GUIDE TO
FLUORESCENCE
LITERATURE

GUIDE TO
FLUORESCENCE
LITERATURE

Richard A. Passwater
**Product Manager of Fluorescence Instrumentation
American Instrument Company
Silver Spring, Maryland**

With the Assistance of
Jarratt G. Bennett
and
Barbara G. Passwater

PLENUM PRESS DATA DIVISION • NEW YORK • 1967

Library of Congress Catalog Card Number 67-18075

ISBN-13: 978-1-4684-6197-8   e-ISBN-13: 978-1-4684-6195-4
DOI: 10.1007/978-1-4684-6195-4

©1967 Plenum Press Data Division
Softcover reprint of the hardcover 1st edition 1967

A Division of Plenum Publishing Corporation
227 West 17 Street, New York, N.Y. 10011
All rights reserved

PREFACE

The major reason for presenting a bibliography on fluorescence and phosphorescence can be summed up in one statement: A recent survey showed that twenty-two percent of all chemical and clinical research was unintentionally duplicated. A comprehensive source book of fluorescence and phosphorescence techniques is therefore needed not only to suggest ideas for future research, but to help decrease needless duplication and expense, and thus to promote the development of both disciplines.

The authors hope that researchers new to fluorescence techniques will appreciate the convenience of this Guide for obtaining data which otherwise could be found only by reviewing dozens of papers, many difficult to find, and that old hands will find it a valuable reference work that will save them many hours and help them to reduce the expense of research (ASTM estimates that more than one billion dollars is spent each year for searching the literature).

Naturally, any work that involves as many technical terms, arbitrary numbers, and proper names as this Guide is subject to error both in the process of organization and in the final transcription. Months of proofreading and correction have been devoted to finding these errors, but inevitably some still remain. The authors would appreciate having these errors called to their attention so that future editions of the Guide can be made even more accurate.

The 4800 papers listed in this volume represent only a fraction of the papers considered. The criterion for inclusion was the presence of substantial information pertinent to fluorescence or phosphorescence techniques, and papers which, for example, only mention that a compound fluorescence under ultraviolet light, or which make only a casual reference to the fluorescence technique were usually rejected. However, occasionally papers of this nature were included because fluorescence methods seem to have unusual potential for the problems discussed. Again, if pertinent papers were missed the authors would be grateful to have these omissions called to their attention.

The abbreviations of journal names employed in this Guide are those used by Chemical Abstracts. Each paper has been given an alpha-numerical identification. Section A contains papers published in the years 1950-1953, section B the years 1954-1956, section C the years 1957-1959, and section D the years 1960-1964. Section E contains papers missed in the original compilation. Within each section the papers are arranged first by year and within each year in alphabetical order of the senior authors. The numeration in each section is continuous and independent of the others.

The indexes employ the alpha-numerical designations rather than page numbers. The author index is a strictly alphabetical listing of all authors, including junior as well as senior authors. The subject index does not contain entries under "Fluorescence," "Phosphorescence," "Energy transfer," and "Spectra," since these would have been so voluminous as to be practically useless. Also, to simplify the index, inorganic compounds have not been indexed as such, and the pertinent papers will be found under the individual elements instead. The indexing of organic compounds is more detailed, but even here it would not have been practical to list all the many different compounds mentioned in article titles. In looking for an organic compound, the reader is advised to seek the specific compound first. If it is not listed, he should next look for a somewhat more general classification of compounds that includes the desired one. This step may be repeated, if necessary, working from the specific compound toward the general class to which it belongs.

The authors hope that this Guide will accomplish its objective and facilitate the retrieval of information now buried in the voluminous literature of fluorescence and phosphorescence and will welcome any suggestions readers may wish to offer to improve its organization or enlarge its scope.

Richard Passwater

CONTENTS

(1950-1953)

A 1 Scintillations in liquids and solutions.
   Ageno, M., Chiozzotto, M., and Querzoli,
   R.
   Phys. Rev. 79: 720 (1950)
   CA 44: 9810h

A 2 Fluorescence method for determination of
   oil fogs.
   Alekseeva, M.V., and Gol'dina, Ts.A.
   Zavodsk. Lab. 16: 35-6 (1950)
   CA 44: 6608e

A 3 Relation between the Franck-Condon fre-
   quencies of absorption and fluorescence
   for some unsaturated hydrocarbons.
   Altmann, S.L.
   Proc. Phys. Soc. 63A: 1234-40 (1950)
   CA 45: 4561d

A 4 Bacterial pyrogenic substances, especially their nature from the viewpoint of
   fluorescence reaction.
   Aoyama, K.
   Bull. Natl. Hyg. Lab., Tokyo 68: 127-37
   (1950)
   CA 48: 13159i

A 5 The phosphorescence of quartz.
   Audubert, R., Bonnemay, M., and Lautout,
   M.
   Compt. Rend. 230: 1771-2 (1950)
   CA 44: 8243a

A 6 Fluorescent reagents. Acyl chlorides and
   acyl hydrazides.
   Baker, W., Haksar, C.N., and McOmie,
   J.F.W.
   J. Chem. Soc. 1950: 170-3 (1950)
   CA 44: 4875b

A 7 Fluorescence studies of some simple ben-
   zene derivatives in the near ultraviolet.
   I. Fluorbenzene and chlorobenzene.
   Bass, A.M., and Sponer, H.
   J. Opt. Soc. Am. 40: 389-96 (1950)
   CA 44: 7153b

A 8 Fluorescence studies of some simple
   benzene derivatives in the near ultra-
   violet. II. Toluene and benzonitrile.
   Bass, A.M.
   J. Chem. Phys. 18: 1403-10 (1950)
   CA 45: 949h

A 9 Fluorescence spectra of natural estrogens
   and their application to biological ex-
   tracts.
   Bates, R.W., and Cohen, H.
   Endocrinology 47: 182-92 (1950)
   CA 45: 6239bd

A 10 Optical bleaches.
   Bayley, C.H.
   Textile J. Australia 25(1): 28, 30, 32, 34,
   and 36 (1950)
   CA 44: 6134b

A 11 Abnormal efficiencies in the
   scintillation-counting of γ-rays.
   Belcher, E.H.
   Nature 166: 742-3 (1950)
   CA 45: 2320h

A 12 Fluorescence of crystalline magnesium
   oxide.
   Bhagavantam, S., and Puranik, P.G.
   Current Sci. 19: 241 (1950)
   CA 45: 453f

A 13 The constituents of Chana (Cicer arieti-
   num, Linn.). III. Chemical examina-
   tion of the fixed oils from Chana and
   Kabuli Chana (ordinary and white varie-
   ties).
   Bhandari, P.R., Bose, J.L., and Siddiqui,
   S.
   J. Sci. Ind. Res. 9B(3): 60-3 (1950)
   CA 44: 7570b

A 14 Origin of fluorescence in diamond.
   Bishui, B.M.
   Indian J. Phys. 24: 441-60 (1950)
   CA 45: 6931e

A 15 The quenching of fluorescence. Deviations from the Stern-Volmer law.
Boaz, H., and Rollefson, G.K.
J. Am. Chem. Soc. 72: 3435-43 (1950)
CA 44: 10519g

A 16 Coriolis perturbation in the ultraviolet spectrum of formaldehyde.
Brand, J.C.D.
Trans. Faraday Soc. 46: 805-12 (1950)
CA 45: 2774h

A 17 Fluorescence and the Beer-Lambert law. A note on the technique of absorption spectrophotometry.
Braude, E.A., Fawcett, J.S., and Timmons, C.J.
J. Chem. Soc. 1950: 1019-21 (1950)
CA 44: 10519b

A 18 Spectrophotometric investigations on the fluorescence of the eye lens in rats given naphthalene.
Brolin, S.E.
Acta Ophthalmol. 28: 163-77 (1950)
CA 45: 6289i

A 19 Quantitative measurements on the elementary process of light excitation from phosphors by single alpha-particles. I.
Broser, I., Kallmann, H., and Reuber, C.
Z. Naturforsch. 5a: 79-85 (1950)
CA 44: 6279a

A 20 Quantitative fluorescence of blood serum in nervous, particularly schizophrenic patients.
Buscaino, G.A.
Acta Neurol. 5: 686-708 (1950)
CA 46: 3150d

A 21 Suitability of inorganic substances as ground material for organophosphors.
Chomse, H.
Forsch. Fortschr. 26: 32 (1950)
CA 46: 5975g

A 22 Organic phosphors with inorganic carrier. V. Arsenic trioxide phosphors.
Chomse, H., and Schone, E.
Z. Anorg. Chem. 261: 153-6 (1950)
CA 44: 7152h

A 23 Photographic emulsion.
Clark, L.E.
U.S. 2,511,462 (1950)
CA 44: 7696f

A 24 Characteristic example of retrapping in phosphorescence.
Curie, D.
Nature 166: 70 (1950)
CA 44: 9804g

A 25 The mechanism of phosphorescence in crystals.
Curie, D.
J. Phys. Radium 11: 179-85 (1950)
CA 44: 7658c

A 26 Experimental proof of electronic recapture in crystalline phosphorescent sulfides.
Curie, D.
Compt. Rend. 230: 1400-2 (1950)
CA 44: 6278f

A 27 Fluorescence of normal human blood serum.
De Lerma, B.
Rend. Accad. Sci. Fis. Mat. (Napoli) 17: 62-7 (1950)
CA 46: 3150c

A 28 Examination of the surface of metallurgical bodies by fluorescence.
Deribere, M.
Rev. Met. 47: 704-5 (1950)
CA 45: 512c

A 29 Effect of temperature on the transient illumination of phosphorescent sulfides submitted to the action of electric fields.
Destriau, G.
Compt. Rend. 230: 205-6 (1950)
CA 44: 3798b

A 30 Recent progress in spectrochemistry of fluorescence from biological products.
Dhere, C.
Fortschr. Chem. Org. Naturstoffe 6: 311-56 (1950)
CA 45: 693f

A 31 Phosphorescence of vapors of phenanthrene.
Dikun, P.P.
Zh. Eksperim. i Teor. Fiz. 20: 193-8 (1950)
CA 44: 6728e

A 32 Variation of the decay time of the fluor-
escence of anthracene and stilbene
with temperature.
Elliot, J.O., Liebson, S.H., and Ravilious,
C.F.
Phys. Rev. 79: 393 (1950)
CA 44: 8775h

A 33 Electron transport and stabilization in
the reaction between dienes and philo-
dienes. I.
Euler, H.V., and Hasselquist, H.
Arkiv Kemi 2: 367-72 (1950)
CA 45: 1989c

A 34 Influence of extraneous molecules on the
absorption and fluorescence spectrum
of magnesium phthalocyanine and of
chlorophyll in solution.
Evstigneev, V.B., Gavrilova, V.A., and
Krasnovskii, A.A.
Dokl. Akad. Nauk SSSR 70: 261-4 (1950)
CA 45: 3720i

A 35 Quenching of the fluorescence of chloro-
phyll and of magnesium phthalocyanin
in their interaction with quenchers.
Evstigneev, V.B., Gavrilova, V.A., and
Krasnovskii, A.A.
Dokl. Akad. Nauk SSSR 74: 315-18 (1950)
CA 45: 1872a

A 36 The fluorescence of silver halides at
low temperatures. I. The pure halides.
Farnell, G.C., Burton, P.C., and Hallama,
R.
Phil. Mag. 41: 157-68 (1950)
CA 44: 6277b

A 37 Fluorescence of silver halides at low
temperatures. II. Mixed crystals of
silver halides.
Farnell, G.C., Burton, P.C., and Hallama,
R.
Phil. Mag. 41: 545-56 (1950)
CA 45: 453d

A 38 The fluorescence of the salts of
8-quinolinol.
Feigl, F., Torok, C., and Zocher, H.
Anais Assoc. Brasil. Quim. 9: 21-7 (1950)
CA 46: 5438e

A 39 Transfers of energy between active
nitrogen and sodium, potassium, and
antimony.
Finkelstein, A.
Compt. Rend. 231: 341-2 (1950)
CA 45: 1425h

A 40 The fluorescence of adrenaline and
adrenochrome.
Fischer, P., and Bacq, Z.M.
Exptl. Med. Surg. 8: 104-12 (1950)
CA 44: 10519d

A 41 Adrenochrome. I. Fluorescence and
cations.
Fischer, P., Derouaux, G., Lambot, H.,
and Lecomte, J.
Bull. Soc. Chim. Belges 59: 72-82 (1950)
CA 44: 7309b

A 42 Dependence of emission spectra of phos-
phors upon activator concentration and
temperature.
Fonda, G.R.
J. Opt. Soc. Am. 40: 347-52 (1950)
CA 44: 7151a

A 43 Fluorometric determination of acetol.
Forist, A.A., and Speck, J.C.
Anal. Chem. 22: 902-4 (1950)
CA 44: 9869c

A 44 Electrolytic dissociation of excited
molecules.
Forster, T.
Z. Elektrochem. 54: 42-6 (1950)
CA 44: 6280c

A 45 Influence of pH on the fluorescence of
naphthalene derivatives.
Forster, T.
Z. Elektrochem. 54: 531-5 (1950)
CA 45: 5515c

A 46 Fluorescent substances for mercury-
vapor electric lamp.
Fujimori, A., and Maekawa, S.
Japan 2823 (1950)
CA 46: 9998c

A 47 Fluorescent reactions. II. Light fluor-
escent reaction of oxidative fluorescent
molecule.
Fujimori, E.
J. Chem. Soc. Japan 71: 491-3 (1950)
CA 45: 6492f

A 48 Fluorescence of magnesium phthalocyanine and of chlorophyll in different
states. Effect of oxygen on the fluorescence of magnesium phthalocyanine and
of chlorophyll in the absorbed state.
Gachkovskii, V.F.
Dokl. Akad. Nauk SSSR 70: 51-4 (1950)
CA 45: 3720c

A 49 Fluorescence of magnesium phthalocya-
nine and chlorophyll in different states.
Complex structure of the main maxi-
mum in the fluorescence spectrum.
Gachkovskii, V.F.
Dokl. Akad. Nauk SSSR 71: 509-11 (1950)
CA 44: 5218e

A 50 Fluorescence of magnesium phthalocya-
nine in complex formation with other
molecules in the adsorbed state.
Gachkovskii, V.F.
Dokl. Akad. Nauk SSSR 73: 963-6 (1950)
CA 45: 41b

A 51 Fluorescence of magnesium phthalocya-
nine and chlorophyll in different states.
Structure of the absorption and fluores-
cence spectra of magnesium porphyrin
and chlorophyll.
Gachkovskii, V.F.
Dokl. Akad. Nauk SSSR 75: 407-10 (1950)
CA 45: 3248b

A 52 Effect of the temperature on the duration
of luminescence of fluorescein solutions.
Galanin, M.D.
Dokl. Akad. Nauk SSSR 70: 989-90 (1950)
CA 44: 4789h

A 53 Measurement of the duration of fluores-
cence with a "phase fluorometer."
Galanin, M.D.
Dokl. Akad. Nauk SSSR 73: 925-7 (1950)
CA 45: 6062b

A 54 Duration of the excited state of a molecule and the properties of fluorescent
solutions.
Galanin, M.D.
Tr. Fiz. Inst. Akad. Nauk SSSR 5: 339-86
(1950)
CA 49: 2873a

A 55 The quantum yields of fluorescence and
phosphorescence of some organic com-
pounds.
Gilmore, E.H., and McClure, D.S.
Phys. Rev. 81: 651 (1950)
CA 46: 6925f

A 56 Color reaction of tryptophan with alde-
hydes.
Giral, J., and Laguna, J.
Ciencia 10: 83-4 (1950)
CA 44: 10605g

A 57 Fluorescence of coumarin derivatives as
a function of pH.
Goodwin, R.H., and Kavanagh, F.
Arch. Biochem. 27: 152-73 (1950)
CA 44: 10519e

A 58 Quenching of fluorescence in liquid
solution.
Grand, S., Collins, F.C., and Kimball, G.E.
Phys. Rev. 82: 338 (1950)
CA 46: 6926a

A 59 Influence of oxidation state of fluores-
cence centers on the luminescent color
of zinc sulfide activated with copper.
Grillot, E., and Bancie-Grillot, M.
Compt. Rend. 231: 966-8 (1950)
CA 45: 3718e

A 60 The luminescence of phosphors of the
aluminum oxide-calcium fluoride type
activated with manganese.
Gunther, G., Anderson, G., and Perlitz, H.
Arkiv Kemi 1: 565-72 (1950)
CA 44: 7152f

A 61 Comparison of the relaxations of photo-
conductivity and of phosphorescence.
Gurevich, D.B., Tolstoi, N.A., and
Feofilov, P.P.
Zh. Eksperim. i Teor. Fiz. 20: 1039-46
(1950)
CA 45: 2770b

A 62 The fluorescence spectra of uranium
minerals in filtered ultraviolet light.
Haberlandt, H., Hernegger, F., and
Scheminsky, F.
Spectrochim. Acta 4: 21-35 (1950)
CA 44: 7152i

A 63 The color and fluorescence of wulfenite in relation to the content of chromium and other trace elements.
Haberlandt, H., and Schroll, E.
Experientia 6: 89-90 (1950)
CA 44: 7719i

A 64 Spectral emission from scintillation solutions and crystals.
Harrison, F.B., and Reynolds, G.T.
Phys. Rev. 79: 732 (1950)
CA 44: 9804g

A 65 Bacterial pyrogenic substances.
Hatta, S., Aoyama, K., and Tanji, S.
Japan Med. J. 3: 125-35 (1950)
CA 46: 569b

A 66 Intensity of fluorescence of solutions.
Heintz, E.
J. Phys. Chim. 47: 676-8 (1950)
CA 45: 4139i

A 67 Fluorescence of mixtures of arterenol, adrenaline, and phosphate.
Heller, J.H., Setlow, R.B., and Mylon, E.
Science 112: 88-9 (1950)
CA 44: 10025h

A 68 Fluorimetric studies of adrenaline and arterenol.
Heller, J.H., Setlow, R.B., and Mylon, E.
Am. J. Physiol. 161: 268-77 (1950)
CA 44: 6901e

A 69 Excitation of fluorescence of organic substances by α-particles and γ-rays.
Herforth, L.
Ann. Physik 7: 312-20 (1950)
CA 44: 9260i

A 70 Emission from oxygen diluted in an atmosphere of xenon.
Herman, R., and Herman, L.
J. Phys. Radium 11: 69-76 (1950)
CA 44: 5703b

A 71 Fluorescence of 1,4-diarylbutadienes and its relation to their configuration.
Hirshberg, Y., Bergmann, E., and Bergmann, F.
J. Am. Chem. Soc. 72: 5117-18 (1950)
CA 45: 2314g

A 72 The fluorescence of cyanine and related dyes in the monomeric state.
Hofer, L.J.E., Grabenstetter, R.J., and Wiig, E.O.
J. Am. Chem. Soc. 72: 203-9 (1950)
CA 44: 4791b

A 73 The ultraviolet spectral absorption of chrysene, its monomethoxy derivatives and 1,2-dimethoxychrysene.
Holiday, E.R., and Jope, E.M.
Spectrochim. Acta 4: 157-64 (1950)
CA 44: 10506b

A 74 Application and preparation of fluorescent materials for electric lamps and tubes.
Holleman, H.C.A.
Chem. Weekblad 46: 33-7 (1950)
CA 44: 4324i

A 75 The photoluminescence of minerals.
Horne, J.E.T.
Bull. Geol. Surv. Gt. Brit. 3: 20-42 (1950)
CA 45: 7471h

A 76 A new standard fluorescent solution to be used in thiochrome method.
Hoshino, M., and Ueno, H.
Ann. Rept. Takeda Res. Lab. 8: 102-5 (1949) (1950)
CA 46: 11296f

A 77 Coloring matters from Aphis fabae.
Human, J.P.E., Johnson, A.W., MacDonald, S.F., and Todd, A.R.
J. Chem. Soc. 1950: 477-85 (1950)
CA 44: 7841h

A 78 Fluorescent substance.
Ide, H.
Japan 3889, Nov. 9 (1950)
CA 46: 9430e

A 79 Fluorescent porcelains.
Iimori, Satoyasu, and Iimori, Shoji
Japan 2866, Sept. 26 (1950)
CA 46: 8978b

A 80 Calibration of proportional counters by the excitation of fluorescence radiation with radioactive sources.
Insch, G.M.
Phil. Mag. 41: 857-62 (1950)
CA 45: 2785e

A 81 Effect of thermal history on color of
zinc beryllium silicate (Mn) phosphors.
Jones, S.
J. Electrochem. Soc. 97: 25-8 (1950)
CA 44: 2377a

A 82 Fluorescence of solutions bombarded
with high-energy radiation (energy
transport in liquids).
Kallmann, H., and Furst, M.
Phys. Rev. 79: 857-70 (1950)
CA 44: 10518c

A 83 Fluorescence of liquids under
γ-bombardment.
Kallmann, H., and Furst, M.
Nucleonics 7: 69-71 (1950)
CA 44: 9810i

A 84 Formation and separation of 2-amino-4-
hydroxy-6- and 2-amino-4-hydroxy-7-
methylpteridine. Methylpteridine Red.
Karrer, P., and Schwyzer, R.
Helv. Chim. Acta 33: 39-45 (1950)
CA 44: 4477i

A 85 Mean lifetime of the fluorescence of
acetone and biacetyl vapors.
Kaskan, W.E., and Duncan, A.B.F.
J. Chem. Phys. 18: 427-31 (1950)
CA 44: 9261a

A 86 Effect of temperature on the lifetime of
fluorescence of solid acetone.
Kaskan, W.E., and Duncan, A.B.F.
J. Chem. Phys. 18: 432-4 (1950)
CA 44: 9261c

A 87 Fluorescence test for copper.
Kedvessy, G.
Magy. Kem. Folyoirat 56: 447-8 (1950)
CA 46: 375f

A 88 Examination of wheat and other flours
by the fluorescence in Wood's light.
Kiger, J.
Ann. Pharm. Franc. 8: 788-90 (1950)
CA 45: 6763c

A 89 The nature of some fluorescing sub-
stances contained in a deep sea mud.
Koe, B.K., Fox, D.L., and Zechmeister, L.
Arch. Biochem. 27: 449-52 (1950)
CA 44: 10394f

A 90 Relation between atmospheric stability
of pigmented paint films and the
pigment-photosensitized formation of
peroxide compounds.
Krasnovskii, A.A., and Gurevich, T.N.
Dokl. Akad. Nauk SSSR 74: 569-72 (1950)
CA 46: 755d

A 91 Examination of meat contaminated by
Cysticercus by means of filtered ultra-
violet light.
Krisilov, D.V.
Gigiena i Sanit. 1950: 52-3 (1950)
CA 44: 8014i

A 92 Fluorescence of magnesium germanate
activated by manganese.
Kroeger, F.A., and van den Boomgaard, J.
J. Electrochem. Soc. 97: 377-82 (1950)
CA 45: 1425e

A 93 The location of dissipative transitions in
luminescent systems.
Kroger, F.A., and Hoogenstraten, W.
Physica 16: 30-2 (1950)
CA 44: 7150i

A 94 Trivalent cations in fluorescent zinc
sulfide.
Kroger, F.A., and Dikhoff, J.
Physica 16: 297-316 (1950)
CA 45: 2315c

A 95 Physical chemistry of the formation of
fluorescence centers in zinc sulfide-
copper.
Kroger, F.A., and Smit, N.W.
Physica 16: 317-28 (1950)
CA 45: 2315d

A 96 Cascaded fluorescent material.
Kroger, F.A., and Voogd, J.
U.S. 2,494,883 (1950)
CA 44: 2856e

A 97 The photovoltaic behavior of organic
substances in solution.
Levin, I., and White, C.E.
J. Chem. Phys. 18: 417-26 (1950)
CA 44: 9242d

A 98 Fluorescent decay of scintillation crys-
tals.
Liebson, S.H., Bishop, M.E., and Elliot,
J.O.
Phys. Rev. 80: 907-8 (1950)
CA 45: 1425f

A 99 Quenching of the fluorescence of chloro-
phyll a solutions.
Livingston, R., and Ke, C.L.
J. Am. Chem. Soc. 72: 909-15 (1950)
CA 44: 4789h

A 100 Colored lubricating oil.
Mastin, R.G.
U.S. 2,508,617 (1950)
CA 44: 9667i

A 101 Singlet-triplet absorption bands in some
organic molecules.
McClure, D.S., Blake, N., and Hanst, P.
Phys. Rev. 81: 651 (1950)
CA 46: 6925f

A 102 Further studies in mercury band fluor-
escence.
McCoubrey, A.D., Alpert, D., and
Holstein, T.
Phys. Rev. 82: 567 (1950)
CA 46: 6926d

A 103 Chronic pulmonary beryllosis in work-
ers using fluorescent powders containing beryllium.
MacMahon, H.E., and Olken, H.G.
Arch. Ind. Hyg. Occupational Med. 1:
195-214 (1950)
CA 44: 4165e

A 104 Observations on exceptional duration of
mineral phosphorescence.
Millson, H.E., and Millson, H.E., Jr.
J. Opt. Soc. Am. 40: 430-5 (1950)
CA 44: 8243b

A 105 The effects of mepacrine hydrochloride
(atebrin) upon the human skin.
Miller, O.B., Herrmann, F., and Rubin, J.
J. Invest. Dermatol. 15: 445-52 (1950)
CA 46: 633e

A 106 Whiter wool.
Moncrieff, R.W.
Textile Mfr. 76(904): 184-6 (1950)
CA 44: 5597c

A 107 Synthetic hydraulic fluids.
Murphy, C.M., and Zisman, W.A.
Prod. Eng. 21: 109-13 (1950)
CA 44: 11076e

A 108 Fluorescent indicators for acid-base
titrations. I.
Neelakantam, K., and Viswanath, G.
Current Sci. (India) 19: 15-16 (1950)
CA 44: 6761b

A 109 Effect of external conditions on the ab-
sorption and fluorescence of vapors of
aromatic compounds.
Neporent, B.S.
Dokl. Akad. Nauk SSSR 72: 35-8 (1950)
CA 44: 6721d

A 110 Stabilizing collisions involving excited
aromatic compounds.
Neporent, B.S.
Zh. Fiz. Khim. 24: 1219-34 (1950)
CA 45: 5518h

A 111 Fluorescent substance.
Ohno, Y.
Japan 1805, June, 21 (1950)
CA 46: 7879c

A 112 Fluorescent substance.
Ohno, Y.
Japan 2865, Sept. 26 (1950)
CA 46: 8978a

A 113 White fluorescent substance.
Ohno, Y.
Japan 3354 (1950)
CA 46: 9998d

A 114 Fluorescent body for use with white
fluorescent electric lamps.
Ohsuga, T., and Hayao, T.
Japan 3355 (1950)
CA 46: 9998e

A 115 Fluorescent substance for white fluor-
escent lamp.
Ohsuga, T.
Japan 3890, Nov. 9 (1950)
CA 46: 9430f

A 116 Intensity relationships in the a and b
bands in the absorption and fluorescence spectra of some uranyl salts at
various temperatures.
Pant, D.D.
Proc. Indian Acad. Sci. 31: 103-6 (1950)
CA 44: 8236e

A 117 Fluorescence changes in solutions of glucose and glycine and of acetaldehyde and ammonia.
Pearce, J.A.
Food Technol. 4: 416-19 (1950)
CA 45: 3719i

A 118 Qualitative analysis of synthetic resins used in leather finishes.
Pektor, V.
Paint Technol. 15(171): 105-7 (1950)
CA 44: 5635c

A 119 Fluorometric determination of gitoxoside.
Pesez, M.
Ann. Pharm. Franc. 8: 746-50 (1950)
CA 45: 3996i

A 120 The determination and the distribution of aneurine in wheat, in flour, and in bread. Apparatus for measuring the fluorescence of solutions.
Petit, L.
Ann. Inst. Natl. Rech. Agron., Ser. A 1: 41-107 (1950)
CA 44: 6979h

A 121 Daylight fluorescence.
Petzold, O.
Paint, Oil Colour J. 118: 988, 990, 992 (1950)
CA 45: 5025a

A 122 Reconnaissance study of Peruvian minerals and ores with an ultraviolet light.
Plaza, G.R.
Bol. Inst. Nacl. Invest. y Fomento Mineros 1(1): 145-50 (1950)
CA 46: 2965e

A 123 Absolute yield of the fluorescence of solutions of anthracene and some of its derivatives.
Polovikov, F.I.
Dokl. Akad. Nauk SSSR 71: 453-6 (1950)
CA 44: 5218a

A 124 Mechanism of phosphorescence in combination of variable compositions.
Pregel, B.
Compt. Rend. 231: 489-90 (1950)
CA 45: 3248g

A 125 The cellulose-dye complex. IV. The polarized fluorescence from dyed fibers.
Preston, J.M., and Ser, Y.F.
J. Soc. Dyers Colourists 66: 357-61 (1950)
CA 44: 8662i

A 126 Striking differences in the activation of potassium and sodium halogen phosphors.
Pringsheim, P.
Acta Phys. Austriaca 3: 396-404 (1950)
CA 44: 7657f

A 127 Phosphorescence of benzene hydrocarbons at the temperature of liquid oxygen.
Pyatnitskii, B.A.
Dokl. Akad. Nauk SSSR 71: 457-60 (1950)
CA 44: 5217c

A 128 Comparison of the quenching actions of nickel and cobalt on luminescent zinc sulfide.
Saddy, J., and Arpiarian, N.
Compt. Rend. 230: 1948-50 (1950)
CA 44: 8775g

A 129 Phosphorescence-chemical investigations on sulfates.
Schikore, W.
Z. Anorg. Chem. 261: 121-9 (1950)
CA 44: 7657h

A 130 Calabash curare alkaloids.
Schmid, H., and Karrer, P.
Helv. Chim. Acta 33: 512-5 (1950)
CA 44: 8060b

A 131 Variation of emission spectrum of manganese-activated zinc beryllium silicate with decay time.
Schulman, J.H., Klick, C.C., and Ginther, R.J.
J. Opt. Soc. Am. 40: 337-8 (1950)
CA 46: 9425c

A 132 Fluorescence and phosphorescence emission spectra of manganese-activated zinc silicate.
Schulman, J.H., and Klick, C.C.
J. Opt. Soc. Am. 40: 622-3 (1950)
CA 46: 6937d

A 133 Polarization of the fluorescence of
vapors of organic dyes.
Shendrick, K.K., and Siryuk, A.G.
Dokl. Akad. Nauk SSSR 75: 665-7 (1950)
CA 45: 3718i

A 134 Fluorescent substance.
Shimizu, E.
Japan 3711, Oct. 27 (1950)
CA 46: 9430e

A 135 Fluorescence spectra of some polycyc-
lic carbohydrates at liquid-air temper-
ature.
Shpol'skii, E.V., Il'ina, A.A., and
Basilevich, V.V.
Izv. Akad. Nauk SSSR, Ser. Fiz. 14: 511
(1950)
CA 46: 1871c

A 136 Nicotine compounds containing univa-
lent copper.
Smith, C.R.
U.S. 2,512,689 (1950)
CA 44: 8592b

A 137 Discussion of the lowest singlet transition in naphthalene as a forbidden transition $A_{1g} \rightarrow A_{1g}$ and remarks on the
higher singlet levels.
Sponer, H., and Nordheim, G.P.
Discussions Faraday Soc. 1950: 19-26
(1950)
CA 46: 3398f

A 138 The evaluation of the antiseptic action
of glycerol solutions with biochemical
methods.
Stepan, J.
Casopis Ceskeho Leharnictva 63: 1-14
(1950)
CA 46: 4172h

A 139 Paper chromatography of digitalis
glycosides.
Svendsen, A.B., and Jensen, K.B.
Pharm. Acta Helv. 25: 241-7 (1950)
CA 45: 1726a

A 140 Influence of the medium on the phos-
phorescence of organic luminophors.
Sveshnikov, B.Y., and Petrov, A.A.
Dokl. Akad. Nauk SSSR 71: 461-4 (1950)
CA 44: 5217g

A 141 Polarization of the phosphorescence of
organoluminophors at the temperature
of liquid air.
Sveshnikov, B.Y., and Ermolaev, V.L.
Dokl. Akad. Nauk SSSR 71: 647-50 (1950)
CA 44: 5711g

A 142 Theory of quenching of the lumines-
cence of organic phosphors.
Sveshnikov, B.Y.
Zh. Eksperim. i Teor. Fiz. 18: 878-85
(1948) (1950)
CA 46: 10908g

A 143 Heterocyclic fluorescent coloring
materials.
Switzer, J.L., and Ward, R.A.
U.S. 2,495,202 (1950)
CA 44: 5501b

A 144 Daylight fluorescent pigment composi-
tions.
Switzer, J.L., and Switzer, R.C.
U.S. 2,498,592 (1950)
CA 44: 5613b

A 145 Daylight fluorescent resinous sheeting
materials.
Switzer, J.L., and Switzer, R.C.
U.S. 2,498,593 (1950)
CA 44: 5613c

A 146 New method for the demonstration of
self-excitation.
Szalay, L., and Szollosy, L.
Acta Univ. Szeged., Acta Phys. Chem. 2:
259-62 (1950)
CA 45: 8360a

A 147 The relationship between the absorption
and emission of the alcoholic solution
of Acridine Orange NO.
Szor, P.
Acta Univ. Szeged. Phys. Chem. 2: 249-55
(1950)
CA 46: 350d

A 148 Browning and the fluorescence of evap-
orated milk.
Tarassuk, N.P., and Semonson, H.D.
Food Technol. 4: 88-92 (1950)
CA 45: 1697h

A 149 Theoretical and practical aspects of
fluorescence, luminescence, and phos-
phorescence.
Taylor, G.G.
J. Soc. Dyers Colourists 66: 181-6 (1950)
CA 44: 4788a

A 150 Reabsorption of fluorescence produced
in an ionization chamber or counter.
Tellez-Plasencia, H., and Theron, P.
J. Phys. Radium 11: 93-6 (1950)
CA 44: 5698d

A 151 Reabsorption of secondary fluorescence
radiation in a photographic emulsion
exposed to X rays.
Tellez-Plasencia, H., and Theron, P.
Sci. Ind. Phot. 21: 361-70 (1950)
CA 45: 6516i

A 152 Interaction of activators in phosphors.
Trapeznikova, Z.A.
Dokl. Akad. Nauk SSSR 74: 465-8 (1950)
CA 45: 454e

A 153 Fluorescent substances.
Uehara, Y., and Kofuya, Y.
Japan 1659, June 7 (1950)
CA 46: 7879b

A 154 The K-centers of potassium chloride
crystals.
Ueta, M.
Busseiron Kenkyu 26:81-95 (1950); Mem.
Coll. Sci., Univ. Kyoto, 26A:69-74 (1950)
CA 46: 10905d

A 155 Emission spectra of some chemolum-
inescent reactions. I. Oxidation of
dimethylbisacridinium nitrate.
Van der Burg, A.S.
Rec. Trav. Chim. 69: 1525-35 (1950)
CA 46: 7437d

A 156 Emission spectra of some chemolum-
inescent reactions. II. Oxidation of
derivatives of 2,3-dihydro-1,4-
phthalazinedione.
Van der Burg, A.S.
Rec. Trav. Chim. 69: 1536-44 (1950)
CA 46: 7437e

A 157 Composition for coating cathode-ray
tubes.
Van Hoorn, A.J.
U.S. 2,496,091 (1950)
CA 44: 4354c

A 158 Photoconductivity of solid anthracene.
Vartanyan, A.T.
Dokl. Akad. Nauk SSSR 71: 641-2 (1950)
CA 44: 5711e

A 159 The Relations between fluoresence and
the nature of the substituents on the molecule of 9, (10H)-acridone. I. Preparation of 9, (10H)-acridone derivatives.
Villemey, L.
Ann. Chim. 5: 570-93 (1950)
CA 45: 2947c

A 160 Relation between fluorescence and the
nature of the substituents on the mole-
cule of 9, (10H)-acridone. II. Fluorometry.
Villemey, L.
Ann. Chim. 5: 642-70 (1950)
CA 45: 3719b

A 161 Spectrographic study of the fluorescent
light from monosubstituted derivatives
of acridone in solution.
Villemey, L.
Compt. Rend. 230: 303-4 (1950)
CA 44: 5218i

A 162 Fluorescent microscopic investigations
of erythrocytes.
Wagner, K,
Wien. Med. Wochschr. 100: 808-9 (1950)
CA 45: 6680b

A 163 Constant quantum efficiency of some
luminescent materials in the vacuum
ultraviolet.
Watanabe, K., Johnson, F.S., and Tousey,
R.
Phys. Rev. 79: 217 (1950)
CA 46: 6923f

A 164 Self-quenching and sensitization of
fluorescence of chlorophyll solutions.
Watson, W.F., and Livingston, R.
J. Chem. Phys. 18: 802-9 (1950)
CA 44: 10520a

A 165 Fluorescence of riboflavin and flavin
adenine dinucleotide.
Weber, G.
Biochem. J. 47: 114-21 (1950)
CA 44: 10757a

A 166 Xanthopterin. IV. Fluorescence of
xanthopterin adsorbates.
Weber, K., and Hojman, J.
Bull. Soc. Chim. Belgrade 15: 27-38 (1950)
CA 46: 5976g

A 167 Proportional counters in a magnetic field: Investigation of the isomerism of Br^{80} and measurement of the fluorescence yield of krypton.
West, D., and Rothwell, P.
Phil. Mag. 41: 873-89 (1950)
CA 45: 2788f

A 168 Fluorometric analysis.
White, C.E.
Anal. Chem. 22: 69-71 (1950)
CA 44: 1843e

A 169 The origin of petroleum: Effects of low-temperature pyrolysis on the organic extract of a recent marine sediment.
Whitehead, W.L., and Breger, I.A.
Science 111: 335-7 (1950)
CA 44: 5285b

A 170 Fluorescence of amino acids, peptides, and amines on filter paper.
Woiwood, A.J.
Nature 166: 272 (1950)
CA 44: 10518h

A 171 Formation of fluorescent films.
Yoshimoto, Y.
Japan 3376 (1950)
CA 46: 9998e

A 172 Fluorescent coloring materials.
Yule, J.A.C.
U.S. 2,503,790 (1950)
CA 44: 5613d

A 173 Optical bleaching agents.
Yura, S.
Kagaku No Ryoiki 4: 364-9 (1950)
CA 44: 9679i

A 174 Polarization spectra of anthracene derivatives.
Zhevandrov, N.D.
Dokl. Akad. Nauk SSSR 74: 25-8 (1950)
CA 45: 36a

A 175 Decay of the phosphorescence of alcoholic solutions of some simple aromatic compounds at the temperature of liquid oxygen.
Zinov'eva, O.G.
Zh. Eksperim. i Teor. Fiz. 20: 132-8 (1950)
CA 44: 6726h

A 176 Luminescence and photoconductivity of crystal phosphors under conditions of weak excitation.
Advirovich, E.I.
Zh. Eksperim. i Teor. Fiz. 21: 275-82 (1951)
CA 46: 2404d

A 177 Mechanism of scintillation in organic chemistry.
Ageno, M., and Querzoli, R.
Rend. Ist. Super. Sanita 14: 779-91 (1951)
CA 47: 408f

A 178 Colorimetric measurement of the yield of fluorescence.
Alentsev, M.N.
Zh. Eksperim. i Teor. Fiz. 21: 133-41 (1951)
CA 45: 7875c

A 179 Measurement of the intrinsic efficiency of ultraviolet luminescence.
Alentsev, M.N., and Vinokurov, L.A.
Izv. Akad. Nauk SSSR, Ser. Fiz. 15: 725-9 (1951)
CA 46: 6493a

A 180 Improved photoelectric fluorometer.
Alford, W.C., and Daniel, J.H.
Anal. Chem. 23: 1130-3 (1951)
CA 46: 5b

A 181 Use of the luminescence method for characterizing the properties of coals.
Ammosov, I.I., and Babinkova, N.I.
Izv. Akad. Nauk SSSR, Otd. Tekhn. Nauk 1951: 341-9 (1951)
CA 45: 8230e

A 182 Taaffeite, a new beryllium mineral, found as a cut gemstone.
Anderson, B.W., Claringbull, G.F., and Hey, M.H.
Mineral. Mag. 29: 765-72 (1951)
CA 46: 1925h

A 183 Mycobacterial fluoromicroscopy and Dubos' cytochemical reaction.
de Androde, L.
Mem. Inst. Oswaldo Cruz 49: 7-32 (1951)
CA 46: 5124d

A 184 Fluorescence spectrophotometry of blood serum in acute yellow atrophy.
Arpino, G., and De Lerma, B.
Progr. Med. 7: 656-60 (1951)
CA 47: 2345b

A 185 Fluorescent substance.
Asoda, T.
Japan 869 (1951)
CA 46: 10914f

A 186 Fluorescence and phosphorescence of
cerium-activated alkaline-earth phosphors.
Avikina, L.I.
Zh. Eksperim. i Teor. Fiz. 21: 310-13
(1951)
CA 45: 7880d

A 187 Ultraviolet absorption spectra of
chromatographic fractions of lignins.
Bailey, A.
J. Am. Chem. Soc. 73: 2325-6 (1951)
CA 45: 9264i

A 188 The fluorescence in diamond excited
by X-rays.
Bishui, B.M.
Indian J. Phys. 25: 575-80 (1951)
CA 46: 6939e

A 189 Scintillations from organic crystals:
specific fluorescence and relative
response to different radiations.
Birks, J.B.
Proc. Phys. Soc. 64: 874-7 (1951)
CA 46: 10002f

A 190 The specific fluorescence of anthracene
and other organic materials.
Birks, J.B.
Phys. Rev. 84: 364-5 (1951)
CA 46: 1879c

A 191 Analysis of uranium solutions by X-ray
fluorescence.
Birks, L.S., and Brooks, E.J.
Anal. Chem. 23: 707-9 (1951)
CA 45: 7915f

A 192 Transfer and transport of energy by
resonance processes in luminescent
solids.
Botden, T.P.J.
Philips Res. Rept. 6: 425-73 (1951)
CA 46: 6492a

A 193 Dried whole-egg powder. XXIX. The
effect of high moisture content on
fluorescence and other chemical
changes.
Boulet, M., and Pearce, J.A.
Can. J. Technol. 29: 153-8 (1951)
CA 45· 7270b

A 194 The quenching of anthracene fluores-
cence.
Bowen, E.J., and Metcalf, W.S.
Proc. Roy. Soc. (London), Ser. A 206:
437-47 (1951)
CA 47: 10351g

A 195 Blackening and fluorescence in potatoes.
Bowman, F., and Hanning, F.
Food. Res. 16: 462-8 (1951)
CA 46: 10910d

A 196 Vibrational analyses of the fluores-
cence spectrum of formaldehyde.
Brand, J.C.D.
J. Chem. Phys. 19: 377-8 (1951)
CA 45: 6931d

A 197 Polarized fluorescence of a single
crystal of pure naphthalene.
Brodersen, S.
Compt. Rend. 233: 1094-6 (1951)
CA 46: 3411b

A 198 Electronic and vibrational levels of the
molecule and the crystal of benzene.
Broude, V.L., Medvedev, V.S., and
Prikhot'ko, A.F.
Zh. Eksperim. i Teor. Fiz. 21: 665-72
(1951)
CA 46: 8958g

A 199 Relations between results of fluorescence analysis and the newest mano-
metric measurements of carbon dioxide
assimilation.
Burk, D., Warburg, O., and Kautsky, H.
Z. Naturforsch. 6b: 292-5 (1951)
CA 46: 3120g

A 200 Quantitative fluorescence spectrum of
the diethylamide of lysergic acid.
Buscaino, G.A.
Ric. Sci. 21: 519-25 (1951)
CA 47: 8518f

A 201 Fluorescence studies of carcinogens in
the rat skin.
Cambel, P.
Cancer Res. 11: 370-5 (1951)
CA 45: 9700c

A 202 Fluorspar in California.
Crosby, J.W., and Hoffman, S.R.
Calif. J. Mines Geol. 47: 619-38 (1951)
CA 46: 3466c

A 203 Theory of luminescence of molecular
crystals.
Davydov, A.S.
Izv. Akad. Nauk SSSR, Ser. Fiz. 15: 605-7
(1951)
CA 46: 8973d

A 204 Spectrophotometric testing of the
fluorescence of estrone and α-estradiol
in binary mixtures.
DeLerma, B., and Bompiani, A.
Boll. Soc. Ital. Biol. Sper. 27: 1699-1701
(1951)
CA 48: 791g

A 205 Spectrophotometric aspects of the
fluorescence of α-estradiol and estrone
after treatment with concentrated sulfuric acid.
DeLerma, B., and Bompiani, A.
Boll. Soc. Ital. Biol. Sper. 27: 233-6 (1951)
CA 48: 791f

A 206 The fluorescence yield of argon.
Dexter, D.L., and Beeman, W.W.
Phys. Rev. 81: 456 (1951)
CA 45: 3710d

A 207 Duration of the phosphorescence of
benzene and its derivatives.
Dikun, P.P., Petrov, A.A., and
Sveshnikov, B.Y.
Zh. Eksperim. i Teor. Fiz. 21: 150-63
(1951)
CA 45: 8357i

A 208 Spectra of the metallo derivatives of
$\alpha,\beta,\gamma,\delta$-tetraphenylporphine.
Dorough, G.D., Miller, J.R., and
Huennekens, F.M.
J. Am. Chem. Soc. 73: 4315-20 (1951)
CA 46: 5968b

A 209 The nature of quenching of fluorescence.
Eisenbrand, J.
Z. Elektrochem. 55: 374-80 (1951)
CA 46: 9430a

A 210 Low-temperature fluorescence of
silver halides and photographic emul-
sions.
Farnell, G.C., and Burton, P.C.
Fundamental Mech. Phot. Sensitivity,
Proc. Symp. Univ. Bristol, Engl. 1950,
61-73 (1951)
CA 45: 8925i

A 211 Fluorescence and photoinactivation of
indoleacetic acid.
Ferri, M.G.
Arch. Biochem. Biophys. 31: 127-31 (1951)
CA 45: 5026a

A 212 Ultraviolet absorption and fluorescence
of 2,6,8,10-tetrahydroxy- and 2,6,10-
trihydroxyhomopurine.
Fischer, F.G., and Neumann, W.P.
Ann. 572: 230-240 (1951)
CA 46: 4021a

A 213 A preliminary report on the energy
distribution and the half life of the
phosphorescent spectra of calcite induced by X-radiation.
Forman, G., and Kruschwitz, W.H.
Phys. Rev. 83: 487 (1951)
CA 47: 6754h

A 214 Fluoreszenz organischer Verbindungen.
Forster, T.
Göttingen: Vandenboeck and Ruprecht,
1951, 312 pp.
CA 45: 4565a

A 215 Nonphosphorescent calcium silicate
phosphor containing indium.
Froelich, H.C.
U.S. 2,545,880 (1951)
CA 45: 6498d

A 216 Fluorescent reactions. III. Fluores-
cence of fluorescein derivatives.
Fujimori, E.
J. Chem. Soc. Japan, Pure Chem. Sect.
72: 315-16 (1951)
CA 46: 825e

A 217 Fluorescent reactions. IV. Light-
fluorescent reaction of reductive
fluorescent molecule.
Fujimori, E.
J. Chem. Soc. Japan, Pure Chem. Sect.
72: 358-60 (1951)
CA 46: 825f

A 218 Fluorescent reactions. V. Fluorescent
reaction of acid amide and protein.
Fujimori, E.
J. Chem. Soc. Japan, Pure Chem. Sect.
72: 417-20 (1951)
CA 46: 1871d

A 219 Phosphorescent effects with high ener-
gy radiation.
Furst, M., and Kallman, H.
Phys. Rev. 82: 964 (1951)
CA 45: 8366h

A 220 Quenching of fluorescence by a light-
absorbing medium.
Galanin, I.M.D., and Frank, I.M.
Zh. Eksperim. i Teor. Fiz. 21: 114-20
(1951)
CA 45: 7883f

A 221 Quenching of the fluorescence of dipoles
by absorbing substances.
Galamin, I.M.D., and Levshin, L.V.
Zh. Eksperim. i Teor. Fiz. 21: 121-5
(1951)
CA 45: 7884a

A 222 Quenching by absorbing means and
sensitized fluorescence in solutions.
Galanin, M.D.
Izv. Akad. Nauk SSSR, Ser. Fiz. 15: 543-9
(1951)
CA 46: 8971b

A 223 Directional properties of single crys-
tal of anthracene regarding fluores-
cence phenomena.
Ganguly, S.C., and Chaudhury, N.K.
J. Chem. Phys. 19: 617-18 (1951)
CA 45: 9373e

A 224 Rise and decay of willemite lumines-
cence.
Gergely, G.
Acta Phys. Acad. Sci. Hung. 1: 197-8
(1951)
CA 46: 3411a

A 225 Fluorescence microscopic observa-
tions on living chloroplasts.
Gickihorn, J.
Mikrochemie Ver. Mikrochim. Acta
36/37: 1128-33 (1951)
CA 45: 5770a

A 226 Spectrophotometric characteristics of
the determination of titanium with
thymol.
Griel, J.V., and Robinson, R.J.
Anal. Chem. 23: 1871-3 (1951)
CA 46: 4425a

A 227 Fluorescence and absorption-spectral
data and other physical characteristics
of xanthopterin-like pigment synthe-
sized by the human tubercle bacillus
and isolated by chromatographic and
fluorescence analysis.
Growe, M.O'L., and Walker, A.
Phys. Rev. 86: 817 (1951)
CA 47: 6758b

A 228 The fluorescence of gases by
α-particle excitation.
Grun, A.E., and Schopper, E.
Z. Naturforsch. 6a: 698-700 (1951)
CA 46: 6939a

A 229 Inhibition of various enzymes by ster-
oids.
Hayano, M., and Dorfman, R.I.
Ann. N.Y. Acad. Sci. 54: 608-18 (1951)
CA 46: 8157d

A 230 Identification of vegetable dyes
painted on ancient Japanese silk by
fluorescence. (A supplement.)
Hayashi, K., and Suzushino, G.
Misc. Rept. Res. Inst. Nat. Resources 23:
59-62 (1951)
CA 47: 870a

A 231 The polarization of fluorescent light of
polystyrenes under the influence of a
fluorescence inhibitor.
Heintz, E.
J. Chim. Phys. 48: 545-51 (1951)
CA 46: 4374b

A 232 The temperature dependence of the
fluorescence spectrum of anthracene
stimulated with light and with cathode
rays.
Hengst, K.
Z. Naturforsch. 6a: 540-3 (1951)
CA 46: 2914d

A 233 Fluorochrome coloring of living proto-
plasm.
Hofler, K.
Mikrochemie Ver. Mikrochim. Acta
36/37: 1146-57 (1951)
CA 45: 5770d

A 234 Fluorescence and phosphorescence spectra of pyrene-type carbohydrates in frozen solutions.
Il'ina, A.A., and Shpol'skii, E.V.
Izv. Akad. Nauk SSSR, Ser. Fiz. 15: 585-95, Discussion 596 (1951)
CA 46: 8973d

A 235 Two types of solvent influence on the fluorescence spectrum and efficiency of salicylic acid and Rhodamine B extra.
Izmailov, N.A., and Kremer, V.A.
Izv. Akad. Nauk SSSR, Ser. Fiz. 15: 565-72 (1951)
CA 46: 8972b

A 236 Excitation with infrared rays of phosphorescence in luminescent substances containing calcium oxide.
Janin, J., Crozet, A., and Clerc, P.
Compt. Rend. 233: 934-6 (1951)
CA 46: 2404h

A 237 Fluorescence efficiencies of some crystals for electrons and heavy particles.
Jentschke, W.K., Elby, F.S., Taylor, C.J., Remley, M.E., and Kruger, P.G.
Phys. Rev. 83: 170-1 (1951)
CA 45: 8902i

A 238 The effect of temperature on emission of fluorescent lamp phosphors.
Jerome, C.W.
J. Electrochem. Soc. 98: 376-83 (1951)
CA 45: 10060i

A 239 Delays in fluorescence applicable to the blue-red binaries T Cor B, Z Andr, R Agar, AX Pers, α Scorp, and RW Hydr.
Johnson, M.
Monthly Notices Roy. Astron. Soc. 111: 490-505 (1951)
CA 46: 5430c

A 240 Investigations in chemical oceanography with the aid of the Pulfrich photometer. VII. The microdeterminations of chlorophyll content and of fluorescence.
Kalle, K.
Deut. Hydrograph. Z. 4: 92-6 (1951)
CA 46: 5422e

A 241 Fluorescence of solutions bombarded with high-energy radiation (energy transport in liquids). II.
Kallmann, H., and Furst, M.
Phys. Rev. 81: 853-64 (1951)
CA 45: 4560i

A 242 Energy storage and light-stimulated phosphorescence in activated NaCl crystals induced by γ-rays.
Kallmann, H., and Furst, M.
Phys. Rev. 83: 674-5 (1951)
CA 45: 8894g

A 243 Quenching of the fluorescence of acridine compounds in an adsorbed state by oxygen.
Karyakin, A.V.
Izv. Akad. Nauk SSSR, Ser. Fiz. 15: 556-64 (1951)
CA 46: 8971h

A 244 Ultraviolet luminescence mechanism of X-rayed alkali-halogen crystals.
Kats, M.L.
Izv. Akad. Nauk SSSR, Ser. Fiz. 15: 667-8 (1951)
CA 46: 8971g

A 245 Fluorescence spectra of petroleum and its fractions in liquid condition and in a chromatographic column.
Kats, M.L., and Sidorov, N.K.
Izv. Akad. Nauk SSSR, Ser. Fiz. 15: 777-81 (1951)
CA 46: 6816f

A 246 Lignin of tropical dicotyledonous woods. I. Comparison of the lignin with that of wood from temperate zones.
Kawamura, I., and Higuchi, T.
Res. Bull. Fac. Agr., Gifu Univ. 1: 39-43 (1951)
CA 50: 3758i

A 247 Fluorescence microscopic examination of lignified cell walls.
Kisser, J., and Wittmann, W.
Mikrochemie Ver. Mikrochim. Acta 36/37: 1134-45 (1951)
CA 45: 5770c

A 248 Fluorescence method for qualitative and quantitative determination of acridine in rivanol and biologic substances.
Kostyakova, A.I.
Zh. Analit. Khim. 6: 251-6 (1951)
CA 45: 10140h

A 249 Origin of multiple bands in luminophors
activated by bivalent manganese.
Kroeger, F.A., and Zalm, P.
J. Electrochem. Soc. 98: 177-82 (1951)
CA 45: 6494a

A 250 A new highly specific test for alumin-
um.
Kul'berg, L.M., and Mustafin, I.S.
Dokl. Akad. Nauk SSSR 77: 285-8 (1951)
CA 46: 2954e

A 251 Photoelectric measurement of fluores-
cence spectra of highly colored
liquids.
Lauer, J.L., and Rosenbaum, E.J.
J. Opt. Soc. Am. 41: 450-3 (1951)
CA 45: 8356c

A 252 Phosphorescence of sodium chloride.
Lautout, M.
Compt. Rend. 232: 2025-7 (1951)
CA 45: 8356g

A 253 The stereochemistry of complex inorganic compounds. XI. The resolution
of bis-(8-quinolino-5-sulfonic acid)
zinc (II).
Liu, J.C.I., and Bailar, J.C., Jr.
J. Am. Chem. Soc. 73: 5432-3 (1951)
CA 46: 1908h

A 254 Fluorescence quenching by colloid
anions and cations.
Lovelock, J.E.
J. Chem. Soc. 1951: 115-20 (1951)
CA 45: 10061a

A 255 The fluorescence of acetone vapor.
Luckey, G.W., and Noyes, W.A., Jr.
J. Chem. Phys. 19: 227-31 (1951)
CA 45: 6492c

A 256 Electrical and optical properties of
zinc oxide.
Miller, P.H., Jr.
Semi-cond. Mater., Proc. Conf., Univ.
Reading 1951: 172-9 (1951)
CA 49: 28e

A 257 Fluorescence test as applied to car-
boxylic acids.
Mulay, L.N., and Mathur, R.M.
Current Sci. 20: 206-7 (1951)
CA 46: 6939c

A 258 Spectra in phosphorescence.
Nagy, E., and Gergely, G.
Acta Phys. Acad. Sci. Hung. 1: 115-25
(1951)
CA 46: 3410h

A 259 Water content of autunite.
Nakanishi, M.
Nat. Sci. Rept. Ochanomizu Univ., Tokyo
1: 71-3 (1951)
CA 48: 4380h

A 260 Vibration energy and luminescence of
complex molecules.
Neporent, B.S.
Usp. Fiz. Nauk 43: 380-402 (1951)
CA 48: 6833h

A 261 Correspondence between absorption
and emission and the origin of broad
bands in the spectra of complex
molecules.
Neporent, B.S.
Zh. Eksperim. i Teor. Fiz. 21: 172-88
(1951)
CA 46: 2398d

A 262 Fluorescent bluing agents as corrosion
catalysts.
Nieuwenhuis, K.J.
Wasindustrie 1: 12-13 (1951)
CA 46: 11719i

A 263 Studies on fluorescent zinc sulfide by
double refraction. II. Relation of
crystal systems, fluorescence, and
phosphorescence.
Niki, E., and Shirai, H.
J. Chem. Soc. Japan 54: 315-17 (1951)
CA 47: 3123b

A 264 Synthesis of organic fluorescent compounds. I. Syntheses of imidazolone
derivatives from acyloins and urea.
Oda, R., and Yoshida, Z.
Bull. Inst. Chem. Res., Kyoto Univ. 26:
89-90 (1951)
CA 46: 8100i

A 265 The synthesis of organic fluorescent
compounds. I. Synthesis of ureides
containing benzene or naphthalene nucleus and the fluorescence of these
compounds.
Oda, R., Yoshida, Z., and Mosuda, H.
J. Chem. Soc. Japan, Ind. Chem. Sect. 54:
540-3 (1951)
CA 48: 1982i

A 266 Theory on the fluorescence and chemical constitution of organic compounds.
Oda, R., and Yoshida, Z.
Mem. Fac. Eng. Kyoto Univ. 13: 108-22 (1951)
CA 46: 6939b

A 267 Fluorescence of Auramine O in the presence of nucleic acid.
Oster, G.
Compt. Rend. 232: 1708-10 (1951)
CA 45: 8359a

A 268 Fluorescence quenching by nucleic acids.
Oster, G.
Trans. Faraday Soc. 47: 660-6 (1951)
CA 45: 10059d

A 269 Fluorescence spectra of uranyl salts.
Pant, D.D.
J. Sci. Res. Benares Hindu Univ. 1: 60-3 (1950-51)
CA 46: 6938i

A 270 Mechanism of the flash of sodium chloride-copper phosphors excited with X-rays.
Parfianovich, I.A.
Zh. Eksperim. i Teor. Fiz. 21: 314-21 (1951)
CA 45: 10062h

A 271 Fluorescence and absorption of crystalline stilbene.
Pesteil, P.
Compt. Rend. 233: 377-9 (1951)
CA 46: 1355a

A 272 Fluorescence and absorption of crystallized stilbene.
Pesteil, P.
Compt. Rend. 233: 1356 (1951)
CA 46: 4909g

A 273 Decay of phosphorescence and duration of the metastable state of organic molecules at low temperature.
Pyatnitskii, B.A.
Izv. Akad. Nauk SSSR, Ser. Fiz. 15: 597-603, Discussion 604 (1951)
CA 46: 8972i

A 274 Quenching of fluorescence in solution. Effect of the structure of the quencher on the efficiency of the reaction.
Rowell, J.C., and LaMer, V.K.
J. Am. Chem. Soc. 73: 1630-4 (1951)
CA 46: 6940b

A 275 A note on the fluorescence of Wyoming bentonite.
Samson, H.R.
Am. Mineralogist 36: 160-61 (1951)
CA 47: 6831f

A 276 Acidity and basicity of molecules in the fundamental state and in an excited state.
Sandorfy, C.
Compt. Rend. 232: 841-3 (1951)
CA 45: 5515b

A 277 Fluorescence test for the detection of argemone oil in mustard oil.
Sarkar, S.N., and Nandi, D.L.
Current Sci. 20: 232-3 (1951)
CA 46: 8394a

A 278 Fluorescence of benzotrifluoride and p-xylene in the near ultraviolet.
Sastri, M.L.N., and Sponer, H.
Phys. Rev. 83: 245 (1951)
CA 47: 6754e

A 279 Term analysis of the emission bands of the sulfide phosphors.
Schenck, R.
Z. Elektrochem. 55: 1-7 (1951)
CA 45: 6062i

A 280 Term analysis of the emission bands of Lenard's sulfide phosphors. II.
Schenck, R.
Z. Elektrochem. 55: 7-12 (1951)
CA 45: 6063b

A 281 The use of optical whiteners in soaps and detergents.
Schlacter, A.
Fette u. Seifen 53: 735-40 (1951)
CA 46: 6854a

A 282 Effect of visible and ultraviolet light and of cathode rays, and X-rays on a fluorescent screen coated with cadmium iodide.
Schlivitch, S.
Compt. Rend. 233: 1023-4 (1951)
CA 46: 2404i

A 283 A vibrational analysis of the fluorescence of naphthalene vapor.
Schnepp, O., and McClure, D.S.
Phys. Rev. 85: 755 (1951)
CA 47: 6756g

A 284 The terms of the activators in the band models of phosphorescent sulfides.
Schon, M.
Z. Naturforsch. 6a: 287-9 (1951)
CA 46: 2405c

A 285 Identification of petroleum products by chromatographic fluorescence methods.
Schuldiner, J.A.
Anal. Chem. 23: 1676-80 (1951)
CA 46: 7311a

A 286 Fluorometric investigations on terramycin.
Serembe, M.
Arch. Ital. Sci. Farmacol. 1: 244-9 (1951)
CA 50: 1970h

A 287 Fluorescence of terramycin: fluorimetric titration.
Serembe, M.
Boll. Soc. Ital. Biol. Sper. 27: 1330-1 (1951)
CA 46: 5112i

A 288 Effect of water of crystallization on the fluorescence spectrum of uranyl nitrate. I.
Sevchenko, A.N., Vdovenko, V.M., and Kovaleva, T.V.
Zh. Eksperim. i Teor. Fiz. 21: 204-11 (1951)
CA 46: 2407d

A 289 Photoluminescence of sublimate phosphors of zinc sulfide and zinc selenide.
Shalimova, K.V.
Dokl. Akad. Nauk SSSR 80: 587-90 (1951)
CA 46: 829d

A 290 Nature of the photoluminescence of silver-activated silver halide sublimate phosphors.
Shalimova, K.V., and Belkina, A.V.
Zh. Eksperim. i Teor. Fiz. 21: 326-37 (1951)
CA 45: 8895i

A 291 The distribution of thiamine and riboflavin in rice grains and of thiamine in parboiled rice.
Simpson, I.A.
Cereal Chem. 28: 259-70 (1951)
CA 45: 8157a

A 292 The size of soap micelles in benzene from osmotic pressure and from the depolarization fluorescence.
Singleterry, L.C., and Weinberger, L.A.
J. Am. Chem. Soc. 73: 4574-9 (1951)
CA 46: 4325h

A 293 The fluorescence and extinction and partition coefficients of estrogens.
Slaunwhite, W.R., Jr., Engel, K.L., Olmstead, P.C., and Carter, P.
J. Biol. Chem. 191: 627-31 (1951)
CA 46: 556i

A 294 The long-lived phosphorescence of potassium iodide-thallium iodide.
Smaller, B., and Avery, E.C.
Phys. Rev. 85: 766-7 (1951)
CA 47: 6757a

A 295 Influence of nucleic acid upon the fluorescence of nuclei and cytoplasm after injection of aromatic diamidines.
Snapper, I., Schneid, B., Lieben, F., Gerber, I., and Greenspan, E.
J. Lab. Clin. Med. 37: 562-74 (1951)
CA 45: 6290d

A 296 Visible fluorescence and chemical constitution of compounds of the benzopyrone group. IV.
Sreerama-Morti, V.V., Rajagopalan, S., and Ramachandra, L.
Proc. Indian Acad. Sci. 34A: 319-23 (1951)
CA 47: 6411i

A 297 Nature of the elementary processes of absorption and fluorescence of uranyl compounds.
Stepanov, B.I.
Zh. Eksperim. i Teor. Fiz. 21: 1153-7 (1951)
CA 46: 10881e

A 298 Width of spectral lines of uranyl salts and its temperature dependence.
Stepanov, B.I.
Zh. Eksperim. i Teor. Fiz. 21: 1158-63 (1951)
CA 46: 10882d

A 299 Isolation and properties of firefly
luciferesceine.
Strehler, B.
Arch. Biochem. Biophys. 32: 397-406
(1951)
CA 46: 1661b

A 300 Light production by green plants.
Strehler, B.L., and Arnold, W.
J. Gen. Physiol. 34: 809-20 (1951)
CA 46: 9672f

A 301 Solvatochromy.
Suhrmann, R., and Perkampus, H.H.
Naturwissenschaften 38: 382 (1951)
CA 46: 3855h

A 302 Effect of temperature and viscosity
on the absorption and emission of
acridine derivatives.
Tarasova, T.M.
Zh. Eksperim. i Teor. Fiz. 21: 189-203
(1951)
CA 46: 2399f

A 303 Sensitized fluorescence in vapors of
organic compounds.
Terenin, A.N., and Karyakin, A.V.
Izv. Akad. Nauk SSSR, Ser. Fiz. 15: 550-5
(1951)
CA 46: 8971d

A 304 Transfer and migration of energy in
biochemical processes.
Terenin, A.N.
Usp. Fiz. Nauk 43: 347-79 (1951)
CA 48: 7120b

A 305 Effect of adsorption of water on the
quenching of the fluorescence of adsorbates.
Terenin, A.N., and Karyakin, A.V.
Zh. Eksperim. i Teor. Fiz. 21: 107-13
(1951)
CA 45: 7882g

A 306 The early stages of the oxidation of
adrenaline in dilute solution.
Trautner, E.M., and Bradley, T.R.
Australian J. Sci. Res. B4: 303-43 (1951)
CA 46: 681h

A 307 The measurement of "daylight fluores-
cent" materials.
Tyler, J.E., and Callahan, F.P., Jr.
J. Opt. Soc. Am. 41: 997-1001 (1951)
CA 46: 2913c

A 308 Use of whiteners (in laundering).
Uhl, O.
Fette u. Seifen 53: 545-8 (1951)
CA 46: 4255a

A 309 Fluorescence and induction phenomena
in photosynthesis.
Van der Veen, R.
Physiol. Plantarum 4: 486-94 (1951)
CA 48: 11572c

A 310 Refractive index and fluorescence of
different samples of milk.
Venkatasubramanian, T.A., and
Ramakrishnan, C.V.
Sci. Cult. 17: 260-1 (1951)
CA 46: 9224f

A 311 The emission of zinc sulfide phosphors.
Vinokurov, L.A., Levshin, V.L., and
Baranova, E.G.
Zh. Eksperim. i Teor. Fiz. 21: 236-51
(1951)
CA 45: 7877g

A 312 Extinction of the chlorophyll fluores-
cence.
Vrbaski, T.
Arhiv Kem. 23: 6-13 (1951)
CA 46: 9995e

A 313 Size effect in the luminescence of zinc
sulfide phosphors.
Wallick, G.C.
Phys. Rev. 84: 375 (1951)
CA 46: 829h

A 314 Chlorophyll fluorescence and photosyn-
thesis.
Wassink, E.C.
Advan. Enzymol. 11: 91-199 (1951)
CA 46: 585i

A 315 Properties of a calcium sulfate: Man-
ganese phosphor under vacuum ultra-
violet excitation.
Watanabe, K.
Phys. Rev. 83: 785-91 (1951)
CA 45: 9372i

A 316 Energy transfer in the photosensitiza-
tion of silver halide photographic emulsions: Optical sensitization, supersen-
sitization, and antisensitization.
West, W., and Carroll, B.H.
J. Chem. Phys. 19: 417-27 (1951)
CA 45: 9405f

A 317 The use of fluorescence in qualitative
analysis.
White, C.E.
J. Chem. Educ. 28: 369-72 (1951)
CA 45: 9417h

A 318 Determination of lithium in rocks.
Fluorometric method.
White, C.E., Fletcher, M.H., and Parks, J.
Anal. Chem. 23: 478-81 (1951)
CA 45: 5564c

A 319 Optical bleaching agents.
Widaly
Seifen-Oele-Fette-Wachse 77: 471-2 (1951)
CA 46: 11720c

A 320 Relations between dye phosphors and
hydrogen bonds of rigid media or sur-
face of adsorbents.
Yamamoto, D., and Iwaki, R.
J. Chem. Phys. 19: 662 (1951)
CA 45: 7882e

A 321 dye gelatin phosphors.
Yamamoto, D., and Iwaki, R.
J. Chem. Soc. Japan, Pure Chem. Sect.
72: 1075-8 (1951)
CA 46: 5974h

A 322 Nonexponential decay of the lumines-
cence of solid aromatic hydrocarbons.
Yastrebov, V.A.
Zh. Eksperim. i Teor. Fiz. 21: 164-71
(1951)
CA 45: 7881f

A 323 Synthesis of organic fluorescent compounds. III. Synthesis of the imidazolone derivatives from acyloins and
urea.
Yoshida, Z., Masuda, H., and Oda, R.
J. Chem. Soc. Japan., Ind. Chem. Sect.
54: 685-6 (1951)
CA 48: 2690g

A 324 Synthesis of organic fluorescent compounds. VI. Sulfonation of triaryl-
imidazolinones.
Yoshida, Z., and Oda, R.
J. Chem. Soc. Japan, Ind. Chem. Sect. 54:
732-3 (1951)
CA 48: 3353g

A 325 Detergent composition containing di-
imidazoles.
Ackerman, F., and Meyer, J.
U.S. 2,604,454 (1952)
CA 46: 10650i

A 326 The fluorescence of solutions by
α-particles.
Ageno, I.M., and Querzoli, I.M.
Rend. Ist. Super. Sanita 15: 28-31 (1952)
CA 46: 8537a

A 327 Fluorescence produced in stilbene by
α-particles.
Ageno, M., and Cortellessa, G.
Nuovo Cimento 9: 196-8 (1952)
CA 46: 9425g

A 328 Kinetics of phosphorescence.
Antonov-Romanovskii, V.V.
Dokl. Akad. Nauk SSSR 85: 517-20 (1952)
CA 46: 9991a

A 329 Spectroscopic examination of fluores-
cence in normal human bile.
Arpino, G.
Riforma Med. 66: 578 (1952)
CA 49: 5622a

A 330 Some characteristics of X-ray-excited
phosphorescence of calcite.
Auxier, J., and Forman, G.
Phys. Rev. 87: 170-1 (1952)
CA 48: 8023g

A 331 Possible form of an infrared-sensitive
multiplier.
Baumgartner, W., and Schaetti, N.
Helv. Phys. Acta 25: 611-15 (1952)
CA 47: 5255e

A 332 Fluorescence lines in the spectrum of
light scattered by calcite.
Bhagavantam, S., and Puranik, P.G.
Nature 169: 37 (1952)
CA 46: 5437i

A 333 Photographic sensitivity as a function
of exposure time and temperature.
Biltz, M.
J. Opt. Soc. Am. 42: 898-903 (1952)
CA 47: 2071h

A 334 The ultraviolet absorption spectra and
the intensity of the fluorescence band
4156 A of diamond.
Bishui, B.M.
Indian J. Phys. 26: 347-56 (1952)
CA 51: 7871c

A 335 Decay time, fluorescent efficiencies,
and energy storage properties for various substances with γ-ray or
α-particle excitation.
Bittman, L., Furst, M., and Kallmann, H.
Phys. Rev. 87: 83-6 (1952)
CA 46: 8981g

A 336 The effect of pyrazolone derivatives on
the fluorescence of dyes.
Bock, J., and Klingenberg, H.
Arch. Ophthalmol. 152: 514-19 (1952)
CA 46: 7661c

A 337 Hygienic evaluation of water by fluor-
escence in filtered ultraviolet light.
Bode, F.
Schweiz. Z. Hydrol. 14: 298-303 (1952)
CA 47: 12711h

A 338 Light absorption and fluorescence of
aromatic amines in superalkaline solu-
tion.
Bonitz, E., and Forster, T.
Z. Elektrochem. 56: 137-40 (1952)
CA 46: 8970e

A 339 Field-dependent fluorescence of vitre-
ous Zn_2SiO_4 phosphor.
Bramley, A., and Rosenthal, J.E.
Phys. Rev. 87: 1125 (1952)
CA 46: 10911b

A 340 Effect of pressure on the sensitivity
and fluorescence of the photographic
emulsion.
Braum, A.
Czech. J. Phys. 1: 171-84 (1952)
CA 47: 7351i

A 341 Dependence of the wavelength of the
fluorescence of organic phosphors in
the X-ray region.
Breitling, G.
Z. Angew. Phys. 4: 401-9 (1952)
CA 47: 11965b

A 342 Electrical conductivity connected with
phosphorescence in cadmium sulfide
crystals.
Broser, I., and Warminsky, R.
Z. Physik 133: 340-61 (1952)
CA 47: 7326b

A 343 Use of ultraviolet light in chromato-
graphy.
Brumberg, E.M.
Izv. Akad. Nauk SSR, Otd. Khim. Nauk
1950: 127-36 (1952)
CA 58: 2479h

A 344 Decreasing fluorescence of synthetic
caffeine.
Buckley, J.S.
U.S. 2,584,839 (1952)
CA 46: 4183a

A 345 Deposition of an impervious metal film
on a granular surface.
Campbell, V.C.
U.S. 2,597,617 (1952)
CA 46: 6943b

A 346 Separation and fluorometric microde-
termination of some ergot alkaloids.
Cerciotti, G.
Farm Sci. e Tec. 7: 298-304 (1952)
CA 46: 10535f

A 347 Molecular conductivity of dyes in solu-
tion. I.
Chaudhury, N.K.
Pakistan J. Sci. Res. 4: 114-17 (1952)
CA 48: 4938e

A 348 Internal absorption of fluorescent
light in large plastic scintillators.
Chou, C.N.
Phys. Rev. 87: 376-7 (1952)
CA 46: 9435b

A 349 Spectroscopic studies in the near ultraviolet of the three isomeric dimethylbenzene vapors. I. Absorption and
fluorescence spectra of p-dimethylben-
zene.
Cooper, C.D., and Sastri, M.L.N.
J. Chem. Phys. 20: 607-13 (1952)
CA 46: 10884h

A 350 Trapping and retrapping of electrons
during phosphorescence.
Curie, D.
Ann. Phys. 7: 749-801 (1952)
CA 47: 9782f

A 351 Experimental measurement of the ef-
fect of a solvent on the rate of a very
fast bimolecular reaction.
Curme, H.G., and Rollefson, G.K.
J. Am. Chem. Soc. 74: 3766-8 (1952)
CA 46: 10806g

A 352 The effect of inert gases on the quench-
ing of fluorescence in the gaseous
state.
Curme, H.G., and Rollefson, G.K.
J. Am. Chem. Soc. 74: 28-31 (1952)
CA 46: 2911e

A 353 Unusual phosphorescence of a diamond.
Custers, J.F.H.
Physica 18: 489-96 (1952)
CA 47: 1495d

A 354 Detection of petroleum products in
spring waters.
Dangl, F., and Nietsch, B.
Mikrochemie Ver. Mikrochim. Acta 39:
336-8 (1952)
CA 46: 11518e

A 355 A note on the phosphorescence of pro-
teins.
Debye, P., and Edwards, J.O.
Science 116: 143-4 (1952)
CA 47: 2043a

A 356 Long-lifetime phosphorescence and the
diffusion process.
Debye, P., and Edwards, J.O.
J. Chem. Phys. 20: 236-9 (1952)
CA 46: 7434i

A 357 Photoelectric testing of different chemical fibers.
de Hauss, J.L.
Reyon, Zellwolle Chemiefasern 30: 248-51
(1952)
CA 46: 8378c

A 358 Biochemistry and th embryonal development of insects. II. Nature of fluorescent substances of eggs of Locusta
migratoria.
DeLerma, B.
Arch. Zool. Ital. 37: 81-91 (1952)
CA 48: 7212d

A 359 Manganese-activated magnesium arsen-
ate phosphors.
Dobrolyubskaya, T.S.
Dokl. Akad. Nauk SSSR 85: 537-8 (1952)
CA 46: 9992a

A 360 Fluorimetric determination of uranium
in ores.
Draganic, I.
Rec. Trav. Inst. Recherches Structure
Matière 1: 89-94 (1952)
CA 47: 7940e

A 361 The "α"-band of formaldehyde and the
relation between the ultraviolet absorp-
tion and fluorescence systems.
Dyne, P.J.
J. Chem. Phys. 20: 811-18 (1952)
CA 47: 2036h

A 362 Fluorescence obtained from formic
acid, carbonyl chloride, and methylene
iodide.
Dyne, P.J., and Style, D.W.G.
J. Chem. Soc. 1952: 2122-4 (1952)
CA 46: 10910h

A 363 Sensitized phosphorescence of organic
molecules at low temperatures.
Ermolaev, V.L., and Terenin, A.N.
Akad. Nauk SSSR, Pamyati, S.
I. Vavilova 1952: 137-46 (1952)
CA 47: 7901h

A 364 Flexible and insoluble phosphorescent
paint for textiles.
Espinosa, M.P.
Span. 205,271, Sept. 24, 1952
CA 48: 9073a

A 365 Luminescent colored plastics.
Espinosa, M.P.
Span. 204,521 (1952)
CA 48: 8585d

A 366 Comparison of the spectral properties
of chlorophyll and pheophytin in differ-
ent solvents.
Evstigneev, V.B., and Gavrilova, V.A.
Dokl. Akad. Nauk SSSR 85: 1073-6 (1952)
CA 47: 1490h

A 367 X-ray spectrum and bond characteris-
tics; K_α fluorescence radiation of sul-
fur.
Faessler, A., and Goehring, M.
Naturwissenschaften 39: 169-77 (1952)
CA 47: 2588h

A 368 Microchemical detection of 8-quinolinol
and its derivatives.
Feigl, F.
Mikrochemie Ver. Mikrochim. Acta 39:
404-8 (1952)
CA 46: 11013b

A 369 New demonstrations on fluorescence.
Feigl, F., and Heisig, G.B.
J. Chem. Educ. 29: 192-4 (1952)
CA 46: 8428e

A 370 The phosphorescence emission of
benzophenone.
Ferguson, J., and Tinson, H.J.
J. Chem. Soc. 1952: 3083-5 (1952)
CA 47: 7903c

A 371 Energy transfers in active nitrogen. I.
Transfers with mercury.
Finkelstein, M.A.
J. Chim. Phys. 49: 185-95 (1952)
CA 49: 8517c

A 371a Energy transfers in active nitrogen.
II. Transfers with cadmium, zinc, sod-
ium, potassium, and antimony.
Finkelstein, M.A.
J. Chim. Phys. 49: 196-203 (1952)
CA 46: 8517c

A 372 Fluorescence measurements of chloro-
phyll solutions.
Forster, L.S.
Dissertation Abstr. 12: 1311-2 (1952)
CA 46: 7419h

A 373 The absolute quantum yields of the
fluorescence of chlorophyll solutions.
Forster, L.S., and Livingston, R.
J. Chem. Phys. 20: 1315-20 (1952)
CA 47: 9780i

A 374 Primary photochemical processes in
polyatomic molecules.
Forster, T.
Z. Elektrochem. 56: 716-22 (1952)
CA 47: 9749a

A 375 Calcination and fluorescence in the
evaluation of samples of the sillimanite
minerals.
Foster, W.R., Riddle, F.H., and Royal, H.F.
Am. Ceram. Soc. Bull. 31: 326-8 (1952)
CA 46: 10566d

A 376 Fluorescence spectra of red algae and
the transfer of energy from phycoeryth-
rin to phycocyanin and chlorophyll.
French, C.S., and Young, V.K.
J. Gen. Physiol. 35: 873-90 (1952)
CA 46: 11342i

A 377 Determination of thiamine in the pres-
ence of fluorescent substances.
Fried, R.
Hoppe-Seylers Z. Physiol. Chem. 291:
57-9 (1952)
CA 48: 12212a

A 378 Investigations of petrolatum from
Emsland crude oil.
Fuchs, W., and Lommerzheim, W.
Erdoel. Kohle 5: 148-51 (1952)
CA 46: 7313c

A 379 High-energy-induced fluorescence in
organic liquid solutions (energy trans-
port in liquids). III.
Furst, M., and Kallmann, H.
Phys. Rev. 85: 816-25 (1952)
CA 46: 5977c

A 380 Change of the fluorescence spectra of
magnesium phthalocyanin and chloro-
phyll as a result of the dark reaction.
Gachkovskii, V.I.
Dokl. Akad. Nauk SSSR 82: 739-42 (1952)
CA 46: 6940d

A 381 The role of the magnesium aton in oxidation reaction, catalyzed by magnesium phthalocyamine and chlorophyll, in
the dark.
Gachkovskii, V.I.
Zh. Fiz. Khim. 26: 1713-15 (1952)
CA 49: 5972i

A 382 Lubricating oils.
Geiser, N.
German 829,163 (1952)
CA 50: 7446f

A 383 Luminescence of alkali halide crystals
subjected to ionizing radiation.
Ghormley, J.A., and Levi, H.A.
J. Phys. Chem. 56: 548-54 (1952)
CA 46: 8527a

A 384 The spectral distribution of energy in
the fluorescence of sodium fluorescein.
Gilmore, E.H.
Phys. Rev. 88: 174 (1952)
CA 48: 8025h

A 385 Absolute quantum efficiencies of lumin-
escence of organic molecules in solid
solution.
Gilmore, E.H., Gibson, G.E., and McClure,
D.S.
J. Chem. Phys. 20: 829-36 (1952)
CA 47: 2043h

A 386 Luminescence of alkali sulfides and
sulfates.
Gobrecht, H., and Hahn, D.
Z. Physik 132: 111-28 (1952)
CA 46: 8969g

A 387 The fluorescence reactions of steroids.
I. Estrogens.
Goldzieher, J.W., Bodenchuk, J.M., and
Nolan, P.
J. Biol. Chem. 199: 621-9 (1952)
CA 47: 2804h

A 388 The fluorescence of coumarin deriva-
tives as a function of pH. II.
Goodwin, R.H., and Kavanagh, F.
Arch. Biochem. Biophys. 36: 442-55
(1952)
CA 47: 408e

A 389 Fluorimetric methods of uranium ana-
lysis.
Grimaldi, F.S,, May, I., and Fletcher,
M.H.
U.S. Geol. Surv., Circ. 199, 200 pp. (1952)
CA 47: 3180d

A 390 The luminescence of some powder
phosphors.
Guminski, K., and Ruziewicz, Z.
Bull. Intern. Acad. Polon. Sci., Classe
Sci. Math. Nat., Ser A 1951: 109-21
(1952)
CA 48: 1159a

A 391 The structure of phosphors of the alum-
inum oxide-calcium fluoride type acti-
vated with manganese.
Gunther, G., and Perlitz, H.
Arkiv Kemi 3: 565-9 (1952)
CA 46: 9426i

A 392 New geochemical studies at Bad-
Gastein.
Haberlandt, H.
Mikrochemie Ver. Mikrochim. Acta 39:
92-100 (1952)
CA 46: 4971f

A 393 The molecular orbital theory of chemical valency. XI. Bond energies, resonance energies, and triplet state energies.
Hall, G.C.
Proc. Roy. Soc. (London), Ser. A 213:
113-23 (1952)
CA 46: 10851c

A 394 Purple fluorescent substances in the
skin and scales of fish.
Hama, T., Goto, K., and Kushibiki, K.
Kagaku 22: 478 (1952)
CA 46: 10465h

A 395 Polarization of fluorescence from dis-
sociated molecules.
Hanle, W., and Scharman, A.
Z. Naturforsch. 7a: 635-6 (1952)
CA 47: 7325i

A 396 A roentgen tube with transparent anode
for secondary excitation of roentgen
spectra.
Herglotz, H.
Oesterr. Ing.-Arch. 6: 135-40 (1952)
CA 46: 8512i

A 397 A halochromic boron-quercetin com-
pound.
Horhammer, L., Hansel, R., and Strasser,
F.
Arch. Pharm. 285: 286-90 (1952)
CA 48: 1192i

A 398 Fluorescence of porcelain wares.
Iimori, S., and Kato, K.
Rept. Sci. Res. Inst. 28: 132-8 (1952)
CA 46: 10910i

A 399 The aspergilli found in canned
marrons glaces.
Iizuka, H.
J. Agr. Chem. Soc. Japan 26: 68-71 (1952)
CA 48: 10113h

A 400 Examination of corrosion of refractories by observing the fluorescence of the surrounding glass.
Jaupain, M.
Verres Refractaires 6: 356-61 (1952)
CA 47: 7181e

A 401 Induced conductivity and light emission in different luminescent-type powders.
Kallmann, H., and Kramer, B.
Phys. Rev. 87: 91-107 (1952)
CA 46: 8970a

A 402 Scintillation counting techniques.
Kallmann, H., Furst, M., and Sidran, M.
Nucleonics 10: 15-17 (1952)
CA 47: 7336h

A 403 Phosphorescence of 2-hydroxyanthraquinone.
Karyakin, A.V., and Kalenichenko, Y.I.
Zh. Fiz. Khim. 26: 103-5 (1952)
CA 47: 4747g

A 404 Collisional perturbation of spin-orbital coupling and the mechanism of fluorescence quenching. A visual demonstration of the perturbation.
Kasha, M.
J. Chem. Phys. 20: 71-4 (1952)
CA 46: 6491e

A 405 Fluorescence of ionic and atomic copper centers in NaCl-Cu phosphors.
Kats, M.L.
Dokl. Akad. Nauk SSSR 85: 757-60 (1952)
CA 46: 10910c

A 406 Luminescence of ionic and atomic silver centers in X-rayed NaCl-Ag phosphors.
Kats, M.L.
Zh. Eksperim. i Teor. Fiz. 23: 720-7 (1952)
CA 47: 3702a

A 407 Luminescence of atomic and ionic silver centers in NaCl-Ag phosphors.
Kats, M.L.
Dokl. Akad. Nauk SSSR 85: 539-42 (1952)
CA 46: 9992b

A 408 The interactions between organic dyes and polymers. I. Influence of potassium polyvinyl sulfate on the absorption spectrum and fluorescence of rhodamine GG aqueous solution.
Koizumi, M., and Mataga, N.
J. Chem. Soc. Japan 73: 814-16, 879-81 (1952)
CA 47: 5797b

A 409 Characteristics of the class of sulfate luminophors.
Konstantinova-Shlezinger, M.A., Gorbacheva, N.A., and Panasyuk, E.I.
Zh. Eksperim. i Teor. Fiz. 23: 588-92 (1952)
CA 47: 3122d

A 410 Quenching of fluorescence by foreign substances.
Kortum, G., Baur, H., and Friedheim, G.
Z. Physik. Chem. 200: 293-301 (1952)
CA 47: 3122i

A 411 Oxyacid phosphors. I. Addition of an oxidizing agent at firing.
Kotera, Y.
Rept. Govt. Chem. Ind. Res. Inst., Tokyo 46: 451-4 (1952)
CA 47: 2042h

A 412 Absorption and fluorescence of solid solutions MgO-NiO.
Kroger, F.A., Vink, H.J., and van den Boomgaard, J.
Physica 18: 77-82 (1952)
CA 46: 5438c

A 413 Germination ability and production of fluorescent material by seeds.
Kugler, I.
Naturwissenschaften 39: 213 (1952)
CA 47: 2838f

A 414 Application of the metallic model to the absorption and fluorescence of some dyes.
Laffitte, E.
Compt. Rend. 235: 36-7 (1952)
CA 47: 44d

A 415 The fundamental polarization of the luminescence of dyes. Fluorescence and α-phosphorescence.
Laffitte, E.
Compt. Rend. 234: 424-6 (1952)
CA 46: 6491d

A 416 Fluorometric determination of traces of beryllium.
Laitinen, H.A., and Kivalo, P.
Anal. Chem. 24: 1467-71 (1952)
CA 46: 11022d

A 417 Trapping of electrons in fused quartz.
Lautout, M.
Compt. Rend. 234: 330-2 (1952)
CA 46: 7878a

A 418 Effect of some reagents on the fluorescence of stilbamidine and 2-hydroxystilbamidine in vitro.
Lieben, F.
J. Mt. Sinai Hosp., N.Y. 19: 217-20 (1952)
CA 46: 10909g

A 419 Temperature effects in organic fluors.
Liebson, S.H.
Nucleonics 10: 41-5 (1952)
CA 46: 10909f

A 420 Absorption and fluorescent properties of the α-estradiol-sulfuric acid complex.
Linford, J.H.
Can. J. Med. Sci. 30: 199-212 (1952)
CA 46: 8520h

A 421 Absorption and fluorescence properties in the visible spectral region of certain steroids in sulfuric acid solutions.
Linford, J.H., and Paulson, O.B.
Can. J. Med. Sci. 30: 213-30 (1952)
CA 46: 8970c

A 422 Fluorescence spectra of some organic solid solutions.
Lipsett, F.R., and Dekker, A.J.
Can. J. Phys. 30: 165-73 (1952)
CA 46: 9428c

A 423 Quenching of the fluorescence of solutions of porphyrins and of chlorophyll.
Livingston, R., Thompson, L., and Ramarao, M.V.
J. Am. Chem. Soc. 74: 1073-5 (1952)
CA 46: 5438g

A 424 The preparation of fluorescent screens from colloidal solutions of polycyclic hydrocarbons in water.
Ludwig, F., and Noach, P.L.
Compt. Rend. 234: 2441-3 (1952)
CA 46: 9994h

A 425 Photochemical reactions in alkali halide crystals.
Martienssen, W.
Z. Physik 131: 488-504 (1952)
CA 46: 6495i

A 426 Influence of the addition of potassium polyvinyl sulfate upon the absorption spectra and fluorescence of the aqueous rhodamine GG solution.
Mataga, N., and Koizumi, M.
J. Inst. Polytech., Osaka City Univ. 3: 21-35 (1952)
CA 47: 8376b

A 427 Spin-orbit interaction in aromatic molecules.
McClure, D.S.
J. Chem. Phys. 20: 682-6 (1952)
CA 46: 10851i

A 428 Some factors influencing fluorescence in minerals.
McDougall, D.J.
Am. Mineralogist 37: 427-37 (1952)
CA 48: 5025i

A 429 The cellular localization of material after the administration of the carcinogen, β-naphthylamine: fluorescence microspectroscopy of epithelial cells of the bladder.
Mellors, R.C., and Hlinka, J.
Cancer 5: 242-8 (1952)
CA 46: 6258g

A 430 Inter- and intramolecular energy-transfer processes.
Moodie, M.M.
J. Chem. Phys. 20: 1212-13, 1510-15 (1952)
CA 47: 9780a

A 431 The clinical value of fluorescent and radioactive tracer methods for the diagnosis and localization of brain tumors.
Moore, G.E., Peyton, W.T., French, L.A., and Caudill, M.
Acta, Unio. Intern. Contra Cancrum 8: 595-8 (1952)
CA 48: 9532h

A 432 The synthesis of organic fluorescent compounds. VIII. Sulfonation of 3,6-dinitrocarbazole.
Oda, R., Yoshida, Z., and Kato, Y.
J. Chem. Soc. Japan, Ind. Chem. Sect. 55: 239-41 (1952)
CA 48: 7599g

A 433 Phosphorescence of alkaline earth sulfides and oxides and of the sulfide and oxide of zinc.
Patrovsky, V.
Chemie 8: 194-6 (1952)
CA 47: 10987e

A 434 Absorption and fluorescence of several organic compounds pure and in solution.
Pesteil, P.
Compt. Rend. 234: 2532-4 (1952)
CA 46: 9427c

A 435 Fluorescence of crystalline naphthalene at 20°.
Pesteil, P.
Compt. Rend. 235: 150-2 (1952)
CA 46: 10910g

A 436 Spectroscopic studies in the rear ultraviolet of the three isomeric dimethylbenzene (xylene) vapors. III. Fluorescence of m-dimethylbenzene (m-xylene).
Rao, V.R., and Sastri, M.L.N.
J. Chem. Phys. 20: 1552-3 (1952)
CA 47: 7326e

A 437 The absorption and fluorescence spectra of p-terphenyl in toluene.
Ravilious, C.F.
Rev. Sci. Instr. 23: 760 (1952)
CA 47: 9772g

A 438 The lowest triplet levels of aromatic hydrocarbons.
Reid, C.
J. Chem. Phys. 20: 1214-15 (1952)
CA 47: 9780c

A 439 The effect of pressure on scintillation phosphors (anthracene).
Reinsch, A.J., and Drichamer, H.G.
J. Appl. Phys. 23: 152-3 (1952)
CA 46: 6491d

A 440 Spectra of ruby.
Ritschl, R., and Muller, R.
Z. Physik 133: 237-43 (1952)
CA 47: 4203e

A 441 Phosphorescent paint.
Rodriquez, C.F., and Espinosa, P.J.P.
Span. 206,368 (1952)
CA 48: 8557e

A 442 Production of monochromatic X-radiation for microradiography by excitation of fluorescent characteristic radiation.
Roger, T.H.
J. Appl. Phys. 23: 881-7 (1952)
CA 46: 10864h

A 443 Properties of the basic blast-furnace slags liable to disintegration.
Sabela, W.
Prace Inst. Met. 4: 371-6 (1952)
CA 47: 8610b

A 444 The fluorescent substances of the cocoon layer in Bombyx mori.
Sakate, S., Kawaguchi, K., and Sato, K.
J. Sericult. Sci. Japan 21: 237-9 (1952)
CA 48: 10942a

A 445 Fluorescence quantitative spectrophotometry of cerebrospinal fluid (CSF) in schizophrenia.
Salvi, P.
Acta Neurol. 7: 49-58 (1952)
CA 47: 11486e

A 446 Fluorescence studies of some simple benzene derivatives in the near ultraviolet. III. Benzotrifluoride.
Sastri, M.L.N., and Sponer, H.
J. Chem. Phys. 20: 1428-31 (1952)
CA 47: 9779i

A 447 Determination of the fluorescing intensity of uranium traces.
Savic, P., and Draganic, I.
Bull. Acad. Serbe Sci., Classe Sci. Math. Nat. 4: 149-50 (1952)
CA 50: 7619d

A 448 Problems of photochemical sensitization.
Scheibe, G.
Z. Elektrochem. 56: 723-8 (1952)
CA 47: 9194g

A 449 Term analysis of the emission bands of the sulfide phosphors. III.
Schenck, R.
Z. Elektrochem. 56: 132-6 (1952)
CA 46: 8970f

A 450 A vibrational analysis of the fluorescence of naphthalene vapor.
Schnepp, O., and McClure, D.S.
J. Chem. Phys. 20: 1375-83 (1952)
CA 47: 9772e

A 451 Lattice defects in surfaces.
Schwab, G.M.
Z. Elektrochem. 56: 297-302 (1952)
CA 46: 8447g

A 452 Fluorescence determinations on filter-
paper strips.
Semm, K., and Fried, R.
Naturwissenschaften 39: 326-7 (1952)
CA 47: 7325h

A 453 Life of the excited state of some phos-
phors.
Shalimova, K.V., and Belikova, T.P.
Dokl. Akad. Nauk SSSR 82: 713-6 (1952)
CA 46: 5976f

A 454 Fluorescence spectrum of coronene in
frozen solutions.
Shpol'skii, E.V., Ll'ina, A.A., and
Klimova, L.A.
Dokl. Akad. Nauk SSSR 87: 935-8 (1952)
CA 47: 4205b

A 455 Fluorescence and associated changes
produced upon storage of evaporated
milk.
Simonson, H.D., and Tarassuk, N.P.
J. Dairy Sci. 35: 166-73 (1952)
CA 46: 4692d

A 456 A new case of molecular light scatter-
ing by completely miscible liquids.
Simova, P.
Izv. Bulgar. Akad. Nauk Otd. Fiz.-Mat. i
Tekh. Nauki, Ser. Fiz. 3: 3-12 (1952)
CA 49: 5128g

A 457 Bond differences between crystalline
and amorphous conditions.
Smekal, A.G.
Naturwissenschaften 39: 505-6 (1952)
CA 47: 9075h

A 458 Titanium-activated calcium magnesium
silicate phosphor.
Smith, A.L.
U.S. 2,589,513 (1952)
CA 46: 4917b

A 459 Fluorescent screens.
Spencer, P.L.
U.S. 2,601,178 (1952)
CA 46: 7880a

A 460 Application of fluorescence X-rays to
metallurgical microradiography.
Splettstosser, H.R., and Seemann, H.E.
J. Appl. Phys. 23: 1217-22 (1952)
CA 47: 1485c

A 461 Fluorescence in ultraviolet light in the
study of boron deficiency in celery.
Spun, A.R.
Science 116: 421-3 (1952)
CA 47: 12535b

A 462 Physics of crystal phosphors.
Stockman, F.
Naturwissenschaften 39: 226-33, 246-54
(1952)
CA 47: 2587i

A 463 Surface-active agents and detergents.
Stupel, H.
SVF Fachorgan Textilveredlung 7: 219-24
(1952)
CA 46: 10647i

A 464 Chlorophyll a. The polarization of
fluorescent light for the determination
of the directions of the transition moments of the absorption bands.
Stupp, R., and Kuhn, H.
Helv. Chim. Acta 35: 2469-82 (1952)
CA 47: 3362h

A 465 Origin of the fluorescence obtained
from formic acid and methylene iodide.
Style, D.W.G., and Ward, J.C.
J. Chem. Soc. 1952: 2125-7 (1952)
CA 46: 9996f

A 466 Solvatochromism.
Suhrmann, R., and Perkampus, H.H.
Z. Elektrochem. 56: 743-5 (1952)
CA 47: 9773a

A 467 The quenching of fluorescence in solutions by additives as a method for investigating the kinetics of bimolecular
reactions in solution.
Sveshnikov, B.Y.
Usp. Fiz. Nauk 46: 331-47 (1952)
CA 48: 10410i

A 468 Kinetics of photoconductivity and kin-
etics of phosphorescence.
Tolstoi, N.A., and Feofilov, P.P.
Izv. Akad. Nauk SSSR, Ser. Fiz. 16: 59-69
(1952)
CA 46: 9426d

A 469 Chemical indicators. I. A study of
fluorescence indicators.
Tomicek, O., and Suk, V.
Chem. Listy 46: 139-44 (1952)
CA 46: 11018g

A 470 Fluorescence spectroscopic investiga-
tions of serum mucoproteins in malig-
nant diseases.
Tomiser, J., Kraus, H., and
Steinbereithner, K.
Z. Krebsforsch. 58: 425-37 (1952)
CA 47: 2868c

A 471 Fluorescence of dye solutions. Visco-
sity and temperature effects.
Tawde, N.R., and Ramanathan, N.
Proc. Phys. Soc. 65: 33-40 (1952)
CA 46: 10910f

A 472 Sensitized phosphorescence of organic
molecules at low temperatures. Intermolecular energy transfer with excitation of the triplet level.
Terenin, A.N., and Ermolaev, V.L.
Dokl. Akad. Nauk SSSR 85: 547-50 (1952)
CA 46: 9994b

A 473 Synthesis and X-ray study of uranium
sulfate minerals.
Traill, R.J.
Am. Mineralogist 37: 394-406 (1952)
CA 48: 4380c

A 474 The luminescence of basic magnesium
arsenate activated by manganese.
Travnicek, M., Kroger, F.A., Botden,
T.P.J., and Zalm, P.
Physica 18: 33-42 (1952)
CA 46: 4914b

A 475 Absorption and fluorescence spectra of
anthracene, phenanthrene, and chrysene
at low temperatures.
Tsujikawa, I., and Kanda, E.
Sci. Rept. Res. Inst., Tohoku Univ. 4A:
471-80 (1952)
CA 47: 9147e

A 476 Chemical studies on some minerals of
Formosan occurrence. I.
Unohara, N., Iimori, H., and Hakomori, S.
J. Chem. Soc. Japan 73: 327-31 (1952)
CA 47: 2651h

A 477 Luminescence of the electronic semi-
conductor, zinc oxide.
Vergunas, F.I., and Konovalov, G.A.
Zh. Eksperim. i Teor. Fiz. 23: 712-19
(1952)
CA 47: 3701e

A 478 Influence of phenylhydrazine on the
fluorescence of chlorophyll in solution.
Watson, W.F.
Trans. Faraday Soc. 48: 526-30 (1952)
CA 46: 10901a

A 479 Polarization of the fluorescence of
macromolecules. II. Fluorescent
conjugates of ovalbumin and bovine
serum albumin.
Weber, G.
Biochem. J. 51: 155-67 (1952)
CA 46: 6939g

A 480 Polarization of the fluorescence of
macromolecules. I. Theory and ex-
perimental method.
Weber, G.
Biochem. J. 51: 145-55 (1952)
CA 46: 6939f

A 481 The fluorescence of solutions of chloro-
phyll and related substances.
Weil, S.A.
Univ. Microfilms 4355, 104 pp. (1952)
CA 47: 3103h

A 482 Fluorescence shifts of naphthols.
Weller, A.
Z. Elektrochem. 56: 662-8 (1952)
CA 47: 3703f

A 483 Fluorometric analysis.
White, C.E.
Anal. Chem. 24: 85-90 (1952)
CA 46: 3895b

A 484 Fluorometric analysis.
White, C.E.
Proc. U.S. Tech. Conf. Air Pollution
1950: 196-200 (1952)
CA 46: 6287d

A 485 Dye phosphors. I. Phosphorescent
bodies suitable for dye phosphors. II.
Quenching of the phosphorescence by
free water in the dye-gelatin phosphors.
Yamamoto, D.
J. Chem. Soc. Japan 73: 739-44, 794-6
(1952)
CA 47: 3702i

A 486 Fluorescent substances in the tissues
of endoparasites.
Yamao, Y.
Zool. Mag. Japan 61: 38-40 (1952)
CA 48: 7804i

A 487 Organic fluorescence and its applica-
tions.
Yoshida, Z.
Kagaku 7: 540-4, 616-23 (1952)
CA 49: 15719f

A 488 Synthesis of organic fluorescent compounds. XIII. Emissivity of fluorescence of naphth-1,2-imidazoles and
perimidines. Fluorescence of the synthetic compounds.
Yoshida, Z., and Oda, R.
J. Chem. Soc. Japan, Ind. Chem. Sect. 55:
565-6 (1952)
CA 48: 6261e

A 489 Synthesis of organic fluorescent compounds. X. Emissivity of fluorescence
of naphth-1,2-imidazoles and perimidines.
Yoshida, Z., Shimada, Y., and Oda, R.
J. Chem. Soc. Japan, Ind. Chem. Sect. 55:
354-6 (1952)
CA 48: 11398c

A 490 Synthesis of organic fluorescent compounds. XII. Emissivity of fluorescence of naphth-1,2-imidazoles and
perimidines. Synthesis of perimidines.
Yoshida, Z., Shimada, Y., and Oda, R.
J. Chem. Soc. Japan 55: 523-4 (1952)
CA 49: 1021d

A 491 The realtion between fluorescence and
chemical constitution of organic compounds.
Yoshida, Z., Shimada, Y., and Oda, R.
Bull. Inst. Chem. Res., Kyoto Univ. 28: 76
(1952)
CA 46: 11185b

A 492 The fluorescence of samarium and
gadolinium in borax.
Zaidel, A.N., and Malakhova, G.P.
Dokl. Akad. Nauk SSSR 85: 591-3 (1952)
CA 46: 99993i

A 493 Quantitative absorption and emission
measurements on the Acridine Orange
cation at room and low temperatures
in an organic solvent and their contribution to the explanation of metachromic fluorescence.
Zanker, V.
Z. Physik. Chem. 200: 250-92 (1952)
CA 47: 3119h

A 494 The proof of definite reversible association of Acridine Orange by absorption
and fluorescence measurements in
aqueous solution.
Zanker, V.
Z. Physik. Chem. 199: 225-58 (1952)
CA 46: 5968c

A 495 Polarization of the fluorescence of
organic crystals.
Zhevandrov, N.D.
Dokl. Akad. Nauk SSSR 83: 677-80 (1952)
CA 46: 7435a

A 496 A multiwavelength fluorescence spec-
trometer (X-ray fluorescence).
Adler, I., and Axelrod, J.M.
J. Opt. Soc. Am. 43: 769-72 (1953)
CA 48: 39d

A 497 Aging of fluorescent liquids.
Ageno, M., Cortellessa, G., and Querzoli,
R.
Rend. Ist Super. Sanita 16: 211-6 (1953)
CA 48: 1161c

A 498 The fluorescence and absorption spec-
tra of estrone, estradiol-M, and estriol
compounds.
Aitken, E.H., and Preedy, J.R.K.
J. Endocrinol. 9: 251-60 (1953)
CA 47: 10352b

A 499 Effect of the emitting·action of the ex-
citing light on the yield of phosphores-
cence.
Anikina, L.I.
Dokl. Akad. Nauk SSSR 88: 41-4 (1953)
CA 47: 5801e

A 500 Effect of the stimulating action of the
exciting light on phosphorescence.
Anikina, L.I.
U.S. At. Energy Comm., Tech. Inform.
Serv., Oak Ridge, Tenn. NSF-tr-23
(1953)
CA 47: 11999c

A 501 Response of a zinc sulfide screen to
α-rays of energy less than 5 MeV.
Anthony, J.P., and Ambrosino, G.
Compt. Rend. 236: 1774-6 (1953)
CA 47: 11014c

A 502 Fluorescent substances.
Aoki, Y.
Japan 978 (1953)
CA 48: 1818g

A 503 Fluorescent substance.
Aoki, Y., and Hanno, M.
Japan 6272 (1953)
CA 48: 11936g

A 504 The spectral sensitivity of infrared-
responsive zinc sulfide phosphors.
Asano, S.
Oyo Butsuri 22: 273-7 (1953)
CA 48: 3145b

A 505 Fluorescent substance.
Awazu, K., and Ide, H.
Japan 1571 (1953)
CA 48: 5660bc

A 505a The determination of the fluorescence
lifetimes of dissolved substances by a
phase-shift method.
Bailey, E.A., and Rollefson, G.K.
J. Chem. Phys. 21: 1315-22 (1953)
CA 47: 11997e

A 505b Fluorescence or organic single crys-
tals at low temperatures. I.
1,1,4,4-tetraphenyl-1,3-butadiene.
Barbaron, M., and Pesteil, P.
Compt. Rend. 236: 1763-4 (1953)
CA 47: 10351c

A 505c Optical bleaching.
Bartkowicz, S.
Przemysl Chem. 9: 112 (1953)
CA 48: 8549c

A 505d New theory concerning the relation
between chemical structure and biolo-
gical activity.
Bertrand, D.
Compt. Rend. 237: 1800-2 (1953)
CA 48: 5236i

A 505e Photofluorescence decay times of
organic phosphors.
Birks, J.B., and Little, W.A.
Proc. Phys. Soc. 66: 921-8 (1953)
CA 48: 6261d

A 505f Absolute scintillation efficiency of
anthracene.
Birks, J.B., and Szendrei, M.E.
Phys. Rev. 91: 197-8 (1953)
CA 47: 10357e

A 505g The decay in fluorescence efficiency
of organic materials on irradiation by
particles and photons.
Black, F.A.
Phil. Mag. 44: 263-7 (1953)
CA 51: 9328h

A 505h Geological features of the Panasqueira
tin-tungsten ore occurrence (Portugal).
Bloot, C., and deWolf, L.C.M.
Bol. Soc. Geol. Port. 11: 1-57 (1953)
CA 48: 9874a

A 506 Spectrophotometry of fluorescence and
phosphorescence.
Bowen, E.J.
Photoelec. Spectrometry Group Bull. 6:
124 (1953)
CA 49: 12135h

A 507 Energy transfer in hydrocarbon solu-
tions.
Bowen, E.J., and Brocklehurst, B.
Trans. Faraday Soc. 49: 1131-3 (1953)
CA 48: 6838c

A 508 Temperature coefficients of fluores-
cence.
Bowen, E.J., and Cook, R.J.
J. Chem. Soc. 1953: 3059-61 (1953)
CA 48: 1160i

A 509 Fluorescence of solutions.
Bowen, E.J., and Wokes, F.
Longmans, Green & Co. 1953, 96 pp. (1953)
CA 47: 9160e

A 510 Fluorescence effects in photoelectric
absorption spectrophotometry.
Braude, E.A., and Timmons, C.J.
Photoelec. Spectrometry Group Bull. 6:
139-40 (1953)
CA 49: 12135f

A 511 Correction for absorption and fluorescence in the determination of molecular weights by light scattering.
Brice, B.A., Nutting, G.C., and Halwer, M.
J. Am. Chem. Soc. 75: 824-8 (1953)
CA 47: 6203g

A 512 The fluorescence spectrum of coronene.
Brocklehurst, B.
J. Chem. Soc. 1953: 3318-9 (1953)
CA 48: 1161a

A 513 Fluorescent screen for electric discharge tubes.
Broos, H.A., and Kroger, F.A.
Ger. 861,897 (1953)
CA 48: 10440h

A 514 Automatic measurement of light absorption and fluorescence on paper chromatograms.
Brown, J.A., and Marsh, M.M.
Anal. Chem. 25: 1865-9 (1953)
CA 48: 2422h

A 515 Preparation and performance of efficient plastic scintillators.
Buck, W.L., and Swank, R.K.
Nucleonics 11: 48-52 (1953)
CA 48: 4325c

A 516 Photoxidation and stabilization of polythene.
Burgess, A.R.
Natl. Bur. Std., Circ. 525: 149-58 (1953)
CA 48: 7337c

A 517 See A 505a

A 518 See A 505b

A 519 See A 505c

A 520 See A 505d

A 521 See A 505e

A 522 See A 505f

A 523 See A 505g

A 524 See A 505h

A 525 Luminescent properties of some of the minerals of Arabia Mountain, DeKalb County, Georgia.
Coffer, H.E., and Renshaw, E.W.
Georgia, Dept. Mines, Mining Geol., Geol. Surv., Bull. No. 60: 312-15 (1953)
CA 48: 4379i

A 526 Colorimetric and fluorometric properties of wheat in relation to germ damage.
Cole, E.W., and Milner, M.
Cereal Chem. 30: 378-91 (1953)
CA 47: 12670i

A 527 The mechanism of phosphorescence in crystal phosphors.
Curie, D.
Phys. Rev. 90: 154-5 (1953)
CA 47: 9782h

A 528 The uranium and fluorescent minerals.
Dake, H.C.
Mineralogist Publ. Co., 70 pp. (1953)
CA 48: 9878a

A 529 Fluorimetric method for the determination of 5-amino-acridine hydrochloride.
Devi, J.G., Khorana, M.L., and Padhye, M.R.
Indian J. Pharm. 15: 3-5 (1953)
CA 47: 6821e

A 530 Molded polystyrene scintillators.
deWaard, H., Prins, W., and Prins, A.
Appl. Sci. Res. B3: 372-6 (1953)
CA 48: 3801h

A 531 Cyanine dyes from β-naphthaquinaldine.
Doja, M.Q., and Sanyal, S.N.
J. Indian Chem. Soc. 30: 261-8 (1953)
CA 48: 479e

A 532 The free amino acids and the fluorescent compound of fish autolyzates.
Drilhon, A.
Bull. Inst. Oceanog. No. 1028, 4 pp. (1953)
CA 48: 3575g

A 533 Coloring of diamonds by neutron and electron bombardment.
Dugdale, R.A.
Brit. J. Appl. Phys. 4: 334-7 (1953)
CA 48: 4332b

A 534 Energy response of some inorganic
scintillators.
Duggal, V.P.
Proc. Indian Acad. Sci. A38: 320-6 (1953)
CA 48: 4990i

A 535 Fluorescence of the condensation pro-
ducts of some aromatic aldehydes and
hippuric acid (azlactones).
Eisenbrand, J.
Arch. Pharm. 286: 441-7 (1953)
CA 49: 7554i

A 536 Fluorimetry. I. Dependence of the
fluorescence intensity upon the dilution
of fluorescing solutions and the confirmation of Beer's law of light absorption
indirectly in high dilutions.
Eisenbrand, J.
Z. Anal. Chem. 140: 401-16 (1953)
CA 48: 4957h

A 537 Butter yellow hepatoma and its origin
(quinaldine).
Euler, H.
Arkiv Kemi 5: 251-9 (1953)
CA 48: 2712f

A 538 Spectral properties of reduced forms
of chlorophyll a and b.
Evstigneev, V.B., and Gavrilova, V.A.
Dokl. Akad. Nauk SSSR 91: 899-902 (1953)
CA 48: 447h

A 539 Studies on the concept of large activation centers in crystal phosphors. Dependence of luminescence efficiency on
concentration of activator.
Ewles, J., and Lee, N.
J. Electrochem. Soc. 100: 392-8 (1953)
CA 48: 451e

A 540 Low-temperature absorption and phos-
phorescence spectra of some iodo
compounds.
Ferguson, J., and Tredale, T.
J. Chem. Soc. 1953: 2959-66 (1953)
CA 48: 3146c

A 541 The kinetics of the polymerization of
acrylonitrile in aqueous solution. II.
Ferroni, E., and Baistrocchi, R.
Ann. Chim. 43: 555-8 (1953)
CA 48: 3770i

A 542 Balanced null-point fluorimeter with
photoelectric multipliers.
Field, W.E., and Drage, O.
Photoelec. Spectrometry Group Bull. 6:
128-32 (1953)
CA 49: 12135a

A 543 A sensitive photometer using modulated
light and its application in a uranium
fluorimeter.
Florida, C.D., and Davey, C.N.
J. Sci. Instr. 30: 409-12 (1953)
CA 48: 4318h

A 544 Emission of molecular complexes of
nitro compounds.
Foster, R., and Mannick, D.L.
Nature 171: 40 (1953)
CA 47: 4203d

A 545 Fluorescence spectra and photochemi-
cal activity of chloroplast pigments.
French, C.S., and Koski, V.M.
Proc. Intern. Botan. Congr., 7th, Stock-
holm 1950: 741-2 (1953)
CA 48: 12928e

A 546 Activation and deactivation of fluorescence in π-electronic systems. I. Various types of activation and deactivation
and the activation of fluorescence in
p-aminosalicylic acid-acetaldehyde
system.
Fujimori, E.
J. Chem. Soc. Japan, Pure Chem. Sect.
74: 911-4 (1953)
CA 48: 3146h

A 547 Activation and deactivation of fluorescence in π-electronic systems. II. Absorption and fluorescence spectra in
the activation of fluorescence in systems of anthrone with aldehyde, ketone,
and alcohol.
Fujimori, E.
J. Chem. Soc. Japan, Pure Chem. Sect.
74: 983-6 (1953)
CA 48: 3147b

A 548 Activation and deactivation of fluorescence in π-electronic systems. III.
Activation and deactivation of fluorescence in systems of anthrone derivatives with alcohol, phenol, and aniline.
Fujimori, E.
J. Chem. Soc. Japan, Pure Chem. Sect.
74: 986-9 (1953)
CA 48: 3147c

A 549 Fluorescent light yields with α-, β-,
and γ-radiations.
Furst, M., and Kallmann, H.
Phys. Rev. 91: 766-7 (1953)
CA 47: 12015e

A 550 Fluorescence, phosphorescence, and
photostimulation of NaCl (AgCl) with
high-energy irradiation.
Furst, M., and Kallmann, H.
Phys. Rev. 91: 1356-67 (1953)
CA 47: 11998g

A 551 Percentage polarization of the fluores-
cence of organic monocrystals.
Ganguly, S.C., and Chaudhury, N.K.
Z. Physik 135: 255-9 (1953)
CA 47: 11006c

A 552 Polarized fluorescence of molecules of
some single organic crystals.
Ganguly, S.C., and Chaudhury, N.K.
J. Chem. Phys. 21: 554-7 (1953)
CA 47: 5802h

A 553 Fluorescence in ceramic research.
Ghose, S.S., van Cott-Bouvier, H., and
Cormier, C.
Bull. Soc. Franc. Ceram. 21: 36-44 (1953)
CA 50: 9706h

A 554 Heavy accessory mineral study in the
Ducktown basin.
Gibson, O.
Georgia, Dept. Mines, Mining, Geol.,
Geol Surv. Bull. 60: 278-88 (1953)
CA 48: 4385e

A 555 Luminescence of thallium and lead
halides at low temperatures.
Gobrecht, H., and Becker, F.
Z. Physik 135: 553-7 (1953)
CA 48: 1158c

A 556 Absorption spectra and fluorescence
spectra of chlorophyll and magnesium
phthalocyanine in the adsorbed state.
Gochkovskii, V.F.
Dokl. Akad. Nauk SSSR 93: 511-4 (1953)
CA 48: 6828a

A 557 Phosphorescent inserts for slippers.
Goldstein, J.
U.S. 2,650,169 (1953)
CA 48: 456f

A 558 Variation in thermal electromotive
force in gelatin-dye phosphors.
Gombay, L.
Kolloid-Z. 133: 40-4 (1953)
CA 48: 4311e

A 559 (Determination of aluminum with salicylaldehyde condensation products.)
Goon, E., Retley, J.E., McMullen, W.H.,
and Wiberly, S.E.
Anal. Chem. 25: 608 (1953)

A 560 The role of oxygen in the Zn(Cu)S
luminophors.
Grillot, E.
J. Chim. Phys. 50: 515-23 (1953)
CA 48: 3795h

A 561 The fluorescence of biacetyl vapor.
Groh, H.J., Jr.
J. Chem. Phys. 21: 674-7 (1953)
CA 47: 6258g

A 562 The mechanism of acetone vapor
fluorescence.
Groh, H.J., Luckey, G.W., and Noyes,
W.A., Jr.
J. Chem. Phys. 21: 115-18 (1953)
CA 47: 3704a

A 563 Chromatographic and fluorometric investigation of biochemical polyphenia
of eye-color genes in Ephestia kuhniella.
Hadorn, E., and Kuhn, A.
Z. Naturforsch. 8: 582-9 (1953)
CA 48: 6035h

A 564 Pteroid fluorescent substances of the
skin and eyes of the frog.
Hama, T.
Experientia 9: 299-300 (1953)
CA 48: 891a

A 565 Fluorescent compositions.
Hanno, M., et al.
Japan 4912 (1953)
CA 48: 10440i

A 566 Organic pigment phosphorescent
paints.
Heidrich, H.
German 873,294 (1953)
CA 52: 19174g

A 567 Fluorescence of organic materials excited by fast electrons and gamma rays with emphasis on composition.
Herforth, L., and Rosahl, D.
Ann. Physik 12: 340-7 (1953)
CA 49: 5131c

A 568 New fluorescence reactions of aluminum.
Holzbecher, Z.
Chem. Listy 47: 680-8 (1953)
CA 48: 3189d

A 569 Measurement of fluorescent spectra of liquids with a modified Beckman DU spectrophotometer.
Huke, F.B., Heidel, R.H., and Fassel, V.A.
J. Opt. Soc. Am. 43: 400-4 (1953)
CA 47: 6771i

A 570 The fluorescence and thermoluminescence of certain marbles.
Ignjatovic, N., and Sljivic, S.
Srpska Akad. Nauk Sb. Radova 33, Geol. Inst. 5: 303-7 (1953)
CA 48: 3867h

A 571 Certain synthetic fluorescent minerals and aluminosilicate phosphors.
Iimori, M., and Iimori, S.
Repts. Sci. Research Inst. 29: 463-7 (1953)
CA 48: 8054d

A 572 Fluorescent bleaching agents of benzidine series. III. Synthesis of benzidine-3,3'-disulfonic acid derivatives containing triazine ring.
Inukai, K., and Maki, Y.
Rept. Govt. Ind. Res. Inst., Nagoya 2: 207-11, 267-71 (1953)
CA 50: 16112d

A 573 White fluorescent substances.
Ite, H.
Japan 6270 (1953)
CA 48: 11936f

A 574 The dye-gelatin phosphors. I. Quenching of the phosphorescence by bound water in the dye-gelatin phosphors.
Iwaki, R.
J. Chem. Soc. Japan, Pure Chem. Sect. 74: 503-6 (1953)
CA 48: 452a

A 575 Fluorescent emission of resonance lines at higher pressures.
Jablonski, A.
Nuovo Cimento 10: 573-80 (1953)
CA 47: 7900h

A 576 Investigations of luminescent antimony oxide.
Janin, J., and Bernard, R.
Compt. Rend. 237: 798-800 (1953)
CA 48: 3145e

A 577 A study of the afterglow of helium excited by centimeter waves.
Janin, J., and Eyrand, I.
Compt. Rend. 237: 1073-5 (1953)
CA 48: 6249h

A 578 Fluorescence test for serotonin and other tryptamines.
Jepson, J.B., and Stevens, B.J.
Nature 172: 772 (1953)
CA 48: 7102e

A 579 The measurement of quantum efficiency of fluorescent lamp phosphors.
Jerome, C.W.
J. Electrochem. Soc. 100: 586-7 (1953)
CA 48: 1815a

A 580 Differential fluorescence in identification of tobacco trashy leaf.
Johanson, R.
Nature 171: 753-4 (1953)
CA 47: 11668c

A 581 Porcellaneous phosphors.
Kato, K., and Iimori, S.
Repts. Sci. Research Inst. 29: 481-7 (1953)
CA 48: 8054f

A 582 Quenching of organic phosphors. II.
Kato, S.
J. Inst. Polytech., Osaka City Univ. 4C: 155-66 (1953)
CA 49: 2873g

A 583 The nature of fluorescent centers and traps in zinc sulfide.
Klasens, H.A.
J. Electrochem. Soc. 100: 72-80 (1953)
CA 47: 11998c

A 584 Fluorescent substance.
Kodera, Y., and Sekine, T.
Japan 1964 (1953)
CA 48: 5660c

A 585 Quenching of organic phosphorescence.
Koizumi, M., and Kato, S.
J. Chem. Phys. 21: 2088-9 (1953)
CA 48: 1816c

A 586 The quenching of organic phosphors
(instrument).
Koizumi, M., and Kato, S.
J. Chem. Soc. Japan, Pure Chem. Sect.,
74: 757-60 (1953)
CA 48: 1160d

A 587 Luminescence and electric properties
of zinc oxide.
Konovalov, G.A.
Tr. Sibirsk. Fiz.-Tekhn. Inst. pri Tomsk.
Gos. Univ. 1953: 32-52 (1953)
CA 50: 13612i

A 588 Liquid settling of fluorescent screens.
Krause, P.W.
U.S. 2,662,829 (1953)
CA 48: 6264g

A 589 Fluorescence and fluorochromy in
biology and medicine.
Krieg, A.
Klin. Wochschr. 31: 350-6 (1953)
CA 47: 7014g

A 590 Variation of the state of polarization of
the fluorescence of dyes with wave-
length of the exciting radiation.
Laffitte, E.
Compt. Rend. 236: 680-2 (1953)
CA 47: 7326f

A 591 Semidirect determination of coefficients
of fluorescence for the K level of certain atoms from measurements of the
width of X-rays.
Laskar, W.
Compt. Rend. 236: 2149-50 (1953)
CA 47: 9761i

A 592 Spectrophotometric determinations of
nicotinic acid and nicotinamide.
LeClerc, A.M., and Douzou, P.
Compt. Rend. 236: 2006-8 (1953)
CA 47: 9558b

A 593 The characteristics of ultraviolet sen-
sitized photographic plates in the vac-
uum ultraviolet.
Lee, P., and Weissler, G.L.
J. Opt. Soc. Am. 43: 512-15 (1953)
CA 47: 7891e

A 594 Phosphorescent effect in lead selenide.
Lee, P.A.
Can. J. Phys. 31: 1023-4 (1953)
CA 48: 6260i

A 595 A study of the α-ray-excited fluores-
cence of zinc sulfide by means of a
photomultiplier.
Leuba, P., and Anthony, J.P.
Compt. Rend. 236: 374-7 (1953)
CA 47: 4747e

A 596 Sugars in the spring sap of birch and
ironwood.
Lohr, E.
Physiol. Plantarum 6: 529-32 (1953)
CA 48: 5303c

A 597 Histological evidence of α-emitter
through fluorescence.
Ludwig, F.
Compt. Rend. 236: 751-3 (1953)
CA 48: 12855a

A 598 Fluorescent substance.
Machida, J.
Japan 6271 (1953)
CA 48: 11936f

A 599 Action of ultrasonics on some sub-
stances present in vegetable oils.
Maffei, F., and Buonsanto, M.
Olearia 7: 132-6 (1953)
CA 47: 12841f

A 600 Low-temperature fluorescence ob-
served in many silver salts.
Makishima, S., and Tomotsu, T.
Kagaku 23: 586 (1953)
CA 48: 452e

A 601 Relation between wavelength of excita-
tion and banded fluorescence of a
tetrahydrofluorocyclene solution.
Malkovskii, Z.
Byull. Polsk. Akad. Nauk, Otd. 3, 1(7): 287-
91 (1953); Referat. Zhur., Fiz. 1955,
No. 3526 (1953)
CA 50: 3090h

A 602 Electronic and vibrational energy of
tetrahydrofluorocyclene in benzene
solution.
Malkovskii, Z.
Bull. Acad. Polon. Sci.,Classe (III) 1:
113-17 (1953)
CA 48: 6253f

A 603 Monochromatic excitation of the fluorescence in tetrahydrofluorocyclene
solutions.
Malkovskii, Z.
Bull. Acad. Polon. Sci., Classe (III) 1:
287-91 (1953)
CA 48: 8656f

A 604 The phosphorescence of thoria.
Mandeville, C.E.
Phys. Rev. 90: 992-3 (1953)
CA 47: 9778i

A 605 The α-particle-induced phosphores-
cence of silver-activated sodium
chloride.
Mandeville, C.E., and Albrecht, H.O.
Phys. Rev. 90: 25-8 (1953)
CA 47: 9155g

A 606 Influence of the addition of alkyl sodium and potassium sulfates upon the
fluorescence and absorption spectra of
aqueous solutions of Rhodamine 6G.
Mataga, N., and Koizumi, M.
J. Inst. Polytech., Osaka City Univ. 4C:
177-88 (1953)
CA 49: 2868b

A 607 Even-odd character of the 2900-3200 A
absorption transition in naphthalene.
McConnell, H., and McClure, D.S.
J. Chem. Phys. 21: 1296-97 (1953)
CA 47: 10999g

A 608 Mineralogical observations of coleman-
ite, inyoite, meyerhofferite, tertschite,
and ulexite from new Turkish borate
deposits.
Meixner, H.
Heidelberger Beitr. Mineral. Petrog. 3:
445-55 (1953)
CA 48: 3856b

A 609 Bis (alkoxybenzamido) stilbenedisulfon-
ate (brightener).
Merner, R.R.
U.S. 2,635,113 (1953)
CA 48: 7636c

A 610 Fluorescence of aesculoside and meth-
ylaesculetol in the presence of horse-
chestnut extract.
Mesnard, P., and Durand, R.
Bull. Soc. Pharm. Bordeaux 92: 162-4
(1953)
CA 48: 5437e

A 611 Luminescent-bitumenological survey in
main deposits.
Moskalev, N.P.
Vestn. Mosk. Univ. 8, No. 3, Ser. Fiz.-Mat.
i Estestven. Nauk, No. 2, 157-60 (1953)
CA 48: 8706d

A 612 Some relative fluorescent efficiencies
in the Schumann region.
Moslen, V.W., White, N.E., and
Williams, S.E.
Brit. J. Appl. Phys. 4: 303-6 (1953)
CA 48: 5655i

A 613 Absorption and fluorescence of styrene
vapor.
Morgan, J.V.
J. Am. Chem. Soc. 75: 5055-7 (1953)
CA 48: 1807i

A 614 Absorption and fluorescence of styrene
vapor.
Morgan, J.V.
Univ. Microfilms 5447, 46 pp. (1953)
CA 47: 11980b

A 615 Effect of plasma on adrenaline fluores-
cence.
Mylon, E., and Roston, S.
Am. J. Physiol. 172: 601-11 (1953)
CA 47: 6475h

A 616 Investigation of the spectra of substitu-
ted phthalimides in vapor and in solu-
tion.
Neporent, B.S., Zelinskii, V.V., and
Klochkov, V.P.
Dokl. Akad. Nauk. SSSR 92: 927-30 (1953)
CA 48: 9810i

A 617 Growth substances in extracts of im-
mature corn grain studied through
their effects on cultures of Jerusalem
artichoke tissue.
Netien, G., and Beauchesne, G.
Compt. Rend. 237: 1026-8 (1953)
CA 48: 4645e

A 618 Measurements of the decay of the phosphorescence of several alkali halides.
Neuert, H., and Retz-Schmidt, T.
Z. Physik 134: 165-72 (1953)
CA 47: 9155d

A 619 The fluorescence of organic compounds through the action of X-rays.
Neunhoeffer, O., and Rosahl, D.
Z. Elektrochem. 57: 81-7 (1953)
CA 47: 11998e

A 620 Synthesis and fluorescence of substituted 2-pyrazolines.
Neunhoeffer, O., and Rosahl, D.
Chem. Ber. 86: 226-31 (1953)
CA 48: 674c

A 621 Chemical determination and metabolism of the steroid hormones of the adrenal cortex.
Neyman, B.
Bull. Soc. Fribourg. Sci. Nat. 43: 174-218 (1953)
CA 49: 5613c

A 622 Sulfide-type fluorescent substances.
Nishikawa, K.
Japan 979 (1953)
CA 48: 1818h

A 623 Fluorescent substance.
Nishikawa, K., and Yamamoto, M.
Japan 5871 (1953)
CA 48: 9208b

A 624 Experimental studies on the pathogenesis and prevention of chlorinated naphthalene poisoning. VIII-2.
Nomura, S.
J. Sci. Labour 29: 57-69 (1953)
CA 48: 5401h

A 625 Synthesis of organic fluorescent compounds. XXVI.
Oda, R., Yoshida, Z., and Yasuda, R.
J. Chem. Soc. Japan 56: 549-51 (1953)
CA 49: 7546gi

A 626 Fluorescent plate.
Ohya, S.
Japan 1170 (1953)
CA 48: 1164e

A 627 Fluorescent substance.
Ohya, S., and Imada, Y.
Japan 1965 (1953)
CA 48: 4320a

A 628 Variations in the decay of phosphorescence with frequency of applied field.
Olson, K.W.
Phys. Rev. 92: 1323 (1953)
CA 48: 2479f

A 629 Ultraviolet absorption spectra of glasses containing titanium and heavy metals.
Osada, K.
J. Phys. Soc. Japan 8: 226-8 (1953)
CA 50: 12646i

A 630 Extent of the fluorescence effect in absorption spectrophotometry.
Ovenston, T.C.J.
Photoelec. Spectrometry Group Bull. 6: 132-8 (1953)
CA 49: 12135df

A 631 Effect of temperature on the fluorescence of uranyl sulfate.
Pant, D.D.
J. Sci. Res. Banaras Hindu Univ. 3: 27-37 (1953)
CA 48: 6262a

A 632 Fluorescence spectra of five samples of uranyl sulfate.
Pant, D.D.
J. Sci. Res. Banaras Hindu Univ. 3: 19-27 (1953)
CA 48: 6261h

A 633 Mechanism of luminescence of potassium chloride-thallium-activated phosphors excited by X-rays.
Parfianovich, I.A.
Zh. Eksperim. i Teor. Fiz. 24: 117-23 (1953)
CA 49: 2193c

A 634 Investigations on the luminescence of minerals. II. Carbonates.
Parsanov, G.P., and Sheveleva, V.A.
Tr. Mineralog. Muzeya, Akad. Nauk SSSR 5: 56-89 (1953)
CA 49: 5130c

A 635 Fluorescence of crystallized acenaph-
thene at 20°.
Pesteil, P.
Compt. Rend. 237: 235-7 (1953)
CA 47: 11998a

A 636 Fluorescence of organic monocrystals
at low temperatures. II. Anthracene,
biphenyl, and acenaphthene.
Pesteil, P., and Barbaron, M.
Compt. Rend. 237: 884-6 (1953)
CA 48: 4311f

A 637 Measurements of scintillation lifetimes.
Phillips, H.B., and Swank, R.K.
Rev. Sci. Instr. 24: 611-6 (1953)
CA 48: 6273f

A 638 Chelating properties of 8-quinolinol,
Mannich bases.
Phillips, J.P., and Ferando, Q.
J. Am. Chem. Soc. 75: 3768-9 (1953)
CA 48: 1873d

A 639 Noncrystalline fluorescent solids for
measurement of radioactivity.
Pichot, L., Pesteil, P., and Clement, J.
J. Chim. Phys. 50: 26-41 (1953)
CA 47: 5808c

A 640 Color bands in fluorspar.
Przibram, K.
Nature 172: 860-1 (1953)
CA 48: 3795f

A 641 Fluorescent screens.
Putnan, C.W.
U. S. 2,660,539 (1953)
CA 48: 4320b

A 642 Effect of concentration on the fluores-
cence of dye solutions.
Ramanathan, N., and Tawde, N.R.
Proc. Natl. Inst. Sci. India, Pt. A 19:
405-10 (1953)
CA 48: 1816a

A 643 Photochemistry of porphyrins.
Randall, G.
U.S. At. Energy Comm., UCRL 2417,
120 pp. (1953)
CA 48: 11197d

A 644 The measurement of the decay of
luminescence by means of the phase
tube.
Rohde, F.
Z. Naturforsch. 8a: 156-161 (1953)
CA 47: 9157d

A 645 Fluorescence spectra and quantum
yields of some organic materials ex-
cited with ultraviolet light.
Rosahl, D.
Ann. Physik 12: 35-44 (1953)
CA 49: 5972f

A 646 Thermal extinction of the fluorescence
of solutions of biacene.
Rosinski, K.
Bull. Acad. Polon. Sci., Classe (III) 1:
55-9 (1953)
CA 48: 452h

A 647 Phosphorescence of cellulose.
Rousset, A., Lochet, R., and Darrine, J.
Compt. Rend. 237: 37-8 (1953)
CA 48: 357h

A 648 Photodielectric effect, ferroelectricity,
and thermoluminescence in ultraviolet-
irradiated zinc oxide.
Roux, J.
Compt. Rend. 236: 2492-4 (1953)
CA 48: 10424e

A 649 Analysis of fluorescent X-radiation by
means of proportional counters.
Rowland, R.E.
J. Appl. Phys. 29: 811-12 (1953)
CA 47: 11964h

A 650 Raman spectra and fluorescence of o-
and p-chlorotoluene in the solid state.
Sanyal, S.B.
Indian J. Phys. 27: 447-51 (1953)
CA 51: 7866c

A 651 Fluorescent substance.
Sato, Y., and Fukumoto, T.
Japan 6020 (1953)
CA 48: 11199f

A 652 Measurements of decay times on liquid
and solid solutions with a new fluoro-
meter.
Schmillen, A.
Z. Physik 135: 294-308 (1953)
CA 41: 11006f

A 653 The decay period of solid solutions of
aromatic hydrocarbons.
Schmillen, A., Schmillen, L., and Rohde,F.
Z. Naturforsch. 8a: 213-14 (1953)
CA 47: 9157f

A 654 Symmetry assignments for the first
two excited singlet electronic states of
the naphthalene molecule.
Schnepp, O., and McClure, D.S.
J. Chem. Phys. 21: 959 (1953)
CA 47: 7890d

A 655 Radiophotoluminescence dosimetry
system of the U.S. Navy.
Schulman, J.H.
Nucleonics 11: 52-6 (1953)
CA 48: 3156e

A 656 Fluorescent substance.
Shawa, S., et al.
Japan 5413 (1953)
CA 48: 11199e

A 657 Change in the fluorescence of deltapine-
15 seed cotton as a result of heat.
Sheehan, W.C.
Textile Res. J. 23: 736-43 (1953)
CA 48: 1008b

A 658 Investigation of the structure of intercomplex compounds by the method of
vibrational spectra in connection with
the question about the nature of the
hydrogen bond.
Shigorin, D.N.
Izv. Akad. Nauk SSSR, Ser. Fiz. 17:
596-603 (1953)
CA 48: 5652a

A 659 Fluorescence and absorption spectra of
estrogens heated in sulfuric acid.
Slaunwhite, W.R., Jr., Engel, L.L., Scott,
J.F., and Ham, C.L.
J. Biol. Chem. 201: 615-20 (1953)
CA 47: 8805f

A 660 Effects of neutron bombardment on a
zinc sulfide phosphor.
Smith, A.W., and Turkevich, H.
J. Chem. Phys. 21: 367-8 (1953)
CA 47: 9155i

A 661 Fluorescence of terramycin and aureo-
mycin.
Soncin, E.
Boll. Soc. Ital. Biol. Sper. 29: 337-8 (1953)
CA 48: 330g

A 662 Reversible association processes of
globular proteins. IV. Fluorescence
methods in studying protein interactions.
Steiner, R.F.
Arch. Biochem. Biophys. 46: 291-311
(1953)
CA 49: 8347c

A 663 Vapor pressures and the heats of sub-
limation of anthracene and of 9,10-
diphenylanthracene.
Stevens, B.
J. Chem. Soc. 1953: 2973-4 (1953)
CA 48: 3781d

A 664 Influence of linear ion density upon the
excitation of ultraviolet and visible
fluorescence by γ-rays and X-rays.
Maier, H.
Strahlentherapie 91: 566-75 (1953)
CA 47: 11996d

A 665 Fluorescent spectra from ethyl nitrate.
I. Supposed emissions from alkoxy
radicals and NO_2.
Style, D.W.G., and Ward, J.C.
Trans. Faraday Soc. 49: 999-1002 (1953)
CA 48: 4971i

A 666 A study of intensity variations in sensi-
tized fluorescence.
Swanson, R.E., and McFarland, R.H.
Phys. Rev. 89: 911-2 (1953)
CA 48: 8026f

A 667 Light-responsive fluorescent media.
Switzer, J.L., Switzer, R.C., and Ward,
R.A.
U.S. 2,653,109 (1953)
CA 48: 1164b

A 668 Tetrakis (carboxyphenyl) thiophene
(dye).
Toland, W.G.
U.S. 2,610,191 (1953)
CA 48: 7641g

A 669 Fluorescent heterocyclic compounds.
Trosken, O.
Ger. 869,490 (1953)
CA 52: 16372c

A 670 The ultraviolet absorption and fluores-
cence spectra of acetophenone.
Vanselow, R.D., and Duncan, A.B.F.
J. Am. Chem. Soc. 75: 829-32 (1953)
CA 47: 5251c

A 671 Titration fluorometer.
Vecerek, B., and Shovronsky, O.
Chem. Listy 47: 272 (1953)
CA 48: 3069e

A 672 Fluorescence of dermatophytes,
Trichophyton meutagrophytes.
Vivancos, G.
Rev. inst. Malbrán 15: 149-53 (1953)
CA 48: 8312b

A 673 Recent applications of fluorescence to
textile fibers.
Wallner, L.
Teintex 18: 69-93 (1953)
CA 47: 5121d

A 674 Polarization of the fluorescence of
labeled protein molecules.
Weber, G.
Discussions Faraday Soc. 1953: 33-9 (1953)
CA 47: 6471f

A 675 Rotational Brownian motion and polari-
zation of the fluorescence of solutions.
Weber, G.
Advan. Protein Chem. 8: 415-59 (1953)
CA 48: 3428a

A 676 Metals in the atomic state in glasses.
Weyl, W.A.
J. Phys. Chem. 57: 753-6 (1953)
CA 48: 968a

A 677 Progress with fluorescent materials.
Williams, A.E.
Ind. Finishing 6: 324-8 (1953)
CA 48: 3704i

A 678 The fluorescence of living, dying, and
dead cells of the Florideae.
Wimmer, C., and Hofler, K.
Oesterr. Akad. Wiss. Math.-Naturw. Kl.,
Sitzber. 162: 625-41 (1953)
CA 49: 4091i

A 679 Fluorescence characteristics of mixed
organic crystals.
Wright, G.T.
Proc. Phys. Soc. 66A: 777-83 (1953)
CA 47: 11997h

A 680 Application of fluorescent coatings by a
precipitation method.
Yamada, M., and Akagi, K.
Japan 6040 (1953)
CA 48: 11199g

A 681 Dye phosphors. IV. Contribution of
conduction band to the dye phosphors.
Yamamoto, D.
J. Chem. Soc. Japan 74: 173-83 (1953)
CA 47: 5252b

A 682 The law of damping of the luminescence
of solid organic substances.
Yastrebov, V.A.
Dokl. Akad. Nauk SSSR 90: 1015-8 (1953)
CA 48: 8054g

A 683 Syntheses of organic fluorescent com-
pounds.
Yoshida, Z., Kato, Y., and Oda, R.
J. Chem. Soc. Japan 56: 411-13 (1953)
CA 49: 6910i

A 684 Synthesis of organic fluorescent compounds. XXIII. Synthesis of 9-
carboxymethyl-3,6-diaminocarbazole-
4,4-disulfonic acid and its fluorescence.
Yoshida, Z., and Oda, R.
J. Chem. Soc. Japan, Ind. Chem. Sect. 56:
261-3 (1953)
CA 48: 10748b

A 685 A new 6100-A band in zinc orthosilicate
activated with manganese.
Zalm, P., and Klasens, H.A.
Philips Res. Rept. 8: 386-92 (1953)
CA 48: 8657a

(1954-1956)

B 1 The determination of quality in fish by
means of fluorescent compounds.
Aker, H., and Appleman, M.D.
Bacteriol. Proc. 54: 18-19 (1954)
CA 51: 15829a

B 2 The storage of energy in some activated
alkali halide phosphors.
Albrecht, H.O., and Mandeville, C.E.
J. Franklin Inst. 257: 353-68 (1954)
CA 54: 11195c

B 3 Lining electron-tube envelopes with
fluorescent materials.
Anderson, J.T., and Ward, H.F.
U.S. 2,676,894, April 27, 1954
CA 48: 9814c

B 4 Fluorescent substance.
Aoki, Y.
Japan 967 (1954)
CA 45: 13444c

B 5 Fluorescence of adsorbed dyes, especial-
ly of Acridine Orange on aluminum
oxide.
Bandow, F.
Z. Physik. Chem. 1: 63-8 (1954)
CA 48: 9205i

B 6 Energy transfer between chromophore
and protein in phycocyanin.
Bannister, T.T.
Arch. Biochem. Biophys. 49: 222-33 (1954)
CA 48: 7093c

B 7 Fluorescence of organic single crystals
at low temperatures. III. Fluorene.
Barbaron, M., and Pesteil, P.
Compt. Rend. 238: 1400-2 (1954)
CA 48: 10438h

B 8 A supposed Raman spectrum of gaseous
bromine.
Barrow, R.F.
J. Chem. Phys. 2: 1775 (1954)
CA 49: 722d

B 9 Fluorescent screens for cathode-ray
tubes.
Bayford, L.J.C., and Seats, P.
U.S. 2,681,293, June 15, 1954
CA 48: 11200a

B 10 Stabilization of the phosphorescence of
the organic fluorescent adsorbates.
Bernanose, A., Comte, M., and Vouaux, P.
J. Chim. Phys. 51: 400-1 (1954)
CA 49: 7390c

B 11 Carbazole fluorescence.
Bernanose, A., and Marquet, G.
J. Chim. Phys. 51: 255-9 (1954)
CA 48: 13438i

B 12 Organic electroluminescence and phos-
phorescence.
Bernanose, A., and Michon, F.
J. Chim. Phys. 51: 622-3 (1954)
CA 49: 4409b

B 13 Temperature dependence of organic
scintillation materials.
Birks, J.B.
Phys. Rev. 95: 277 (1954)
CA 48: 9824c

B 14 Energy transfer in organic phosphors.
Birks, J.B.
Phys. Rev. 94: 1567-73 (1954)
CA 48: 9205b

B 15 Raman and fluorescence spectra of o-
and p-bromotoluene in the solid state
at low temperatures.
Biswas, D.C.
Indian J. Phys. 28: 423-30 (1954)
CA 51: 7865i

B 16 Fluorescence spectra of organic crys-
tals.
Birks, J.B., and Wright, G.T.
Proc. Phys. Soc. 67B: 657-63 (1954)
CA 49: 728f

B 17 The fluorescence of some phenolic steroid urinary extracts in connection with the hydrolytic procedure (hydrochloric acid or glucuronidase).
Bompiani, A.
Boll. Soc. Ital. Biol. Sper. 30: 865-8 (1954)
CA 49: 12563b

B 18 Spectrophotometry of the fluorescence of 11-deoxycorticosterone, 11-hydroxycorticosterone, and 17-hydroxy-11-dehydrocorticosterone in color reactions in various solvents.
Bompiani, A.
Boll. Soc. Ital. Sper. 30: 868-71 (1954)
CA 49: 12563c

B 19 The spectrophotometric character of the fluorescence of 16-ketoestradiol in color reactions in various solvents.
Bompiani, A.
Boll. Soc. Ital. Biol. Sper. 30: 871-3 (1954)
CA 49: 12563d

B 20 The yield of fluorescence of the vapors and solutions of substituted phthalimides.
Borisevich, H.A., Zelinskii, V.V., and Neporent, B.S.
Dokl. Akad. Nauk SSSR 94: 37-9 (1954)
CA 50: 6196f

B 21 Action of foreign gases on the fluorescence of vapors of aromatic compounds.
Borisevich, N.A.
Dokl. Akad. Nauk SSSR 99: 695-8 (1954)
CA 49: 15499i

B 22 Fluorescence quenching in solution and in the vapor state.
Bowen, E.J.
Trans. Faraday Soc. 50: 90-102 (1954)
CA 48: 12560e

B 23 Fluorescence.
Bowen, E.J.
J. Oil Colour Chemists' Assoc. 37: 264-7 (1954)
CA 48: 11903e

B 24 The fluorescence spectra of coronene and 1,12-benzoperylene at low temperatures.
Bowen, E.J., and Brocklehurst, B.
J. Chem. Soc. 1954: 3875-8 (1954)
CA 49: 3667h

B 25 An experimental study of the transfer of energy of excitation between unlike molecules in liquid solution.
Bowen, E.J., and Livingston, R.
J. Am. Chem. Soc. 76: 6300-4 (1954)
CA 49: 5131i

B 26 Polarization correction in the deceleration of fast electrons of 3 to 15 MeV.
Breitling, G., and Glocker, R.
Naturwissenschaften 41: 471-2 (1954)
CA 49: 10759d

B 27 Emission of the forbidden bands ($^3\Sigma_u^+ \rightarrow {}^3\Sigma_g^-$) of O_2.
Broida, H.P., and Gaydon, A.G.
J. Phys. Radium 15: 385-7 (1954)
CA 48: 11187e

B 28 Phosphorescence of atoms and molecules of solid nitrogen at 4.2°K.
Broida, H.P., and Pellam, J.R.
Phys. Rev. 95: 845-6 (1954)
CA 48: 11932h

B 29 The fluorescence of liquid egg. I. The relation between the fluorescence and the mustiness of frozen whole egg.
Brooks, J.
Food Technol. 8: 400-5 (1954)
CA 50: 11548i

B 30 Fluorescence analytical investigation on secondary uranium minerals.
Bultemann, H.W.
Neues Jahrb. Mineral., Abhandl. 86: 155-62 (1954)
CA 49: 2957g

B 31 Energy transfer from solvents to solute in liquid organic solutions under ultraviolet excitation.
Cohen, S.G., and Weinreb, A.
Phys. Rev. 93: 1117 (1954)
CA 48: 6838e

B 32 Luminescent substances.
Compagnie des Lampes.
Fr. 1,077,637 (1954)
CA 53: 13801h

B 33 Luminescence characteristics of some scintillating crystals.
Cook, J.R., and Mahmoud, K.A.
Proc. Phys. Soc. 67B: 817-24 (1954)
CA 49: 1439d

B 34 The fluorescence of biacetyl vapor at
4358 A.
Coward, N.A., and Noyes, W.A.
J. Chem. Phys. 22: 1207-10 (1954)
CA 48: 13439d

B 35 The luminescence of calcium oxide.
Crozet, A.
Ann. Univ. Lyon, Sci., Sect. B 7 (1954)
CA 49: 6732c

B 36 Stimulation of calcium oxide phosphores-
cence.
Crozet, A., and Janin, J.
Compt. Rend. 239: 1031-4 (1954)
CA 49: 5130e

B 37 Type IIb diamonds.
Custers, J.F.H.
Physica 20: 183-4 (1954)
CA 52: 5121c and CA 48: 11932g

B 38 Kinetics of the coloration and luminescence of vitreous silica induced by irradiation with X- and γ-rays, with observations on related phenomena.
Dainton, F.S., and Rowbottom, J.
Trans. Faraday Soc. 50: 480-93 (1954)
CA 49: 728b

B 39 A browning reaction between thiamine
and glucose.
deLange, P.
Nature 173: 1040-1 (1954)
CA 48: 11662i

B 40 Spectrophotometric analysis of the
fluorescence of the pigmented material
of the "residual body" (corpora lutea)
of Onthoptera.
DeLerma, B.
Boll. Soc. Ital. Biol. Sper. 30: 1311-14
(1954)
CA 49: 12737a

B 41 Microspectrography of the fluorescence
of the pigmented material of the "residual body" (corpora lutea) of insects.
DeLerma, B.
Boll. Soc. Ital. Biol. Sper. 30: 1309-11
(1954)
CA 49: 12736i

B 42 Phenomenon of luminescence in mineral-
ogy.
Deribere, M.
Bull. Soc. Franc. Mineral. Crist. 77:
939-52 (1954)
CA 49: 61b

B 43 Emission spectra of electroluminescent
substances with multiple activators.
Destriau, G.
J. Phys. Radium 15: 13-15 (1954)
CA 48: 6260e

B 44 Theory of concentration quenching in
inorganic phosphors.
Dexter, D.L., and Schulman, J.H.
J. Chem. Phys. 22: 1063-70 (1954)
CA 48: 12559g

B 45 Luminescence of photographic silver
halide emulsions at low temperatures.
Dorfner, K.R., and Joos, G.
Sitzber. Math.-Naturw. Kl. Bayer. Akad.
Wiss. Muenchen. 1953: 63-7 (1954)
CA 49: 8715f

B 46 Derivatives of o-hydroxyquinoline.
Dorier, P., Tronche, P., and Blanquet, P.
Trav. Soc. Pharm. Montpellier 14: 152-5
(1954)
CA 50: 1011a

B 47 A twin-beam null-point fluorimeter for
the analysis of liquid samples.
Dowdall, J.P., and Stretch, H.
Analyst 79: 651-5 (1954)
CA 49: 771d

B 48 Fluorescence phenomena during germ-
ination of Brassiceae.
Eifrig, H.
Ber. Deut. Botan. Ges. 67: 300-10 (1954)
CA 49: 5597c

B 49 A comparison of the intensities of the
Raman effect and fluorescence.
Eisenbrand, J.
Optik 11: 557-61 (1954)
CA 49: 5965f

B 50 X-ray-induced photostimulated conduc-
tivity in magnesium oxide.
Eisenstein, A.S.
Phys. Rev. 93: 1017-18 (1954)
CA 48: 6818c

B 51 The longlived phosphorescent compo-
nents of thallium-activated sodium
iodide.
Emigh, C.R., and Megill, L.R.
Phys. Rev. 93: 1190-4 (1954)
CA 48: 6837g

B 52 Quenching and degree of polarization of
fluorescence in solutions.
Epple, R., and Forster, T.
Z. Elektrochem. 58: 783-7 (1954)
CA 49: 7987g

B 53 The photooxidation of 6,13-diphenylpent-
acene and 5,7,12,14-tetraphenylpenta-
cene.
Etienne, A., and Beauvois, C.
Compt. Rend. 239: 64-6 (1954)
CA 50: 3375c

B 54 The phosphorescence spectra of naph-
thalene and some simple derivatives.
Ferguson, J., Iredale, T., and Taylor, J.A.
J. Chem. Soc. 1954: 3160-5 (1954)
CA 49: 2869b

B 55 A fluorometer for solutions.
Fletcher, M.H., and Warner, E.R.
U. S. Geol. Surv., Circ. No. 311, 9 pp.
(1954)
CA 48: 13278i

B 56 Basic theory and fundamentals of fluor-
escent X-ray spectrographic analysis.
Friedman, H., Birks, L.S., and Brooks,
E.J.
Symp. Fluorescent X-ray Spectrographic
Anal., ASTM Spec. Tech. Publ. 157:
3-26 (1954)
CA 48: 9257h

B 57 Activation and deactivation of fluorescence in electronic systems. IV. Activation and deactivation of fluorescence
in fluorescein derivatives.
Fujimori, E.
J. Chem. Soc. Japan 75: 24-7 (1954)
CA 48: 4982a

B 58 Energy transfer by means of collision in
liquid organic solutions under high-
energy and ultraviolet excitations.
Furst, M., and Kallmann, H.
Phys. Rev. 94: 503-7 (1954)
CA 48: 8058b

B 59 New fluorimeter for the determination of
uranium.
Galvanek, P., and Morrison, T.J.
U.S. At. Energy Comm. ACCO-47 (1954)
CA 49: 6b

B 60 Anisotrophy of fluorescence of some
organic crystals.
Ganguly, S.C., and Chaudhury, N.K.
Phys. Rev. 95: 1148-52 (1954)
CA 48: 12560d

B 61 Energy transfer in irradiated solutions
of mixed phosphors.
Germann, F.E.E., Brown, F.T., Wissell,
R., and Waugh, T.D.
Science 120: 540-2 (1954)
CA 49: 4410i

B 62 Luminescent screens.
Gier, J.D.
U.S. 2,665,220 (1954)
CA 48: 6264b

B 63 Behavior of optical bleaching agents on
cellulosic materials.
Glarum, S.N., and Penner, S.E.
Am. Dyestuff Reptr. 43, Proc. Am. Assoc.
Textile Chemists Colorists P310-14
(1954)
CA 48: 9072c

B 64 The fluorescence reactions of steroids.
Goldzieher, J.W., Bodenchuk, J.M., and
Nolan, P.
Anal. Chem. 26: 853-6 (1954)
CA 48: 11196f

B 65 Roots. I. Properties and distribution of
fluorescent constituents in Avena roots.
Goodwin, R.H., and Pollock, B.M.
Am. J. Botany 41: 516-20 (1954)
CA 48: 12255d

B 66 Investigation of polarizational characteristics of luminescence of complex
organic compounds by the photoelectric
method.
Gribkov, V.I., and Zhevandrov, N.D.
Dokl. Akad. Nauk SSSR 98: 565-8 (1954)
CA 50: 1471c

B 67 Fluorescence and thermoluminescence
of ice.
Grossweiner, L.I., and Matheson, M.S.
J. Chem. Phys. 22: 1514-26 (1954)
CA 49: 60h

B 68 The fluorescence of gases on impact of fast particles.
Grun, A.F.
Z. Naturforsch. 9a: 55-63 (1954)
CA 48: 8058c

B 69 Investigation of energy transfer and quenching processes in gases on excitation with fast particles.
Grun, A.E., and Schopper, E.
Z. Naturforsch. 9a: 134-47 (1954)
CA 48: 8657h

B 70 Luminescence studies on fluorite and other minerals.
Haberlandt, V.H.
Oesterr. Akad. Wiss., Math.-Naturw. Kl., Sitzber., Abt. I 163: 375-99 (1954)
CA 49: 12969f

B 71 Slow component in decay of fluors.
Harrison, F.B.
Nucleonics 12: 24-5 (1954)
CA 48: 6839c

B 72 Salt effects on the rates of fast reactions in aqueous solutions.
Harty, W.E., and Rollefson, G.K.
J. Am. Chem. Soc. 76: 4811-15 (1954)
CA 49: 2841a

B 73 Identification of cardiac glycosides and aglycones in Strophanthus seeds by paper chromatography.
Heftmann, E., Berner, P., Hayden, A.L., Miller, H.K., and Mosettig, E.
Arch. Biochem. Biophys. 37: 329-39 (1954)
CA 48: 12931e

B 74 Molecular-weight determination of high polymer chains by fluorescence and spectroscopic methods.
Heintz, E.
J. Phys. Radium 15: 219-21 (1954)
CA 48: 7986h

B 75 Secondary radiation effects in the photographic action of α-rays.
Herz, R.H.
Strahlentherapie 94: 455-9 (1954)
CA 48: 13451e

B 76 Elimination of the fluorescence of estrogens in urinary extracts by hydrogen peroxide.
Heusghem, C.
Nature 173: 1043-4 (1954)
CA 48: 11535d

B 77 The luminescence of high polymers.
Hinrichs, I.H.
Z. Naturforsch. 9a: 617-24 (1954)
CA 49: 2873b

B 78 Fluorescence spectra of the condensation products of salicylaldehyde and their salts.
Holzbecher, Z.
Chem. Listy 48: 1160-6 (1954)
CA 48: 13543d

B 79 Fluorimetry.
Indemans, A.W.M.
Chem. Weekblad. 50: 236-40 (1954)
CA 48: 8691d

B 80 Fluorometric analyses. I. Trial construction of fluorophotometer and fluorocolorimetric determination of aluminum with morin.
Ishibashi, M., Shigematsu, T., and Nakegawa, Y.
Japan Analyst 3: 294-6 (1954)
CA 49: 14561a

B 81 Organic phosphors having metallic compounds as the phosphorescent bodies.
Iwaski, R.
J. Chem. Soc. Japan, Pure Chem. Sect. 75: 524-30 (1954)
CA 48: 8057c

B 82 Dye-gelatin phosphors. II. The lifetime of phosphorescence.
Iwaki, R.
J. Chem. Soc. Japan, Pure Chem. Sect. 75: 843-8 (1954)
CA 48: 11196d

B 83 Yield of anti-Stokes fluorescence of dye solutions.
Jablonski, A.
Acta Phys. Polon. 13: 239-42 (1954)
CA 49: 8705a

B 84 Measurement of fluorescent spectra.
Jatkar, S.K.K., and Mattoo, B.N.
J. Univ. Poona, Sci. Technol. 6: 67-73 (1954)
CA 49: 15498g

B 85 The Raman and fluorescence spectra of allyl diglycol carbonate.
Jeppesen, M.A.
Phys. Rev. 98: 1211 (1954)
CA 50: 10513b

B 86 Influence of condensing oxygen on the fluorescence and absorption spectra of anthraquinone derivatives in the adsorbed state.
Karyakin, A.V., and Terenin, A.N.
Dokl. Akad. Nauk SSSR 97: 479-82 (1954)
CA 49: 12970h

B 87 Organic phosphorescence. I.
Kato, S., and Koizumi, M.
Bull. Chem. Soc. Japan 27: 189-94 (1954)
CA 49: 7390e

B 88 Fluorescence spectra of crude oils and their fractions in the liquid state and on a chromatographic column.
Kats, M.L., and Sidorov, N.K.
Uch. Zap. Saratovsk. Gos Univ. Vypusk Fiz. 40: 3-59 (1954)
CA 54: 10291f

B 89 Fluorophotometer for determination of uranium in fused sodium fluoride pellets.
Kelley, M.T., Hemphill, H.L., and Collier, D.M.
U.S. At. Energy Comm. ORNL–1445, 15 pp. (1954)
CA 48: 8596i

B 90 Spectro-photoelectric colorimetry.
Ketelaar, J.A.A.
Chem. Weekblad. 50: 225-30 (1954)
CA 48: 8689i

B 91 Model VI transmission fluorimeter for determination of uranium.
Kinser, C.A.
U.S. Geol. Surv., Circ. No. 330, 9 pp. (1954)
CA 48: 9813f

B 92 The theory of colored glass.
Kocik, J.
Sklar Keram. 4: 255-62 (1954)
CA 50: 10359f

B 93 Influence of the addition of high-molecular electrolytes upon the absorption spectra and fluorescence of organic dyestuffs. II.
Koizumi, M., and Mataga, N.
Bull. Chem. Soc. Japan 27: 194-7 (1954)
CA 49: 7382c

B 94 Light-filter combination for observation of fluorescence in entire spectral range.
Kolbel, H.
Naturwissenschaften 41: 550 (1954)
CA 49: 12135i

B 95 Photochromy of dehydrobianthrones.
Kortum, G., Theilacker, W., and Braun, V.
Z. Physik. Chem. 2: 179-96 (1954)
CA 49: 722b

B 96 Studies on the oxy-acid phosphors. II. Vanadate phosphors.
Kotera, Y., and Sekine, T.
Bull. Chem. Soc. Japan 27: 13-18 (1954)
CA 49: 7986e

B 97 The origin of the fluorescence in self-activated zinc sulfide, cadmium sulfide, and zinc oxide.
Kroger, F.A., and Vink, H.J.
J. Chem. Phys. 22: 250-2 (1954)
CA 48: 5655a

B 98 Investigation on zinc sulfide crystals.
Krumbiegel, J.
Z. Naturforsch. 9a: 903-4 (1954)
CA 49: 5130g

B 99 Fluorescent film.
Kubota, C., and Higashide, F.
Japan 8440, Dec. 21, 1954
CA 50: 9162g

B 100 Optical properties of cadmium sulfide in glass.
Kuwabara, G.
J. Phys. Soc. Japan 9: 992-6 (1954)
CA 50: 1462d

B 101 Absorption and fluorescence of crystal violet, methyl violet, and malachite green.
Laffitte, E., and Dubreuil, Y.
Compt. Rend. 238: 787-9 (1954)
CA 48: 8651h

B 102 Luminescence of vitreous silica.
Lautout, M.
Compt. Rend. 238: 2409-10 (1954)
CA 48: 13438d

B 103 Surface states of cadmium sulfide.
Liebson, S.H.
J. Electrochem. Soc. 101: 359-62 (1954)
CA 50: 51i

B 104 Spectroscopic investigation of the
fluorescence of nitro compounds.
Lippert, E.
Z. Physik. Chem. 2: 328-35 (1954)
CA 49: 8704i

B 105 The influence of the solvent on the
electronic spectra of intramolecular
ionic aromatic compounds.
Lippert, E., and Moll, F.
Z. Elektrochem. 58: 718-24 (1954)
CA 49: 7976c

B 106 Fluorescent spectra of solid solutions
of naphthalene with added anthracene.
Lipsett, F., and Dekker, A.J.
Nature 173: 736-7 (1954)
CA 48: 9813a

B 107 Absorption spectra and fluorescence of
extraction products from blood and in-
testine of the normal or diabetic animal.
Loubatieres, A., and Boayard, P.
J. Physiol. 46: 437-41 (1954)
CA 49: 7104d

B 108 Vibration temperature in the phosphor-
escence spectrum of the Swan bands of
the C_2 molecule.
Lukacs, G., and Herman, L.
Compt. Rend. 239: 640-2 (1954)
CA 49: 2184b

B 109 Internal vibrational structure found in
the low-temperature fluorescence of
silver nitrate crystals.
Makishima, S., and Tomotsu, T.
Bull. Chem. Soc. Japan 27: 476 (1954)
CA 49: 10057a

B 110 A problem of solid-state physics. Fun-
damental absorption and recombination.
Makishima, S., and Tomotsu, T.
Kagaku 24: 292-4 (1954)
CA 48: 9205e

B 111 Interferometric analysis of the fluor-
escent spectrum of iodine vapor.
Malamond, C., and Boiteux, H.
Compt. Rend. 238: 778-80 (1954)
CA 48: 6838f

B 112 Influence of the addition of sodium and
potassium alcohol-sulfonic esters upon
the absorption spectra and fluorescence
of aqueous dye solutions.
Mataga, N., and Koizumi, M.
J. Chem. Soc. Japan 75: 273-6 (1954)
CA 48: 5648b

B 113 Fluorescence measurements with the
Beckamn Model DC1 spectrophotometer.
McAnally, J.S.
Anal. Chem. 26: 1526 (1954)
CA 49: 1382h

B 114 Excited states of the naphthalene mole-
cule. I. Symmetry properties of the
first two excited singlet states.
McClure, D.S.
J. Chem. Phys. 22: 1668-75 (1954)
CA 49: 719f

B 115 The band fluorescence of mercury
vapor.
McCoubrey, A.O.
Phys. Rev. 93: 1249-60 (1954)
CA 48: 6838g

B 116 Quenching of fluorescence by flavonoids
and other substances.
McLaughlin, J.A., and Szent-Gyorgyi, A.
Enzymologia 16: 384-9 (1954)
CA 49: 10066g

B 117 Quenching of the fluorescence of an-
thracene. Transition from strong to
weak quenching.
Melhuish, H.W., and Metcalf, W.S.
J. Chem. Soc. 1954: 976-9 (1954)
CA 48: 6839d

B 118 Fluorescence of adrenaline compounds.
Mesnard, P., Romain, P., and Marzat, J.
Bull. Soc. Pharm. Bordeaux 92: 121-5
(1954)
CA 50: 1472c

B 119 Quenching of the fluorescence of an-
thracene gas.
Metcalf, W.S.
J. Chem. Soc. 1954: 2485-6 (1954)
CA 48: 11196c

B 120 Resonance fluorescence with nuclei:
Tl^{203}.
Metzger, F.R., and Todd, W.B.
Phys. Rev. 95: 627 (1954)
CA 50: 10545c

B 121 Nuclear resonance fluorescence in
mercury and the lifetime of the 411 keV
excited state of mercury.
Metzger, F.R., and Todd, W.B.
Phys. Rev. 95: 853-4 (1954)
CA 48: 11922a

B 122 Physical and chemical consequences of
advanced spontaneous heating in stored
soybeans.
Milner, M., and Thompson, J.B.
J. Agr. Food Chem. 2: 303-9 (1954)
CA 48: 5396g

B 123 Inter- and intramolecular energy transfer processes. III. Phosphorescence
bands of some carcinogenic aromatic
hydrocarbons.
Moodie, M.M., and Reid, C.
J. Chem. Phys. 22: 252-4 (1954)
CA 48: 5655d

B 124 Spectrometric studies of the persis-
tence of fluorescent derivatives of
carcinogens in mice.
Moodie, M.M., Reid, C., and Wallick, C.A.
Cancer Res. 14: 367-71 (1954)
CA 48: 11627e

B 125 Design and use of an integrating-type
electrophotometer on the colorimetric
and fluorometric analysis.
Musha, S., and Ito, M.
Japan Analyst 3: 316-20 (1954)
CA 49: 14392g

B 126 Fluorescence spectra of steroids,
especially corticosteroids.
Nakao, T., Aizawa, Y., and Ui, H.
Jikeikai Med. J. 1: 81-94 (1954)
CA 49: 15484a

B 127 A study of anti-Stokes fluorescence of
vapors of aromatic compounds.
Neporent, B.S., and Borisevich, N.A.
Dokl. Akad. Nauk SSSR 94: 447-50 (1954)
CA 50: 6925i

B 128 Fluorescence and phosphorescence
spectra of phthalimide and its deriva-
tives in frozen solutions.
Neporent, B.S., and Inyushin, A.I.
Dokl. Akad. Nauk SSSR 98: 197-200 (1954)
CA 49: 12970d

B 129 Fluorescence of nitrogen dioxide.
Neuberger, D., and Duncan, A.B.F.
J. Chem. Phys. 22: 1693-6 (1954)
CA 49: 728i

B 130 Effect of sedormid (allylisopropylace-
tyl carbamide) on the developing chick
embryo.
Frisch, A.W., Talman, E.L., Aldrich, R.A.,
Neve, R.A., and Case, J.D.
Proc. Soc. Exptl. Biol. Med. 85: 573-5
(1954)
CA 48: 10232c

B 131 Influence of foreign ions on intensity of
fluorescence and afterglow of calcium
wolframate phosphor for X-rays.
Nishikawa, K.
J. Chem. Soc. Japan 57: 795-7 (1954)
CA 49: 10747b

B 132 Fluorescent substance.
Nishikawa, K., and Yamamoto, M.
Japan 110 (1954)
CA 48: 12562e

B 133 Electrical conductivity of some con-
densed aromatic hydrocarbons.
Northrop, D.C., and Simpson, O.
Proc. Phys. Soc. 67B: 892-4 (1954)
CA 49: 5058i

B 134 A treatment of chemical kinetics with
special applicability to diffusion-
controlled reactions.
Noyes, R.N.
J. Chem. Phys. 22: 1349-59 (1954)
CA 48: 13372f

B 135 Spectrophotometric measurements on
high-pressure mercury vapor lamps.
Parolini, G.
Ingegnere 28: 253-8 (1954)
CA 48: 11930d

B 136 Fluorescent lakes.
Pecquery, R.
Fr. 1,065,432 (1954)
CA 52: 19177a

B 137 The passivity of the 11-keto steroids
with respect to the phenomena of halo-
chromism and halofluorescence.
Pesez, M.
Bull. Soc. Chim. France 1954: 1070-2
(1954)
CA 49: 5125c

B 138 Fluorescence of crystalline fluorene.
Pesteil, P., and Pesteil, L.
Compt. Rend. 238: 75-7 (1954)
CA 48: 6839b

B 139 Fluorescence of several aromatic
crystals. Discussion of the results.
Pesteil, P., and Pesteil, L.
Compt. Rend. 238: 226-8 (1954)
CA 48: 9182i

B 140 Fluorescence of single crystals at low
temperatures. IV. Benzophenone.
Pesteil, P., and Barbaron, M.
Compt. Rend. 238: 1789-90 (1954)
CA 48: 9812e

B 141 Fluorescence spectra of aromatic
crystals at low temperatures.
Pesteil, P., and Barbaron, M.
J. Phys. Radium 15: 92-8 (1954)
CA 48: 9812g

B 142 Polarization of the fluorescence of
several organic single crystals.
Pesteil, P.
J. Phys. Radium 15: 407-9 (1954)
CA 49: 2873f

B 143 The triplet state in fluid solvents.
Porter, G., and Windsor, M.W.
Discussions Faraday Soc. 1954: 178-86
(1954)
CA 49: 15469i

B 144 Polystyrene scintillators.
Prins, W., and de Waard, H.
Plastica 7: 240-4 (1954)
CA 48: 14289f

B 145 Fluorescence-photometrical identifica-
tion of the rectified olive oils.
Provvedi, F.
Olii Minerali, Grassi Saponi, Colori
Vernici 31: 139-43 (1954)
CA 49: 5005i

B 146 Measurement of organic fluorescence
decay times.
Ravilious, C.F., Farrar, R.T., and
Libson, S.H.
J. Opt. Soc. Am. 44: 238-41 (1954)
CA 48: 6273eg

B 147 Decomposition in liquid scintillation
systems.
Reid, C.
J. Chem. Phys. 22: 1947 (1954)
CA 49: 2194i

B 148 The aromatic carbonium ions.
Reid, C.
J. Am. Chem. Soc. 76: 3264-8 (1954)
CA 48: 11924b

B 149 Pressure effects in phosphorescence.
Reiffel, L.
Phys. Rev. 94: 856 (1954)
CA 48: 9204b

B 150 Fluorometric determination of
N-methylnicotinamide.
Rosenthal, H.L.
Science 120: 231 (1954)
CA 48: 13546e

B 151 Fluorescent tungstates for X-ray in-
tensifying screens.
Rothschild, S.
Brit. 705,024, Mar. 3 (1954)
CA 48: 9208c

B 152 Photodielectric effect.
Roux, J.
J. Phys. Radium 15: 176-88 (1954)
CA 48: 8643i

B 153 Phosphorescence spectra of potassium
chloride-thallium powder phosphor.
Ruziewicz, Z.
Roczniki Chem. 28: 295-6 (1954)
CA 48: 11195d

B 154 Decay of phosphorescence of long dura-
tion in magnesium oxide.
Saksena, B. D., and Pant, L. M.
Current Sci. 23: 393-4 (1954)
CA 49: 12136b

B 155 Cathodo-luminescence of MgO.
Saksena, B.D., and Pant, L.M.
Proc. Phys. Soc. 67B: 811-16 (1954)
CA 49: 1435g

B 156 Some experiments on the damaged
Burma rice.
Sakurai, Y., Shiroishi, M., Fukamachi, C.,
and Hayakawa, S.
Shokuryo Kenkyush Kenkyu Hokoku 9:
181-6 (1954)
CA 53: 11684g

B 157 Fluorescent substance.
Sato, Y.
Japan 259 (1954)
CA 48: 12562e

B 158 Fluorescence and Joshi effect.
Saxena, A.P., and Ramanamurti, M.V.
Agra Univ. J. Res., Science 3: 233-40
(1954)
CA 48: 12544c

B 159 The relationship of absorption bands to
oscillators of fixed position in chromo-
phores and its application to determining the position of molecular parts in
macromolecules.
Scheibe, G.
Z. Naturforsch. 9b: 85-9 (1954)
CA 48: 9193f

B 160 Fluorescence of cadmium iodide.
Schlivitch, S., and Monod-Herzen, G.
Compt. Rend. 238: 2071 (1954)
CA 48: 11196a

B 161 Behavior of bituminous substances in
ultraviolet light.
Schmidt, H.
Erdoel Kohle 7: 428-33 (1954)
CA 49: 1313f

B 162 The quenching of the fluorescence of
fluorescein.
Schmillen, A.
Z. Angew. Phys. 6: 260-2 (1954)
CA 49: 61c

B 163 The decay of the Acridine Orange
fluorescence.
Schmillen, A.
Z. Naturforsch. 9a: 1036-9 (1954)
CA 49: 5972g

B 164 The chemistry of experimental chloro-
ma. I. Porphyrins and peroxidases.
Schultz, J., Shay, H., and Gruenstein, M.
Cancer Res. 14: 157-62 (1954)
CA 48: 11625f

B 165 Paper chromatography of plant-growth
regulators and allied compounds.
Sen, S.P., and Leopold, A.C.
Physiol. Plantarum 7: 98-108 (1954)
CA 49: 13574g

B 166 A method of determining the dissocia-
tion energy of nitrogen.
Sheehan, W.F.
J. Chem. Phys. 22: 1461 (1954)
CA 48: 13397a

B 167 Fine structure of fluorescence spectra
of aromatic hydrocarbons in frozen
solutions.
Shpol'skii, E.V., and Klimova, L.A.
Izv. Akad. Nauk SSSR., Ser. Fiz. 18: 673
(1954)
CA 50: 7585i

B 168 The use of optical brighteners in the
paper industry.
Siegrist, A.E.
Das Papier 8: 109-20 (1954)
CA 48: 9062e

B 169 Estimation of the component cardiac
glycosides in digitalis plant samples.
II. Deglucoglycosides and ultraviolet
fluorescence with trichloroacetic acid.
Silberman, H., and Thorp, R.H.
J. Pharm. Pharmacol. 6: 546-51 (1954)
CA 49: 1279g

B 170 Fluorescence transfer in condensed
aromatic hydrocarbons.
Simpson, O., and Northrop, D.C.
Physica 20: 1122-5 (1954)
CA 49: 5045h

B 171 Fluorescence of quinoline in acid solu-
tions.
Sljivic, S.
Z. Anal. Chem. 143: 113-14 (1954)
CA 49: 61d

B 172 Quenching of potassium resonance
radiation by hydrogen and deuterium.
Smith, W.M., Stewart, J.A., and Taylor,
G.W.
Can. J. Chem. 32: 961-8 (1954)
CA 49: 5119i

B 173 The formation of absorption lines with
incoherent scattering of light.
Sobolev, V.V.
Astron. Zh. 31: 231-48 (1954)
CA 48: 11930a

B 174 The measurement of the lifetime of an
excited state by optical resonance.
Soleillet, P.
Compt. Rend. 239: 698-700 (1954)
CA 49: 5120d

B 175 Influence of a magnetic field on the polarization of the 2139 A resonance radiation emitted by a zinc atom.
Spitzer, M.
Compt. Rend. 239: 696-8 (1954)
CA 49: 5120b

B 176 Simultaneous observation of oxidation-reduction potentials and chlorophyll fluorescence of chlorella suspensions.
Spruit, C.J.P., and Wassink, E.C.
Biochim. Biophys. Acta 15: 357-66 (1954)
CA 49: 4806d

B 177 Reversible association processes of globar proteins. VI. The combination of trypsin with soybean inhibitor (polarization).
Steiner, R.F.
Arch. Biochem. Biophys. 49: 71-92 (1954)
CA 48: 7091h

B 178 Possibilities of electron-vibration transformations and a study of the laws of the extinguishing fluorescent radiation of complex molecules.
Stepanov, B.I.
Vestsi Akad. Navuk Belarusk. SSR 1954: 60-9 (1954)
CA 49: 15500b

B 179 Color photographs of fluorescent objects in ultraviolet light.
Stoll, S.
Chim. Anal. 36: 39-40 (1954)
CA 48: 4355d

B 180 Temperature dependence of light absorption and fluorescence intensity in the solvatochromy of 1,2,4-trimethyl-3-hydroxyphenazine.
Suhrmann, R., and Perkampus, H.H.
Z. Physik Chem. 2: 290-311 (1954)
CA 49: 6726i

B 181 Quenching of the fluorescence of the eosin ion.
Svirbely, W.J., and Sharpless, N.E.
J. Am. Chem. Soc. 76: 1404-9 (1954)
CA 48: 6261b

B 182 An adaptation of the Beckman spectrophotometer for use as a fluorimeter.
Swann, R.V.
Analyst 79: 176-8 (1954)
CA 48: 6169b

B 183 Sulfuric acid-induced fluorescence of corticosteroids.
Sweat, M.L.
Anal. Chem. 26: 773-6 (1954)
CA 48: 10096b

B 184 Lanthanum oxychloride phosphors.
Swindells, F.E.
J. Electrochem. Soc. 101: 415-18 (1954)
CA 50: 64e

B 185 Light transducers.
Szegho, C.S., and Pakswer, S.
U.S. 2,680,213, June 1 (1954)
CA 48: 9814d

B 186 Photoconductivity effect in indium antimonide.
Tauc, J., Smirous, K., and Abraham, A.
Czech. J. Phys. 4: 255 (1954)
CA 49: 7370d

B 187 Energy transfer in sensitized fluorescence of mixtures of vapors of organic compounds.
Terenin, A.N., and Karyakin, A.V.
Dokl. Akad. Nauk SSSR 96: 269-72 (1954)
CA 49: 10067c

B 188 Kinetics and equilibria in flavoprotein systems. I. A fluorescence recorder and its application to a study of the dissociation of the old yellow enzyme and its resynthesis from riboflavine phosphate and protein.
Theorell, H., and Nygaard, A.P.
Acta Chem. Scand. 8: 877-88 (1954)
CA 49: 4036i

B 189 The linkages between flavine mononucleotide (FMN) and the protein of the old yellow enzyme studied by fluorescence measurements.
Theorell, H., and Nygaard, A.P.
Arkiv Kemi 7: 205-9 (1954)
CA 48: 11519a

B 190 Apparent phosphorescence of cypridina luciferin solution.
Tsuji, F.I., and Harvey, E.N.
Arch. Biochem. Biophys. 52: 285-6 (1954)
CA 48: 13764i

B 191 Fluorescence spectra of cuprous
halides at low temperatures.
Tsujikawa, I., and Kanda, E.
Sci. Rept. Res. Inst., Tohoku Univ., Ser. A
6: 220-3 (1954)
CA 49: 5131e

B 192 The hyperfine structure of magnetic
resonance in some phosphorescent
materials.
Uebersfeld, J.
J. Phys. Radium 15: 126-7 (1954)
CA 48: 6248c

B 193 Fluorescent substance.
Uehara, Y., and Kofuya, Y.
Japan 109 (1954)
CA 48: 12562d

B 194 Fluorescent substances.
Uehara, Y., and Kofuya, Y.
Japan 7574 (1954)
CA 50: 5414f

B 195 Synthetic studies on minerals. II. The
reaction of sodium tungstate and cal-
cium carbonate.
Umegaki, Y., Kashiwagi, H., and Habara,
T.
J. Sci. Hiroshima Univ., Ser. C, 1: 79-84
(1954)
CA 50: 117f

B 196 Capillary (paper chromatography) and
luminescence analysis of fern extracts
and their components.
Vaverane, V.
Latvijas PSR Zinatnu Akad. Vestis
1954(10): 85-91 (1954)
CA 49: 7188i

B 197 Comparison of various phenomena connected with photosynthesis (fluorescence, redox potentials, phosphate ex-
changes, gas exchange, and others)
with special reference to induction effects in chlorella.
Wassink, E.L., and Spruit, C.J.P.
Congr. Intern. Botan., Paris 8: 1-2 (1954)
CA 48: 13842f

B 198 Fluorescence of spectrum of triphenyl-
methyl at 4°K.
Weissman, S.I.
J. Chem. Phys. 22: 155 (1954)
CA 48: 4311h

B 199 General base catalysis for the electro-
lytic dissociation of excited naphthol.
Weller, A.
Z. Elektrochem. 58: 849-53 (1954)
CA 49: 9388h

B 200 Reaction of chlorophyll in amines.
Weller, A., and Livingston, R.
J. Am. Chem. Soc. 76: 1575-8 (1954)
CA 48: 6796e

B 201 Dependence of the polarization of the
fluorescence on the concentration.
Weber, G.
Trans. Faraday Soc. 50: 552-5 (1954)
CA 49: 1435e

B 202 Sensitized fluorescence in organic
mixed crystals.
Wolf, H.C.
Z. Physik 139: 318-27 (1954)
CA 50: 3902d

B 203 Cellular-physiological study of pH_n and
fluorescence of colored plants.
Yablokova, V.A.
Dokl. Akad. Nauk SSSR 98: 145-8 (1954)
CA 49: 1155e

B 204 The photoelectric fluorometer and its
application to biochemical experiments.
Yagi, K., and Tabata, T.
Kagaku No Ryoiki 8: 45-53 (1954)
CA 48: 8851f

B 205 Synthesis of organic fluorescent com-
pounds. XXXI.
Yoshida, Z.
J. Chem. Soc. Japan 57: 241-3 (1954)
CA 49: 11620g

B 206 Low temperature spectra of some
homologous six-membered nitrogen-
containing heterocyclic compounds and
their development to the spectra of
acridine dyes.
Zanker, V.
Z. Physik. Chem. 2: 52-78 (1954)
CA 49: 55h

B 207 The fluorescence of self-activated
zinc sulfide.
Addamiano, A.
J. Chem. Phys. 23: 1541 (1955)
CA 49: 14487h

B 208 Absorption and fluorescence spectra of
    polyatomic molecules.
    Agranovich, V.M., and Davydov, A.S.
    Nauk. Zap., Kiivs'k. Derzh. Univ. 14, No.
    8; Sb. Fiz. Fak. 7: 15-20 (1955)
    CA 51: 9309c

B 209 Yield of resonance fluorescence of
    atoms.
    Alentsev, M.N., Antonov, Romanovskii,
    V.V., Stepnaov, B.I., and Fok, M.V.
    Zh. Eksprim. i Teor. Fiz. 28: 253-4 (1955)
    CA 50: 2283d

B 210 Application of fluorescent coatings to
    discharge tubes.
    Anderson, J.T., and Ward, H.F.
    U.S. 2,726,966 (1955)
    CA 50: 6934i

B 211 Photomultiplier study of scintillations
    produced in a zinc sulfide screen by
    α-particles.
    Anthony, J.P.
    J. Phys. Radium 16: 182-90 (1955)
    CA 49: 12980d

B 212 Luminescence ZnS (Au). Action of Fe.
    Arpiarian, N.
    Compt. Rend. 240: 1202-5 (1955)
    CA 49: 12967d

B 213 The spectral distribution of infrared
    stimulated phosphorescence of lead-
    activated zinc sulfide type phosphors.
    Asano, S.
    J. Phys. Soc. Japan 10: 903-5 (1955)
    CA 50: 12657e

B 214 Quenching of biacetyl fluorescence in
    solution.
    Backstrom, H.L.J., and Sandros, K.
    J. Chem. Phys. 23: 2197 (1955)
    CA 50: 2302a

B 215 Fluorescent brightening agents.
    Badische Anilin- & Soda-Fabrik, A.-G.
    Brit. 741,798 (1955)
    CA 50: 11026h

B 216 Luminescence spectroscopy of
    porphyrin-like molecules including the
    chlorophylls.
    Becker, R.S., and Kasha, M.
    J. Am. Chem. Soc. 77: 3669-70 (1955)
    CA 49: 12969h

B 217 Anomalous light emission of azulene.
    Beer, M., and Loaguet-Higgins, H.C.
    J. Chem. Phys. 23: 1390-1 (1955)
    CA 49: 15499f

B 218 The principles of identification and
    measurements of vulvar fluorescence.
    Benson, R.C., and Vogel, M.J.
    J. Clin. Endocrinol. Metab. 15: 784-800
    (1955)
    CA 49: 13444i

B 219 Quenching of the long-lifetime compo-
    nent of positron annihilation in benzene.
    Berko, S., and Zuchelli, A.J.
    Phys. Rev. 99: 1653 (1955)
    CA 51: 11089g

B 220 Organic phosphorescence. Urinary
    elimination of acriflavine.
    Bernanose, A., Comte, M., and Vouaux, P.
    Bull. Soc. Pharm. Nancy 26: 5-11 (1955)
    CA 51: 5893i

B 221 Relation between organic electrolumin-
    escence and concentration of active
    product.
    Bernanose, A., and Vouaux, P.
    J. Chim. Phys. 52: 509-10 (1955)
    CA 50: 16341h

B 222 Glycosides and aglycons. CLIII. Iden-
    tification of glycosides and aglycons of
    the sulla-bufo type in paper chromato-
    graphy by direct photocopies.
    Bernasconi, R., Sigg, H.P., and
    Reichstein, T.
    Helv. Chim. Acta 38: 1767-75 (1955)
    CA 50: 13063c

B 223 Raman spectra and fluorescence of a
    few substituted toluenes in the solid
    state at different low temperatures.
    Biswas, D.C.
    Indian J. Phys. 29: 257-71 (1955)
    CA 51: 7865h

B 224 Raman spectra of a few monosubstituted
    benzene compounds in the solid state at
    different low temperatures.
    Biswas, D.C.
    Indian J. Phys. 29: 503-17 (1955)
    CA 51: 7865b

B 225 Stabilization of energy-rich molecules.
   I. Energy transfer with hydrogen.
   Boudart, M., and Dubois, J.T.
   J. Chem. Phys. 23: 223-9 (1955)
   CA 49: 6717e

B 226 The emission spectra of aromatic hydrocarbons in crystalline paraffins at
   -180°.
   Bowen, E.J., and Brocklehurst, B.
   J. Chem. Soc. 1955: 4320-31 (1955)
   CA 50: 8334a

B 227 Solvent quenching of the fluorescence
   of anthracene.
   Bowen, E.J., and West, K.
   J. Chem. Soc. 1955: 4394-5 (1955)
   CA 50: 8334f

B 228 Photochemistry of anthracenes. III.
   Interrelations between fluorescence
   quenching, dimerization, and photo-
   oxidation.
   Bowen, E.J., and Tanner, D.W.
   Trans. Faraday Soc. 51: 475-81 (1955)
   CA 49: 13787h

B 229 Energy transfer in rigid solvents.
   Bowen, E.J., and Brocklehurst, B.
   Trans. Faraday Soc. 51: 774-7 (1955)
   CA 50: 65g

B 230 Spectrophotofluorometric assay in the
   visible and ultraviolet.
   Bowman, R.L., Caulfield, P.A., and
   Udenfriend, S.
   Science 122: 32-3 (1955)
   CA 49: 15290f

B 231 Law of extinction of phosphorescence
   in organoluminophors.
   Bredel, V.V.
   Dokl. Akad. Nauk SSSR 103: 787-90 (1955)
   CA 50: 8334c

B 232 Agreement of dose measurement for
   rapid electron rays (3-15-MeV) with
   X-ray dose measurement.
   Breitling, G., and Glocker, R.
   Naturwissenschaften 42: 11-12 (1955)
   CA 49: 12147h

B 233 Fluorescence and average lifetime of
   excited OH $(^2\Sigma^+)$ in flames.
   Broida, H.P., and Carrington, T.
   J. Chem. Phys. 23: 2202 (1955)
   CA 50: 2301e

B 234 Fluorescence of the serum in rats with
   alloxan diabetes and cataract. I. Intensity determinations.
   Brolin, S.E.
   Acta Physiol. Scand. 33: 359-69 (1955)
   CA 49: 15013a

B 235 Fluorescence yield of chlorophyll in
   chlorella as a function of light intensity.
   Brugger, J.E.
   Natl. Acad. Sci. — Natl. Res. Council
   1955: 113-17 (1955)
   CA 52: 10303c

B 236 The spectral fluorescence and absorption of 3,4-benzopyrene in the visible
   region.
   Berg, N.O., and Norden, G.
   Acta Pathol. Microbiol. Scand. 36: 193-204
   (1955)
   CA 49: 7381e

B 237 Inhibition of horse-liver esterase by
   Rhodamine B.
   Burch, J.
   Biochem. J. 59: 97-110 (1955)
   CA 49: 6330e

B 238 Absorption and luminescence investigation of some mesoaryl and mesoalkyl
   anthracene derivatives.
   Cherkasov, A.S.
   Zh. Fiz. Khim. 29: 2209-17 (1955)
   CA 50: 13614h

B 239 Lingual fluorescence under Wood's
   light in psychiatry. Observations on
   411 patients.
   Colombo, C., and Verga, G.
   Acta Vitaminol. 9: 157-62 (1955)
   CA 50: 8861f

B 240 Fluorescence and phosphorescence.
   Dammers-de Klerk, A.
   Chem. Weekblad 51: 741-50 (1955)
   CA 50: 2301c

B 241 Sedimentation of phosphorescent
   screens in cathode-ray tubes.
   de Boer, F., and Emmens, H.
   Philips' Tech. Rundschau 16: 272-6 (1955)
   CA 52: 19494a

B 242 The fluorescence of chromoactive sub-
stances of crustaceans and insects.
DeLerma, B., Dupont-Raabe, M., and
Knowles, F.
Compt. Rend. 241: 995-8 (1955)
CA 50: 7335i

B 243 The low temperature luminescence of
silver bromide with silver sulfide addi-
tion.
Dorfner, K.R.
Ann. Physik 6: 331-60 (1955)
CA 50: 7598b

B 244 Preparation of 1,4,6,11-tetraphenyl-
naphthacene, an isomer of rubene.
Douris, R.G.
Compt. Rend. 240: 1113-15 (1955)
CA 50: 3369f

B 245 Yield of anti-Stokes fluorescence of
very viscous dye solutions.
Drabent, R., and Frackowiak, D.
Acta Phys. Polon. 14: 447-54 (1955)
CA 50: 16405c

B 246 Fluorescent derivative of guanine
formed during the hydrolysis of
deoxyribonucleic acid.
Dunn, D.B.
Biochim. Biophys. Acta 18: 317-8 (1955)
CA 50: 2708d

B 247 Comparison of the luminescence of
calcium silicate (Mn,Pb) and zinc
beryllium silicate (Mn).
Dziergwa, H., and Lange, H.
Z. Physik 140: 359-69 (1955)
CA 50: 6925d

B 248 Fluorescent compounds in oat roots.
Eberhardt, F.
Z. Botan. 43: 405-22 (1955)
CA 50: 2752e

B 249 Distribution of fluorescing islets,
adrenaline, and noradrenaline in the
adrenal medulla of the hamster.
Eranko, O.
Acta Endocrinol. 18: 174-9 (1955)
CA 49: 8435i

B 250 Distribution of fluorescing islets,
adrenaline, and noradrenaline in the
adrenal medulla of the cat.
Eranko, O.
Acta Endocrinol. 18: 180-8 (1955)
CA 49: 8436a

B 251 Effect of insulin on chromaffin reaction, fluorescing islets, and catechol
amines in the adrenal medulla of the
rat.
Eranko, O.
Acta Pathol. Microbiol. Scand. 36: 219-23
(1955)
CA 49: 7697g

B 252 The extinction and the luminescence
duration changes during sensitized
phosphorescence of organic compounds.
Ermolaev, V.L.
Dokl. Akad. Nauk SSSR 102: 925-8 (1955)
CA 50: 3091e

B 253 Photomagnetism of the triplet states of
organic molecules (including fluores-
cein).
Evans, D.F.
Nature 176: 777-8 (1955)
CA 50: 12659c

B 254 A simple procedure for the photography
of fluorescent substances.
Fasella, P., and Baglioni, C.
Boll. Soc. Ital. Biol. Sper. 31: 554-7 (1955)
CA 50: 709d

B 255 Electrophoretic study of some trans-
aminase reactions.
Fasella, P., Lis, H., and Siliprandi, N.
Riv. Ist. Sieroterap. Ital. 30: 206-10 (1955)
CA 50: 415a

B 256 Polarization theory of resonant emis-
sion and fluorescence of atoms and
diatomic molecules.
Feofilov, P.P.
Dokl. Akad. Nauk SSSR 104: 846-9 (1955)
CA 50: 12658i

B 257 Influence of fluorescent compounds on
rooting of cuttings. II.
Ferri, M.G., and de Morretes, B.L.
Rev. Brasil. Biol. 15: 321-8 (1955)
CA 50: 6589g

B 258 Calculation of errors in spectral analy-
sis by current measurements. II.
Fishman, I.S., and Stolov, A.L.
Uch. Zap. Kazansk. Gos. Univ. 115: 57-71
(1955)
CA 52: 13517i

B 259 Concentration reversal of the fluorescence of pyrene.
Forster, T., and Kasper, K.
Z. Elektrochem. 59: 976-80 (1955)
CA 50: 3902h

B 260 Correlation of structure with fluorescence of some metal-organic chelates.
Dissertation, University of Maryland (1955)
CA 50: 2349h

B 261 Fluorescence spectra of protochlorophyll, chlorophylls C and D and their pheophytins.
French, C.S., Smith, J.H.C., and Virgin, H.I.
Natl. Acad. Sci. — Natl. Res. Council 1955: 17-18 (1955)
CA 52: 10302g

B 262 Fluorescence in the photographic emulsion.
Friedman, J.S., and Horwitz, L.
Phot. Sci. Tech. (2) 2: 42-7 (1955)
CA 49: 4429b

B 263 Radiation spectrograms of fluorescence of radioelements.
Frilley, M.
J. Phys. Radium 16: 630-4 (1955)
CA 49: 15475b

B 264 Effective cross section of secondary collisions in the sensitized fluorescence.
Frish, S.E., and Kraulinya, E.K.
Dokl. Akad. Nauk SSSR 101: 837-40 (1955)
CA 50: 6931e

B 265 Chlorophyll-photosensitized reduction of triphenyltetrazolium chloride by hydrazine hydrate.
Fujimori, E.
J. Am. Chem. Soc. 77: 6495-8 (1955)
CA 50: 5412i

B 266 Organic fluorescent and photosensitive substances.
Fujimori, E.
Rept. Inst. Ind. Sci., Univ. Tokyo 4: 93-151 (1955)
CA 49: 14545g

B 267 Enhancement of fluorescence in solutions under high-energy irradiation.
Furst, M., and Kallmann, H.
Phys. Rev. 97: 583-7 (1955)
CA 49: 6733f

B 268 Fluorescent behavior of solutions containing more than one solvent.
Furst, M., and Kallmann, H.
J. Chem. Phys. 23: 607-12 (1955)
CA 49: 10747f

B 269 Increasing fluorescence efficiency of liquid-scintillation solutions.
Furst, M., Kallmann, H., and Brown, F.H.
Nucleonics 13: 58, 60 (1955)
CA 49: 10067b

B 270 2,3,2',3'-Imidazole-(1,2,4)-triazine compounds.
Fusco, R., and Rossi, S.
Ital. 536,121 (1955)
CA 53: 2264f

B 271 Biochemistry of microorganisms. XCVI. The coloring matter of Penicillium herquei.
Galarraga, J.A., Neill, K.G., and Raistrick, H.
Biochem. J. 61: 456-64 (1955)
CA 50: 3537c

B 272 Chlorophyll spectra and molecular structure.
Goedheer, J.C.
Nature 176: 928-9 (1955)
CA 50: 6931g

B 273 Water-soluble paint.
Gomez, A.M.
Span. 213,231 (1955)
CA 50: 4527i

B 274 Simple adaptation of the Beckman DCL spectrophotometer as a spectrofluorometer.
Gornall, A.G., and Kalant, H.
Anal. Chem. 27: 474-5 (1955)
CA 49: 8636c

B 275 Radiation micromethods in the study of crystal-structure defects.
Grillot, E.
Compt. Rend. Congr. Soc. Savantes Paris Dept., Sect. Sci. 1955: 71-87 (1955)
CA 52: 13433g

B 276 Changes in the fluorescent substances
(of tissues) during development of
amphibians.
Hama, T., and Goto, T.
Compt. Rend. Soc. Biol. 149: 859-60 (1955)
CA 50: 3656c

B 277 Quenching of sodium iodide fluores-
cence by hydrogen, hydrogen chloride,
carbon dioxide, and water.
Hanson, H.G.
J. Chem. Phys. 23: 1391-7 (1955)
CA 49: 15500e

B 278 Identification of pyrimidines in the
fluorescing fractions of the teeth of the
sperm whale.
Hartles, R.L., and Leaver, A.G.
J. Dental Res. 34: 820-30 (1955)
CA 50: 4339c

B 279 The interaction of deoxyribonucleic
acid with acriflavine.
Heilweil, H.G., and Van Winkle, Q.
J. Phys. Chem. 59: 939-43 (1955)
CA 50: 1101i

B 280 Effect of ultrasound on conductivity
and fluorescence of zinc sulfide and
cadmium sulfide crystals.
Herforth, L., and Krumbiegel, I.
Naturwissenschaften 42: 39 (1955)
CA 49: 15457b

B 281 Fine structure study with X-rays by
fluorescence measurement.
Herforth, L.
Naturwissenschaften 42: 412 (1955)
CA 51: 2387d

B 282 Some new laboratory results for the
spectrum of carbon.
Herman, R., and Herman, L.
Mem. Soc. Roy. Sci. Liege 15: 352-9 (1955)
CA 49: 14480e

B 283 Fluorescence spectra of the condensation products of salicylaldehyde and
their salts.
Holzbecher, Z.
Collection Czech. Chem. Commun. 20:
59-66 (1955)
CA 49: 7989f

B 284 Fluorescence spectra of salicylaldehyde condensation products and their
salts. II. o-Salicylideneamino phenol.
Holzbecher, Z.
Collection Czech. Chem. Commun. 20:
1292-6 (1955)
CA 50: 5411g

B 285 Fluorescence spectra of salicylaldehyde
condensation products and their salts.
Holzbecher, Z.
Chem. Listy 49: 1030-3 (1955)
CA 49: 12971c

B 286 Fluorescence spectra of salicylaldehyde
condensation products and their salts.
III.
Holzbecher, Z.
Chem. Listy 49: 1162-8 (1955)
CA 49: 14489i

B 287 Fluorescence of zinc sulfide.
Hoogenstraaten, W.
Ned. Tijdschr. Natuurk. 21: 150-9 (1955)
CA 62: 8749e

B 288 Analyses of flavones. VIII. Further
properties of halochromic boron com-
plexes.
Horhammer, L., and Hansel, R.
Arch. Pharm. 288: 315-21 (1955)
CA 51: 6619h

B 289 Fluorescence quenching by blue and
yellow light in a sensitized photograph-
ic emulsion.
Horwitz, L., and Friedman, J.
J. Opt. Soc. Am. 45: 798 (1955)
CA 50: 95d

B 290 Studies on organic phosphors containing
metallic compounds as phosphorescent
bodies. II. Crystals of metallic salts
suitable for the phosphorescent bodies
and the effects of metallic ions on the
phosphorescent color. 2.
Iwaki, R.
J. Chem. Soc. Japan, Pure Chem. Sect.
76: 605-10 (1955)
CA 50: 3902a

B 291 The optical properties of calcium
metaantimonate.
Janin, J., and Bernard, R.
Compt. Rend. 240: 614-15 (1955)
CA 49: 7986h

B 292 Long time phosphorescence of organic
crystals.
Kallmann, H., Kramer, B., and Sucov, E.
J. Chem. Phys. 23: 1043-7 (1955)
CA 49: 12970e

B 293 Studies on the quenching of organic
phosphors. III. Energy states of organic phosphors as revealed by their
quenching phenomena.
Kato, S., Kimura, K., and Koizumi, M.
J. Chem. Soc. Japan 76: 262-7 (1955)
CA 49: 15500g

B 294 Determination of citrimin contained in
rice by fluorometry.
Kawashiro, I., Tanabe, H., Takeuchi, H.,
and Nishimura, C.
Bull. Natl. Hyg. Lab., Tokyo 73: 191-6
(1955)
CA 50: 7229c

B 295 Thermoluminescence and fluorescence
in alkali halide crystals induced by
soft X rays.
Keller, S.P., and Clemmons, J.J.
J. Chem. Phys. 23: 586-7 (1955)
CA 49: 7986i

B 296 Infrared phosphorescence of germani-
um at low temperatures.
Kessler, R.
Z. Naturforsch. 10a: 87-8 (1955)
CA 49: 7389g

B 297 Modified calcium pyrophosphate phos-
phors.
Kinney, D.E.
J. Electrochem. Soc. 102: 676-81 (1955)
CA 50: 6926g

B 298 Borazole and boron nitride scintillators.
Kirkbride, J.
Proc. Phys. Soc. 68B: 253 (1955)
CA 49: 13800h

B 299 The effect of light from a Hg arc on
flame emission.
Kishko, S.M.
Nauchn. Zap. Uzhgorodsk. Inst. 12: 59-63
(1955)
CA 55: 3190i

B 300 Influence of the light of a mercury arc
on flame radiation.
Kishko, S.M., and Miluyanchuk, V.S.
Izv. Akad. Nauk SSSR, Ser. Fiz. 19: 19
(1955)
CA 50: 3885i

B 301 Characteristic fluorescence in solids.
Klasens, H.A.
Ned. Tijdschr. Natuurk. 21: 131-41 (1955)
CA 50: 63f

B 302 Fluorescence spectrum of gaseous
nitric oxide and the effect of some
gases on the spectrum.
Kleinberg, A.V., and Terenin, A.N.
Dokl. Akad. Nauk SSSR 101: 445-7 (1955)
CA 50: 2301fh

B 303 Semimicro butyric acid values, semimicro total values, and semimicro
residual values of cocoa butter and
cocoa-butter-substitute fats.
Kleinert, J.
Rev. Intern. Chocolat. 10: 449-54 (1955)
CA 50: 8231i

B 304 The dependence of continuous absorption and fluorescence spectra of vapors
and solutions of substituted phthalimides from the temperature and the
solvent.
Klochkov, V.P.
Zh. Fiz. Khim. 29: 1432-41 (1955)
CA 51: 851g

B 305 Porphyrins in relation to the develop-
ment of the nervous system.
Kluver, H.
Biochem. Develop. Nervous Systems,
Proc. Intern. Neurochem. Symp. 1st,
Oxford 1954: 137-44 (1955)
CA 50: 6532b

B 306 Fluorescent substance.
Kodera, Y., and Sekine, T.
Japan 417 (1955)
CA 50: 11830h

B 307 Mechanism of quenching of fluores-
cence.
Kortum, G., and Wilski, H.
Trans. Faraday Soc. 51: 1620-3 (1955)
CA 50: 11827i

B 308 Oxyacid phosphors. V. Antimony oxide phosphor with red emission.
Kotera, Y., and Sekine, T.
Rept. Govt. Chem. Ind. Res. Inst., Tokyo 50: 259-66 (1955)
CA 50: 1119b

B 309 Energy transfer in polystyrene-anthracene.
Krenz, F.H.
Trans. Faraday Soc. 51: 172-83 (1955)
CA 49: 12136h

B 310 Electrical quadrupole moments of nuclei rubidium-85 and rubidium-87 by determination of the high-frequency transitions in the 6^2P 3/2 term of the rubidium atom.
Kruger, H., and Meyer-Berkhout, U.
Naturwissenschaften 42: 94-5 (1955)
CA 50: 57i

B 311 Fluorometer for paper chromatograms.
Kuhn, A.
Naturwissenschaften 42: 529-30 (1955)
CA 51: 10961g

B 312 Light absorption of organic dyestuffs.
Kuhn, H.
Chimia (Switz.) 9: 237-49 (1955)
CA 51: 839h

B 313 The luminescence of colored materials.
Laffitte, E.
Ann. Phys. 10: 71-127 (1955)
CA 49: 8704f

B 314 Ultraviolet absorption spectra of some mono- and disubstituted (meso) anthracene derivatives.
Laffitte, E., and Lalande, R.
Bull. Soc. Chim. France 1955: 531-4 (1955)
CA 49: 13777h

B 315 The polarization of the fluorescence of diatomic gaseous molecules.
Laffitte, E.
J. Phys. Radium 16: 66-71 (1955)
CA 49: 5103h

B 316 Emission spectra of fused quartz irradiated by X-rays. The role of impurities and of the vitreous state.
Lautout, M.
J. Chim. Phys. 52: 176-8 (1955)
CA 49: 10748a

B 317 Photostimulation and coloration of fused quartz irradiated by X- or γ-rays.
Lautout, M.
J. Chim. Phys. 52: 267-71 (1955)
CA 49: 11425c

B 318 Halochromy of tertiary alicyclic and aliphatic carbinols.
Lavrushin, V.F., Verkhovod, N.N., and Movchan, P.K.
Dokl. Akad. Nauk SSSR 105: 1723-6 (1955)
CA 50: 11256h

B 319 Decay of phosphorescence with activators and traps arising from the same impurity in different valence states.
Lehovec, K.
J. Opt. Soc. Am. 45: 219-22 (1955)
CA 49: 5971b

B 320 N-substituted 2,3,5-triphenylpyrroles.
Lespagnol, A., Dumont, J.M., Mercier, J., and Etzensperger, M.
Bull. Soc. Pharm. Lille 1955(1): 87-9 (1955)
CA 50: 3399a

B 321 Structure of chlorophyll. Analytical and synthetic evidence.
Linstead, R.P., Eisner, U., Ficken, G.E., and Johns, R.B.
Chem. Soc., Spec. Publ. 3: 83-97 (1955)
CA 50: 10822a

B 322 Factors affecting the fluorescence of cotton.
Lourigan, G.H.
Am. Dyestuff Reptr. 44: 348-9, 359 (1955)
CA 49: 11287f

B 323 New luminiferous substance for fluorescent tubes made from domestic raw materials.
Lucatu, E.
Comun. Acad. Rep. Populare Romine 5: 333-7 (1955)
CA 50: 10544a

B 324 Recombined mechanism of afterglow of some crystallophosphors.
Lushchik, C.B.
Tr. Inst. Fiz. i Astron., Akad. Nauk Est. SSR 1955(1): 57-71 (1955)
CA 51: 87f

B 325 Fluorescent spots in raw cotton associated with the growth of microorganisms.
Marsh, P.B., Bollenbacher, K., San Antonio, J.P., and Merola, G.V.
Textile Res. J. 25: 1007-16 (1955)
CA 50: 2177e

B 326 Fluorescence of compounds of the adrenaline series. II.
Marzat, J., Romain, P., and Mesnard, P.
Bull. Soc. Pharm. Bordeaux 93: 18-21 (1955)
CA 50: 8334f

B 327 Influence of the addition of high-molecular electrolytes on the absorption spectra and fluorescence of organic dyes. IV.
Mataga, N., and Koizumi, M.
Bull. Chem. Soc. Japan 28: 51-4 (1955)
CA 52: 2537f

B 328 The solvent effect on fluorescence spectrum. Change of solute-solvent interaction during the lifetime of excited solute molecule.
Mataga, N., Kaifu, Y., and Koizumi, M.
Bull. Chem. Soc. Japan 28: 690-1 (1955)
CA 50: 6930f

B 329 Equilibrium of hydrogen-bond formation in the excited state.
Mataga, N., Kaibe, Y., and Koizumi, M.
Nature 175: 731-2 (1955)
CA 49: 15499b

B 330 Photoelectric end-point determination in the titration of fluorides with thorium nitrate.
Mavrodineanu, R., and Gwirtsman, J.
Contrib. Boyce Thompson Inst. 18: 181-6 (1955)
CA 51: 1665a

B 331 Measurement of absolute quantum efficiencies of fluorescence.
Melhuish, W.H.
New Zealand J. Sci. Technol. 37: 142-9 (1955)
CA 50: 12659e

B 332 Nuclear resonance fluorescence in germanium-74 and praseodymium-141.
Metzger, F.R.
Phys. Rev. 99: 613 (1955)
CA 51: 11029a

B 333 Angular distribution of the resonance fluorescence radiation from the 411-keV excited state of mercury[198].
Metzger, F.R.
Phys. Rev 97: 1258-60 (1955)
CA 49: 7375c

B 334 A photoelectric fluorimeter of simple construction.
Mezincesu, M.D., and Popescu-Stefanescu, A.
Rev. Chim. 6: 659-63 (1955)
CA 50: 16189a

B 335 Dependence of fluorescence on the structure of the solvent molecule.
Miller, E.J.
Bull. Inst. Nucl. Sci. "Boris Kidrich" 5: 107-11 (1955)
CA 50: 682c

B 336 Fluorescent substance.
Nagy, E., Mende, L., and Gazda, I.
Ger. 933,645 (1955)
CA 52: 19517h

B 337 Change of fluorescence spectra of derivatives of phthalimide on transition from gases to solutions through the critical state.
Neporent, B.S., Klochkov, V.P., and Motovilov, O.A.
Zh. Fiz. Khim. 29: 305-13 (1955)
CA 50: 12659a

B 338 Alkali titanates.
Neuhans, A., Schmitz-DuMont, O., and Reckhard, H.
Forschungsber. Wirtsch. Verkehrsministeriums Nordrhein-Westfalen No. 190, 48 pp. (1955)
CA 53: 6857g

B 339 Emission of elliptically polarized fluorescence radiation by optically active compounds.
Neunhoeffer, O., and Ulrich, H.
Z. Elektrochem. 59: 122-6 (1955)
CA 49: 12970g

B 340 Synthesis and fluorescence of quinoline-substituted 1,3,5-triphenylpyrazolines.
Neunhoeffer, P., and Ulrich, H.
Chem. Ber. 88: 1123-33 (1955)
CA 50: 13881h

B 341 Metallic complexes of isonicotinyl
hydrazide.
Neuzil, E., and Segonne, J.
Bull. Soc. Pharm. Bordeaux 93: 49-66
(1955)
CA 50: 9200g

B 342 Effect of oxidizing or reducing agents
on the fluorescence intensity of calcium
wolframate.
Nishikawa, K., and Yamamoto, M.
J. Chem. Soc. Japan 58: 471-2 (1955)
CA 49: 14487i

B 343 Photoeffects in condensed aromatic
hydrocarbons.
Northrop, D.C., and Simpson, O.
Proc. Phys. Soc. 68: 974-7 (1955)
CA 50: 6181h

B 344 Phosphorescence of solids containing
the manganous or ferric ions.
Orgel, L.E.
J. Chem. Phys. 23: 1958 (1955)
CA 50: 681f

B 345 Slow disappearance of fluorescence in
thionine on addition of abietic acid.
Palit, S.R., and Biswas, B.
Nature 176: 214-15 (1955)
CA 49: 15501a

B 346 Intensity measurements in absorption
and fluorescence of uranyl acetate
solution.
Pant, D.D., and Khandelwal, D.P.
Current Sci. 24: 376 (1955)
CA 50: 5401i

B 347 Action of X-rays on dilute solutions of
Uranine S.
Patti, F.
J. Chim. Phys. 52: 38-40 (1955)
CA 49: 8712d

B 348 The fluorescent reaction in some
1-(3,4-dihydroxyphenyl)2-aminoetha-
nols.
Pavolini, T.
Ann. Chim. 45: 380-6 (1955)
CA 52: 9010g

B 349 Properties of three oxonaphthacenes
derived from the oxide of rubrene
(5,6,11,12-tetraphenylnaphthacene).
Perronnet, J.
Compt. Rend. 241: 1474-7 (1955)
CA 50: 14685f

B 350 The absorption and fluorescence of
several organic molecules in the
crystalline state.
Pesteil, P.
Ann. Phys. 10: 128-84 (1955)
CA 49: 8704e

B 351 Vibrational structure of the phosphor-
escence spectra of aromatic crystals
at very low temperatures.
Pesteil, P., and Zmerli, A.
Ann. Phys. 10: 1079-97 (1955)
CA 50: 6930h

B 352 Luminescence of crystals at low tem-
peratures. VI. T → S transition of
hexachlorobenzene.
Pesteil, P., Pesteil, L., and Kara, R.
Compt. Rend. 240: 960-2 (1955)
CA 49: 10747d

B 353 Spectroscopic studies of the phosphor-
escent states of aromatic hydrocarbons.
Porter, G., and Windsor, M.W.
Mol. Spectr., Rept. Conf., London 1954:
6-19 (1955)
CA 50: 7600d

B 354 Anaglyphic X-ray pictures.
Prell, R.
Ger. 928,994 (1955)
CA 52: 1784b

B 355 Fluorescence of human lymphatic and
cancer tissues following high doses of
intravenous hematoporphyrin.
Rasmussen-Taxdal, D.S., Ward, G.E., and
Figge, F.H.J.
Cancer 8: 78-81 (1955)
CA 49: 4867b

B 356 Fluorescence spectroscopy with
roentgen rays.
Regler, F.
Mikrochim. Acta 1955: 671-83 (1955)
CA 49: 12134g

B 357 Vitamin B_{12}.
Robinson, F.M., Miller, I.M., McPherson,
J.F., and Folkers, K.
J. Am. Chem. Soc. 77: 5192 (1955)
CA 50: 1107b

B 358 The fluorescence yields of the L levels
of bismuth.
Ross, M.A.S., Cochran, A.J., Hughes, J.,
and Feather, N.
Proc. Phys. Soc. 68A: 612-24 (1955)
CA 49: 13786d

B 359 Changes in the behavior of fluorescent
screens under electron bombardment
in cathode tubes.
Rottgardt, K.H.J., and Berthold, W.
Z. Naturforsch. 10a: 736-40 (1955)
CA 51: 3294d

B 360 Fluorescent centers in uranium-
activated sodium fluoride.
Runciman, W.A.
Nature 175: 1082 (1955)
CA 49: 14488b

B 361 Utilization of phosphorescence at room
temperature as a method of chemical
analysis. Application to amino acids.
Rybak, B., Lochet, R., and Rousset, A.
Compt. Rend. 241: 1278-80 (1955)
CA 50: 7013h

B 362 Relative intensities of anthracene
and naphthacene emissions under
cathode-ray, ultraviolet, and X-ray
excitations.
Saksena, B.D., and Pant, L.M.
J. Chem. Phys. 23: 987-8 (1955)
CA 49: 10747e

B 363 Fluorescence and fluorescence polari-
zation of myelotropic nerve fibers af-
ter fluorochromation by fluorescein.
Scharf, J.H.
Z. Naturforsch. 10b: 355-6 (1955)
CA 50: 3531d

B 364 Decrease of intensity of fluorescence
of cigaret smoke caused by irradiation.
Schmahl, D., and Schneider, H.
Arzneimittel-Forsch. 5: 348-50 (1955)
CA 49: 12785a

B 365 Fluorescence as a measure of brown
substances in soybean lecithin.
Scholfield, C.R., and Dutton, H.J.
J. Am. Oil Chemists' Soc. 32: 169-70
(1955)
CA 49: 7268g

B 366 Ratio of effective cross section of recombination and capture of electrons
and the concentration of ionic vacancies
in crystals of KCl-Tl.
Shchukin, I.P.
Dokl. Akad. Nauk SSSR 104: 211-14 (1955)
CA 50: 11099i

B 367 Photochemical studies. II. Photochem-
istry of biacetyl at 3650 and 4358 A and
its relation to fluorescence.
Sheats, G.F., and Noyes, W.A., Jr.
J. Am. Chem. Soc. 77: 1421-6 (1955)
CA 49: 7988e

B 368 Long-wave photochemistry of biacetyl
and its correlation with fluorescence
at temperatures over 100°.
Sheats, G.F., and Noyes, W.A., Jr.
J. Am. Chem. Soc. 77: 4532-3 (1955)
CA 49: 15503d

B 369 Estimation of fluorescein in dilute
solutions.
Shotton, E., and Habeeb, A.F.S.A.
J. Pharm. Pharmacol. 7: 456-62 (1955)
CA 49: 15499g

B 370 Electronic and vibrational states of
biacetyl and biacetyl-d_6. I. Electronic
states.
Sidman, J.W., and McClure, D.S.
J. Am. Chem. Soc. 77: 6461-70 (1955)
CA 50: 5402c

B 371 Luminescence spectra of petroleum
products in frozen solutions.
Sidorov, N.K., and Kirillov, L.A.
Nauchn. Ezhegodnik Saratovsk. Univ.
1954: 596-8 (1955)
CA 54: 19159d

B 372 Fluorescence of drying oils.
Sims, R.P.A., and Cooper, F.P.
J. Am. Oil Chemists' Soc. 32: 381-4 (1955)
CA 49: 13663h

B 373 Method of recording the decay of phos-
phorescence.
Skarsvag, K.
Rev. Sci. Instr. 26: 397-8 (1955)
CA 49: 12136a

B 374 Dependence of energetic release of fluorescence on temperature.
Sokolova, V.S.
Vestn. Akad. Nauk Kaz. SSR 11: 73-7 (1955)
CA 49: 16083f

B 375 Synthesis of fused heterocyclics. I.
Somasekhara, S., and Phadke, R.
J. Indian Inst. Sci. 37A: 120-9 (1955)
CA 50: 3444b

B 376 Radiation control by photographic recording of fluorescence.
Sommermeyer, K., and Heiner, G.
Naturwissenschaften 42: 508 (1955)
CA 51: 4150g

B 377 Grain-storage studies. XX. Relation of viability, fat, acidity, germ damage, fluorescence value, and formazan value of commercial wheat samples.
Sorger-Domenigg, H., Cuendet, L.S., and Geddes, W.F.
Cereal Chem. 32: 499-506 (1955)
CA 50: 3662b

B 378 Removing adherent materials.
Stankey, J.A.
U.S. 2,726,180 (1955)
CA 50: 4430c

B 379 Influence of alternating electric fields on the light emission of some phosphors.
Steinberger, I.T., Low, W., and Alexander, E.
Phys. Rev. 99: 1217-22 (1955)
CA 49: 15500i

B 380 Fluorescence quenching of anthracene, 9-phenylanthracene, and 9,10-diphenylanthracene in the vapor phase.
Stevens, B.
Trans. Faraday Soc. 51: 610-19 (1955)
CA 49: 15500c

B 381 Determination of the relative yield and the intensity of phosphorescence in organoluminophors.
Sveshnikov, B.Y.
Dokl. Akad. Nauk SSSR 105: 1208-11 (1955)
CA 50: 13614f

B 382 Temperature sensitivities of the sensitized fluorescence spectrum of thallium.
Swanson, R.E., and McFarland, R.H.
Phys. Rev. 98: 1063-7 (1955)
CA 49: 11425e

B 383 The fluorimetric determination of ergot alkaloids.
Syenes, I., and Szasz, K.
Magy. Kem. Folyoirat 61: 393-8 (1955)
CA 52: 11361b

B 384 Fluorescent globulin of the lens.
Szent-Gyorgi, A.
Biochim. Biophys. Acta 16: 167 (1955)
CA 49: 6340g

B 385 Estimation of fluorescent brightening agents.
Taylor, G.G.
J. Soc. Dyers Colourists 71: 697-704 (1955)
CA 50: 2981g

B 386 Fluorescent substance.
Torio, K.
Japan 1223 (1955)
CA 50: 16414g

B 387 An investigation of the $\gamma \rightarrow \alpha\,Al_2O_3$ polymorphic transformation by the luminescence spectra.
Trofimov, A.K., and Tolkachov, S.S.
Dokl. Akad. Nauk SSSR 104: 54-5 (1955)
CA 50: 7597g

B 388 Fluorescence characteristics of 5-hydroxytryptamine (seratonin).
Udenfriend, S., Bogdanski, D.F., and Weissbach, H.
Science 122: 972-3 (1955)
CA 50: 6197c

B 389 The nature, properties, and significance of fluorescing substances formed in the processes of adrenaline and noradrenaline oxidation.
Vtevskii, A.M., and Osinskaya, V.O.
Ukr. Biokhim. Zh. 27: 401-7 (1955)
CA 50: 1948b

B 390 Intensity calculations for molecular electronic spectra.
van Dranen, J.
Chem. Weekblad 51: 735-40 (1955)
CA 50: 2273e

B 391 Apparatus for the measurement of the life of phosphorescent phenomena.
Van Roggen, A., and Vroom, R.A.
J. Sci. Instr. 32: 180-3 (1955)
CA 52: 16894e

B 392 Scintillation phenomena in sodium iodide and cesium fluoride.
Van Sciver, W.J.
Dissertation, Stanford University, California (1955)
CA 50: 6201b

B 393 Scintillation and luminescence in un-activated sodium iodide.
Van Sciver, W., and Hofstadter, R.
Phys. Rev. 97: 1181 (1955)
CA 49: 7402d

B 394 The forbidden oxygen I lines and phos-phorescence bands emitted from inert gases containing traces of oxygen.
Vegard, L., and Kvifte, G.
Skrifter Norske Videnskaps-Akad. Oslo, I, Mat.-Naturv. Kl. 1955, No. 2, 20 pp. (1955)
CA 51: 9303f

B 395 The mechanism of the photosensitiza-tion of zinc oxide under the formation of hydrogen peroxide, and the fluores-cent properties of zinc oxide.
Veselovskii, V.I., and Shub, D.M.
Probl. Kinetiki i Kataliza, Akad. Nauk SSSR 8: 43-52 (1955)
CA 50: 1471e

B 396 Isolation of fluorescent materials from Drosophila melanogaster.
Viscontini, M., Schoeller, M., Loeser, E., Karrer, P., and Hadorn, E.
Helv. Chim. Acta 38: 397-401 (1955)
CA 50: 8918i

B 397 Fluorescent substances from Droso-phila melanogaster. III.
Viscontini, M., Loeser, E., Karrer, P., and Hadorn, E.
Helv. Chim. Acta 38: 2034-5 (1955)
CA 50: 5922f

B 398 Isolation of fluorescent material from Astacus fluviatilis.
Viscontini, M., Schmid, H., and Hadorn, E.
Experientia 11: 390-2 (1955)
CA 50: 4411g

B 399 Studies in the naphthalene series. I. Synthesis of 3-amino-1-naphthoic acid.
Vondracek, M., and Vecerek, B.
Chem. Listy 49: 772-5 (1955)
CA 50: 3353d

B 400 Fluorescence of salicylic acid and related compounds.
Weller, A.
Naturwissenschaften 42: 175-6 (1955)
CA 50: 5411h

B 401 Quenching of fluorescence by some hemin proteins.
Werth, G., and Eisenbrand, J.
Hoppe-Seylers Z. Physiol. Chem. 299: 156-67 (1955)
CA 49: 6334g

B 402 2-Pyrones. XV. Substituted 3-cinnamoyl-4-hydroxy-6-methyl-2-pyrones from dehydroacetic acid.
Wiley, R.H., Jarboe, C.H., and Ellert, H.G.
J. Am. Chem. Soc. 77: 5102-5 (1955)
CA 50: 8619g

B 403 The vibrational structure of the fluor-escence spectrum of naphthalene in solution and in crystalline form.
Wolf, H.C.
Z. Naturforsch. 10a: 3-9 (1955)
CA 50: 10530f

B 404 The vibrational structure of the fluor-escence spectrum of naphthalene. II. Naphthalene in durol.
Wolf, H.C.
Z. Naturforsch. 10a: 244-8 (1955)
CA 50: 11110c

B 405 Spectroscopic behavior of the methyl derivatives of naphthalene; spectro-scopic analysis of the position isomer-ic derivatives of aromatic hydrocar-bons.
Wolf, H.C.
Z. Naturforsch. 10a: 270-8 (1955)
CA 51: 3289c

B 406 Fluorescence and absorption spectrum of 1,8-dimethyl-naphthalene.
Wolf, H.C.
Z. Naturforsch. 10a: 800-1 (1955)
CA 51: 3289f

B 407 Band width in the fluorescence spectrum of organic molecular crystals. Investigations on mixed crystals of naphthalene.
Wolf, H.C.
Z. Physik. 143: 266-73 (1955)
CA 50: 15239i

B 408 The photographic fixation of fluorescence demonstrations on paper chromatograms in ultraviolet light.
Wolff, K.
Pharmazie 10: 371 (1955)
CA 49: 15301i

B 409 Absolute quantum efficiency of photoflourescence of anthracene crystals.
Wright, G.T.
Proc. Phys. Soc. 68B: 241-8 (1955)
CA 49: 13785i

B 410 Fluorescence excitation spectra and quantum efficiencies of organic crystals.
Wright, G.T.
Proc. Phys. Soc. 68: 701-12 (1955)
CA 50: 680i

B 411 Absolute scintillation efficiency of anthracene crystals.
Wright, G.T.
Proc. Phys. Soc. 68: 929-37 (1955)
CA 50: 6195i

B 412 Fluorescence excitation spectrum of anthracene.
Wright, G.T.
Phys. Rev. 100: 587-8 (1955)
CA 49: 3091h

B 413 Fluorometric analysis of vitamins. III. Fluorescence spectra of thiochrome and other similar fluorescent substances.
Yagi, K., Tabata, T., Kotaki, E., and Arakawa, T.
Vitamins 9: 391-2 (1955)
CA 50: 15674e

B 414 Ratio of quantum yields of phosphorescence and fluorescence of substituted phthalimides.
Zelinski, V.V., and Kolobkov, V.P.
Dokl. Akad. Nauk SSSR 101: 241-4 (1955)
CA 50: 2302c

B 415 Effect of the migration of energy on the polarization of fluorescent monocrystals.
Zhevandrov, N.D.
Dokl. Akad. Nauk SSSR 100: 455-8 (1955)
CA 50: 1470b

B 416 Relation of the polarization of luminescence and other optical properties of the anthracene derivatives to their structure.
Zhevandrov, N.D.
Tr. Fiz. Inst., Akad. Nauk SSSR, Fiz. Inst. 6: 123-98 (1955)
CA 50: 7955h

B 417 Scintillation response of anthracene to low-energy protons and helium ions.
Zimmerman, E.J.
Phys. Rev. 99: 1199-1203 (1955)
CA 49: 15498a

B 418 Luminescence of crystals at low temperatures. VIII. Transition $T \rightarrow S$ of hexachlorobenzene at 90°K.
Zmerli, A., and Pesteil, P.
Compt. Rend. 240: 2217-19 (1955)
CA 49: 15499d

B 419 Long-lived states in photochemical reactions. II. Photoreduction of fluorescein and its halogenated derivatives.
Adelman, A.A., and Oster, G.
J. Am. Chem. Soc. 78: 3977-80 (1956)
CA 51: 1739c

B 420 Effect of the expelling action of excitation light on the photoluminescence yield of crystal phosphors.
Anikina, L.I.
Tr. Fiz. Inst., Akad. Nauk SSSR, Fiz. Inst. 7: 5-46 (1956)
CA 51: 4148g

B 421 The polarization of fluorescence and energy transfer in grana.
Arnold, W., and Meek, E.S.
Arch. Biochem. Biophys. 60: 82-90 (1956)
CA 50: 6600h

B 422 Changes in the fluorescence of some minerals from Serbian deposits.
Arsenijevic, M.
Glasnik Prirod. Muzeia Srpske Zemlye, Ser. A 7: 151-6 (1956)
CA 51: 13661a

B 423 Emission bands of benzonitrile.
Asundi, R.K., and Joshi, B.D.
Current Sci. 25: 150-1 (1956)
CA 50: 16384e

B 424 Fluorescence of steroids.
Arrhenius, S.
Acta Chem. Scand. 10: 154 (1956)
CA 50: 12659h

B 425 Investigation of the luminescence of
cadmium sulfide activated with silver.
Bancie-Grillot, M., and Grillot, E.
J. Chim. Phys. 53: 521-6 (1956)
CA 50: 16360f

B 426 Stereoisomerism and excited states of
simple polymethine dyes.
Baumgartner, F., Gunther, E., and
Scheibe, G.
Z. Elektrochem. 60: 570-2 (1956)
CA 50: 16405e

B 427 Electronic spectra of polyacetylenes.
Beer, M.
J. Chem. Phys. 25: 745-50 (1956)
CA 51: 2392d

B 428 Fluorescence of cerium in sodium
fluoride melts.
Belegisanin, N.
Glasnik Hem. Drushtva, Beograd 21: 271-6
(1956)
CA 52: 15263a

B 429 Phosphorescence of the crystal phos-
phor ZnS-Cu during excitation by an
electron beam.
Belikova, T.P.
Soviet Phys. JETP 2: 776-7 (1956)
CA 51: 6356d

B 430 Fluorimetry employing a photoelectric
cell.
Bernanose, A., and Rene, M.
Bull. Soc. Pharm. Nancy 28: 16-18 (1956)
CA 51: 9867h

B 431 Radiation-induced luminescence. II.
Effect of oxygen and bromobenzene.
Berry, P.J., Lipsky, S., and Burton, M.
Trans. Faraday Soc. 52: 311-20 (1956)
CA 50: 14371f

B 432 Fluorescence spectra of some organic
crystals.
Birks, J.B., and Cameron, A.J.W.
S. African J. Sci. 53: 16-19 (1956)
CA 51: 1735i

B 433 Fluorescence of *p*-chlorotoluene in the
solid state at low temperature.
Biswas, D.C.
Indian J. Phys. 30: 143-50 (1956)
CA 51: 9329b

B 434 Dependence of the intensity of the
fluorescence of *p*-chlorotoluene on the
wavelength of the exciting radiation.
Biswas, D.C.
Indian J. Phys. 30: 255-7 (1956)
CA 51: 9329c

B 435 Fluorescence of *p*-bromotoluene,
o-bromo-, and *o*-chlorotoluene in the
solid state at low temperature.
Biswas, D.C.
Indian J. Phys. 30: 407-14 (1956)
CA 51: 9329d

B 436 Fluorescence spectra of methyl benzo-
ate *m*-chlorotoluene, and *m*-bromotolu-
ene.
Biswas, D.C.
Indian J. Phys. 30: 565-9 (1956)
CA 51: 9329f

B 437 The yield of resonance fluorescence of
sodium in a flame.
Boers, A.L., Alkemade, C.T.J., and Smit,
J.A.
Physica 22: 358-60 (1956)
CA 52: 9752i

B 438 Measurement of fluorescence duration
by means of a phase fluorometer.
Bonch-Bruevich, A.M.
Bull. Acad. Sci. USSR, Phys. Ser. 20:
536-8 (1956)
CA 51: 11859e

B 439 Measurement of fluorescence duration
by means of a phase fluorometer.
Bonch-Bruevich, A.M.
Izv. Akad. Nauk SSSR, Ser. Fiz. 20: 591-5
(1956)
CA 51: 1737c

B 440 A new phase fluorometer.
Bonch-Bruevich, A.M., Molchanov, V.A.,
   and Shirokov, V.I.
Izv. Akad. Nauk SSSR, Ser. Fiz. 20:
   596-600 (1956)
CA 51: 1737d

B 441 Development of the fluorometric meth-
   od for studying the duration of the ex-
   cited state of molecules.
Bonch-Bruevich, A.M.
Usp. Fiz. Nauk 58: 85-110 (1956)
CA 50: 16337f

B 442 The effect of foreign gases on the
   spectra and on the fluorescence yield
   for vapors of aromatic compounds.
Borisevich, N.A., and Neporent, B.S.
Opt. i Spektroskopiya 1: 536-45 (1956)
CA 51: 5568e

B 443 Absolute photoluminescence yield of
   anthracene and naphthalene crystals.
Borisov, M.D., and Vishnevskii, V.N.
Bull. Acad. Sci. USSR, Phys. Ser. 20:
   459-61 (1956)
CA 51: 11080h

B 444 The enhancement of the fluorescence of
   vapors.
Bowen, E.J., and Veljkovic, S.
Proc. Roy. Soc.(London), Ser. A 236: 1-6
   (1956)
CA 50: 15241g

B 445 Change in reduced diphosphopyridine
   nucleotide (DPNH) fluorescence upon
   combination with liver alcohol dehydrogenase.
Boyer, P.D., and Theorell, H.
Acta Chem. Scand. 10: 447-50 (1956)
CA 52: 9237d

B 446 Misassignment of the multiplicity for-
   bidden transitions in pyridine.
Brealey, G.J.
J. Chem. Phys. 24: 571-3 (1956)
CA 50: 9152b

B 447 Measurement of fluorescence and qual-
   ity of eggs.
Brooks, J., and Hale, H.P.
Bull. Inst. Intern. Froid, Suppl. 1956(1):
   169-75 (1956)
CA 55: 3865f

B 448 Oxygen as activator of zinc sulfide
   phosphors.
Brundel, A.A.
Zh. Fiz. Khim. 30: 2469-77 (1956)
CA 51: 9327a

B 449 Influence of secondary fluorescence on
   the emission spectra of luminescent
   solutions.
Budo, A., and Ketskemety, I.
J. Chem. Phys. 25: 595-6 (1956)
CA 51: 88i

B 450 Determination of the absolute quantum
   yield of fluorescent solutions.
Budo, A., Dombi, J., and Szollosy, L.
Acta Univ. Szeged., Acta Phys. Chem. 2:
   18-27 (1956)
CA 51: 15278d

B 451 Fluorescence of thallium-activated
   halide phosphors.
Butler, K.H.
J. Electrochem. Soc. 103: 508-12 (1956)
CA 50: 16399d

B 452 (Fluorometric determination of urani-
   um.)
Centanni, F.A., Ross, A.M., and Desesa,
   M.A.
Anal. Chem. 28: 1651 (1956)

B 453 Fluorescence analysis for polycyclic
   aromatic hydrocarbons.
Chaudet, J.H., and Kaye, W.I.
Am. Chem. Soc. Div. Petrol. Chem.,
   Preprints 1, No. 4, Polycyclic Hydro-
   carbons, 147-54 (1956)
CA 53: 13775e

B 454 Separation of natural coumarins by
   circular chromatography.
Chakraborty, D.P., and Chakraborty, H.C.
Sci. Culture 22: 117-19 (1956)
CA 53: 4019c

B 455 Absorption spectra, fluorescence
   spectra, and fluorescence quantum
   yields of some meso derivatives of
   anthracene.
Cherkasov, A.S.
Izv. Akad. Nauk SSSR, Ser. Fiz. 20: 478-81
   (1956)
CA 51: 870g

B 456 Duration of fluorescence for the meso
      derivatives of anthracene.
      Cherkasov, A.S., Molchanov, V.A.,
      Vember, T.M., and Voldaikina, K.G.
      Dokl. Akad. Nauk SSSR 109: 292-4 (1956)
      CA 51: 9329i

B 457 Duration of fluorescence for the meso
      derivatives of anthracene.
      Cherkasov, A.S., Molchanov, V.A.,
      Vember, T.M., and Voldaikina, K.G.
      Soviet Phys. Doklady 1: 427-9 (1956)
      CA 51: 13580b

B 458 Absorption spectra, fluorescence
      spectra, and fluorescence quantum
      yields of some meso derivatives of
      anthracene.
      Cherkasov, A.S.
      Bull. Acad. Sci. USSR, Phys. Ser. 20:
      436-9 (1956)
      CA 51: 11080g

B 459 Light-quenching effect of organic
      phosphors.
      Chomse, H., Hoffman, W., and Seidel, P.
      Naturwissenschaften 43: 12 (1956)
      CA 51: 16110b

B 460 Light scattering measurements of
      cellulose solutions in concentrated
      acids.
      Choudhury, P.K., and Frank, H.P.
      J. Polymer Sci. 20: 218-23 (1956)
      CA 51: 7710h

B 461 Aromatic hydrocarbons. LXXII. The
      relationships between chemical reacti-
      vity, phosphorescence, and α-absorp-
      tion bands and the "hydrogen similari-
      ty" of the upper level of the p-bands in
      the absorption spectra of aromatic
      hydrocarbons.
      Clar, E., and Zander, M.
      Chem. Ber. 89: 749-62 (1956)
      CA 51: 346a

B 462 Energy transfer from solvent to solute
      in liquid organic solutions under ultra-
      violet excitation.
      Cohen, S.G., and Weinreb, A.
      Proc. Phys. Soc. 69: 593-605 (1956)
      CA 50: 16384c

B 463 Photoconductivity and semiconductivity
      of anthracene: effect of nitrogen dioxide
      and chlorine.
      Compton, D.M.J., and Waddington, T.C.
      J. Chem. Phys. 25: 1075-6 (1956)
      CA 51: 3283d

B 464 Towards better scintillation counting.
      Cooper, D.I., and Morton, G.A.
      Nucleonics 14: 46-8 (1956)
      CA 50: 12677a

B 465 Detection of crude oil in subterranean
      formations.
      Cross, C.F., and Wayo, S.J.
      U.S. 2,740,758 (1956)
      CA 50: 10390b

B 466 Fluorescence spectrum of plutonium in
      lanthanum chloride.
      Cunningham, B.B., Gruen, D.M., Conway,
      J.G., and McLaughlin, R.D.
      J. Chem. Phys. 24: 1275 (1956)
      CA 50: 12643c

B 467 Theoretical and experimental study of
      some properties of electron traps and
      luminescence centers in sulfides.
      Curie, D.
      J. Phys. Radium 17: 699-702 (1956)
      CA 51: 865g

B 468 Sensitized phosphorescence and charge
      transfer-fluorescence in organic
      molecular compounds.
      Czekalla, J.
      Naturwissenschaften 43: 467-8 (1956)
      CA 53: 21165e

B 469 Affinities of vat dyes.
      DeCort, W.J.
      J. Soc. Dyers Colourists 72: 439 (1956)
      CA 50: 17455h

B 470 Characterization of some bile consti-
      tuents separated by electrophoresis.
      Dessi, P., and Pellegrini, R.
      Giorn. Biochim. 5: 146-52 (1956)
      CA 50: 15830g

B 471 Spectral repetition of the electroen-
      hancement effect of mixtures of CdS-
      ZnS activated by manganese and silver.
      Destriau, G.
      J. Phys. Radium 17: 734-6 (1956)
      CA 51: 1734h

B 472 Absorption, fluorescence, and energy
levels of the dysprosium ion.
Dieke, G.H., and Singh, S.
J. Opt. Soc. Am. 46: 495-9 (1956)
CA 50: 11812d

B 473 The concentration dependence of the
spectral effect of secondary fluorescence.
Dombi, J., and Horvai, R.
Acta Univ. Szeged., Acta Phys. Chem. 2:
9-17 (1956)
CA 51: 15278b

B 474 Transfer of electronic energy.
Quenching of fluorescence by oxygen
and nitric oxide.
Dubois, J.T.
J. Chem. Phys. 25: 178 (1956)
CA 50: 13615d

B 475 Phosphor synthesis by recrystallization
and migration of electrons.
D'yulai, Z.
Izv. Akad. Nauk SSSR, Ser. Fiz. 20: 1569-70
(1956)
CA 51: 10242i

B 476 Quantitative comparison of the fluorescent patterns of the eggs of three genotypes of Ephestia kuhniella (Anagasta
kuhniella).
Egelhaaf, A.
Naturwissenschaften 43: 165-6 (1956)
CA 52: 14017b

B 477 Sensitized phosphorescence of aromatic compounds (energy transport from
one triplet level to another).
Ermolaev, V.L.
Bull. Acad. Sci. USSR Phys. Ser. 20: 471-5
(1956)
CA 51: 11859g

B 478 Sensitized phosphorescence of aromatic compounds (energy transport from
one triplet level to another).
Ermolaev, V.L.
Izv. Akad. Nauk SSSR, Ser. Fiz. 20:
514-19 (1956)
CA 51: 1738c

B 479 The connection between the dark and
light photochemical reduction of
chlorophyll and its analogs.
Evstigneev, V.B., and Gavrilova, V.A.
Dokl. Akad. Nauk SSSR 108: 507-10 (1956)
CA 50: 16409b

B 480 Metabolite inhibitors. I. 6,7-Dimethyl-
9-formylmethylisoalloxazine, 6,7-
dimethyl-9-(2-hydroxyethyl)isoallox-
azine and derivatives.
Fall, H.H., and Petering, H.G.
J. Am. Chem. Soc. 78: 377-80 (1956)
CA 50: 13039d

B 481 Fluorescent compounds.
Farbenfabriken Bayer A.-G.
Brit. 738,884 (1956)
CA 50: 15246d

B 482 Migration of excitation energy in or-
ganic crystals. I. Tetracene included
in anthracene.
Ferguson, J.
Australian J. Chem. 9: 160-71 (1956)
CA 50: 11110a

B 483 Polarization of anthracene crystal
fluorescence.
Ferguson, J., and Schneider, W.G.
J. Chem. Phys. 25: 780 (1956)
CA 51: 870f

B 484 Phosphorescence and fluorescence of
some aromatic nitroamines.
Foster, R., Hammick, D.L., Hood, G.M.,
and Sanders, A.C.E.
J. Chem. Soc. 1956: 4865-8 (1956)
CA 51: 3295h

B 485 Pigment production by a strain of
Pseudomonas aeruginosa isolated from
the fleece of sheep.
Fraser, I.E.B., and Mulcock, A.P.
J. New Zealand Assoc. Bacteriologists
11: 2-7 (1956)
CA 52: 11168h

B 486 The structure and characteristics of
the fluorescent metal chelates of
O,O'-dihydroxyazo compounds.
Freeman, D.C., Jr., and White, C.E.
J. Am. Chem. Soc. 78: 2678-82 (1956)
CA 50: 14369i

B 487 Fluorescence-spectrum curves of
chlorophylls, pheophytins, and hypericin.
French, C.S., Smith, J.H.C., Virgin, H.I.,
and Airth, R.L.
Plant Physiol. 31: 369-74 (1956)
CA 51: 2131d

B 488 Absorption, action, and fluorescent
spectra of photosynthetic pigments in
living cells and in solutions.
French, C.S., and Young, V.M.K.
Radiation Biol. 3: 343-91 (1956)
CA 50: 7196g

B 489 The structure of magnesium phthalocy-
anine adsorbates on MgO, ZnO, Al_2O_3,
and glass, and their fluorescence.
Gachkovskii, V.F.
Dokl. Akad. Nauk SSSR 110: 408-10 (1956)
CA 51: 13511d

B 490 Absolute luminescence yield of γ-
scintillations in naphthalene crystals
containing anthracene.
Galanin, M.D., and Grishin, A.P.
Zh. Eksperim. i Teor. Fiz. 30: 33-41
(1956)
CA 50: 14386b

B 491 Modification of Beckman DK-1 spectro-
photometer for use as a recording
spectrofluorometer.
Gemmill, C.L.
Anal. Chem. 28: 1061-3 (1956)
CA 50: 12656i

B 492 The transfer of excitation energy from
p-terphenyl to riboflavine.
Gemmill, C.L.
Radiation Res. 5: 216-24 (1956)
CA 50: 16403i

B 493 Two measuring procedures for fluores-
cence decay processes.
Glaser, F.
Z. Naturforsch. 11a: 1030-6 (1956)
CA 51: 7157g

B 494 The azole series. III. The structure
of 2-oxazolone and 2-oxazolethione:
ultraviolet and fluorescence spectra
of these and related compounds.
Gompper, R., and Herlinger, H.
Chem. Ber. 89: 2816-24 (1956)
CA 51: 11854e

B 495 Selection techniques in screening for
coumarin-deficient sweet-clover
plants.
Goplen, B.P., Greenshields, J.E.R., and
White, W.J.
Can. J. Botany 34: 711-19 (1956)
CA 50: 14060f

B 496 The lowest optical excited state of the
naphthalene crystal.
Griessbach, D., Will, G., and Wolf, H.C.
Z. Naturforsch. 11a: 791-6 (1956)
CA 51: 4140h

B 497 Fluorescence of the exciton in pure
cadmium sulfide.
Grillot, E.
J. Phys. Radium 17: 822-3 (1956)
CA 51: 9309g

B 498 Quasi-hydrogenoid (exciton) fluores-
cence spectrum of pure cadmium sul-
fide at 20°K.
Grillot, E., Grillot, M., Pesteil, P., and
Zmerli, E.
Compt. Rend. 242: 1794-6 (1956)
CA 50: 11818g

B 499 Radiation spectrum of the exciton.
Gross, E.F., and Yakobson, M.A.
Zh. Tekhn. Fiz. 26: 1369-71 (1956)
CA 50: 13607b

B 500 Fluorescence spectrum of americi-
um $^{+++}$ in $LaCl_3$.
Gruen, D.M., Conway, J.G., McLaughlin,
R.D., and Cunningham, B.B.
J. Chem. Phys. 24: 1115-16 (1956)
CA 50: 11119f

B 501 The extinction and sensitization effects
of electric fields.
Gumlich, H.E.
J. Phys. Radium 17: 117-21 (1956)
CA 50: 9158g

B 502 Biochemical polyphenyl and material
distribution in the body of various eye
colored genotypes of Ephestia kuhniella.
Hadorn, E., and Egelhaaf, A.
Z. Naturforsch. 11b: 21-5 (1956)
CA 50: 9639c

B 503 Active center of chymotrypsin. I.
Labeling with a fluorescent dye.
Hartley, B.S., and Massey, V.
Biochim. Biophys. Acta 21: 58-70 (1956)
CA 50: 15643c

B 504 Fluorescent organic materials and
chemical constitution.
Herforth, L.
Wiss. Ann. 5: 744-53 (1956)
CA 52: 10861d

B 505 Fluorescent substances in skin trans-
plants.
Herforth, L., and Schafer, P.
Naturwissenschaften 43: 110-11 (1956)
CA 52: 12170b

B 506 Origin of phosphorescence of short
duration in the nitrogen molecule.
Herman, L., Salmona, A., and Lucas, G.
Compt. Rend. 243: 1029-32 (1956)
CA 51: 12665i

B 507 Fluorescence studies with uranin.
Hofler, K.
Protoplasma 47: 322-66 (1956)
CA 55: 2813i

B 508 Fluorescence of azulene in the vapor
phase.
Hunt, G.R., and Ross, I.G.
Z. Naturforsch. 11a: 1043 (1956)
CA 51: 12667i

B 509 Blue luminescent substance in
berberine-containing plants. IV. Blue
luminescent substance from Berberis
thunbergii var. maximowizii and B.
amurensis var. joponica forma. Bret-
schnuderi.
Ishida, Y., and Okamura, T.
J. Pharm. Soc. Japan 76: 223-4 (1956)
CA 50: 7957e

B 510 Phenylcinchoninic acid analogs. I. Relation between their fluorescence and
chemical structures. II. Capillary
images.
Ito, H., and Fukushima, H.
Ann. Rept. Pharm. Tokushima Univ. 5:
12-14, 15-17 (1956)
CA 53: 3877e

B 511 Organic phosphors containing metallic
compounds as the phosphorescent bodies. III. Effect of metallic compounds
on the fluorescence and phosphorescence spectra.
Iwaki, R.
J. Chem. Soc. Japan 77: 26-31 (1956)
CA 50: 5411a

B 512 Organic phosphors containing metallic
compounds as the phosphorescent bodies. IV. Phosphorescence of uranin-
lead acetate phosphors.
Iwaki, R.
J. Chem. Soc. Japan 77: 31-6 (1956)
CA 50: 5411c

B 513 The dye gelatin phosphors. III. Satura-
tion characteristics of phosphorescence.
Iwaki, R.
J. Chem. Soc. Japan, Pure Chem. Sect.
77: 801-4 (1956)
CA 50: 16405e

B 515 Absorption and fluorescence spectra of
chromones, flavones, flavonols.
Jatkar, S.K.K., and Mattoo, B.N.
J. Indian Chem. Soc. 599-604, 623-9,
641-9 (1956).
CA 51: 9313i

B 516 Absorption and fluorescence spectra of
benzylidene coumaranones.
Jatkar, S.K.K., and Mattoo, B.N.
J. Indian Chem. Soc. 33: 647-50 (1956)
CA 51: 9315a

B 517 Identification of azotobacter species
by fluorescence and cell analysis.
Johnstone, D.B., and Fishbein, J.R.
J. Gen. Microbiol. 14: 330-5 (1956)
CA 50: 11429d

B 518 The disappearance of electrons during
phosphorescence in argon.
Joslet, C., Weniger, S., and Herman, R.
Compt. Rend. 242: 2538-9 (1956)
CA 50: 12658h

B 519 Quantitative evaluation of fluorescent
paper chromatographs.
Kaiser, H., and Wildermann, L.
Intern. Z. Vitaminforsch. 27: 131-9 (1956)
CA 51: 11156e

B 520 Number of traps and behavior of ex-
cited electrons in luminescent mater-
ials.
Kallmann, H., and Spruch, G.M.
Phys. Rev. 103: 94-102 (1956)
CA 50: 15240c

B 521 Color reactions of rosin derivatives.
Kamath, N.R., and Shetye, G.D.
Paintindia 6: 29-34 (1956)
CA 50: 16134g

B 522 The nature of the centers of luminosity
of the phosphors NaCl-Ag and NaCl-Cu.
Kats, M.L., Grigor'eva, N.A., Mironenko,
L.A., and Smigirev, B.N.
Uch. Zap. Saratovsk. Gos. Univ. 44: 115-
29 (1956)
CA 54: 20516e

B 523 The influence of various gases on the luminescence of X-irradiated crystals of alkali halides.
Kats, M.L.
Uch. Zap. Saratovsk. Gos. Univ. 44: 131-5 (1956)
CA 54: 4169b

B 524 Atomic centers of nickel in sodium chloride-nickel phosphors.
Kats, M.L., and Semenov, B.Z.
Dokl. Akad. Nauk SSSR 106: 415-18 (1956)
CA 51: 2402b

B 525 Several new results from the measurement of fluorescence in plants.
Kautsky, H.
Z. Naturforsch. 11b: 116-7 (1956)
CA 50: 16992d

B 526 The electrolytic reduction of diphosphopyridine nucleotide.
Ke, B.
Arch. Biochem. Biophys. 60: 505-6 (1956)
CA 50: 7179i

B 527 The scattering of the vibrational quanta of an excited nitric oxide molecule upon collision with other molecules.
Kleinberg, A.V.
Opt. i Spektroskopiya 1: 469-77 (1956)
CA 51: 5569i

B 528 The relation of the spectra of organic compounds to the dielectric constant of the medium.
Klochkov, V.P.
Opt. i Spektroskopiya 1: 546-53 (1956)
CA 51: 5568d

B 529 Stability of internal complex salts in the excited state.
Kokubun, H., Ono, N., and Inamura, Y.
Naturwissenschaften 43: 105 (1956)
CA 51: 16094f

B 530 The behavior of dye ions in detergent solutions. II. The behavior of fluorescent dye ions as revealed by fluorescence measurements.
Kondo, T.
Nippon Kagaku Zasshi 77: 1281-4 (1956)
CA 51: 10187h

B 531 Fluorometer.
Korol'kov, S.I., and Kudryavtsev, V.I.
U.S.S.R. 104,009 (1956)
CA 51: 4067h

B 532 The occurrence of porphyrin in the planarian.
Krugelis-Macrae, E.
Biol. Bull. 110: 69-76 (1956)
CA 50: 8926c

B 533 The phosphorescence spectra of organolumiphors in anti-Stokesian excitation.
Kudryashov, P.I., and Sveshnikov, B.Ya.
Opt. i Spektroskopiya 1: 554-9 (1956)
CA 51: 6355i

B 534 Influence of dye concentration on the luminescence of Acridine Orange in alcohol solution at -183°.
Kuznetsova, L.A., and Sveshnikov, B.Ya.
Izv. Akad. Nauk SSSR, Ser. Fiz. 20: 433-41 (1956)
CA 51: 869d

B 535 Electronic properties of aluminum nitride.
Lagrenaudie, J.
J. Chim. Phys. 54: 222-5 (1956)
CA 50: 11758f

B 536 Fluorescence and scattering of light by plant pigments.
Latimer, P.
Univ. Microfilms No. 18165, 141 pp. (1956)
CA 51: 2403e

B 537 Quantum yields of fluorescence of plant pigment.
Latimer, P., Bannister, T.T., and Rabinowitch, E.
Science 124: 585-6 (1956)
CA 51: 4505i

B 538 Influence of the existence of several trapping systems and of the phenomenon of recapture on the decay law of phosphorescence.
Levchin, V.
J. Phys. Radium 17: 684-7 (1956)
CA 51: 5567e

B 539 Apparatus for spectral analysis of molecular phosphorescence at ordinary temperatures.
Lochet, R., Valentin, F., and Rousset, A.
J. Phys. Radium 17: 307-8 (1956)
CA 51: 2404c

B 540 Effect of crystal structure upon the luminescence of manganese-activated lithium titanate.
Lorenz, M.R., and Prener, J.S.
J. Chem. Phys. 25: 1013-15 (1956)
CA 51: 4147i

B 541 The influence of degree of dispersion of some fluorescent substances on the fluorescence intensity. I. Fluorescein in water, in alcohols, in acetone, and in ether.
Lucatu, F.
Comun. Acad. Rep. Populare Romine 6: 35-42 (1956)
CA 51: 871c

B 542 Anomalous fluorescence in torbernite from Rum Jungle, N. T., Australia.
Lyon, R.J.P.
Am. Minerologist 41: 789-92 (1956)
CA 51: 7957f

B 543 Paper chromatographic comparative study on various strains of medicago sativa.
Manunta, C.
Genet. Agrar. 6: 377-84 (1956)
CA 52: 17423d

B 544 Physical properties of chymotrypsin and chymotrypsinogen using the depolarization of fluorescence techniques.
Massey, V., Harrington, W.F., and Hartley, B.S.
Discussions Faraday Soc. 20: 24-32 (1955) (Pub. 1956)
CA 50: 16922b

B 545 Influence of high-molecular cations on the fluorescence and absorption spectra of dye anions.
Mataga, N.
J. Inst. Polytech., Osaka City Univ., Ser. C, 74-84 (1956)
CA 52: 5971i

B 546 Hydrogen-bonding effect on the fluorescence of π-electron system.
Mataga, N., Kaifu, Y., and Koizumi, M.
Bull. Chem. Soc. Japan 29: 115-22 (1956)
CA 51: 7158b

B 547 The base strength of some nitrogen heterocycles in the excited state.
Mataga, N., Kaifu, Y., and Koizumi, M.
Bull. Chem. Soc. Japan 29: 373-9 (1956)
CA 50: 11820c

B 548 Solvent effects upon the fluorescence spectra and the dipole moments of excited molecules.
Mataga, N., Kaifu, Y., and Koizumi, M.
Bull. Chem. Soc. Japan 29: 465-70 (1956)
CA 51: 89c

B 549 Fluorescence spectra of natural and irradiated diamonds.
Matthews, I.G.
J. Phys. Radium 17: 649 (1956)
CA 51: 5566e

B 550 Absorption and fluorescence spectra of coumarins.
Mattoo, B.N.
Trans. Faraday Soc. 52: 1184-94 (1956)
CA 51: 5552b

B 551 Excited states of the naphthalene molecule. II. Further studies on the first singlet-singlet transition.
McClure, D.S.
J. Chem. Phys. 24: 1-12 (1956)
CA 50: 4637h

B 552 Nuclear resonance fluorescence in nickel-60.
Metzger, F.R.
Phys. Rev. 103: 983-7 (1956)
CA 50: 15266h

B 553 Fluorescence spectra of illuminating oils.
Mihul, C., Ruscior, C., and Pop, V.
Analele Stiint. Univ. "A. I. Cuza" Iasi, Sect. I. 2: 199-209 (1956)
CA 53: 3877h

B 554 Preparation of 10-vinyl acridone.
Mikhant'ev, B.I., and Sklyarov, V.A.
Zh. Obshch. Khim. 26: 784-5 (1956)
CA 50: 14760h

B 555 Fluorescence of cadmium iodide.
Monod-Herzen, G.
Compt. Rend. 242: 2830-1 (1956)
CA 50: 16405b

B 556 The continuous spectrum structure of complex molecules.
Neporent, B.S.
Zh. Fiz. Khim. 30: 1048-61 (1956)
CA 50: 16391g

B 557 Spectra and yields of anti-Stokes and
Stokes fluorescence of the vapors of
aromatic compounds.
Neporent, B.S., and Borisevich, N.A.
Opt. i Spektroskopiya 1: 143-54 (1956)
CA 50: 16404e

B 558 The effect of helium on the intensity of
the spectra for the vapors of complex
aromatic compounds.
Neporent, B.S., and Solodovnikov, A.A.
Opt. i Spektroskopiya 1: 951-2 (1956)
CA 51: 12652i

B 559 A spectrometric installation of high
light power for the investigation of
luminescence.
Neporent, B.S., and Klochkov, V.P.
Izv. Akad. Nauk SSSR, Ser. Fiz. 20: 601-4
(1956)
CA 51: 1737e

B 560 Observation of the products from the
thermal dissociation of the vapors of
polyatomic molecules from their
fluorescence spectra.
Neuimin, G.G.
Opt. i Spektroskopiya 1: 463-8 (1956)
CA 51: 5569g

B 561 Electroluminescence and thermolumin-
escence of zinc sulfide single crystals.
Neumark, G.F.
Phys. Rev. 103: 41-6 (1956)
CA 50: 15239a

B 562 Detection of petroleum products in nat-
ural waters by fluorescence.
Nietsch, B.
Mikrochim. Acta 1956: 171-8 (1956)
CA 50: 9724i

B 563 Development of high-visibility paints.
Noonan, F.M., and Cowling, J.E.
Rept. NRL Progr. 1956: 6-11 (1956)
CA 51: 4019f

B 564 Electronic properties of aromatic hydrocarbons. II. Fluorescence transfer
in solid solutions.
Northrop, D.C., and Simpson, O.
Proc. Roy. Soc. (London), Ser. A 234:
136-49 (1956)
CA 50: 11118h

B 565 Interaction of pyridine nucleotide
linked enzymes.
Nygaard, A.P., and Rutter, W.J.
Acta Chem. Scand. 10: 37-48 (1956)
CA 50: 12132g

B 566 Fluorescent materials associated
with discoloration in aspen.
Obert, J.C., Hossfeld, R.L., and Kaufert,
F.H.
Tappi 39: 470-1 (1956)
CA 50: 13439f

B 567 Fluorometric determination of thallium
and indium with Rhodamine B.
Onishi, A.
Bull. Chem. Soc. Japan 29: 945 (1956)

B 568 Phosphorescence of sodium and potas-
sium acetates.
Osada, K.
J. Phys. Soc. Japan 11: 425-9 (1956)
CA 50: 15239f

B 569 The effect of absorbed water on the
phosphorescence of sodium acetate.
Osada, K.
J. Phys. Soc. Japan 11: 1014-15 (1956)
CA 51: 7872d

B 570 Fluorescence and internal rotation:
their dependence on viscosity of the
medium.
Oster, G., and Nishijima, Y.
J. Am. Chem. Soc. 78: 1581-4 (1956)
CA 50: 10538d

B 571 Inherent inconsistencies in fluorescence
and scintillation spectra.
Ott, D.G., Hayes, F.N., Kerr, V.N., and
Benz, R.W.
Science 123: 1071 (1956)
CA 50: 16403e

B 572 Lowest triplet state of anthracene.
Padhye, M.R.
J. Chem. Phys. 24: 588-94 (1956)
CA 50: 9153b

B 573 Analytical studies of the fluorescence
of samarium in calcium tungstate.
Peattie, C.G., and Rogers, L.B.
Spectrochim. Acta 7: 321-48 (1956)
CA 50: 8383f

B 574 Luminescence of crystals at low temperatures. Naphthalene at 20°K.
Pesteil, P., and Zmerli, A.
Compt. Rend. 242: 1876-8 (1956)
CA 51: 88e

B 575 Detection of ultrasound with phosphorescent materials.
Peterman, L.A., and Oncley, P.B.
Proc. Natl. Electronics Conf. 11: 481-9 (1955) (Pub. 1956)
CA 50: 7527i

B 576 Hydrolytic cleavage products of boron trifluoride complexes of β-carotene, some dehydrogenated carotenes and anhydrovitamin A_1.
Petracek, F.J., and Zechmeister, L.
J. Am. Chem. Soc. 78: 3188-9 (1956)
CA 50: 14585e

B 577 The role of sulfur in the life of plants. III. Fluorescent substances in the bleeding sap of maize.
Potapov, N.G., and Feier, D.
Agrokem. Talajtan 5: 37-46 (1956)
CA 50: 15747i

B 578 Associated donor-acceptor luminescent centers.
Prener, J.S., and Williams, F.E.
Phys. Rev. 101: 1427 (1956)
CA 50: 6927c

B 579 Color and luminescence of feldspars (With an addendum: anhydrite fluorescence reversible by tempering).
Przibram, K.
Oesterr. Akad. Wiss., Math.-Naturw. Kl., Sitzber., Abt. II. 165: 281-311 (1956)
CA 51: 13579g

B 580 Temperature quenching of the phosphorescence of some aromatic acids.
Pyatnitskii, B.A., and Fadeeva, M.S.
Bull. Acad. Sci. USSR, Phys. Ser. 20: 479-82 (1956)
CA 51: 11859h

B 581 Phosphorescence spectra of some aromatic acids at liquid-air temperatures.
Pyatnitskii, B.A.
Dokl. Akad. Nauk SSSR 109: 503-6 (1956)
CA 51: 9330a

B 582 Temperature quenching of the phosphorescence of some aromatic acids.
Pyatnitskii, B.A., and Fadeeva, M.S.
Izv. Akad. Nauk SSSR, Ser. Fiz. 20: 524-8 (1956)
CA 51: 1737h

B 583 Analysis of the absorption and fluorescence spectra of uranyl salts. I. Uranyl acetate (absorption).
Rao, V.R., and Narasimham, K.V.
Indian J. Phys. 30: 334-47 (1956)
CA 51: 9308g

B 584 Reaction products of aminoalkylated cellulosic textiles with 8-hydroxyquinoline-sulfonic acids and metals.
Reeves, W.A., and Guthrie, J.D.
U.S. 2,753,240 (1956)
CA 50: 13468f

B 585 A study of dihydroxy-monoethynylanthracene.
Rio, G., and Cornu, P.J.
Compt. Rend. 242: 523-6 (1956)
CA 50: 14682e

B 586 Particle-size distribution and number of particles per unit mass of fluorescent powders.
Rosinski, J., Glaess, H.E., and McCully, C.R.
Anal. Chem. 28: 486-90 (1956)
CA 50: 10543e

B 587 Dichrosim and difluorescence of chloroplasts.
Ruch, F.
Cytochemical methods with quantitative aims. Biophysical and biochemical approaches. Symposium held at Karolinska Institutet, Stockholm, Sweden, Sept. 1956.
CA 52: 5518a

B 588 The luminescence of uranium-activated sodium fluoride.
Runciman, W.A.
Proc. Roy. Soc. (London) Ser. A 237: 39-47 (1956)
CA 51: 7872e

B 589 Absorption and fluorescence spectra of ions in crystals.
Runciman, W.A.
Rept. Progr. Phys. 19: 30-58 (1956)
CA 54: 1085f

B 590 Hydroxylation of anhydromethyltetra-
hydroberberine-A. 13-Hydroxydihydro-
allocryptopine.
Russell, P.B.
J. Am. Chem. Soc. 78: 3115-21 (1956)
CA 50: 14788d

B 591 Gas emission of vacuum-tube mater-
ials.
Saito, N.
Shinku Kogyo 3: 273-81 (1956)
CA 52: 14337c

B 592 Fluorescence of riboflavine.
Sakai, K.
Nagoya J. Med. Sci. 18: 245-51 (1956)
CA 50: 11828b

B 593 The relation between fluorescence and
the unfermentable reducing substances
in blackstrap molasses.
Sattler, L.
Intern. Sugar J. 58: 194-5, 215-18 (1956)
CA 51: 2314f

B 594 Equipment and method of microspec-
trographic analysis of emissions of
fluorescence applicable to minerals.
Sandrea, A.P.
Bull. Soc. Franc. Mineral Crist. 79: 325-8
(1956)
CA 51: 17618h

B 595 Organic scintillators.
Sangster, R.C., and Irvine, J.W.,Jr.
J. Chem. Phys. 24: 670-715 (1956)
CA 50: 9873h

B 596 Polymethine dyes and their practical
application. I. Anils and flourescence.
Sassi, L.
Arch. Inst. Pasteur Tunis 33: 91-103
(1956)
CA 55: 2361c

B 597 Spectra and quantum states of the
europic ion in crystals. I. Fluorescence and absorption spectra of single
crystals of europic ethyl sulfate mono-
hydrate.
Sayre, E.V., and Freed, S.
J. Chem. Phys. 24: 1213-19 (1956)
CA 50: 12643a

B 598 Measurement of the fluorescent decay
with a fluorometer with variable path.
Scharmann, A.
Z. Naturforsch. 11a: 398-402 (1956)
CA 51: 6355a

B 599 Photoelectric emission measurement
of Acridine Orange-fluorochromed
tissue sections.
Scheibe, O., and Eder, M.
Acta Histochem. 3: 6-18 (1956)
CA 51: 1356c

B 600 Effect of gases on the photoconductivity
of anthracene.
Schneider, W.G., and Waddington, T.C.
J. Chem. Phys. 25: 358 (1956)
CA 50: 14353c

B 601 Different forms of porphyria in whites
and Bantus in South Africa.
Scott, F.P., and Grotepass, W.
Med. Klin. (Munich) 51: 679-82 (1956)
CA 50: 10897i

B 602 Composite phosphor screens.
Schultz, W.W.
U.S. 2,740,050 (1956)
CA 50: 8338b

B 603 Porphyrin metabolism. I. Modified
procedure for the quantitative determination of the urinary coproporphyrin
isomers (I and III).
Schwartz, S., Cohen, S., and Watson, C.J.
U.S. At. Energy Comm. TID-5220, 183-4
(1956)
CA 51: 3695a

B 604 Fluorescent quenching by nitrobenzene.
Seelentag, H.
Z. Physik. Chem. 9: 373-92 (1956)
CA 51: 2403c

B 605 Intensity ratio of sodium D lines in
fluorescence.
Seiwert, R.
Ann. Physik 18: 35-53 (1956)
CA 51: 75bc

B 606 Gaging of thin nickel coatings by X-ray
fluorescence.
Sellers, W.W., and Carroll, K.G.
Tech. Proc. Am. Electroplaters' Soc.,
43rd Ann. Conv., Washington 1956:
97-100 (1956)
CA 51: 5594i

B 607 Fluorescence in diamond excited by
    X rays.
    Sen, S.N., and Bishui, B.M.
    Indian J. Phys. 30: 620-5 (1956)
    CA 51: 9328d

B 608 Luminescence of uranyl salt solutions.
    Sevchenko, A.N., and Volod'ko, L.V.
    Izv. Akad. Nauk SSSR, Ser. Fiz. 20: 464-70
    (1956)
    CA 51: 870c

B 609 Distribution of activators in alkali
    halide phosphors.
    Shamovskii, L.M., and Zhvanko, Ya.N.
    Dokl. Akad. Nauk SSSR 111: 140-3 (1956)
    CA 51: 14425d

B 610 Index of refraction effect on absolute
    fluorescence measurements.
    Shepp, A.
    J. Chem. Phys. 25: 579 (1956)
    CA 51: 88d

B 611 Fluorescence of some proteins, nucleic
    acids, and related compounds.
    Shore, V.G., and Pardee, A.B.
    Arch. Biochem. Biophys. 60: 100-7 (1956)
    CA 50: 6543h

B 612 Emission spectrum of coronene in sol-
    utions at low temperatures.
    Shpol'skii, E.V., and Klimova, L.A.
    Dokl. Akad. Nauk SSSR 111: 1227-31 (1956)
    CA 52: 6926a

B 613 Influence of the solvent on the lumines-
    cence spectrum of aromatic carbohy-
    drates at low temperatures.
    Shpol'skii, E.V., and Klimova, L.A.
    Izv. Akad. Nauk SSSR, Ser. Fiz. 20: 471-5
    (1956)
    CA 51: 870e

B 614 Electronic and vibrational states of
    anthracene.
    Sidman, J.W.
    J. Chem. Phys. 25: 115-21 (1956)
    CA 50: 14362g

B 615 Electronic and vibrational states of
    tetracene(naphthacene).
    Sidman, J.W.
    J. Chem. Phys. 25: 122-4 (1956)
    CA 50: 14362i

B 616 Phosphorescence spectra of the β-halo-
    naphthalenes.
    Sidman, J.W.
    J. Chem. Phys. 25: 229-37 (1956)
    CA 50: 15239g

B 617 Polarized absorption and fluorescence
    spectra of crystalline anthracene at
    4°K.: spectral evidence for trapped
    excitons.
    Sidman, J.W.
    Phys. Rev. 102: 96-101 (1956)
    CA 50: 12649g

B 618 Electronic and vibrational states of the
    nitrite ion.
    Sidman, J.W.
    J. Am. Chem. Soc. 78: 2911 (1956)
    CA 50: 11818i

B 619 Electronic and vibrational states of
    pleiadienes.
    Sidman, J.W.
    J. Am. Chem. Soc. 78: 4217-25 (1956)
    CA 51: 854b

B 620 Mechanism of oxygen-quenching of
    trypaflavine phosphorescence in silica
    gel adsorbates.
    Sjoblom, J.J.
    Dissertation, University of Minnesota
    (1956)
    CA 50: 10540h

B 621 Metabolism of folic acid. III. Trans-
    formations of leucovorin.
    Slavik, K., and Slavikova-Matoulkova, V.
    Chem. Listy 50: 1141-6 (1956)
    CA 50: 15628e

B 622 Vavilov's rule (luminescence yield).
    Stepanov, B.I.
    Usp. Fiz. Nauk 58: 3-36 (1956)
    CA 50: 16339i

B 623 First-order deactivation of excited
    2-naphthylamine molecules with low
    vibrational energy reserve.
    Stevens, B.
    J. Chem. Phys. 24: 488-9 (1956)
    CA 50: 6197a

B 624 Effect of time on fluorescing power of
    estrogenic steroids.
    Strickler, H.S., Grauer, R.C., and
    Caughey, M.R.
    Anal. Chem. 28: 1240-3 (1956)
    CA 50: 14913a

B 625 The fluorescence of *p*-chlorotoluene in the solid state at low temperatures.
Sukar, S.C., and Biswas, D.C.
J. Chem. Phys. 24: 470 (1956)
CA 50: 6196e

B 626 Near ultraviolet fluorescence spectra of isomeric fluorotoluenes.
Suryanarayana, V., and Rao, V.R.
J. Sci. Ind. Res. 15B: 662 (1956)
CA 51: 7143f

B 627 Theory of the concentration quenching of the fluorescence of solutions.
Sveshnikov, B.Y.
Dokl. Akad. Nauk SSSR 111: 78-81 (1956)
CA 52: 3505h

B 628 Possibility of the transition from one kind of concentration quenching of the fluorescence to another one.
Sveshnikov, B.Y., Kuznetsova, L.A., and Molchanov, V.A.
Dokl. Akad. Nauk SSSR 109: 746-9 (1956)
CA 52: 5125a

B 629 The possibility of a transition from one kind of concentration quenching of fluorescence to another.
Sveshnikov, B.Y., Kuznetsova, L.A., and Molchanov, V.A.
Soviet Phys. Doklady 1: 484-7 (1956)
CA 52: 6944i

B 630 Mechanism of concentrated extinction of fluorescence in anthracene in solutions.
Sveshnikov, B.Y., and Tishchenko, G.A.
Opt. i Spektroskopiya 1: 155-60 (1956)
CA 50: 16404b

B 631 Lanthanum oxychloride phosphors.
Swindells, F.E.
U.S. 2,729,604 (1956)
CA 50: 4652ab

B 632 Effective cross section of capture and the recombination of thermal electrons in ZnS-Cu (Co) phosphor.
Syuiyun, S.
Opt. i Spektroskopiya 1: 264-70 (1956)
CA 50: 16400e

B 633 Studies on the metabolic products obtained from mouse skin after painting with 3,4-benzopyrene.
Tarbell, D.S., Brooker, E.G., Seifert, P., Vanterpool, A., Claus, C.J., and Conway, W.
Cancer Res. 16: 37-47 (1956)
CA 50: 14952g

B 634 Effects of solvents on the phosphorescence of aromatic compounds at low temperatures.
Teplyakov, P.A.
Opt. i Spektroskopiya 1: 896-900 (1956)
CA 51: 2403h

B 635 The effect of the concentration and the solvent on the phosphorescence of aromatic compounds at low temperatures.
Teplyakov, P.A., and Pyatnitskii, B.A.
Bull. Acad. Sci. USSR, Phys. Ser. 20: 476-8 (1956)
CA 51: 11859g

B 636 The effect of the concentration and the solvent on the phosphorescence of aromatic compounds at low temperatures.
Teplyakov, P.A., and Pyatnitskii, B.A.
Izv. Akad. Nauk SSSR, Ser. Fiz. 20: 520-3 (1956)
CA 51: 1737g

B 637 Sensitized phosphorescence in organic solutions at low temperature. Energy transfer between triplet states.
Terenin, A.N., and Ermolaev, V.L.
Trans. Faraday Soc. 52: 1042-52 (1956)
CA 51: 5567g

B 638 New concepts of photoconductivity mechanism and phosphorescence.
Tolstoi, N.A.
Radiotekhn, i Elektron. 1: 1135-43 (1956)
CA 51: 7840i

B 639 Some methods and results of a kinetic study on luminescence and photoconductivity.
Tolstoi, N.A.
J. Phys. Radium 17: 801-5 (1956)
CA 51: 869b

B 640 Formal analysis of the theory of the two-stage excitation of phosphorescence and photoconductivity. I. Stationary relations.
Tolstoi, N.A., and Shatilov, A.V.
Opt. i Spektroskopiya 1: 216-29 (1956)
CA 51: 4148c

B 641 Energy migration in a system of molecular assemblies.
Tomita, G., Nakajima, T., Takeyama, N., and Mizunoya, T.
Kyushu Daigaku Seisankagaku Kenkyusho Hokoku 19: 6-15 (1956)
II. Reaction mechanism.
Tomita, G., and Takeyama, N.
Ibid 20: 43-9 (1956)
IV. Reaction mechanism.
Ibid 57-61 (1956)
V. Reaction mechanism.
Tomita, G.
Ibid 21: 1-5 (1956)
CA 54: 1046f

B 642 Formation of substituted dibenzothiophene dioxides by sulfonation of *m*- and *p*-terphenyls.
Van Allan, J.A.
J. Org. Chem. 21: 1152-5 (1956)
CA 52: 1993b

B 643 Participation of B-vitamins in nonenzymic browning reactions.
Van der Poel, G.H.
Voeding 14: 452-5 (1956)
CA 54: 3770b

B 644 The photochemically active form of chlorophyll of the leaves of plants.
Vorob'eva, L.M., and Krasnovskii, A.A.
Biokhimiya 21: 126-36 (1956)
CA 50: 10201f

B 645 Some notes on the fluorescence spectra of plants in vivo.
Virgin, H.I.
Physiol. Planatarum 9: 594-81 (1956)
CA 52: 13897c

B 646 Isolation of fluorescent substances from Ephestia kuhniella (Anagasta kuhniella).
Viscontini, M., Kuhn, A., and Egelhaaf, A.
Z. Naturforsch. 11b: 501-4 (1956)
CA 52: 10435f

B 647 Confirmation of the anomalous fluorescence of azulene.
Viswananth, G., and Kasha, M.
J. Chem. Phys. 24: 574-7 (1956)
CA 50: 9153d

B 648 Kinetics and equilibria in flavoprotein systems. V. Effect of pH, anions, and partial structural analogs of the coenzyme on the activity of D-amino acid oxidase.
Walaas, E., and Walaas, O.
Acta Chem. Scand. 10: 122-33 (1956)
CA 50: 12133e

B 649 Photoelectric method for the measurement of the polarization of the fluorescence of solutions.
Weber, G.
J. Opt. Soc. Am. 46: 962-70 (1956)
CA 51: 88g

B 650 Intramolecular proton transfer in excited states.
Weller, A.
Z. Elektrochem. 60: 1144-7 (1956)
CA 51: 6335g

B 651 Fluorometric analysis.
White, C.E.
Anal. Chem. 28: 621-5 (1956)
CA 50: 7000i

B 652 Fluorescence of suspensions of green sulfur bacteria.
Williams, A.M.
Biochim. Biophys. Acta 19: 571 (1956)
CA 50: 11427i

B 653 Fluorescence in cockroaches.
Willis, E.R., and Roth, L.M.
Ann. Entomol. Soc. Am. 49: 495-7 (1956)
CA 51: 16980d

B 654 Mechanism of the energy transmission in the sensitized fluorescence of organic mixed crystals.
Wolf, H.C.
Z. Physik 145: 116-24 (1956)
CA 50: 16403c

B 655 Lead- and manganese-activated cadmium fluorophosphate phosphors.
Wollentin, R.W.
J. Electrochem. Soc. 103: 17-23 (1956)
CA 50: 6927a

B 656 Analysis of lard. The importance of neutral red fat test with regards to natural fluorescence.
Wurziger, J., and Lindemann, E.
Fleischwirtschaft 8: 675-9 (1956)
CA 51: 1497f

B 657 An apparatus for the measurement of fluorescence on filter paper.
Yagi, K., and Tabata, T.
Seikagaku 27: 779-80 (1956)
CA 55: 1093g

B 658 On the fluorescent substances found in the skin of several fishes.
Yamao, Y.
Chiba Daigaku Bunri Gakuba Kiyo, Shizen Kagaku 2: 73-8 (1956)
CA 52: 15759g

B 659 Carassius-purple, a fluorescent substance obtained from the skin of the Japanese crucian carp, Carassius auratus.
Yamao, Y., Daigaku, C., Gakuba, B., and Shizen, K.
Kagaku 2: 79-81 (1956)
CA 52: 17541g

B 660 Treatment of wheat with ionizing radiations. II. Effect on respiration and other indexes of storage deterioration.
Yen, Y., Milner, M., and Ward, H.T.
Food Technol. 10: 411-15 (1956)
CA 50: 15980h

B 661 Low-temperature absorption of the phosphorescence state of the cation of Acridine Orange and its concentration dependence.
Zanker, V.
Z. Physik. Chem. 8: 20-31 (1956)
CA 50: 16405g

B 662 The spectral relation for luminescence yields.
Zelinskii, V.V., Kolobkov, V.P., and Pikulik, L.G.
Opt. i Spektroskopiya 1: 161-7 (1956)
CA 51: 4148b

B 663 The increase in the quantum yield of phosphorescence under the influence of potassium iodide.
Zelinskii, V.V., and Kolobkov, V.P.
Opt. i Spektroskopiya 1: 560-70 (1956)
CA 51: 5568b

B 664 Investigation of the property of complex organic molecules to fluoresce and to phosphoresce.
Zelinskii, V.V., Emets, N.P., Kolobkov, V.P., and Pikulik, L.G.
Bull. Acad. Sci. USSR, Phys. Ser. 20: 465-70 (1956)
CA 51: 11859f

B 665 Relation between the fluorescence and phosphorescence yields of phthalimide derivatives and the temperature.
Zelinskii, V.V., and Kolobkov, V.P.
Dokl. Akad. Nauk SSSR 106: 1042-5 (1956)
CA 50: 15240a

B 666 Investigation of the property of complex organic molecules to fluoresce and to phosphoresce.
Zelinskii, V.V., Emets, N.P., Kolobkov, V.P., and Pikulik, L.G.
Izv. Akad. Nauk SSSR, Ser. Fiz. 20: 507-13 (1956)
CA 51: 1737i

B 667 Polarization spectra of some naphthylamines and polyenes.
Zhevandrov, N.P.
Izv. Akad. Nauk SSSR, Ser. Fiz. 20: 570-3 (1956)
CA 51: 1738d

B 668 Phosphorescence of crystalline naphthalene at 20°K. Effects of surface.
Zmerli, A., Pesteil, L., and Pesteil, P.
Compt. Rend. 242: 2822-5 (1956)
CA 50: 13612f

B 669 Recent developments in brightening agents.
Zussman, H.W., Lennon, W., and Tobin, W.
Soap Chem. Specialties 32: 35-9, 81 (1956)
CA 50: 15090e

B 670 Migration of excitation energy in organic crystals. II. Solid solution of anthracene and tetracene in naphthalene.
Ferguson, J.
Australian J. Chem. 9: 172-9 (1956)
CA 50: 11110c

B 671 Vitamin B_2 photolysis. IX. Effect of amino acids and other compounds on the intensity of fluorescence of riboflavine and of other fluorescent substances.
Sakurai, Y., and Kuroki, Y.
Vitamins 11: 473-9 (1956)
CA 51: 18027c

(1957-1959)

C 1 Effects of the introduction of oxygen into
     calcium fluoride.
     Adler, H., and Kueta, I.
     Oesterr. Akad. Wiss., Math.-Naturw. Kl.,
        Sitzber., Abt. II 166: 199-243 (1957)
     CA 53: 7778i

C 2 Luminescence of sodium chloride.
     Adler, H., and Stegmuller, F.
     Acta Phys. Austriaca 11: 31-58 (1957)
     CA 51: 12663h

C 3 Organic fluorescent materials.
     Akamatsu, H., and Iguchi, N.
     Japan 2012, Mar. 30, 1957
     CA 52: 15270ab

C 4 Population of the vibrational levels of the
     electronic state $B^2\Sigma^+$ of the CN mole-
     cule excited in the presence of active
     nitrogen.
     Akriche, J., and Herman, L.
     Compt. Rend. 244: 1024-6 (1957)
     CA 51: 9308a

C 5 Radiation decomposition products of or-
     ganic materials at 77°K.
     Alger, R.S., and Anderson, T.H.
     Am. Chem. Soc., Div. Petrol. Chem.,
        Preprints 2(4): C71-C79 (1957)
     CA 54: 20442i

C 6 Optical and electrical properties of silver
     chloride.
     Aline, P.G.
     Phys. Rev. 105: 406-12 (1957)
     CA 51: 8527g

C 7 Evaluation of whitening efficiency of
     fluorescent whitening agents.
     Allen, E.
     Soap Chem. Specialties 33: 40-4, 93, 95,
        97, 55-6 (1957)
     CA 51: 15134b

C 8 Synthetic fluorescent substances.
     Anderson, J.T., and Wells, R.S.
     Ger. 963,541 (1957)
     CA 54: 9514a

C 9 Diffusional theory of phosphorescence.
     Antonov-Romanovskii, V.V.
     Opt. i Spektroskopiya 3: 592-601 (1957)
     CA 52: 6944i

C 10 New results in the field of phosphores-
      cence research.
      Antonov-Romanovskii, V.V.
      Izv. Akad. Nauk SSSR, Ser. Fiz. 21: 484-93
         (1957)
      CA 52: 6942e

C 11 Mixtures and chemical impurities in
      Yugoslav scheelites.
      Arsenijevic, M.
      Zbornik Radov Geol. Inst. "Jovan
         Zhujovic" 9: 177-208 (1957)
      CA 53: 996a

C 12 Combined spectrofluorimeter and double
      monochromater spectrophotometer.
      Bartholomew, R.J., Dalgliesh, C.E., and
         Wootton, I.D.P.
      Biochem. J. 65, 27 pp. (1957)
      CA 53: 4829i

C 13 Fluorescence spectrum of sodium salt
      of 4,4'-diaminostilbene-2,2'-disulfonic
      acid and of Blankofor A.
      Bartowicz, S.
      Zeszyty Nauk. Politech. Lodz., Chem. 5:
         17-20 (1957)
      CA 52: 15246b

C 14 A method for accurate determination of
      the relative yields of the fluorescence
      of solutions.
      Bauer, R., and Frackowiak, D.
      Bull. Acad. Polon. Sci., Classe (III) 5:
         729-32 (1957)
      CA 52: 876h

C 15 Measurement of incident energy by sil-
ver halide emulsions containing fluor-
escent substances.
Becker, K., Klein, E., and Zeitler, E.
Ger. 1,082,120 (C157b) (1957)
CA 56: 11112e

C 16 Photoreduction of eosin in the bound
state.
Bellin, J.S., and Oster, G.
J. Am. Chem. Soc. 79: 2461-4 (1957)
CA 51: 11861e

C 17 Infrared fluorescence of simple mole-
cules.
Benson, S.W., and Porter, G.B.
J. Chem. Phys. 26: 714 (1957)
CA 51: 10245a

C 18 Canning of evaporated and fresh milk.
Berg, E.M.
U.S. 2,776,213 (1957)
CA 51: 4597h

C 19 Optical properties of calcium metaanti-
monate activated with lead.
Bernard, R.
Ann. Univ. Lyon, Sci., Sect. B, 10: 49-62
(1957)
CA 52: 9782h

C 20 Induced fluorescence in mammalian
gametes with Acridine Orange.
Bishop, M.W.A., and Smiles, J.
Nature 179: 307-8 (1957)
CA 51: 12193i

C 21 Solvent quenching of fluorescence.
Bowen, E.J., and Stebbens, D.M.
J. Chem. Soc. 1957: 360-3 (1957)
CA 51: 6357e

C 22 Luminescent materials based on alkaline
earth halides.
Braunholz, F., and Krah, H.
Ger. (East) 21,194 (Cl. 57b) (1957)
CA 56: 5528e

C 23 Direct-reading fluorimeter.
Brealey, L., and Ross, R.E.
Analyst 82: 769-73 (1957)
CA 52: 3414i

C 24 Fluorescence lifetimes of photosynthetic
pigments.
Brody, S.S.
Univ. Microfilms Publ. No. 19802, 94 pp.
(1957)
CA 51: 10244f

C 25 Excitation lifetime of photosynthetic
pigments in vitro and in vivo.
Brody, S.S., and Rabinowitch, E.
Science 125: 555 (1957)
CA 51: 10674c

C 26 Fluorescence reactions of steroids with
chlorosulfonic acid.
Bruno, S., and Scialpi, E.
Farmaco, Ed. Sci. 12: 940-5 (1957)
CA 52: 11360d

C 27 The spectral influence of secondary
fluorescence.
Budo, A., Dombi, J., and Horvai, R.
Acta Univ. Szeged., Acta Phys. Chem. 3:
3-15 (1957)
CA 52: 19446a

C 28 Concerning the determination of the true
degree of polarization of the fluorescent
radiation from solution.
Budo, A., Ketskemety, I., Salkovits, E.,
and Gargya, L.
Acta Phys. Acad. Sci. Hung. 8: 181-93
(1957)
CA 53: 13774i

C 29 Influence of secondary fluorescence on
the emission spectra of fluorescing
solutions.
Budo, A., and Ketskemety, I.
Acta Phys. Acad. Sci. Hung. 7: 207-23
(1957)
CA 51: 16111a

C 30 Synthesis of 7,8-benzonaphtho(2',1'-3,4)
fluorene.
Buu-Hoi, N.P., and Saint-Ruf, G.
J. Chem. Soc. 1957: 3806-7 (1957)
CA 52: 4589g

C 31 Fluorometric uranium analyzer.
Byrne, J.T.
Anal. Chem. 29: 1408-12 (1957)
CA 52: 1690i

C 32 The systematic error of spectrometric
measurements in the study of fluores-
cence.
Chechan, C., Audran, R., and Verain, A.
Chim. Anal. 39: 59-61 (1957)
CA 51: 8534g

C 33 Fluorometric determination of trace
quantitites of tungsten.
Chen, K.C., and Chen, C.T.
Hsia Men Ta Hsueh Hsueh Pao, She Hii
K'o Hsueh 1957: 121-9 (1957)
CA 56: 2885i

C 34 Ursilite — a new uranium silicate.
Chernikov, A.A., Krutetskaya, O.U., and
Sidelnikova, V.D.
At. Energ., Vopr. Geol. Urana, Suppl.
1957: 73-7 (1957)
CA 53: 8955h

C 35 Simultaneous fluorimetric determination
of adrenaline and noradrenaline in
plasma. I. Fluorescence characteristics of adrenolutine and noradrenolutine
and their simultaneous determination in
mixtures.
Cohen, G., and Goldenberg, M.
J. Neurochem. 2: 58-70 (1957)
CA 52: 6468c

C 36 Study of the complexes formed by some
flavones with gallium (III) and antimony
(III).
Constantinescu, D.G., Oteleanu, R., and
Baiulescu, G.
Acad. Rep. Populare Romine, Filiala Iasi,
Studii Cercetari Stiint., Chim. 8: 89-
100 (1957)
CA 54: 24693f

C 37 Fluorescence spectra of uranium, nep-
tunium, and curium.
Conway, J.C., Wallmann, J.C., Cunning-
ham, B.B., and Shalimoff, G.V.
J. Chem. Phys. 27: 1416-7 (1957)
CA 52: 5121a

C 38 Porous ceramic ware from materials
containing naphthalene and water.
Cramer, F.W., and Cramer, E.
Ger. 965,987 (1957)
CA 54: 8021i

C 39 Fluorescence and absorption spectra of
molecular compounds at low tempera-
tures. Energy transitions in molecular
compounds.
Czekalla, J., Briegleb, G., Herre, W., and
Glier, R.
Z. Elektrochem. 61: 537-46 (1957)
CA 51: 17465c

C 40 New technique for staining tubercle bac-
teria for the investigation of fluores-
cence.
Degommier, J.
Ann. Inst. Pasteur 92: 692-4 (1957)
CA 51: 15678h

C 41 The U isotope effect and other features
in the absorption and fluorescence
spectra of uranyl compounds.
Dieke, G.H.
U.S. At. Energy Comm. A-3227, 174 pp.
(1957)
CA 55: 18293e

C 42 Fluorescence lifetimes of rare earth
salts and ruby.
Dieke, G.H., and Hall, L.A.
J. Chem. Phys. 27: 465-7 (1957)
CA 52: 890i

C 43 Absorption, fluorescence, and magnetic
properties of gadolinium chloride
($GdCl_3 \cdot 6H_2O$).
Dieke, G.H., and Leopold, L.
J. Opt. Soc. Am. 47: 944-54 (1957)
CA 51: 17440h

C 44 The nature of the edge of emission in
cadmium sulfide.
Diemer, G., van Gurp, G.J., and Meyer,
H.J.G.
Physica 23: 987-8 (1957)
CA 52: 5968a

C 45 Direct lifetime measurements of excited
molecules of chlorophyll and analogous
pigments.
Dmitvevskii, O.D., Ermolaev, V.C., and
Terenin, A.N.
Doklady Akad. Nauk SSSR, 114: 751-3
(1957)
CA 52: 11191i

C 46 Fluorescent derivatives of 1,2,3-triazole. VI. 2-Styrylnaphtho(1,2)triazolesulfonic acids.
Dobas, J., and Pirkl, J.
Chem. Listy 51: 2330-3 (1957)
CA 52: 6367e

C 47 Anti-Stokes fluorescence yield in glucose and collodion rigid solutions.
Drabent, R.
Bull. Acad. Polon. Sci., Classe (III) 5: 1131-6 (1957)
CA 62: 6934a

C 48 Chitin and melanoidins, intermediate products of the melanoidin reaction.
Drozdova, T.V.
Biokhimiya 22: 487-94 (1957)
CA 52: 2120a

C 49 A spectrophotofluorometric study of compounds of biological interest.
Duggan, D.E., Bowman, R.L., Brodie, B.B., and Udenfriend, S.
Arch. Biochem. Biophys. 68: 1-14 (1957)
CA 51: 11860f

C 50 Phosphorescence of amino acids excited by forbidden absorption bands.
Dumartin, M., Lochet, R., Rybak, B., and Rousset, A.
Compt. Rend. 294: 2905-7 (1957)
CA 51: 14427b

C 51 Fluorescence spectrophotometry of pyridine nucleotide in photosynthesizing cells.
Duysens, L.N.M., and Sweep, G.
Biochim. Biophys. Acta 25: 13-16 (1957)
CA 51: 14857f

C 52 Fluorescence spectrum of the complex of reduced phosphopyridine nucleotide and alcohol dehydrogenase from yeast.
Duysens, L.N.M., and Kronenberg, G.H.M.
Biochim. Biophys. Acta 26: 437-8 (1957)
CA 52: 3022f

C 53 Pertubation of singlet-triplet transitions of aromatic molecules by oxygen under pressure.
Evans, D.F.
J. Chem. Soc. 1957: 1351-7 (1957)
CA 51: 11852f

C 54 Fluorescent material.
Fabriques renvies des lampes electriques
Fr. 1,139,056 (1957)
CA 53: 19588b

C 55 Identification of different types of unused film.
Societa per Azioni Ferrania
Ger. 1,008,569 (1957)
CA   : 1142d

C 56 Prospects of utilizing pine bark for tanning.
Filipek, Z.
Sylwan 101: 60-72 (1957)
CA 54: 20263d

C 57 Elementary photoprocesses in solution.
Forster, T.
Photochem. Liquid Solid State, Papers Symp., Dedham, Mass. 1957: 10-15 (1957)
CA 54: 23659a

C 58 Absorption spectra and fluorescence properties of concentrated solutions of organic dyes.
Forster, T., and Konig, E.
Z. Elektrochem. 61: 344-8 (1957)
CA 51: 11066e

C 59 Absorption and fluorescence spectra of aryl-alkali amides.
Forster, T., and Renner, H.
Z. Elektrochem. 61: 340-3 (1957)
CA 51: 11066b

C 60 Decay of phosphorescence of rigid solutions.
Frackowiak, M.
Acta Phys. Polon. 16: 63-78 (1957)
CA 52: 2557i

C 61 Further investigation on the decay of phosphorescence of rigid solutions.
Frackowiak, M.
Bull. Acad. Polon. Sci., Classe (III) 5: 809-12 (1957)
CA 52: 3531i

C 62 Collisions between excited cadmium and cesium atoms.
Friedrich, H., and Seiwert, R.
Ann. Physik 20: 215-29 (1957)
CA 51: 17461b

C 63 Fluorescence efficiencies of organic
    compounds.
    Furst, M., Kallmann, H., and Brown, F.H.
    J. Chem. Phys. 26: 1321-32 (1957)
    CA 51: 13579i

C 64 Complex structure and the nature of the
    absorption and fluorescence spectra of
    magnesium phthalocyanine and chloro-
    phyll.
    Gachkovskii, U.F.
    Fiz. Sb., L'vovsk. Gos. Univ. 1957: 372-5
    (1957)
    CA 55: 17209a

C 65 Changes in the type of quenching for
    zinc sulfide phosphors after excluding
    the exciting source.
    Gasting, N.L.
    Dokl. VII Nauch. Konf., Posvyashch.
    40-Letiyer Velikoi Oktyabr'sh Sots.
    Revolyutsii, Tomsk. Univ. 1957(2):
    125 (1957)
    CA 52: 2994b

C 66 Fluorescing color pigments.
    Gaunt, T.N.
    Ger. 961,575 (1957)
    CA 54: 9320d

C 67 The differential excitation ability of
    α-particles in luminescent plastic
    films.
    Geck, F.W., and Hanle, W.
    Ann. Physik 20: 142-3 (1957)
    CA 51: 17460d

C 68 (Indozyl.)
    Gehauf, B., and Goldenson, J.
    Anal. Chem. 29: 276 (1957)

C 69 Fluorescent stilbyl ditriazole com-
    pounds.
    Geigy, J.R.
    Brit. 779,505, July 24, 1957
    CA 52: 446d

C 70 Marking of photographic material (with
    fluorescing material).
    Ossenbrunner, A. and Schulte, W.
    Ger. 1,009,022 (1957)
    CA 54: 8385g

C 71 Coherent scattering in the determina-
    tions of effective X-ray energy.
    Glocker, R., and Messner, D.
    Z. Physik 149: 480-5 (1957)
    CA 53: 8798f

C 72 Student-built spectrofluorometer.
    Goldstein, J.M., McNabb, W.M., and
    Hazel, J.F.
    J. Chem. Educ. 34: 604-6 (1957)
    CA 52: 4260f

C 73 Absorption and fluorescence spectra of
    the aniline-nitrobenzene system.
    Gol'tsev, V.D.
    Izv. Vysshikh Uchebn. Zavedenii, Fiz.
    1957: 91-102 (1957)
    CA 55: 21791b

C 74 Luminescence of the halides of silver.
    Golob, S.I.
    Materially V Soveshch. po Lyuminest.
    (Kristallofosfory), Akad. Nauk Est.
    SSR, Tartu 1956: 108-24 (1957)
    CA 55: 6167c

C 75 Assimilation of berberine and chelidox-
    anthine by bacteria.
    Gray, P.H.H., and Lachand, R.A.
    Plant Soil 8: 354-66 (1957)
    CA 52: 12071e

C 76 Photometry in the extreme ultraviolet
    using fluorescence sensitized photo-
    graphic material.
    Greiner, H.
    Z. Naturforsch. 12a: 735-8 (1957)
    CA 52: 9825g

C 77 Synthesis of C^{14}-labeled anthracene,
    9-methylanthracene, and 1,2-benzan-
    thracene.
    Hadler, H.I., and Raha, C.R.
    J. Org. Chem. 22: 433-5 (1957)
    CA 52: 1129g

C 78 Fluorescent response of CsI (Tl) to
    energetic nitrogen ions.
    Halbert, M.L.
    Phys. Rev. 107: 647-9 (1957)
    CA 52: 1796f

C 79 Fluorescent lifetimes of uranyl salts at
    different temperatures.
    Hall, L.A., and Dieke, G.H.
    J. Opt. Soc. Am. 47: 1092-6 (1957)
    CA 52: 2556a

C 80 Photofluorescence decay time of organic
    phosphors.
    Hamilton, T.D.S.
    Proc. Phys. Soc. 70B: 144-5 (1957)
    CA 51: 17460c

C 81 (Detection of cyanide.)
Hanker, J.S., and Gamson, R.M.
Anal. Chem. 29: 879 (1957)

C 82 Fluorescent derivatives of 1,2,3-
triazole. IV. Color and fluorescence
of some derivatives of 2-phenylnaph-
tho(1,2)triazole.
Hanousek, V., and Dobas, J.
Chem. Listy 51: 1127-35 (1957)
CA 51: 15506e

C 83 Fluorescence intensity ratio of sodium
doublet observed in the optical disso-
ciation of sodium iodide vapor.
Hanson, H.G.
J. Chem. Phys. 27: 491-4 (1957)
CA 52: 873c

C 84 Measurements of fluorescence in an-
thracene vapor.
Hardtl, K.H., and Scharmann, A.
Z. Naturforsch. 12a: 715-19 (1957)
CA 52: 8730b

C 85 Role of collision transfer in fluorescent
solutions.
Hardwick, E.R.
J. Chem. Phys. 26: 323-4 (1957)
CA 51: 7873a

C 86 Investigation of the scintillation process.
Hardwick, E.R., and McMillan, W.G.
J. Chem. Phys. 26: 1463-71 (1957)
CA 54: 18066g

C 87 Possibility of fluorescence in comets
excited by the emission line hyman and
in the sun spectrum.
Haser, L., and Swings, P.
Ann. Astrophys. 20: 52 (1957)
CA 51: 16089e

C 88 Riboflavine and other fluorescent sub-
stances in the skin of several animals.
Hashimoto, A.
Bitamin 13: 419-22 (1957)
CA 54: 3634h

C 89 Some factors affecting fluorescence
maxima.
Hercules, D.M.
Science 125: 1242-3 (1957)
CA 51: 15277h

C 90 Widening of the arc lines of sodium under the influence of the intermolecular
stark effect.
Herman, L., Weniger, S., and Herman, R.
Spectrochim. Acta, Suppl. 1957: 333-7
(1957)
CA 54: 1062h

C 91 Molecular spectra emitted by radiative
recombination.
Herman, L., Lucas, G., and Hernian, R.
Spectrochim. Acta, Suppl. 1957: 325-8
(1957)
CA 54: 1062g

C 92 Disappearance of H^+ ions in hydrogen
ionized at atmospheric pressure.
Herman, R., Weniger, S., and Herman, L.
Compt. Rend. 244: 1179-82 (1957)
CA 51: 11059c

C 93 Use of interpolation theory in the study
of phosphorescence.
Honig, J.M.
J. Chem. Phys. 26: 1454-62 (1957)
CA 51: 16109d

C 94 Lower excited states and the phosphor-
escent state of biphenyl.
Iguchi, K.
J. Phys. Soc. Japan 12: 1250-5 (1957)
CA 52: 3414b

C 95 Fluorometric analysis. V. Determina-
tion of gallium with 8-quinolinol.
Ishibashi, M., Shgematsu, T., and
Nishikawa, Y.
Nippon Kagaku Zasshi 78: 1139-42 (1957)
CA 52: 11656b

C 96 Products of interaction of acetylacetone-
p-benzoquinones and pyridine.
Islam, A.M., and Selim, M.I.
J. Org. Chem. 22: 1641-3 (1957)
CA 52: 7282e

C 97 Decay of photoluminescence of solutions.
Jablonski, A.
Acta Phys. Polon. 16: 471-9 (1957)
CA 52: 7854d

C 98 Quinoline derivatives.
Jacob, R.M., Robert, J.G., and Liakhoff, L.
U.S. 2,816,893, Dec. 17, 1957
CA 52: 2934a

C 99 Differentiation of spring- and summer-
wood by means of secondary fluores-
cence.
Jayme, G., and Bauer, G.
Holzforschung 11: 16-18 (1957)
CA 51: 17161b

C 100 Decrease in fluorescence intensity of a
solution of fresh cigaret smoke pro-
ducts on exposure to light.
Johnston, H.
Nature 180: 1350 (1957)
CA 52: 8470h

C 101 Absorption spectra of dysprosium (III),
holmium (III), and erbium (III) aquo
ions.
Jorgensen, C.K.
Acta Chem. Scand. 11: 981-9 (1957)
CA 52: 19444e

C 102 Phosphorescing and fluorescing
enameled plates.
Kaiser, G.
Ger. 1,007,142 (1957)
CA 54: 7099i

C 103 Decay times of fluorescent substances
excited by high-energy radiation.
Kallman, H., and Brucker, G.J.
Phys. Rev. 108: 1122-30 (1957)
CA 52: 5967g

C 104 Pterin-like fluorescent substances
produced by aspergillus fungi.
Kaneko, Y.
Nippon Nogeikagaku Kaishi 31: 118-21
(1957)
CA 52: 12993bc

C 105 Correlation between fluorescence and
phosphorescence in esculin- and
uranin-activated boron phosphors.
Kantardzhyan, L.T.
Opt. i Spektroskopiya 2: 378-81 (1957)
CA 51: 11080a

C 106 Long distance energy transfer by
resonance in biology.
Karreman, G., and Steele, R.A.
Biochim. Biophys. Acta 25: 280-91 (1957)
CA 51: 16615d

C 107 The use of infrared absorption spectra
in the investigation of sensitization of
the photooxidation of organic substances
with anthraquinone derivatives.
Karyakin, A.V., and Shablya, A.V.
Dokl. Akad. Nauk SSSR 112: 688-91 (1957)
CA 51: 16106a

C 108 The viscosity dependence and extinction
of the fluorescence of pyrene.
Kasper, K.
Z. Physik. Chem. 12: 52-67 (1957)
CA 51: 14427a

C 109 Organic phosphorescence. II. A revi-
sion of Lewis' mechanism for the
phosphorescence.
Kato, S., and Koizumi, M.
Bull. Chem. Soc. Japan 30: 27-33 (1957)
CA 51: 9326e

C 110 Atomic absorption and emission centers in alkali halide phosphors activated
with heavy-metal ions and their formation under hard radiation.
Kats, M.L.
Izv. Akad. Nauk SSSR, Ser. Fiz. 21: 550-1
(1957)
CA 52: 13438g

C 111 Atomic centers in the absorption and
luminescence of alkali halide phosphors
activated with ions of heavy metals and
their formation by means of ultraviolet
radiation.
Kats, M.L.
Opt. i Spektroskopiya 3: 602-9 (1957)
CA 52: 19492e

C 112 Infrared stimulable phosphors.
Keller, S.P., Mapes, J.E., and Cheroff, G.
Phys. Rev. 108: 663-76 (1957)
CA 52: 5138h

C 113 Effect of extinction on phosphorescence
spectra of boron luminophors.
Khalupovskii, M.D.
Opt. i Spektroskopiya 3: 385-7 (1957)
CA 52: 2559e

C 114 Separation of amaryllidaceous alkaloids
by paper chromatography.
Kincl, F.A., Troncoso, V., and Rosen-
kranz, G.
J. Org. Chem. 22: 574-6 (1957)
CA 51: 17091d

C 115 Ultraviolet fluorescence of some ter-
nary silicates activated with lead.
Klaseno, H.A., Hoekstra, A.H., and Cox,
A.P.M.
J. Electrochem. Soc. 104: 93-100 (1957)
CA 51: 6334d

C 116 Relation of the absorption and fluores-
cence spectra to the solvent.
Klochkov, V.P.
Fiz. Sb. L'vovsk. Gos. Univ. 1957: 71-5
(1957)
CA 55: 16139h

C 117 Energy transfer during fluorescence of
organic solutions.
Knau, H.
Z. Naturforsch. 12a: 881-6 (1957)
CA 52: 5125b

C 118 Cesium iodide as a scintillation phos-
phor.
Knopfe, H., Loepfe, E., and Stoll, P.
Z. Naturforsch. 12a: 348-50 (1957)
CA 51: 11859c

C 119 Basicity of excited acridones.
Kokubun, H.
Naturwissenschaften 44: 233-4 (1957)
CA 51: 14427c

C 120 Intermolecular proton transition in the
excited state.
Kokubun, H.
Z. Physik. Chem. 13: 386-8 (1957)
CA 52: 5123a

C 121 Behavior of dyes in detergent solutions.
III. Investigation of the interaction of
dyes and detergents by means of dye-
detergent complex.
Kondo, T.
Nippon Kagaku Zasshi 78: 1093-6 (1957)
CA 52: 5933f

C 122 Fluorescent thin layers to glass.
Kramer, K., and Kreudenstein, H.S.V.
Ger. 1,011,123 (1957)
CA 54: 8021e

C 123 The effect of long wave ultraviolet
light on the fluorescence of carcinogenic
hydrocarbons in various solvents. II.
The fluorescence behavior of solutions
at various radiation intensities.
Kriegel, H., and Herforth, L.
Z. Naturforsch. 12b: 41-5 (1957)
CA 51: 11860d

C 124 Diffusion theory of quenching fluores-
cence in solutions by foreign substances.
Kuznetsova, L.A., Sveshnikov, B.Y., and
Shirokov, V.U.
Opt. i Spektroskopiya 2: 578-86 (1957)
CA 51: 16110e

C 125 Low-temperature activation of pre-
cipitated ZnS-Cu phosphor.
Kynev, K.D.
Opt. i Spektroskopiya 3: 652-4 (1957)
CA 52: 19492e

C 126 The use of radioactive isotopes in
wood processing.
Lakatosh, B.K.
Derevoobrabatyvayushchaya Prom. 6:
9-10 (1957)
CA 52: 11413c

C 127 Symmetry properties of the V center.
Lambe, J., and West, E.J.
Phys. Rev. 108: 634-7 (1957)
CA 52: 5139a

C 128 Aseptic culture of Arabidopsis thaliana.
Langridge, J.
Australian J. Biol. Sci. 10: 243-52 (1957)
CA 52: 2176b

C 129 Hydrogenation of kerosine to remove
color and fluorescence.
Lanning, W.C.
U.S. 2,793,986 (1957)
CA 51: 15110f

C 130 Effect of energy migration on fluores-
cence in dye solutions.
Lavorel, J.
J. Phys. Chem. 61: 864-9 (1957)
CA 51: 17465f

C 131 Influence of concentration on the ab-
sorption spectrum and the action spec-
trum of fluorescence of dye solutions.
Lavorel, J.
J. Phys. Chem. 61: 1600-5 (1957)
CA 52: 4318f

C 132 Halochromy. Absorption spectra of
tertiary alcohols and aromatic substituted methane derivatives in acid solution.
Larrushin, V.F.
Uch. Zap., Khar'kovsk. Gos. Univ., Tr.
Khim. Fak. i Nauch.-Issled. Inst. Khim.
95(18): 179-206 (1957)
CA 54: 17042i

C 133 Modification of the Raman spectrum of
a substance when the existing frequency
comes very close to a resonance frequency.
Lennuier, R.
Compt. Rend. 244: 1022-4 (1957)
CA 51: 9318d

C 134 The carotenoids, steroids, and higher
fatty acids of Polytoma uvella.
Links, J., Verloop, A., and Havinga, E.
Congr. Intern. Botan., 8[e], Paris, Compt.
Rend. Rappt. Commun. Sect. 17: 35-6
(1957)
CA 52: 7446i

C 135 Effect of surface condition on the
fluorescence and surface conductivity
of anthracene.
Lipsett, F.R., Compton, D.M.J., and
Waddington, T.C.
J. Chem. Phys. 26: 1444-5 (1957)
CA 51: 16109f

C 136 Intermolecular transfer of electronic
excitation.
Livingston, R.S.
J. Phys. Chem. 61: 860-4 (1957)
CA 51: 17439d

C 137 The role of the triplet state in the
photochemical autoxidation of aryl
hydrocarbons.
Livingston, R.S.
Photochem. Liquid Solid State, Papers
Symp., Dedham, Mass. 1957: 76-82
(1957)
CA 54: 19108d

C 138 Absorption and fluorescence spectra
of radicals obtained by high-frequency
discharge in some aromatic vapors,
and stabilized at low temperatures.
Lortie, Y.
J. Phys. Radium 18: 520-2 (1957)
CA 53: 5865e

C 139 Aromatic hydrocarbons from vehicular
exhausts.
Lyons, M.J., and Johnston, H.
Brit. J. Cancer 11: 60-6 (1957)
CA 51: 15102h

C 140 Individual and group marketing methods
for fly population studies.
MacLeod, J., and Donnelly, J.
Bull. Entomol. Res. 58: 585-92 (1957)
CA 52: 8389h

C 141 Analysis of the aromatic fractions from
kerosine by phosphorescence spectra.
Mamedov, K.I.
Opt. i Spektroskopiya 3: 587-91 (1957)
CA 53: 6587i

C 142 Reactions of aldehydes and
m-phenylenediamine.
Maruta, S., and Suzuki, Y.
Nippon Kagaku Zasshi 78: 1604-8 (1957)
CA 54: 1371a

C 143 Formation of a cyclic recurring unit in
free radical polymerization.
Marvel, C.S., and Vest, R.D.
J. Am. Chem. Soc. 79: 5771-3 (1957)
CA 52: 7220e

C 144 Application of chromatography. XXXIII.
V-compound isolated from the myceli-
um of Eremothecium ashbyii.
Masuda, T., Kishi, T., and Asai, M.
Pharm. Bull. 5: 598-606 (1957)
CA 52: 16358h

C 145 Effect of environment on the fluores-
cence and absorption spectra of some
π-electron systems.
Mataga, N.
Bunko Kenkyu 6: 6-19 (1957)
CA 53: 9899f

C 146 Hydrogen bonding effect on the fluores-
cence of some nitrogen heterocycles. I.
Mataga, N., and Tsumo, S.
Bull. Chem. Soc. Japan 30: 368-74 (1957)
CA 51: 17464g

C 147 Influence of hydrogen bonding on the
fluorescence of some π-electron sys-
tems.
Mataga, N., and Tsuno, S.
Naturwissenschaften 44: 304-5 (1957)
CA 51: 15277f

C 148 Fluorimeter for the determination of
uranium.
Mathe, G., and Szalay, S.
Magy. Fiz. Folyoirat 5: 247-50 (1957)
CA 53: 5766i

C 149 Luminescence of silver halides.
Matyas, Z.
Abhandl. Deut. Akad. Wiss. Berlin, Kl.
Chem., Biol. Geol. 1955: 89-92 (1957)
CA 52: 2555h

C 150 Artifact in spectrophotometry caused
by fluorescence.
Mehler, A.H., Bloom, B., Ahrendt, M.E.,
and Stetten, D.
Science 126: 1285-6 (1957)
CA 52: 12554h

C 151 Energy dependence of the fluorescence
of polycrystalline phosphors excited by
electron rays or X rays.
Messner, D.
Z. Physik 147: 24-42 (1957)
CA 51: 6357a

C 152 Fluorescence spectra of kerosine.
Michul, C., Ruscior, C., and Pop, V.
Spectrochim. Acta 1957, Suppl. 562-4
(1957)
CA 54: 9267b

C 153 The fluorescence of Romanian diesel
oils.
Mihul, C., Ruscior, C., Pop, V., Schwartz,
F.R., and Radulescu, G.A.
Analele Stiint Univ. "A. I. Cuza," Iasi,
Sect. I 3: 243-56 (1957)
CA 53: 13567b

C 154 Fluorescent substances.
Mizuno, H., and Kamiya, S.
Japan 3358, June 6 (1957)
CA 52: 15269i

C 155 Effect of ultraviolet radiation on car-
cinogenic hydrocarbons by addition of
benzoyl peroxide.
Monig, H., and Kriegel, H.
Naturwissenschaften 49: 115-16 (1957)
CA 53: 18604h

C 156 Fading and tendering activity in anthra-
quinonoid rat dyes. I. Electronic ab-
sorption spectra of dye solutions.
Moran, J.J., and Stonehill, H.I.
J. Chem. Soc. 1957: 765-78 (1957)
CA 51: 7721f

C 157 Thallium-activated sodium chloride
recrystallization-phosphors.
Morlin, Z.
Nature 180: 89-90 (1957)
CA 52: 11588e

C 158 After-glow of NaCl recrystallized
phosphors activated with thallous
chloride.
Morlin, Z.
Acta Phys. Acad. Sci. Hung. 7: 341-56
(1957)
CA 52: 1778b

C 159 XII. On the fluorescence of thiophene
compounds.
Motoyama, R., Nishimura, S., Imoto, E.,
Murakami, Y., Hari, K., and Ogawa, J.
Nippon Kagaku Zasshi 78: 950-4 (1957)
CA 54: 14224i

C 160 Fluorescence spectrum of various
steroids in phosphoric acid.
Nakamura, R.
Tokyo Jikeikai Ika Daigaku Zasshi 72:
505-8 (1957)
CA 52: 15246f

C 161 Phosphate phosphors. II. Luminescent
color of pyrophosphate phosphors ex-
cited by cathode ray.
Nakano, E., and Takagi, K.
Nippon Kagaku Zasshi 78: 1146-50 (1957)
CA 52: 13438f

C 162 Hydrous uranyl phosphate (uramphite)
$(NH_4UO_2)(PO_4) \cdot 3H_2O$.
Nekrasona, Z.A.
At. Energ., Vopr. Geol. Urana, Suppl.
1957: 67-72 (1957)
CA 53: 8955a

C 163 High-voltage electrophoresis of the
urine of a child with Wilson's disease.
Noller, H.G., Stelgens, P., and Kieser,
H.J.
Monatsschr. Kinderheilk. 105: 343-6 (1957)
CA 52: 546g

C 164 Fluorescence analysis of skin of guinea
pig and rabbit after application of
3,4-benzpyrene.
Norden, G.
Acta Pathol. Microbiol. Scand. 40: 296-302
(1957)
CA 51: 13220i

C 165 Fluorescence of uranyl salts in solu-
tions.
Novak, M.
Jaderna Energie 3: 44-7 (1957)
CA 54: 10505g

C 166 Competition of unimolecular and bimolecular processes with special applications to the quenching of fluorescence in solution.
Noyes, R.M.
J. Am. Chem. Soc. 79: 551-5 (1957)
CA 51: 9328i

C 167 Relative intensities of fluorescence and
phosphorescence in biacetyl vapor.
Okake, H., and Noyes, W.A., Jr.
J. Am. Chem. Soc. 79: 801-6 (1957)
CA 51: 7872g

C 168 Fluorometric determination of thallium
and indium with Rhodamine B.
Onishi, A.
Bull. Chem. Soc. Japan 30: 567 (1957)

C 169 Fluorometric determination of thallium
and indium with Rhodamine B.
Onishi, A.
Bull. Chem. Soc. Japan 30: 827 (1957)

C 170 Phosphorescence of acetic acid.
Osada, K.
J. Phys. Soc. Japan 12: 1420 (1957)
CA 52: 8730a

C 171 Fluorescence spectra of aqueous solu-
tions of uranyl nitrate at room temper-
ature (10°C).
Pant, D.D., and Khandelwal, D.P.
Current Sci. 26: 282-3 (1957)
CA 52: 5968g

C 172 Fluorescence and absorption spectra
of uranyl salts.
Pant, D.D., and Pandey, B.C.
J. Sci. Ind. Res. 16B: 280-5 (1957)
CA 52: 2536f

C 173 Effect of intermittent light excitation
on alkali-halide phosphors.
Parfianovich, I.A.
Opt. i Spektroskopiya 2: 392-5 (1957)
CA 51: 11077a

C 174 Effect of growth of luminescence of
crystal phosphors after cessation of the
action of deexciting light.
Parfianovich, I.A.
Tr. Irkutsk. Gornomet. Univ. 15:
13-21 (1957)
CA 55: 13080c

C 175 Spectrofluorimeters and filter fluor-
imeters.
Parker, C.A., and Barnes, W.J.
Analyst 82: 606-18 (1957)
CA 52: 893a

C 176 Model '54 transmission and reflection
fluorimeter for determination of uranium with adaptation to field use.
Parshall, E.E., and Rader, L.F., Jr.
U.S. Geol. Surv., Bull. No. 1036-M,
221-51 (1957)
CA 52: 2463d

C 177 Behavior of cell autofluorescence in
activated thyroid. Experimental research in activation states from hypo-
physeal thyrotropic hormone (TSH).
Pende, G., and Romano, P.M.
Arch. "E. Maragliano" Pathol. Clin. 13:
743-52 (1957)
CA 52: 3974d

C 178 Quantum yield of F fluorescence.
Perlin, Y.E.
Opt. i Spektroskopiya 3: 328-33 (1957)
CA 52: 3523c

C 179 Luminescence of microporous glass
activated by salts of heavy metals.
Pershina, E.V., and Terenin, A.N.
Materialy V-go [Pyatogo] Soveshch. po
Lyuminests. (Kristallofosfory), Nauk
Est. SSR, Tartu 1956: 139-43 (1957)
CA 54: 20518e

C 180 Vital fluorescence and histochemistry.
Peters, T.
Acta Histochem. 4: 250-59 (1957)
CA 52: 3897e

C 181 Concerning spectra and duration of
afterglow in cement phosphors.
Pfahnl, A.
Acta Phys. Austriaca 11: 252-68 (1957)
CA 52: 890h

C 182 Luminescence in mineralogy.
Pfeiffer, L.
Urania 20: 176-8 (1957)
CA 51: 11941d

C 183 Fluorescent tungstates for X-ray inten-
sifying screens.
Philips, N.V.
Ger. 1,001,440 (1957)
CA 53: 12026f

C 184 Semiconductor properties of colored
potassium chloride crystals.
Politov, N.G.
Tr. Inst. Fiz. Akad. Nauk Gruz. SSR 5:
77-178 (1957)
CA 55: 16166a

C 185 Free radicals and triplet states in
aromatic vapors.
Porter, G.
Chem. Soc. (London), Spec. Publ. No. 9:
139-49, 150 (1957)
CA 53: 11986f

C 186 Metastable states in photochemistry.
Porter, G.
Threshold Space, Proc. Conf. Chem.
Aeronomy, Cambridge, Mass. 1956:
94-8 (1957)
CA 52: 13441d

C 187 Fluorometric researchers on some
fats.
Provvedi, F.
Olii Minerali, Grassi Saponi, Colori
Vernici 34: 373 (1957)
CA 51: 18372h

C 188 Reversibility of fluorescence by an-
nealing.
Przibram, K.
Nature 179: 319-20 (1957)
CA 51: 15278fh

C 189 Fluorescence of skin.
Przibram, K.
Naturwissenschaften 44: 393-4 (1957)
CA 52: 5605b

C 190 Fluorescence of minerals and chemi-
cals reversible by annealing.
Przibram, K.
Oesterr. Akad. Wiss., Math.-Naturw. Kl.,
Sitzber., Abt. II 166: 111-23 (1957)
CA 53: 7779f

C 191 Phosphorescence spectra of some aro-
matic acids at liquid-air temperature.
Pyatnitskii, B.A.
Soviet Phys. "Doklady" 1: 451-4 (1957)
CA 51: 13580c

C 192 Photosynthesis and energy transfer.
Rabinowitch, E.
J. Phys. Chem. 61: 870-8 (1957)
CA 51: 18143e

C 193 Quenching law for zinc sulfide phos-
phors.
Rabotkina, L.R.
Dokl. VII Nauch. Konf., Posvyashch. 40-
Letiyer Velikoi Oktyabr'sh Sots.
Revolyutsii, Tomsk. Univ. 1957: 126-7
(1957)
CA 51: 1035i

C 194 Triplet states of biologically active
molecules.
Reid, C.
Science 125: 396-7 (1957)
CA 51: 9330c

C 195 Response of crystal phosphors to
nuclear radiation.
Reiffel, L.
U.S. At. Energy Comm. OSR-TN-57-313,
56 pp. (1957)
CA 55: 18362c

C 196 Note on heterocyclic substituted pyr-
azolines.
Ried, W., and Dankert, G.
Chem. Ber. 90: 2707-11 (1957)
CA 53: 2207g

C 197 Spectroscopic studies on dyes. IV. The
fluorescence spectra of thioindigo dyes.
Rogers, D.A., Margerum, J.D., and
Wyman, G.M.
J. Am. Chem. Soc. 79: 2464-8 (1957)
CA 51: 15271h

C 198 Ultraviolet-visible absorption spectra
of quinoxaline derivatives.
Sawicki, E., Chastain, B., Bryant, H., and
Carr, A.
J. Org. Chem. 22: 625-9 (1957)
CA 52: 2026i

C 199 Symmetries of electric fields about
ions in solutions. Absorption and fluorescence spectra of europic chloride in
water, methanol, and ethanol.
Sayre, E.V., Miller, D.G., and Freed, S.
J. Chem. Phys. 26: 109-13 (1957)
CA 51: 6333i

C 200 Portable self-powered reader for DT-
60 glass dosimeter.
Schaffert, J.C.
Nucleonics 15: 60-2 (1957)
CA 53: 7800g

C 201 The difference in fluorescence between "light" and "dark" pseudo-unipolar ganglion cells in the semilunar ganglion of cattle.
Scharf, J.H., and Oster, K.
Acta Histochem. 4: 65-89 (1957)
CA 51: 18063i

C 202 Fluorescence and fluorescent polarization of nerve fibers after staining with fluorescein derivatives. Attempt at an interpretation.
Scharf, J.H.
Mikroskopie 11: 261-319, 349-97 (1957)
CA 54: 5798i

C 203 A new class of solid photoluminescents.
Schlivitch, S.
Compt. Rend. 245: 2047-8 (1957)
CA 52: 5124a

C 204 Near-ultraviolet spectrum of crystalline hexamethylbenzene and the structure of the hexamethylbenzene molecule.
Schnepp, O., and McClure, D.S.
J. Chem. Phys. 26: 83-92 (1957)
CA 51: 6335b

C 205 Self-activation of zinc sulfide.
Schwager, E.A., and Fischer, A.
Z. Physik 149: 345-6 (1957)
CA 53: 7779e

C 206 Thermal quenching in α- and γ-excited fluorescent solutions.
Selinger, H.H., and Ziegler, C.A.
J. Res. Natl. Bur. Std. 58: 125-6 (1957)
CA 51: 11859i

C 207 Fluorescence of naphthalene and anthracene vapors under β-ray excitation.
Shepp, A.
J. Chem. Phys. 27: 816-17 (1957)
CA 52: 1796e

C 208 A possible new type of fluorescence of the earth's atmosphere.
Shklovskii, I.S.
Astron. Zh. 34: 127-30 (1957)
CA 51: 12644i

C 209 Fluorescent substances from Mycobacterium.
Shoda, T.
Bitamin 13: 334-43 (1957)
CA 54: 4752a

C 210 Emission spectra of aromatic hydrocarbons at low temperatures.
Shpol'skii, E.V., Girdzhiyauskaite, E.A., and Klimova, L.A.
Materialy Desyatogo Vses. Soveshch. po Spektroskopii, L'vovsk. Gos. Univ., L vov, 1956, Fiz. Sb. 1: 24-36 (1957)
CA 53: 21146f

C 211 Concentration dependence of the absorption and fluorescence spectra of mixed crystals of anthracene with phenanthrene at 77°K.
Sidman, J.W.
J. Am. Chem. Soc. 79: 305-7 (1957)
CA 51: 6336f

C 212 Electronic and vibrational states of the nitrite ion. I. Electronic states.
Sidman, J.W.
J. Am. Chem. Soc. 79: 2669-75 (1957)
CA 51: 17440d

C 213 Maillard reaction in dried grass.
Sjollemo, A.
Landbouwk. Tijdschr. 69: 261-9 (1957)
CA 52: 12109e

C 214 Spectrophoto fluorometric studies of 5-hydroxyindoles and related compounds.
Sprince, H., Rowley, G.R., and Jameson, D.
Science 125: 442-3 (1957)
CA 51: 8534h

C 215 Excitation of biological substances.
Steele, R.H., and Szent-Gyorgyi, A.
Proc. Natl. Acad. Sci. U.S. 43: 477-91 (1957)
CA 52: 19461g

C 216 Fluorescent insulin conjugates.
Steiner, R.F., and McAlister, A.J.
J. Colloid Sci. 12: 80-98 (1957)
CA 51: 7873b

C 217 Use of the fluorescence techniques as an absolute method for obtaining mean relaxation times of globular proteins.
Steiner, R.F., and McAlister, A.J.
J. Polymer Sci. 24: 105-23 (1957)
CA 51: 9731c

C 218 Vibrational energy transfer from collisional deactivation of simple molecules and fluorescence stabilization of complex molecules.
Stevens, B., and Boudart, M.
Ann. N.Y. Acad. Sci. 67: 570-99 (1957)
CA 51: 16073f

C 219 Fluorescence and emission spectra of the three isometric fluorotoluenes.
Suryanaryana, V., Rao, I.A., and Rao, V.R.
Trans. Faraday Soc. 53: 1570-7 (1957)
CA 53: 146685i

C 220 Luminescence of uranium-activated sodium fluoride.
Sverdlov, Z.M.
Opt. i Spektroskopiya 3: 356-60 (1957)
CA 52: 2559d

C 221 Nuclear resonance fluorescence studies in oxygen.
Swann, C.P.
Univ. Microfilms Publ. No. 18015, 120 pp. (1957)
CA 51: 9306f

C 222 Derivatives of 4,4'-diaminostilbene.
Swiss 324,183 (1957)
CA 53: 10134d

C 223 Detection of chromatographic spots in paper.
Szent-Gyorgyi, A.E.
Science 126: 751 (1957)
CA 52: 1837h

C 224 Excitations and polymerization.
Szent-Gyorgyi, A.E.
Proc. Natl. Acad. Sci. U.S. 43: 151-2 (1957)
CA 51: 7872i

C 225 Effect of gas adsorption on zinc oxide luminescence.
Tagantsev, K.V., and Terenin, A.N.
Dokl. Akad. Nauk SSSR 112: 241-4 (1957)
CA 51: 16109a

C 226 Ultraviolet fluorescence of the aromatic amino acids.
Teale, F.W.J., and Weber, G.
Biochem. J. 65: 476-82 (1957)
CA 51: 7158d

C 227 Influence of concentration on phosphorescence of solutions of aromatic compounds at low temperature.
Teplyakov, P.A.
Opt. i Spektroskopiya 2: 269-71 (1957)
CA 51: 9329g

C 228 Infrared spectra (fluorescence) of phthalocyanines with different central metal atoms.
Terenin, A.N., and Sidorov, A.N.
Spectrochim. Acta, Suppl. 1957: 573-8 (1957)
CA 54: 6755e

C 229 Luminescent inclusions in micas.
Tolstikhina, K.I.
Tr. Inst. Geol. Rudn. Mestorozhd., Petrogr. Mineralog. i Geokhim. 1957(17): 53-6 (1957)
CA 53: 21446a

C 230 Ultra-τ-meter.
Tolstoi, N.A., Tkachuk, A.M., and Tkachuk, N.N.
Izv. Akad. Nauk SSSR, Ser. Fiz. 21: 595-611 (1957)
CA 52: 15142c

C 231 A green-blue fluorescent (ultraviolet light) substance in lichens xanthoria, caloplaca, and teloschistes.
Tomaselli, R.
Ist. Botan. Univ., Lab. Crittogam., Pavia, Atti 14: 144-50 (1957)
CA 52: 11196f

C 232 The green 2-band in the ultraviolet luminescence of zinc sulfide.
Tomlinson, T.B., and White, E.A.D.
J. Electron. 2: 404-5 (1957)
CA 51: 6356f

C 233 Metachromasy of polyvinylpyrrolidone.
Tsubomura, I., Yoshioka, J., and Torii, K.
Bunko Kenkyu 6: 10-14 (1957)
CA 53: 8107d

C 234 Duration of the excited state and quantum yield of fluorescence from chlorophyll in vitro and in vivo.
Tumerman, L.A.
Soviet Phys. "Doklady" 2: 525-7 (1957)
CA 53: 20301g

C 235 Fluorescence in planetary atmospheres.
Urey, H.C., and Brewer, A.W.
Proc. Roy. Soc. (London), Ser. A 241: 37–43 (1957)
CA 51: 15261d

C 236 Fluorescence of Argentinian woods.
Valente, E., and Pardo, L.L.
Rev. Invest. Forestales 1: 47–51 (1957)
CA 53: 3689h

C 237 Red shifts in the spectra of anthracene derivatives.
Veljkovic, S.R.
Trans. Faraday Soc. 53: 1181–5 (1957)
CA 52: 9766g

C 238 Fluorescence of aromatic amino acids.
Vladimirov, Y.A.
Dokl. Akad. Nauk SSSR 116: 780–3 (1957)
CA 52: 19461d

C 239 Zirconium oxide, its crystal polymorphism and its suitability as a material for high temperatures.
Weber, B.C., and Schwartz, M.A.
Ber. Deut. Keram. Ges. 34: 391–6 (1957)
CA 54: 9424a

C 240 Determination of the absolute quantum yield of fluorescent solutions.
Weber, G., and Teale, F.W.J.
Trans. Faraday Soc. 53: 646–55 (1957)
CA 52: 5124f

C 241 The quenching of fluorescence. III. Quenching of the fluorescence of optical bleaches.
Weber, K., and Skuric, Z.
Croat. Chem. Acta 29: 115–25 (1957)
CA 52: 10728a

C 241a Polarization of the fluorescence of tetraphenylporphine.
Weigl, J.W.
J. Mol. Spectry. 1: 133–8 (1957)
CA 51: 17465d

C 242 Discrimination between "radiative" and "nonradiative" transfer of molecular excitation energy in liquid systems.
Weinreb, A.
J. Chem. Phys. 27: 133–6 (1957)
CA 51: 17458c

C 243 Protolytic reactions of excited acridine.
Weller, A.
Z. Elektrochem. 61: 956–61 (1957)
CA 52: 4297i

C 244 I. Phosphorimetric method of analysis. II. Fluorometric analysis of rare earth by complex formation.
Wentworth, W.E.
Dissertation, Florida State University, Dissertation Abstr. 17: 2852–3 (1957)
CA 52: 4387g

C 245 Derivatives of polyvinyl alcohol.
Werssermel, K., and Starck, W.
Ger. 1,020,791 (1957)
CA 54: 10401g

C 246 Fluorometry.
White, C.E.
Trace Anal., Papers Symp. Trace Anal., N.Y. 1955: 211–28 (1957)
CA 52: 4381f

C 247 Fluorescence emission spectra, fluorescence excitation spectra, and absorption spectra of some metal chelates.
White, C.E., Hoffman, D.E., and Magee, J.S.
Spectrochim. Acta 9: 105–12 (1957)
CA 51: 14426h

C 248 Heterocyclic analogs of terphenyl. 3,6-Diaryl-1,2,4,5-tetrazines.
Wiley, R.H., Jarboe, C.H., Jr., and Hayes, F.N.
J. Org. Chem. 22: 835–6 (1957)
CA 52: 2870i

C 249 Luminescence of synthetic zeolites activated with copper.
Wilke, K.T.
Z. Physik. Chem. 207: 45–59 (1957)
CA 51: 17464f

C 250 Luminescence in the aluminum fluoride group.
Wilke, K.T., and Mannheim, R.
Naturwissenschaften 44: 631–2 (1957)
CA 52: 8730c

C 251 Aminco-Bowman spectrophotofluorometer.
Williams, R.T.
Biochem. J. 65, 26 pp. (1957)
CA 53: 4829h

C 252 Observation of the relative polarizations of electronic transitions.
Williams, R.
J. Chem. Phys. 26: 1186-8 (1957)
CA 51: 13569h

C 253 New phenomena in the fluorescence spectrum of a diphosphopyridine nucleotide-linked dehydrogenase.
Winer, A.D., Novoa, W.B., and Schwert, G.W.
J. Am. Chem. Soc. 79: 6571-2 (1957)
CA 52: 6929c

C 254 Absorption alteration and chlorophyll fluorescence in the primary process of photosynthesis.
Witt, H.T., and Moraw, R.
Z. Physik. Chem. 12: 393-5 (1957)
CA 51: 18127a

C 255 Fluorescence and dyeing characteristics of fluorescent whitening dyes of triazinylstilbene series. I. Diamino and dyhydroxy derivatives.
Yabe, A., and Hayashi, M.
Kogyo Kagaku Zasshi 60: 740-5 (1957)
CA 53: 8635b

C 256 Fluorescence of nitrogen under the influence of short-wave radiation.
Yakovleva, A.V., and Gromoua, I.I.
Fiz. Sb., L'vovsk. Gos. Univ. 1957: 308-10 (1957)
CA 55: 217787i

C 257 The fluorescence and phosphorescence of organic compounds.
Yamamoto, D.
Kagaku (Tokyo) 27: 384-90 (1957)
CA 51: 15245a

C 258 Composition for fluorescent-penetrant inspection of materials.
Yoshida, K., Nishikawa, K., Tsuchida, S. and Sakata, M.
Japan 1809, Mar. 20, 1957
CA 52: 9484a

C 259 Fluorescent substances.
Yoshida, K., Nishikawa, K., Tsuchida, S., and Sakada, E.
Japan 1810 (1957)
CA 52: 5474d

C 260 Fluorescence analysis of the compounds containing amino group and related substances.
Yoshida, T.
Igaku Kenkyu 27: 443-55 (1957)
CA 52: 10721i

C 261 Relation between fluorescence frequencies and the effectiveness of certain quenchers.
Zelinskii, V.V., Kolobkov, V.P., and Kondaraki, N.I.
Soviet Phys. "Doklady" 2: 501-4 (1957)
CA 53: 18626d

C 262 Determination of molecular volumes in solutions by the use of polarized luminescence.
Zhevandrov, N.D., and Nikolaev, V.P.
Soviet Phys. "Doklady" 2: 175-8 (1957); Dokl. Akad. Nauk SSSR 113: 1025-8 (1957)
CA 53: 33b

C 263 Determination of the oscillator strength of the 3720-A iron line from the decay time of the resonance fluorescence.
Ziock, K.
Z. Physik 147: 99-112 (1957)
CA 51: 6330i

C 264 Luminescence of benzene and hexadeuteriobenzene at 20°K.
Zmerli, A.
Compt. Rend. 245: 1911-3 (1957)
CA 52: 5123c

C 265 Fluorescence spectra of the products resulting from high-frequency electric discharges in benzene, toluene, o-, m-, and p-xylene vapors.
Agerbiceanu, I., Haglescu-Miriste, M., and Weissman, I.
Comun. Acad. Rep. Populare Romine 8: 359-64 (1958)
CA 53: 2781e

C 266 Naphthalene derivatives in inorganic analysis. VI. Reagent for the fluorimetric detection of tin.
Anderson, J.R.A., Garnett, J.L., and Lock, L.C.
Anal. Chim. Acta. 19: 256-9 (1958)
CA 54: 2090f

C 267 Dyes derived from imidazole. I. Preparation and chromatographic separation of dyes from 1,4,5,8-naphthalenetetracarboxylic and 1,8-naphthalenedicarboxylic acids.
Arient, J., and Franc, J.
Chem. Listy 52: 1946-50 (1958)
CA 53: 1722e

C 268 A highly sensitive chemical dosimeter for ionizing radiation.
Armstrong, W.A., and Grant, D.W.
Nature 182: 747 (1958)
CA 53: 6803h

C 269 Transitions without emission in zinc sulfides at low temperature.
Arpiarian, N.
J. Chim. Phys. 55: 667-71 (1958)
CA 53: 4923h

C 270 Quenching of the long-lived fluorescence of biacetyl in solution.
Backstrom, H.L.J., and Sandros, K.
Acta Chem. Scand. 12: 823-32 (1958)
CA 53: 21095c

C 271 Polarization of photoluminescence of organophosphors.
Baczynski, A., and Czajkowski, M.
Bull. Acad. Polon. Sci., Ser. Sci., Math., Astron. Phys. 6: 271-4 (1958)
CA 52: 13438d

C 272 Formation of aromatic hydrocarbons at high temperatures.
Bodger, G.M., Buttery, R.G., Kimber, R.W.L., Lewis, G.E., Moritz, A.G., and Napier, I.M.
J. Chem. Soc. 1958: 2449-52 (1958)
CA 52: 20094d

C 273 The variation of fluorescence by addition of benzene (benzene effect), investigated in case of Acridine Orange-aluminum oxide.
Bandow, F., and Banderet, E.
Z. Physik. Chem. 18: 201-5 (1958)
CA 55: 3192e

C 274 Influence of oxygen on the transfer effi ciency and fluorescence yield of organic solutions.
Bar, V., and Weinreb, A.
J. Chem. Phys. 29: 1912-14 (1958)
CA 53: 6755d

C 275 Ultraviolet fluorescence of crystals of aromatic amino acids.
Barskii, I.Ya., and Brumberg, E.M.
Biokhimiya 23: 791-2 (1958)
CA 53: 2813i

C 276 Relation between exciting wave length and the ratio of the phosphorescence and fluorescence yields.
Bauer, R., and Baczynski, A.
Bull. Acad. Polon. Sci., Ser. Sci., Math., Astron. Phys. 6: 113-7 (1958)
CA 52: 13438b

C 277 Excitation energy transfer from dye molecules in the metastable state.
Bauer, R., Baczynski, A., and Czajkowski, M.
Bull. Acad. Polon. Sci., Ser. Sci., Math., Astron. Phys. 6: 653-8 (1958)
CA 53: 5867a

C 278 The connection of Raman dispersion, adsorption, and fluorescence (resonance Raman effect).
Behringer, J.
Z. Elektrochem. 62: 544-67 (1958)
CA 52: 16872b

C 279 The Elks peroxydisulfate oxidation in the pyridine series: a new synthesis of 2,5-dihydroxypyridine.
Behrman, E.J., and Pih, B.M.
J. Am. Chem. Soc. 80: 3717-18 (1958)
CA 53: 2223c

C 280 Delayed light emission from rare gas coronas.
Bemerl, W.F., and Fetz, H.
Naturwissenschaften 45: 381-2 (1958)
CA 53: 9809e

C 281 The color and fluorescence of Russulas.
Bernanose, A., and Vincent, A.M.
Bull. Soc. Pharm. Nancy 37: 20-2 (1958)
CA 52: 15665h

C 282 Energy transfer in organic systems. I. Photofluorescence of terphenyl-toluene solutions.
Birks, J.B., and Cameron, A.J.W.
Proc. Phys. Soc. 72: 53-64 (1958)
CA 54: 23511i

C 283 Phosphorescence studies of o-, m-, and
      p-xylene at low temperatures.
      Blackwell, L.A.
      Univ. Microfilms, L.C. Card No. Mic 58-
         2728, 64 pp. (1958)
      CA 53: 2782d

C 284 Delayed singlet-singlet emission from
      molecular crystals.
      Blake, N.W., and McClure, D.S.
      J. Chem. Phys. 29: 722-4 (1958)
      CA 53: 2772e

C 285 Carotenoids in man. III. The micro-
      scopic pattern of fluorescence in ather-
      omas and its relation to their growth.
      Blankenhorn, D.H., and Braunstein, H.
      J. Clin. Invest. 37: 160-5 (1958)
      CA 52: 8345d

C 286 Photoconductivity and absorption and
      fluorescence spectra of 9,10-dichloro-
      anthracene.
      Bock, E., Ferguson, J., and Schneider,
         W.G.
      Can. J. Chem. 36: 507-12 (1958)
      CA 52: 8743i

C 287 The photographic effect of charged
      particles at low temperatures.
      Bogomolov, K.S., Razorenova, I.F., and
         Sirotinskaya, A.A.
      Phot. Corpusculaire, Colloq. Intern., 1er
         Strasbourg 1957: 203-11 (1958)
      CA 54: 24041i

C 288 Quenching of the fluorescence of
      phthalimide vapors by oxygen.
      Borisvich, N.A.
      Tr. Inst. Fiz. i Mat., Akad. Nauk Belor-
         ussk. SSR 1956: 94-101 (1958)
      CA 55: 21793g

C 289 Fluorescence of the toxin of clostridi-
      um botulinum and its relation to toxici-
      ty.
      Boroff, D.A., and Fitzgerald, J.E.
      Nature 181: 751-2 (1958)
      CA 52: 13009f

C 290 Energy-transfer mechanism in poly-
      styrene phosphors.
      Bothe, H.K., and Herforth, L.H.
      Semicond. Phosphors, Proc. Intern. Colloq.
         Garmisch-Partenkirchen 1956: 439-44
         (1958)
      CA 54: 17045c

C 291 Fluorescence of acridine and acridone
      solutions.
      Bowen, E.J., and Sahu, J.
      J. Chem. Soc. 1958: 3716-18 (1958)
      CA 53: 8805g

C 292 Energy transfer and fluorescence spec-
      tra in Porphyridium cruentum.
      Brody, S.S.
      J. Chim. Phys. 55: 942-51 (1958)
      CA 54: 12268f

C 293 Energy transfer in liquid and rigid or-
      ganic systems.
      Brown, F.A., Furst, M., and Kallmann, H.P.
      J. Chim. Phys. 55: 688-97 (1958)
      CA 53: 4900c

C 294 Fluorescent decay times and their re-
      lation to the scintillation process.
      Brucker, G.J.
      Dissertation Abstr., New York Univ.
         (1958)
      CA 52: 16894d

C 295 Fluorescence spectrum and low levels
      of neodymium chloride.
      Carlson, E., and Dieke, G.H.
      J. Chem. Phys. 29: 229-30 (1958)
      CA 52: 19440b

C 296 Constitution of the leucoanthocyanidin
      peltogynol.
      Chan, W.R., Forsyth, W.G.C., and
         Hassall, C.H.
      J. Chem. Soc. 1958: 3174-9 (1958)
      CA 53: 3206c

C 297 Respiratory enzymes in oxidative phos-
      phorylation. VII. Binding of intramito-
      chondrial reduced pyridine nucleotide.
      Chance, B., and Baltscheffsky, H.
      J. Biol. Chem. 233: 736-9 (1958)
      CA 53: 2312e

C 298 Fluorescent substances produced by
      dermatophytes.
      Chattaway, F.W., and Barlow, A.J.E.
      Nature 181: 281 (1958)
      CA 52: 9297h

C 299 Measurement of degree of polarization
      of fluorescence of pure anthracene
      crystalline film.
      Chaudhury, N.K.
      Z. Physik 151: 93-105 (1958)
      CA 52: 14344c

C 300 Kinetics of the photochemical transformations and the concentration quenching of the fluorescence of 9-monoalkyl-
substituted anthracenes.
Cherkasov, A.S., and Vember, T.M.
Opt. i Spektroskopiya 4: 203-10 (1958)
CA 52: 13440h

C 301 (Fluorometric detection of aluminum
and beryllium with 2-hydroxy-3-
naphthoic acid.)
Cherkesov, A.I.
Dokl. Akad. Nauk SSSR 118: 309 (1958)
CA 52: 8840c

C 302 Organophosphors. VIII.
Chomse, H.
Festkoerperphys. Leuchtstoffe 1958: 254-9
(1958)
CA 55: 18341i

C 303 Determination of molecular weight by
scattered light from fluorescent solu-
tions.
Ciferri, A., and Weill, G.
Ric. Sci. 28: 765-9 (1958)
CA 52: 17906i

C 304 Use of a coumarin derivative as a
secondary solute in liquid scintillators.
Coche, A., Henck, R., and Laustriat, G.
Compt. Rend. 247: 2123-6 (1958)
CA 53: 12870f

C 305 The electronic spectra of crystalline
toluene, bibenzyl, diphenylmethane, and
biphenyl in the near ultraviolet.
Coffman, R., and McClure, D.S.
Can. J. Chem. 36: 48-58 (1958)
CA 52: 19445b

C 306 Photoconductivity of anthracene. The
effect of neutron bombardment.
Compton, D.M.J., Schneider, W.G., and
Waddington, T.C.
J. Chem. Phys. 28: 741-2 (1958)
CA 52: 12566i

C 307 Glycosides. III. Glycosides of
4-methylumbelliferone.
Constantzas, N., and Kocourek, J.
Chem. Listy 52: 1629-32 (1958)
CA 53: 1318c

C 308 Acid-stable organic azides and the
Schmidt reaction with heterocyclic
ketones.
Coombs, M.M.
J. Chem. Soc. 1958: 4200-2 (1958)
CA 53: 10202g

C 309 Concentration quenching in fluorescent
acene solutions.
Dammers-de-Klerk, A.
Mol. Phys. 1: 141-50 (1958)
CA 53: 2782e

C 310 Fluorescence quenching of organic
molecules in solution.
Dammers-de-Klerk, A.
Chem. Weekblad 54: 281-8 (1958)
CA 52: 19445h

C 311 Introduction to spectrophotofluorimetry,
description of a type of spectrophoto-
fluorimeter.
De Francesco, F.
Ann. Chim. 48: 390-9 (1958)
CA 52: 15249d

C 312 Fluorescence spectrum of PrCl$_3$ and
the levels of the Pr^{+++} ion.
Dieke, G.H., and Sarup, R.
J. Chem. Phys. 29: 741-5 (1958)
CA 53: 2772f

C 313 Analysis of mixed ambipolar and ex-
citon diffusion in cadmium sulfide
crystals.
Diemer, G., VanGurp, G.J., and
Hoogenstraaten, W.
Philips Res. Rept. 13: 458-84 (1958)
CA 53: 6782g

C 314 Stilbenes.
Drefahl, G., and Plotner, G.
Chem. Ber. 91: 1274-80 (1958)
CA 52: 20044d

C 315 Photoluminescence of esters of
phthalic and benzoic acids.
Dubinskii, I.B.
Izv. Krymsk. Ped. Inst. 1957: 29, 321-35
(1958)
CA 54: 16174e

C 316 The spectrum of comet Mrkos (1957d)
in the near infrared.
Dufay, J., and Swings, P.
Ann. Astrophys. 21: 260-72 (1958)
CA 53: 6750h

C 317 Griseofulvin. XIII. Homologs of gris-
    eofulvin and 7-chloro-4,4',6-trimethyl-
    gris-3'-ene-3,2'-dione.
    Duncanson, L.A., Grove, J.F., and Jeffs,
    P.W.
    J. Chem. Soc. 1958: 2929-33 (1958)
    CA 53: 326c

C 318 Two simplified fluorimeters for urani-
    um determination.
    Dutra, C.V.
    Inst. Pesquisas Radioativas, Publ. 10: 1-
    10 (1958)
    CA 56: 8265f

C 319 Thermoluminescence and phosphores-
    cence spectra of some pure and
    impurity-activated alkali halides.
    Dutta, B.C., and Ghosh, A.K.
    Indian J. Phys. 32: 578-9 (1958)
    CA 53: 10994h

C 320 Effect of an activator on phosphores-
    cence quenching.
    Dvorovenko, V.K.
    Nauchn. Zap., Odessk. Gos. Ped. Inst.,
    Kafedry Mate., Fiz. i Estestvoz. 22:
    32-4 (1958)
    CA 55: 11106g

C 321 Organic oxide luminophors.
    Dvorovenko, V.K.
    Nauchn. Zap., Odessk. Gos. Ped. Inst.,
    Fiz.-Mat. Fak. 22: 43-6 (1958)
    CA 52: 2996i

C 322 Damping of phosphorescence of ce-
    mented phosphors activated by tereph-
    thalic acid.
    Dvorovenko, V.K.
    Nauchn. Zap., Odessk. Gos. Ped. Inst.,
    Fiz.-Mat. Fak. 22: 47-52 (1958)
    CA 52: 4217e

C 323 Fluorescence of diamond.
    Dyer, H.B., and Mathews, I.G.
    Proc. Roy. Soc. (London), Ser. A 243:
    320-35 (1958)
    CA 52: 7870a

C 324 Development of a single fired phos-
    phorescent enamel.
    Eichbaum, B.R.
    Am. Ceram. Soc. Bull. 37: 148-51 (1958)
    CA 52: 8492f

C 325 Fluorescent colored porcelain enamel.
    Eichbaum, B.R.
    Ceram. Ind. 70: 92-3 (1958)
    CA 52: 12352e

C 326 Absorption and fluorescence spectra of
    acridone and its dependence on hydro-
    gen ion concentration and solvent.
    Kokubun, H.
    Z. Elektrochem. 62: 599-607 (1958)
    CA 52: 16869e

C 327 Phosphorescence of lubricating oils at
    the temperature of liquid oxygen.
    Elin, L.V., Korobtsov, I.M., and
    Khalupovskii, M.D.
    Nauchn. Zap., Odessk. Gos. Ped. Inst.,
    Fiz.-Mat. Fak. 22(1): 63-5 (1958)
    CA 57: 8799d

C 328 Determination of X-ray absorption co-
    efficients of inhomogeneous materials.
    Ergun, S., and Tiensuu, V.H.
    J. Appl. Phys. 29: 946-9 (1958)
    CA 53: 7756e

C 329 Energy transfer between triplet levels.
    Ermolaev, V.L., and Terenin, A.N.
    J. Chim. Phys. 55: 698-704 (1958)
    CA 53: 4899d

C 330 The polarized fluorescence of very
    thin anthracene crystals.
    Ferguson, J., and Schneider, W.G.
    Can. J. Chem. 36: 1070-80 (1958)
    CA 53: 854f

C 331 Spectral response of photoconduction in
    thin single crystals of anthracene.
    Ferguson, J., and Schneider, W.G.
    Can. J. Chem. 36: 1633-9 (1958)
    CA 53: 6783f

C 332 Methylenedeoxybenzoins. VII. The
    condensation of deoxybenzoin and simi-
    lar ketones with cyanoacetate.
    Fiesselman, H., and Ehmann, W.
    Chem. Ber. 91: 1706-13 (1958)
    CA 53: 1329g

C 333 Identification of glass fragments by
    their physical properties.
    Finch, J., and Williams, P.P.
    Analyst 83: 698-9 (1958)
    CA 53: 9597i

C 334 External quenching, activator recipro-
city, and hole migration in ZnSCu and
ZnSCuCo phosphors.
Fok, M.V.
Semicond. Phosphors, Proc. Intern.
Colloq., Garmisch-Partenkirchen 1956:
593-601 (1958)
CA 54: 14971i

C 335 Phosphorescence spectra and analysis
of some indole derivatives.
Freed, S., and Salmre, W.
Science 128: 1341-2 (1958)
CA 55: 6138c

C 336 Proteins in solutions at low tempera-
tures.
Freed, S., Turnbull, J.H., and Salmre, W.
Nature 181: 1731-2 (1958)
CA 53: 982i

C 337 Photographic duplicating process using
fluorescence.
Friedman, J.S., and Horwitz, L.
U.S. 2,865,744 (1958)
CA 53: 21313e

C 338 Oxidative decomposition of riboflavine.
Fujisawa, T.
Nagoya J. Med. Sci. 21: 69-83 (1958)
CA 53: 5276i

C 339 Cross quenching of fluorescence in
organic solutions.
Furst, M., and Kallmann, H.P.
Phys. Rev. 109: 646-51 (1958)
CA 52: 11567e

C 340 Fluorescent pigments based on tetra-
zaindenes.
Fusco, F., and Rossi, S.
U.S. 2,837,520, June, 1958
CA 52: 16388c

C 341 The experimental investigation of the
rotational depolarization of the fluorescence of solutions.
Gati, L., and Szalay, L.
Acta Univ. Szeged., Acta Phys. Chem. 4:
90-3 (1958)
CA 53: 19684f

C 342 Oxygen-containing heterocycles as
liquid scintillator solutes.
Gilman, H., Weipert, E.A., Dietrich, J.J.,
and Hayes, F.N.
J. Org. Chem. 23: 361-2 (1958)
CA 53: 9225e

C 343 Measurements of X-ray doses by means
of thin, fluorescent screens.
Glocker, R., and Meissner, D.
Z. Physik. 152: 538-45 (1958)
CA 53: 849h

C 344 Effect of light on fluorescence of ethyl-
enediamine derivatives of adrenaline
and norepinephrine.
Goldfien, A., and Karler, R.
Science 127: 1292-3 (1958)
CA 53: 15761d

C 345 Dependence of the ratios of the yields
of phosphorescence and fluorescence on
the position of the fluorescence spectra.
Golikova, L.E., Zelinskii, V.V., and
Kolobkov, V.P.
Opt. i Spektroskopiya 5: 480-2 (1958)
CA 53: 10956b

C 346 Synthesis and reactions of 2-chlorooxa-
zoles.
Gompper, R., and Effenberger, F.
Angew. Chem. 70: 628 (1958)
CA 53: 7141d

C 347 Energy transfer by excitons in pure
cadmium sulfide.
Grillot, E.
J. Chim. Phys. 55: 642-9 (1958)
CA 53: 4923b

C 348 Extinction factors of the fluorescence
of CdS (Ag).
Grillot-Bancie, M.
Semicond. Phosphors, Proc. Intern.
Colloq., Garmisch-Partenkirchen 1956:
610-13 (1958)
CA 54: 10543f

C 349 Heinrichite and metaheinrichite, hy-
drated barium uranyl arsenate minerals.
Gross, E.B., Corey, A.S., Mitchell, R.S.,
and Walenta, K.
Am. Mineralogist 43: 1134-43 (1958)
CA 53: 5027c

C 350 The fluorescence of air excited by fast
electrons: light yield as a function of
pressure.
Grun, A.E.
Can. J. Phys. 36: 858-62 (1958)
CA 52: 15259c

C 351 Dependence of the state of polarization on the wavelength of fluorescence.
Gurinovich, G.P., and Sevcheniko, A.N.
Izv. Akad. Nauk SSSR, Ser. Fiz. 22: 1407–11 (1958)
CA 53: 2783d

C 352 Inhibition of the fading dyes. II.
Hajo's, Z., and Fodor, J.
Acta Chim. Acad. Sci. Hung. 16: 291–9 (1958)
CA 53: 722h

C 353 Theory of excitons and their role in energy transfer in the solid state.
Haken, H.
J. Chim. Phys. 55: 613–20 (1958)
CA 53: 4922i

C 354 The fluorescence of tetracycline in rats treated with dihydrotachysterol.
Hakkinen, I.P.T.
Acta Physiol. Scand. 42: 282–7 (1958)
CA 52: 16611c

C 355 Thermoluminescence spectra of X-ray-colored potassium chloride crystals.
Halperin, A., and Kristianpoller, N.
J. Opt. Soc. Am. 48: 996–1000 (1958)
CA 53: 3897c

C 356 The nature of some fluorescent substances of pterin type in the adult skin of the toad Bufo vulgaris formosus.
Hama, T., and Obika, M.
Experientia 14: 182–4 (1958)
CA 52: 15755b

C 357 Isolation and structure of the fluorescent compound formed from adrenaline and ethylenediamine.
Harley-Mason, J., and Laird, A.H.
Biochem. J. 69: 59–60 (1958)
CA 52: 18543e

C 358 The calcium silicate-tungstate phosphor. Phase relationships and fluorescent properties.
Harrison, D.E., and Hummel, F.A.
J. Electrochem. Soc. 105: 34–7 (1958)
CA 52: 4332h

C 359 Photochemical reduction of thionine. I. The thionine-ferrous ion complex.
Havemann, R., and Pietsch, H.
Z. Physik. Chem. 208: 98–108 (1958)
CA 52: 9785b

C 360 Phosphorescence studies of some heterocyclic and related organic compounds.
Heckman, R.C.
J. Mol. Spectry. 2: 27–41 (1958)
CA 52: 10721i

C 361 Phosphorescence in liquid scintillation counting of proteins.
Herberg, R.J.
Science 128: 199–200 (1958)
CA 52: 18588b

C 362 See C 677a

C 363 Emission processes in the phosphorescence of argon.
Herman, L., Seguier, J., and Herman, R.
J. Phys. Radium 19: 463–74 (1958)
CA 53: 7757b

C 364 Tetramethylrhodamine as an immuno-histochemical fluorescent label in the study of chronic thyroditis.
Hiramoto, R., Engel, K., and Pressman, D.
Proc. Soc. Exptl. Biol. Med. 97: 611–14 (1958)
CA 52: 14809e

C 365 Steroid epoxides.
Hoehn, W.M.
J. Org. Chem. 23: 929–30 (1958)
CA 53: 1412f

C 366 Color reactions for the identification of some derivatives of progesterone.
Hohensee, F., and Huttenrauch, R.
Z. Physiol. Chem. 310: 19–22 (1958)
CA 52: 12329g

C 367 4,6-Dimethyl-7-methylaminocoumarin.
H.B. Holliday & Co., Ltd.
Ger. 1,033,218, July 3,1958
CA 55: 1660f

C 368 Paramagnetic resonance absorption in naphthalene in its phosphorescent state.
Hutchinson, C.A., Jr., and Mangum, B.W.
J. Chem. Phys. 29: 952–3 (1958)
CA 53: 3887i

C 369 Lower excited states and the phosphorescent state of biphenyl. II. Scheme of modification of molecular orbitals.
Iguchi, K.
J. Phys. Soc. Japan 13: 1186–9 (1958)
CA 53: 2765i

C 370 Naphthyridines. III. Synthesis of 2,10-
diazaanthracene and 1,7-naphthyridine.
Ikekawa, N.
Chem. Pharm. Bull. 6: 401-4 (1958)
CA 53: 3227c

C 371 The reaction of methyl β-anilinovinyl
ketone with aldehydes.
Inoue, G.
Nippon Kagaku Zasshi 79: 1243-6 (1958)
CA 54: 24716d

C 372 Mercury-vapor discharge lamp.
Istvan, G., and Kardos, F.
Austrian 198,835, July 25, 1958
CA 52: 16048d

C 373 Metastable state of dye molecules.
Jablonski, A.
Bull. Acad. Polon. Sci., Ser. Sci., Math.,
Astron. Phys. 6: 589-93 (1958)
CA 53: 5866i

C 374 Apparatus for the automatic registra-
tion of fluorescence spectra.
Jokl, J.
Chem. Listy 52: 1370-2 (1958)
CA 53: 5873h

C 375 Energy transfer in liquid organic sys-
tems.
Kallman, H.P., Furst, M., and Brown, F.A.
Semicond. Phosphors, Proc. Intern.
Colloq., Garmisch-Partenkirchen 1956:
269-84 (1958)
CA 53: 8773g

C 376 Triplet-singlet emission spectra of
solid toluene at 4K and 77K and in EPA
solution at 77K.
Kanda, Y., and Sponer, H.
J. Chem. Phys. 28: 798-806 (1958)
CA 52: 16032f

C 377 Oxygen quenching of fluorescence of
chlorophyll and its analogs in the ab-
sorbed state.
Karayakin, A.V., and Shablya, A.V.
Opt. i Spektroskopiya 5: 655-62 (1958)
CA 53: 3877c

C 378 Synthesis of riboflavine by microorganisms. II. The green and violet fluorescent compounds produced in the culture
filtrate of clostridium acetobutylicum.
Katagiri, H., Takeda, J., and Ionai, K.
J. Vitaminol. 4: 207-10 (1958)
CA 54: 12259i

C 379 Absorption and luminescence spectra
of alkali halide crystals activated by
nickel.
Kats, M.L., and Semenov, B.Z.
Opt. i Spektroskopiya 4: 637-42 (1958)
CA 53: 2812h

C 380 Micellar behavior of half ester soaps in
benzene.
Kaufman, S., and Singleterry, C.R.
J. Phys. Chem. 62: 1257-9 (1958)
CA 53: 2744b

C 381 Effect of concentration on the polariza-
tion of the fluorescence of rigid solu-
tions.
Kawski, A.
Bull. Acad. Polon. Sci., Ser. Sci., Math.,
Astron. Phys. 6: 533-9 (1958)
CA 53: 2781a

C 382 The dependence of emission anisotropy
of fluorescence on the concentration of
luminescent molecules in plexiglass
luminophors.
Kawski, A.
Bull. Acad. Polon. Sci., Ser. Sci., Math.,
Astron. Phys. 6: 671-5 (1958)
CA 54: 14970c

C 383 Fluorescence spectra, term assign-
ments, and crystal field splittings of
rare earth activated phosphors.
Keller, S.P.
J. Chem. Phys. 29: 180-7 (1958)
CA 52: 17980b

C 384 Quenching, stimulation, and exhaustion
studies on some infrared stimulable
phosphors.
Keller, S.P., and Pettit, G.D.
Phys. Rev. 111: 1533-9 (1958)
CA 53: 2815b

C 385 Determination of the relative fluorescence intensity and fluorescence yield
by means of polarization measurements.
Ketskeméty, I.
Acta Univ. Szeged., Acta Phys. Chem. 4:
18-20 (1958)
CA 53: 13775b

C 386 The relation between the absorption and
polarization spectra of a fluorescent
morin compound.
Ketskeméty, I., Marek, N., and Sarkany, B.
Acta Univ. Szeged., Acta Phys. Chem. 4:
21-9 (1958)
CA 53: 9811h

C 387 Ultraviolet fluorescence of pyridoxine, indole, and some of their derivatives.
Khan-Magometova, I.D.
Biofizika 3: 558-61 (1958)
CA 53: 1918b

C 388 Reaction of quinoline and iodine with quinaldine and with 2,6-dimethylquinoline.
King, C., and Abrarmo, S.V.
J. Org. Chem. 23: 1926-8 (1958)
CA 53: 11382d

C 389 Dependence of the excited-state duration of organoluminophors on the wave length of the exciting light.
Kislyak, G.M.
Opt. i Spektroskopiya 5: 297-301 (1958)
CA 53: 1917g

C 390 Dependence of the lifetime of the excited state of trypaflavine in formic acid on the wave length of the excitation light.
Kislyak, G.M.
Izv. Krymsk. Ped. Inst. 1957, 29: 358-69 (1958)
CA 55: 14057d

C 391 Results on the fluorescence of gases excited by high-energy charged particles.
Koch, L., and Lesuer, R.
Comm. Energie At. (France), Rappt. 869, 7 pp. (1958)
CA 53: 21140a

C 392 Determination of benzanthrone in air.
Kogan, I.B.
Zavodsk. Lab. 24: 291-3 (1958)
CA 54: 11859e

C 393 Fluorescence of powdered vegetable drugs under ultraviolet radiation.
Kokoski, C.J., Kokoski, R.J., and Slama, F.J.
J. Am. Pharm. Assoc. 47: 715-17 (1958)
CA 53: 1639b

C 394 Influence of polar molecules on the electronic spectra of acridone.
Kokubun, H.
Z. Physik. Chem. 17: 281-91 (1958)
CA 55: 2272d

C 395 Synthesis of polyarylenalkyls. IX. Synthesis and transarylation of monofluoro- and monochloro-diphenylmethanes.
Kolesnikov, G.S., Korshak, V.V., and Smirnova, T.V.
Izv. Akad. Nauk SSSR, Otd. Khim. Nauk 1958: 1123-6 (1958)

C 396 Effect of preliminary annealing on the heat luminescence of X-rayed sodium chloride crystals.
Kolevatykh, G.V., Polonskii, A.M., and Solononyok, R.E.
Nauchn. Zap., Odessk. Gos. Ped. Inst., Fiz.-Mat. Fak. 22: 29-42 (1958)
CA 55: 26601b

C 397 Fluorescent reactions with 8-quinolinol.
Korenman, I.M., and Avrova, N.F.
Tr. po Khim. i Khim. Tekhnol. 1: 138-43 (1958)
CA 54: 6393g

C 398 Fluorescent reactions of beryllium and aluminum.
Korenman, I.M., and Grishin, I.A.
Tr. po Khim. i Khim. Tekhnol. 1(2): 383-8 (1958)
CA 54: 15079c

C 399 Characteristics of fluorescein complexons.
Korbl, J., Vydra, F., and Pribil, R.
Talanta 1: 138-41 (1958)
CA 53: 5952i

C 400 Hydroxyanthraquinones as reagents for germanium.
Korenman, I.M., Kurina, N.V., and Emelin, E.A.
Tr. po Khim i Khim. Tekhnol. 1: 134-7 (1958)
CA 53: 19692e

C 401 Measurement of absorption and fluorescence spectra on paper chromatograms. Determination of specific activities of carbon-14-labeled compounds on chromatography paper.
Korte, F., and Weitkamp, H.
Angew. Chem. 70: 434-7 (1958)
CA 53: 12830b

C 402 Calculation of the influence of reabsorption in luminescence spectra of pheophytin and chlorophyll.
Kravtsov, L.A., and Ivanov, N.P.
Inzh.-Fiz. Zh., Akad. Nauk Belorussk. SSR 1958: 45-52 (1958)
CA 52: 11567d

C 403 Fluorescence yield of sodium salicylate in the ultraviolet and X-ray spectrum.
Krokowski, E.
Naturwissenschaften 45: 509-10 (1958)
CA 53: 7778a

C 404 The ultraviolet absorption and fluorescence spectra of various heterocyclic N-oxides.
Kubota, T., and Miyazaki, H.
Nippon Kagaku Zasshi 79: 916-37 (1958)
CA 53: 15760fhi

C 405 Synthesis of dienic acids from α-pyrones.
Kudryashov, L.I., and Kochetkov, N.K.
Zh. Obshch. Khim. 28: 2448-52 (1958)
CA 53: 3202h

C 406 Application of chromatography. XXXVI. Biosynthesis of riboflavine. 2. Riboflavine-synthesizing enzyme extracted from Eremothecium oshbyii.
Kuwada, S., Masuda, T., Kishi, T., and Osai, M.
Chem. Pharm. Bull. 6: 618-24 (1958)
CA 54: 24768d

C 407 Dependence of the quenching of fluorescence in solution by absorbing substances on the viscosity of the solvent.
Kuznetsova, L.A., and Sveshnikov, B.Y.
Opt. i Spektroskopiya 4: 55-9 (1958)
CA 52: 15246e

C 408 Influence of concentration on the absorption spectrum and the fluorescence spectrum of organic substances in solution.
Lavorel, J.
J. Chim. Phys. 55: 905-10 (1958)
CA 53: 21165f

C 409 Synthesis of 6-hydroxyfluoran, B-D-galactopyranoside; its use as substrate for B-galactosidase; design of a microfluorophotometer.
Lazo, J.S.
Anales. Fac. Quim. Farm., Univ. Chile 10: 158-65 (1958)
CA 54: 6707d

C 410 The synthesis of secondary phosphates and arsenates of uranium.
Leonova, E.N.
Tr. Inst. Geol. Rudn. Mestorozhd., Petrogr., Mineralog. i Geokhim. 1958 (30): 37-55 (1958)
CA 52: 21472g

C 411 Association of molecules of Rhodamine 6G and crystal violet in concentrated water solutions.
Levshin, L.V., and Suvorov, V.S.
Opt. i Spektroskopiya 4: 678-81 (1958)
CA 53: 2780d

C 412 β-γ correlations with resonance fluorescence.
Lewis, R.R., and Curtis, R.B.
Phys. Rev. 110: 910-14 (1958)
CA 52: 16939a

C 413 Reciprocal energy and reciprocal energy gradients between highly polar molecules and dipole migration.
Lippert, E.
Z. Physik. Chem. 6: 125-8 (1956); Chem. Zentr. 129: 2975 (1958)
CA 54: 1986d

C 414 The angular distribution of intensity and polarization ratio of the fluorescence of anthracene crystals.
Lipsett, F.R., and Tardiff, L.
Can. J. Phys. 36: 1438-41 (1958)
CA 53: 51d

C 415 Effect of the degree of dispersion of some fluorescent substances on the fluorescence intensity. II. Uranium in acetone-water mixtures.
Lucatu, E.
Comun. Acad. Rep. Populare Romine 8: 1021-6 (1958)
CA 52: 67a

C 416 Field emission from excited color centers.
Luty, F.
Z. Physik 153: 247-56 (1958)
CA 53: 2804a

C 417 The efficiency of latent image formation from the low-temperature luminescence of silver chloride.
Makishima, S., Tomotsu, T., and Hayakawa, S.
Photographic Sensitivity, Tokyo Symp., 125-53 (1958)
CA 52: 16100h

C 418 Electronic spectra of quinoline and isoquinoline and the mechanism of fluorescence quenching in these molecules.
Mataga, N.
Bull. Chem. Soc. Japan 31: 459-62 (1958)
CA 53: 1916g

C 419 Solvent effects on the fluorescence of some α-electron systems — naphthylamine and naphthol.
Mataga, N.
Bull. Chem. Soc. Japan 31: 481-6 (1958)
CA 54: 18066h

C 420 Fluorescence quenching due to the interaction between π-electron systems in the excited state and acid-base relation: nitrogen heterocycles, naphthylamine, and naphthol in aqueous and alcoholic solution.
Mataga, N.
Bull. Chem. Soc. Japan 31: 487-91 (1958)
CA 53: 8805d

C 421 Fluorescence of diamonds excited by X rays.
Matthews, I.G.
Proc. Phys. Soc. 72: 1074-80 (1958)
CA 54: 19175b

C 422 Yeast uridine diphosphogalactose-4-epimerase, correlation between activity and fluorescence.
Maxwell, E.S., de Robichon-Szulmajster, H., and Kalckar, H.M.
Arch. Biochem. Biophys. 78: 407-15 (1958)
CA 53: 8297c

C 423 Energy transfer in molecular complexes of sym-trinitrobenzene with polyacenes. I. General considerations.
McGlynn, S.P., and Boggus, J.D.
J. Am. Chem. Soc. 80: 5096-101 (1958)
CA 53: 3877b

C 424 Delayed light action spectra of several algae in visible and ultraviolet light.
McLeod, G.C.
J. Gen. Physiol. 42: 243-50 (1958)
CA 53: 8318i

C 425 Incorporation of the carbon 14 of adenine into a pteridine derivative by Eremothecium ashbyii.
McNutt, W.S., and Forrest, H.S.
J. Am. Chem. Soc. 80: 951-2 (1958)
CA 52: 9311f

C 426 Enhancement of phosphorescence ability upon aggregation of dye molecules.
McRae, E.G., and Kasha, M.
J. Chem. Phys. 28: 721-2 (1958)
CA 52: 11567f

C 427 Infrared fluorescence of copper-activated zinc sulfide phosphors.
Meijer, G.
Phys. Chem. Solids 7: 153-8 (1958)
CA 53: 6784g

C 428 The quenching of the fluorescence of anthracene.
Melhuisch, W.H., and Metcalf, W.S.
J. Chem. Soc. 1958: 480-2 (1958)
CA 52: 5124b

C 429 1,2,4-triazines. VIII. A new synthesis of the fluorescent 1,2,4-triazines.
Metze, R.
Chem. Ber. 91: 1863-6 (1958)
CA 53: 3237b

C 430 Luminescence measurements on various Acridine Orange derivatives.
Methke, E., and Zanker, V.
Z. Physik. Chem. 18: 375-90 (1958)
CA 55: 5121e

C 431 Fluorescence of tetracycline antibiotics in bone.
Milch, R.A., Rall, D.P., and Tobie, J.E.
J. Bone Joint Surg. 40A: 897-910 (1958)
CA 52: 17530e

C 432 The activation of CdI_2 by PbI_2.
Monod-Herzen, G.
Compt. Rend. 246: 2605-7 (1958)
CA 52: 19496e

C 433 Zinc oxide — a reactive pigment.
Morley-Smith, C.T.
J. Oil Colour Chemists' Assoc. 41: 85-97 (1958)
CA 53: 8658g

C 434 Phosphorescence spectrum of benzo-3, 4-pyrene in solution gelled at 180°.
Muel, B., and Hubert-Habart, M.
J. Chim. Phys. 55: 377-83 (1958)
CA 52: 19513d

C 435 Fluorescence spectrophotometry of crude oils.
Mukashev, Z.A., and Shmais, I.I.
Uch. Zap., Kazakhsk. Gos. Univ., Geol. i Geogr. 37: 140-4 (1958)
CA 53: 1281a

C 436 Photometric titration with photoelectric spectrophotometer.
Murakami, K., and Kimura, S.
Bunko Kenkyu 6(4): 36-41 (1958)
CA 53: 7681g

C 437 Fluorescent derivatives from the reaction of ethylenediamine with adrenaline and norepinephrine.
Nadeau, G., and Joly, L.P.
Nature 182: 180-1 (1958)
CA 53: 1638b

C 438 Fluorometric determination of sulfate ion and spectrophotometric determinations of thorium with the help of trihydroxyfluorene derivatives.
Nazarenko, V.A., and Shustova, M.B.
Zavodsk. Lab 24: 1344-6 (1958)
CA 54: 13985f

C 439 Fluorometry of vegetable oils.
Nemirovshy, B.
Rev. Fis. 10(3): 5-14 (1958)
CA 55: 4993c

C 440 Particularities of difurylpolyene fluorescence.
Neopchatykh, P.F.
Izv. Akad. Nauk SSSR, Ser. Fiz. 22: 1417-19 (1958)
CA 53: 2784b

C 441 Intensities in the spectra of polyatomic molecules.
Neporent, B.S., and Bakhshiev, N.G.
Opt. i Spektroskopiya 5: 634-45 (1958)
CA 53: 3867f

C 442 The confirmation of existence of persistent phosphorescence of calcium tungstate phosphor containing alkali or alkali-earth oxide impurities.
Nishikawa, K.
Kogyo Kagaku Zasshi 61: 266-7 (1958)
CA 53: 19587b

C 443 Fluorometric analysis. X. Fluorescence of metal quinolinolates and their adaptability in fluorometric analysis. XI. Fluorescence of metal salts of 8-quinolinol derivatives.
Nishikawa, N.
Nippon Kagaku Zasshi 79: 1003-7, 1007-10 (1958)
CA 53: 8917d

C 444 Fluorometric analysis. IX. Determination of gallium with 5,7-dihalo-8-quinolinol.
Nishikawa, Y.
Nippon Kagaku Zasshi 79: 631-7 (1958)
CA 53: 6896e

C 445 Fluorometric analysis. VII. Determination of gallium in silicate rocks with 8-hydroxyquinaldine.
Nishikawa, Y.
Nippon Kagaku Zasshi 79: 236-8 (1958)
CA 52: 13531c

C 446 Fluorimetric determination of gallium with Eriochrome Red B.
Nishikawa, Y.
Bunseki Kagaku 7: 549-53 (1958)
CA 54: 16277e

C 447 Identity tests for petroleum products. II.
Nojima, S., and Hiroshi, A.
Kagaku To Sosa 11(2): 59-66 (1958)
CA 53: 7563e

C 448 Application of fluorescence to microscopic observation of ore.
Okuda, T.
Suiyokaishi 13: 660 (1958)
CA 53: 12097f

C 449 Fluorometric identification of pyridine nucleotide changes in photosynthetic bacteria and algae.
Olson, J.M.
U.S. At. Energy Comm. BNL 512(C-28), 316-24 (1958)
CA 53: 22201d

C 450 Diffusion of activators in luminescent ZnS.
Ortmann, H.
Semicond. Phosphors, Proc. Intern. Colloq., Garmisch-Pertenkirchen 1956: 535-7 (1958)
CA 54: 10546d

C 451 Mechanism of reduction of nicotinamide
derivatives.
Paiss, Y., and Stein, G.
J. Chem. Soc. 1958: 2905-9 (1958)
CA 53: 921e

C 452 Fluorescence spectra of uranyl acetate
solutions.
Pant, D.D., and Khandelwal, D.P.
Current Sci. 27: 242 (1958)
CA 53: 5866f

C 453 Direct recording of fluorescence ex-
citation spectra.
Parker, C.A.
Nature 182: 1002-4 (1958)
CA 53: 4898i

C 454 Comparative electrical and optical
measurements on natural calcium
fluoride crystals.
Peibst, H., and Lemke, H.
Z. Physik. Chem. 208: 188-209 (1958)
CA 52: 13432g

C 455 Pteridines. VII. Methylations of hy-
droxpteridines.
Pfleiderer, W.
Chem. Ber. 91: 1671-80 (1958)
CA 53: 1365i

C 456 The possibility that several phosphor-
escent levels exist in organoluminofors.
Piliporvich, V.A., and Sveshnikov, B.Y.
Dokl. Akad. Nauk SSSR 119: 59-61 (1958)
CA 53: 2781d

C 457 Setup for detailed investigation of decay
curves of phosphorescence with dura-
tion period >0.1 second.
Pilipovich, V.A.
Opt. i Spektroskopiya 4: 116-18 (1958)
CA 52: 6924f

C 458 Decay law for the phosphorescence of
organoluminophors.
Pilipovich, V.A., and Sveshnikov, B.Y.
Opt. i Spektroskopiya 5: 290-6 (1958)
CA 53: 10994a

C 459 New data on uranium minerals in the
U.S.S.R.
Polikarpova, V.A., and Ambartsumyan,
Z.L.
Proc. U.N. Intern. Conf. Peaceful Uses
At. Energy, 2nd, Geneva, 1958 2: 286-
309 (1958)
CA 53: 2946e

C 460 The analysis of food dyes through
fluorescence.
Popovici, M.
Rev. Ind. Aliment., Produse Vegetale 1958:
7-8, 19-24 (1958)
CA 53: 11682h

C 461 Energy transfer from the triplet state
in solution.
Porter, G., and Wright, M.R.
J. Chim. Phys. 55: 705-12 (1958)
CA 53: 4899b

C 462 Autunite in the syenite of Biella.
Potenza, M.F.
Rend. Soc. Mineral. Ital. 14: 215-23 (1958)
CA 53: 2947c

C 463 Fluorescence of adsorbed water.
Przibram, K.
Nature 182: 520 (1958)
CA 53: 2813g

C 464 Phosphorescence spectra of some aro-
matic acids at liquid-air temperature.
Pyatnitskii, B.A.
Izv. Akad. Nauk SSSR, Ser. Fiz. 22: 1304-6
(1958)
CA 53: 2782b

C 465 Fluorescence reaction for the detection
of boric acid.
Raju, N.A., and Neelakantam, K.
Current Sci. (India) 27: 482 (1958)
CA 53: 12934h

C 466 Geochemistry of the fluorite occur-
rences of the Thuringian Forest.
Rentzsch, J.
Geologie 7: 924-34 (1958)
CA 53: 6929f

C 467 Semiquantitative determination of co-
proporphyrin in urine.
Ribeiro, B.A., and Stettiner, H.M.A.
Arquiv. Fac. Hig. Saude Publica Univ. Sao
Paulo 12: 165-80 (1958)
CA 54: 12244b

C 468 Anatomical and fluorescence-optical
investigations of seeds of Popaveraceae.
Roder, I.
Oesterr. Botan. Z. 104: 370-81 (1958)
CA 53: 8315f

C 469 Natural antioxidants in the leaves of
         olive trees.
      Roncero, A.V., and Vela, F.M.
      Grosas y Aceites 8: 247-9 (1958)
      CA 52: 12090h

C 470 Interaction of riboflavine, flavine mono-
         nucleotide, and flavine adenine di-
         nucleotide with various metal ions:
         riboflavine-catalyzed photochemical
         reduction of ferric ion and photooxida-
         tion of ferrous ion.
      Rutter, W.J.
      Acta Chem. Scand. 12: 438-46 (1958)
      CA 53: 14165e

C 471 Fluorescent lamp.
      Schwing, J., and Schiazzano, G.
      U.S. 2,838,707, June 10, 1958
      CA 52: 15270b

C 472 Energy transfer in organic crystals.
      Schmillen, A.
      Semicond. Phosphors, Proc. Intern.
         Colloq., Garmisch-Partenkirchen 1956:
         445-50 (1958)
      CA 54: 23774g

C 473 Energy transfer in polyacene solid
         solutions.
      Schmillen, A.
      Z. Physik 150: 123-33 (1958)
      CA 53: 21205c

C 474 Experiments with 4-thiopyrones and
         with 2,2',6,6'-tetraphenyl-4,4'-dipyryl-
         ene. The piezochromism of diflavylene.
      Schonberg, A., Elkaschef, M., Nosseir, M.,
         and Sidky, M.M.
      J. Am. Chem. Soc. 80: 6312-15 (1958)
      CA 53: 11359d

C 475 Influence of the sulfonic acid group on
         chemical reactions. V. Desulfonability
         and fluorescence colors of aminonaph-
         thalenesulfonic acid as a function of the
         sulfonic acid group position.
      Schriever, K., Bamann, E., and Kraus, C.
      Chem. Ber. 91: 414-7 (1958)
      CA 52: 18331e

C 476 Fluorescence of extracts from Cheli-
         donium majus.
      Semenova, M.N.
      Aptechn. Delo 7: 26-7 (1958)
      CA 53: 20698h

C 477 Fluorimetric determination of telluri-
         um with Rhodamine.
      Shcherbov, C.D.P., and Irankova, A.I.
      Zavods. Lab 24: 1346-9 (1958)
      CA 54: 13979f

C 478 Fluorometric determination of gallium
         in germanium.
      Shigematsu, T.
      Bunseki Kagaku 7: 787-8 (1958)
      CA 53: 19693h

C 479 Nature of the inactive absorption during
         anti-Stokes excitation of fluorescence.
      Shirokov, V.I.
      Opt. i Spektroskopiya 5: 478-9 (1958)
      CA 53: 1918d

C 480 Elementary processes in the upper at-
         mosphere and their manifestation in
         emissions.
      Shklovski, I.S.
      Ann. Geophys. 14: 414-24 (1958)
      CA 53: 4892a

C 481 Simultaneous observation of the Raman
         effect and fluorescence.
      Shorygin, P.P., and Ivanova, T.M.
      Dokl. Akad. Nauk SSSR 121: 70-3 (1958)
      CA 54: 23785c

C 482 Luminescence and absorption of pyrene
         and 3,4-benzopyrene in frozen solutions
         of normal paraffins.
      Shpol'skii, E.V., and Girdzhiyanskaite,
         E.A.
      Opt. i Spektroskopiya 4: 620-30 (1958)
      CA 53: 27811

C 483 Diimidazole derivatives.
      Siegrist, A.E., and Ackermann, F.
      Swiss 332,807 (1958)
      CA 53: 576g

C 484 Some color and fluorescence detection
         reactions of steroids.
      Siblikova, O., and Hais, I.M.
      Cesk. Farm. 7: 1-13 (1958)
      CA 52: 10495f

C 485 Spin-orbit coupling in the $^3A_2$-$^1A_1$
         transition of formaldehyde.
      Sidman, J.W.
      J. Chem. Phys. 29: 644-52 (1958)
      CA 53: 2765h

C 486 Fluorescent color effects.
Siegrist, A.E.
U.S. 2,837,485 (1958)
CA 52: 12576c

C 487 The possibility of occurrence of fluor-
escence in molecules.
Simon, Z.
Acad. Rep. Populare Romine, Studii
Cercetari Fiz. 9: 469-81 (1958)
CA 54: 14935h

C 488 Oxidation of aromatic amines. VI.
Persulfate oxidation of carcinogenic
aromatic amines.
Sims, P.
J. Chem. Soc. 1958: 44-7 (1958)
CA 52: 10979g

C 489 Photosorption of oxygen on silica gel
and crystalline quartz.
Solonitsyn, Y.P.
Zh. Fiz. Khim. 32: 1241-7 (1958)
CA 52: 16898e

C 490 Delayed fluorescence in naphthalene
crystals at 4°K.
Sponer, H., Kanda, Y., and Blackwell, L.A.
J. Chem. Phys. 29: 721 (1958)
CA 53: 2781c

C 491 Spectral changes accompanying binding
of Acridine Orange by polyadenylic
acid.
Steiner, R.F., and Beers, R.F.
Science 127: 335-6 (1958)
CA 52: 10717f

C 492 Some effects of intramolecular vibra-
tional energy transfer in complex
fluorescent molecules.
Stevens, B.
Can. J. Chem. 36: 96-101 (1958)
CA 52: 19442d

C 493 Notes on a blue fluorescent substance
in urine.
Stevens, D.S.
Naturwissenschaften 45: 212 (1958)
CA 52: 16532g

C 494 Structure and tautomerism of the esters
of several substituted pyruvic acids.
Stock, A.M., Donahue, W.E., and Amstutz,
E.D.
J. Org. Chem. 23: 1840-8 (1958)
CA 53: 10219a

C 495 Investigations of fluorescence of tissue
cultures. I. Metachromic vital fluor-
escence with Acridine Orange.
Stockinger, L.
Z. Naturforsch. 13: 407-9 (1958)
CA 52: 20354a

C 496 Fluorescent decay of CsI (Tl) for part-
icles of different ionization density.
Storey, R.S., Jack, W., and Ward, A.
Proc. Phys. Soc. 72: 1-8 (1958)
CA 54: 23963c

C 497 Reactions in the chlorophyll series.
Glaucorhodine.
Strell, M., Kalojanoff, A., and Koniger, M.
Ann. Chem. 614: 205-11 (1958)
CA 53: 4296b

C 498 Photochemical behavior of alkali-
organic addition compounds, dissolved
in organic liquids. II. Increase in the
electrical conductivity on irradiation.
Suhrmann, R., and Matejec, R.
Z. Physik. Chem. 263-75 (1958)
CA 52: 8752d

C 499 Diffusion theory for the kinetics of bi-
molecular reactions in solutions.
Sveshnikov, B.Y., and Kuznetsova, L.A.
Dokl. Akad. Nauk SSSR 121: 1045-7 (1958)
CA 54: 23643a

C 500 Kinetics of fluorescence quenching of
solutions with foreign substances.
Sveshnikov, B.Y., Shirokov, V.M.,
Kuznetsova, L.A., and Kodryashov, P.I.
Izv. Akad. Nauk SSSR, Ser. Fiz. 22: 1047-
50 (1958)
CA 53: 856a

C 501 Dependence of the long afterglow polarization of organic substances on the
viscosity of the solution.
Sveshnikov, B.Y., and Kudryashov, P.I.
Izv. Akad. Nauk SSSR, Ser. Fiz. 22: 1403-6
(1958)
CA 53: 2783c

C 502 Mechanical printing of daylight fluores-
cent compositions.
Switzer, J.L., and Switzer, R.C.
U.S. 2,845,023, July 29, 1958
CA 52: 17753d

C 503 Washing of powdered material with a
      minimum amount of liquid.
      Tarbes, P.
      Fr. 1,163,806 (1958)
      CA 54: 19050e

C 504 Nonenzymic browning.
      Telegdy-Kovats, L., and Rajky, A.
      Nahrung 2: 893-909 (1958)
      CA 53: 20856i

C 505 Indole. XI. Conversion of heteroauxin
      (β-phenylethylamide) into a pyrrocoline
      derivative with polyphosphoric acid.
      Thesing, J., and Funk, F.H.
      Chem. Ber. 91: 1546-51 (1958)
      CA 53: 368c

C 506 9-Phenanthrylboric acid, a new lumin-
      escent organoboron compound.
      Thielens, G.
      Naturwissenschaften 45: 543 (1958)
      CA 53: 10148i

C 507 An antibiotic produced by a Bacillus
      sp. active against hemophilus pertussis
      (Bordetella pertussis).
      Tiffin, A.I.
      Nature 181: 907-8 (1958)
      CA 52: 16471i

C 508 Dependence of the fluorescence spectra
      of solutions of meso-substituted anthra-
      cene on the concentration of a dissolved
      substance.
      Tishchenko, G.A., Sveshnikov, B.Y., and
      Cherkasov, A.S.
      Opt. i Spektroskopiya 4: 631-6 (1958)
      CA 53: 2782g

C 509 The applications of fluorescence spec-
      tra.
      Titeica, R.
      Analele Acad. Rep. Populare Romine, Vol.
      VII, Anexa. Prima Consfatuire Tara
      Spectroscopie Apl., Bucharest 1957:
      77-92 (1958)
      CA 53: 13763c

C 510 Formal analysis of the theory of two-
      step excitation of phosphorescence and
      photoconductivity. II. Relaxation rela-
      tions.
      Tolstoi, N.A., and Shatilov, A.V.
      Opt. i Spektroskopiya 5: 590-600 (1958)
      CA 53: 2816g

C 511 Separation of uranium from thorium by
      means of ion exchange and the fluori-
      metric determination of uranium.
      Tomic, E., Ladenbauer, I.M., and Pollack,
      M.
      Z. Anal. Chem. 161: 28-38 (1958)
      CA 53: 129c

C 512 Activation of the fluorescence by poly-
      mers and photooxidation-reduction.
      Tomita, G., and Takeyama, H.
      Kagaku 28: 528 (1958)
      CA 53: 6755e

C 513 Phosphoryl chloride enhancement of
      fluorescence of steroids in sulfuric
      acid.
      Touchstone, J.C., Keisman, R.A.,
      Marcantonis, A.F., and Greene, J.W.
      Anal. Chem. 30: 1707 (1958)
      CA 53: 3346b

C 514 Acridine derivatives. I. Synthesis of
      1-nitro-2-methoxy-9-chloroacridine.
      Troshchenko, A.T.
      Zh. Obshch. Khim. 28: 2207-13 (1958)
      CA 53: 2230h

C 515 Oxidation of lysergic acid derivatives
      in the 2,3-position.
      Troxler, F.
      Planta Med. 6: 399-401 (1958)
      CA 53: 16105g

C 516 Pteridines. XV. Improved synthesis
      and constitution of erythropterin.
      Tschesche, R., and Ende, H.
      Chem. Ber. 91: 2074-81 (1958)
      CA 53: 4289g

C 517 Quantum efficiency of F-center fluores-
      cence in potassium chloride.
      van Doorn, C.Z.
      Philips Res. Rept. 13: 296-300 (1958)
      CA 52: 15258i

C 518 Fluorescence and photoconduction of
      silver-activated cadmium sulfide.
      Van gool, W.
      Philips Res. Rept. 13: 157-66 (1958)
      CA 52: 11578g

C 519 Marking of proteins with fluorescent
      dye.
      Vehleke, H.
      Naturwissenschaften 45: 87 (1958)
      CA 52: 12045c

C 520 Fluorescence spectra and polarization
of glyceraldehyde-3-phosphate and
lactic dehydrogenase coenzyme complexes.
Velick, S.F.
J. Biol. Chem. 233: 1455-67 (1958)
CA 53: 6754i

C 521 Fluorescent substances from Drosophila melanogaster. VII. Synthesis and
properties of L-erythro- and D-threo-
2-amino-G-hydroxy-8(dihydroxypropyl)
pteridine.
Viscontini, M., and Raschig, H.
Helv. Chim. Acta 41: 108-13 (1958)
CA 52: 13738b

C 522 The polarization of the fluorescence of
conjugated polyethylenimine.
Wahl, P.
J. Polymer Sci. 29: 375-80 (1958)
CA 52: 19444g

C 523 Fluorescence excitation spectrum of
organic compounds in solution. I. Systems with quantum yield independent of
the exciting wavelength.
Weber, G., and Teale, F.W.J.
Trans. Faraday Soc. 54: 640-8 (1958)
CA 53: 1917e

C 524 Kinetics of fluorescence conversion.
Weller, A.
Z. Physik. Chem. 15: 438-53 (1958)
CA 52: 11590d

C 525 The investigation of fast reactions of
excited molecules on the basis of
fluorescence conversion and quenching.
Weller, A.
Z. Physik. Chem. 18: 163-80 (1958)
CA 55: 6112i

C 526 Fluorometric analysis.
White, C.E.
Anal. Chem. 30: 729-34 (1958)
CA 52: 9847b

C 527 Mechanism of formate activation.
Whiteley, H.R., Osborn, M.J., and
Huennekens, F.M.
J. Am. Chem. Soc. 80: 757-8 (1958)
CA 52: 9311h

C 528 Fluorescence activation spectra of a
diphosphopyridine nucleotide-dependent
dehydrogenase.
Winer, A.D., and Schwert, G.W.
Science 128: 660-1 (1958)
CA 53: 3320a

C 529 Fluorescence spectra of ternary complexes of dehydrogenases with reduced
diphosphopyridine nucleotide (DPNH)
and reduced substrates.
Winer, A.D., and Schwert, G.W.
Biochim. Biophys. Acta 29: 424-25 (1958)
CA 53: 469i

C 530 Synthesis of 2-azaindolizidine and
2,3-substituted derivatives.
Winterfeld, K., and Schuler, H.
Naturwissenschaften 45: 492 (1958)
CA 53: 16133h

C 531 The lowest electronic excitation state
of anthracene crystals.
Wolf, H.C.
Z. Naturforsch. 13: 414-19 (1958)
CA 52: 19495g

C 532 Fluorescence and dyeing characteristics of fluorescent whitening dyes of the
triazinylstilbene series. III. Combined
derivatives of different amino or hydroxyl groups.
Yabe, A., and Hayashi, M.
Kogyo Kagaku Zasshi 61: 78-82 (1958)
CA 53: 16542fi

C 533 Determination of fluorescence of fla-
vines on paper strips.
Yagi, K., and Okuda, J.
Chem. Pharm. Bull. 6: 659-62 (1958)
CA 54: 16741d

C 534 Fluorescence of naphthalimide derivatives. Consideration of the mirror-
image relation between fluorescence
and absorption spectra.
Yasuda, K., Inukai, K., and Ito, K.
Nippon Kagaku Zasshi 79: 897-9 (1958)
CA 53: 48f

C 535 Advances in applications of organic
fluorescent compounds.
Yoshida, Z.
Kagaku No Ryoiki 12: 198-204 (1958)
CA 52: 18959h

C 536 Measurements of the relative quantum yields of the fluorescence of acridine and fluorescein dyes.
Zanker, V., Rammensee, H., and Haibach, T.
Z. Angew. Phys. 10: 357-61 (1958)
CA 52: 19461b

C 537 Effect of temperature on the fluorescence spectra of phthalimide derivatives.
Zelinskii, V.V., and Kolobkov, V.P.
Opt. i Spektroskopiya 5: 423-7 (1958)
CA 53: 10955i

C 538 Ratio of phosphorescence and fluorescence quantum yields and the fluorescence band frequency.
Zelinskii, V.V., and Kolobkov, V.P.
Soviet Phys. "Doklady" 3: 361-4 (1958)
CA 53: 18626e

C 539 Effect of the structure of organic molecules on the probability of a transition into the metastable state.
Zelinskii, V.V., Kolobkov, V.P., and Reznikova, I.I.
Dokl. Akad. Nauk SSSR 121: 315-8 (1958)
CA 54: 23512g

C 540 See C910a

C 541 Procedure for the study of X-ray fluorescence spectra of alkali-halide crystal phosphors in the visible spectrum.
Zolotavev, G.K.
Uch. Zap., Odessk. Gos. Ped. Inst., Fiz.-Mat. Fak. 22: 67-70 (1958)
CA 55: 25454e

C 542 Electronic spectrum of 4,4'-bis(dimethyl-amino)fuchsone and related triphenylmethane dyes.
Adam, F.C., and Simpson, W.T.
J. Mol. Spectry. 3: 363-80 (1959)
CA 55: 9043f

C 543 The absorption and fluorescence of thin layers of anthracene.
Agubiceanu, I., and Gheorghita-Oancea, C.
Comun. Acad. Rep. Populare Romine 9: 545-9 (1959)
CA 54: 5262i

C 544 Luminescence of dyes absorbed on certain semiconductors.
Akimov, I.A.
Zh. Nauchn. i Prikl. Fotogr. i Kinematog. 4: 64-6 (1959)
CA 53: 15761f

C 545 The fluorescence of (Zn,Co) S phosphors.
Albers, K.
Monatsber. Deut. Akad. Wiss. Berlin 1: 93-5 (1959)
CA 54: 8308i

C 546 Quenching effects and the afterglow of chlorophyll.
Albrecht, A.O., Denison, W.C., Livingston, L.G., and Mandeville, C.E.
J. Franklin Inst. 268: 278-82 (1959)
CA 55: 15116d

C 547 Why a limit to optical bleach fluorescence?
Allen, E.
Soap Chem. Specialties 35: 51-3 (1959)
CA 53: 16543h

C 548 Effects of added gases on the intensities of thallium-sensitized fluorescence.
Anderson, R.A.
Univ. Microfilms Mic 59-1520 (1959)
CA 53: 18621i

C 549 Naphthalene derivatives in inorganic analysis. VII. Nitronaphthols as fluorometric reagents for stannous tin.
Anderson, J.R.A., Costoulas, A.J., and Garnett, J.L.
Anal. Chim. Acta 20: 236-42 (1959)
CA 54: 2090h

C 550 Action of phenylhydrazine on Mannich bases from benzyl-ideneacetone.
Andrisano, R., and Chieriei, L.
Gazz. Chim. Ital. 89: 505-16 (1959)
CA 54: 12116e

C 551 Analysis and recovery of uranium from low-grade ores.
Arden, T.V.
At. Energy Res. Estab., Rept. R2862, 41 pp. (1959)
CA 53: 21422h

C 552 Differential fluorescence in living rat
eggs treated with Acridine Orange.
Austin, C.R., and Bishop, M.W.H.
Exptl. Cell Res. 17: 35-43 (1959)
CA 55: 2755f

C 553 Effect of Al on the green emission of
CdS.
Avinor, M.
Physica 25: 1095-6 (1959)
CA 54: 20516g

C 554 The metastable state of dye molecules.
Baczynski, A., and Czajhowski, M.
Bull. Acad. Polon. Sci., Ser. Sci., Math.,
Astron. Phys. 7: 357-60 (1959)
CA 54: 10509e

C 555 Crystal structure of hurlbutite.
Bakakin, V.V., and Belov, N.V.
Dokl. Akad. Nauk SSSR 125: 383-5 (1959)
CA 53: 19710i

C 556 The internal field and the position of the
electronic absorption and emission
bands of polyatomic organic molecules
in solution.
Bakhshiev, N.G.
Opt. i Spektroskopiya 7: 52-61 (1959)
CA 54: 23815e

C 557 Rays of fluorescent emission and of
light absorption in pure cadmium sul-
fide crystals frozen to 4.2°K.
Bancie-Grillot, M., Gross, E.F., Grillot,
E., and Razbirine, B.S.
Compt. Rend. 248: 86-9 (1959)
CA 53: 12013d

C 558 The ultraviolet fluorescence of guanine
in solution and on paper chromatograms.
Barskii, I.Y.
Biokhimiya 24: 823-5 (1959)
CA 54: 15505c

C 559 Demonstration of chloroquine or chlor-
oquine derivatives in tissues.
Baumer, A., Par, H., and Conrads, H.
Z. Rheumaforsch. 18: 433-40 (1959)
CA 54: 6959a

C 560 A new type of fluorometer. Measure-
ments of decay periods of fluorescence
of Acridine Yellow solutions as a func-
tion of concentration.
Bauer, R., and Rozwadowski, M.
Bull. Acad. Polon. Sci., Ser. Sci., Math.,
Astron. Phys. 7: 365-8 (1959)
CA 54: 10509d

C 561 Spectrophotofluorometric assay of
griseofulvin.
Bedford, C., Child, K.J., and Tomick, E.G.
Nature 184, Suppl. No. 6, 364-5 (1959)
CA 54: 15512g

C 562 Effect of electron irradiation on natural
organic substances in the electron
microscope and electron diffractograph.
Belavtseva, E.M.
Kristallografiya 4: 421-2 (1959)
CA 57: 4134d

C 563 Delayed luminescence of dibenzyl and
diphenylamine crystals for photo- and
β-excitation.
Belikova, T.P.
Opt. i Spektroskopiya 6: 117-18 (1959)
CA 53: 12856e

C 564 New method for the study of the chemi-
luminescence at simultaneous fluores-
cence.
Bersis, D.S.
Z. Physik. Chem. 22: 328-35 (1959)
CA 54: 7335i

C 565 Reaction of cis- and trans-stilbene-2-
carboxylic acids with peroxy acids.
Berti, G.
J. Org. Chem. 24: 934-8 (1959)
CA 54: 22609e

C 566 Contact photocopying of fluorescent
fractions in ultraviolet light. (A photographic technique for the detection of
fluorescent light.)
Betke, K., Clotten, R., and Schiebe, G.
Klin. Wochschr. 37: 403 (1959)
CA 54: 11128e

C 567 Estrogenic factor.
Bickoff, E.M., and Booth, A.N.
U.S. 2,890,116 (1959)
CA 53: 1554a

C 568 Sulfur compounds in the kerosine boil-
ing range of Middle East distillates.
Occurrence of a bicyclic thiophene and
a thienyl sulfide.
Birch, S.F., Cullum, T.V., Dean, R.A., and
Redford, D.G.
Tetrahedron 7: 311-18 (1959)
CA 54: 7678f

C 569 Crystal fluorescence of carcinogens and related organic compounds.
Birks, J.B., and Cameron, A.J.W.
Proc. Roy. Soc. (London), Ser. A 249: 297-317 (1959)
CA 54: 17045e

C 570 Energy transfer in fluorescent plastic solutions.
Birks, J.B., and Kuchela, K.N.
Discussions Faraday Soc. No. 27: 57-63 (1959)
CA 54: 14936d

C 571 Polarization of fluorescence in cadmium sulfide and zinc sulfide single crystals.
Birman, J.L.
Phys. Rev. Letters 2: 157-9 (1959)
CA 53: 8826g

C 572 Fluorescence and photochemical action in uranyl nitrate solution.
Bist, H.D.
J. Sci. Ind. Res. 18B: 387-8 (1959)
CA 54: 18065h

C 573 The glow discharge in a mixture of butane and bromine.
Bodarea, E., and Popovici, C.
Comun. Acad. Rep. Populare Romine 9: 1249-56 (1959)
CA 54: 21952g

C 574 Determination of the critical molecule distance for concentration depolarization of fluorescence.
Bojarski, L., and Kawsky, A.
Ann. Physik 5: 31-4 (1959)
CA 53: 4147e

C 575 Fluorescence spectra of frozen crystalline solutions of simple aromatic hydrocarbons.
Bolotnikova, T.N.
Izv. Akad. Nauk SSSR, Ser. Fiz. 23: 29-31 (1959)
CA 53: 11988d

C 576 Interpretation of the fluorescence spectrum of naphthalene.
Bolotnikova, T.N.
Opt. i Spektroskopiya 7: 44-51 (1959)
CA 54: 23815d

C 577 Spectroscopy of some simple aromatic hydrocarbons in frozen crystalline solutions.
Bolotnikova, T.N.
Opt. i Spektroskopiya 7: 217-22 (1959)
CA 54: 8276e

C 578 Experimental sample of a phase fluorometer.
Bonch-Bruevich, A.M., Karazin, V.A., Molchanov, V.A., and Shirokov, V.I.
Pribory i Tekhn. Eksperim. 1959: 53-6 (1959)
CA 53: 21142g

C 579 A new type of fluorometer.
Borisov, A.Y., and Tumerman, L.A.
Izv. Akad. Nauk SSSR, Ser. Fiz. 23: 97-101 (1959)
CA 53: 11990d

C 580 Fluorescence decay of thallium-activated inorganic scintillation crystals with particles of various ionization density.
Bormann, M., Anderson-Lindstrom, G., Neuert, H., and Pollehn, H.
Z. Naturforsch. 14a: 681-2 (1959)
CA 53: 21200c

C 581 The effect of temperature on fluorescence of solutions.
Bowen, E.J., and Sahu, J.
J. Phys. Chem. 63: 4-7 (1959)
CA 53: 7761a

C 582 Viscosity and temperature effects in fluorescence.
Bowen, E.J.
Discussions Faraday Soc. No. 27, 40-2 (1959)
CA 54: 14930e

C 583 The effect of viscosity on the fluorescence yield of solutions.
Bowen, E.J., and Miskin, S.F.A.
J. Chem. Soc. 1959: 3172-3 (1959)
CA 54: 16140d

C 584 The relation between the fluorescence and the structure of luminescent indicators and reagents.
Bozhevol'nov, E.A.
Tr. Vses. Nauchn.-Issled. Inst. Khim. Reaktivov No. 23, 147-65 (1959)
CA 54: 23723b

C 585 Determination of zinc and cadmium.
Bozhevol'nov, E.A., Dziomko, V.M., and Serebyakova, G.U.
U.S.S.R. 120,029 (1959)
CA 54: 20676a

C 586 Solar Lyman-β fluorescence mechanism in the upper atmosphere.
Brandt, J.C.
Astrophys. J. 130: 228-40 (1959)
CA 53: 21135f

C 587 Hydroxytetracenequinones.
Brockman, H., and Muller, W.
Chem. Ber. 92: 1164-70 (1959)
CA 53: 17076i

C 588 Medium with negative absorption coefficient.
Butaeva, F.A., and Fabukant, V.A.
Issled. po Eksperim. i Teor. Fiz., Akad. Nauk SSSR, Fiz. Inst. 1959: 62-70 (1959)
CA 53: 1056g

C 589 Solution techniques in fluorescent X-ray spectrography.
Campbell, W.J., Leon, M., and Thatcher, J.W.
U.S. Bur. Mines Rept. Invest. No. 5497, 24 pp. (1959)
CA 53: 21369i

C 590 Preparation of 6H-6-oxo-5,10,11-triozaphenanthrene.
Carboni, S., and Pardi, M.
Ann. Chim. 49: 1220-7 (1959)
CA 54: 8831i

C 591 Electronic quenching of OH $(^2\Sigma^+)$ in flames and its significance in the interpretation of rotational relaxation.
Carrington, T.
J. Chem. Phys. 30: 1087-95 (1959)
CA 53: 16687b

C 592 Rotational transfer in the fluorescence spectrum of OH $(^2\Sigma^+)$.
Carrington, T.
J. Chem. Phys. 31: 1418-19 (1959)
CA 54: 7334i

C 593 Fluorometer CISE-CR1, a device for determinations on solids and liquids.
Cerrai, E., and Rossi, G.
Energia Nucl. (Milan) 6: 399-408 (1959)
CA 54: 1942e

C 594 Changes in fluorescence in a frog sartorius muscle following a twitch.
Chance, B., and Jobsis, F.
Nature 184, Suppl. No. 4, 195-6 (1959)
CA 54: 6982d

C 595 Differential microfluorimeter for the localization of reduced pyridine nucleotide in living cells.
Chance, B., and Legallais, V.
Rev. Sci. Instr. 30: 732-5 (1959)
CA 54: 13757e

C 596 Fluorescence measurements of mitochondrial pyridine nucleotide in aerobiosis and anaerobiosis.
Chance, B., and Thorell, B.
Nature 184: 931-4 (1959)
CA 54: 13218a

C 597 Synthesis of furano compounds. XVII. Synthesis of coumestrol.
Chatterjea, J.N.
J. Indian Chem. Soc. 36: 254-6 (1959)
CA 54: 10987i

C 598 Concentration quenching of fluorescence in solutions.
Chandhuri, K.D.
Z. Physik 154: 34-42 (1959)
CA 53: 5866a

C 599 Reactions of hexamethylenetetramine.
Checchi, S.
Gazz. Chim. Ital. 89: 2151-62 (1959)
CA 55: 5499e

C 600 Fluorescence of some salicyloyl hydrazones.
Chen, P.S.
Anal. Chem. 31: 296-8 (1959)
CA 53: 12923h

C 601 The effect of p-toluidene on the quantum yield of the photooxidation and the photodimerization of some anthracene derivatives.
Cherkasov, A.S., and Vember, T.M.
Opt. i Spektroskopiya 7: 321-5 (1959)
CA 54: 8237a

C 602 Absorption and fluorescence spectra and fluorescence quantum yield of some methyl- and methylmesoarylanthracenes.
Cherkasov, A.S.
Opt. i Spektroskopiya 7: 326-31 (1959)
CA 54: 8277e

C 603 Effect of subsituting groups on the position of absorption and fluorescence spectra of anthracene derivatives.
Cherkasov, A.S.
Opt. i Spektroskopiya 6: 496-502 (1959)
CA 55: 10056c

C 604 Influence of substituting groups on the position of the absorption and fluorescence spectra of derivatives of anthracene.
Cherkasov, A.S.
Opt. Spectry. 6: 315-18 (1959)
CA 55: 15115a

C 605 Effect of oxygen on the photochemical transformation and the concentration quenching of fluorescence of some anthracene derivatives.
Cherkasov, A.S., and Vember, T.M.
Opt. i Spektroskopiya 6: 503-11 (1959)
CA 55: 11076c

C 606 Effect of the conjugating of the anthracene nucleus with the double bind of an alkenyl substituent on the fluorescence and absorption spectra.
Cherkasov, A.S.
Dokl. Akad. Nauk SSSR 125: 848-51 (1959)
CA 55: 10068f

C 607 Transpositions in the series of the 5-hydroxyflavanones. I. 5,6,7- and 5,7,8-substituted flavanones.
Chopin, J., Chadenson, M., Grenier, G., and Bouillant, M.
Bull. Soc. Chim. France 1959: 1585-96 (1959)
CA 54: 11009b

C 608 Transformation of trans-o-hydroxy-cinnamic acids into the corresponding coumarins.
Cingolani, E.
Gaz. Chim. Ital. 89: 999-1008 (1959)
CA 54: 22617d

C 609 Effects of alternating and continuous electric fields on the luminescence of certain zinc sulfides excited by α-radiation.
Coche, A., and Henck, R.
J. Phys. Radium 20: 827-9 (1959)
CA 54: 2974i

C 610 α-Cyanostyrene.
Colonge, J., Dreux, J., and Regeand, J.P.
Bull. Soc. Chim. France 1959: 1244-7 (1959)
CA 54: 6628c

C 611 Green fluorescence of guanidinium compounds with ninhydrin.
Conn, R.B., and Davis, R.B.
Nature 183: 1053-5 (1959)
CA 53: 15854h

C 612 Synthesis of 2,3:6,7-dibenzodiphenylene.
Curtis, R.F., and Viswanath, G.
J. Chem. Soc. 1959: 1670-6 (1959)
CA 53: 21845g

C 613 Intermolecular bonds by resonance mesomery. XI. Fluorescence spectra, reflection spectra, and fluorescence relaxation times in crystalline molecular compounds.
Czekalla, J., Schmillen, A., and Mager, K.J.
Z. Elektrochem. 63: 623-6 (1959)
CA 53: 21148f

C 614 Correction: Phosphorescence spectra and relaxation times of aromatic hydrocarbons and their donor-acceptor complexes.
Czekalla, J., Briegleb, G., Herre, W., and Vahlensieck, H.J.
Z. Elektrochem. 63: 1197 (1959)
CA 53: 5243f

C 615 Luminous spots on electrodes in insulating oil gaps.
Dakin, T.W., and Berg, D.
Nature 184: 120 (1959)
CA 54: 2047h

C 616 Theory of the temporal growth of ionization in gases involving the action of metastable atoms and trapped radiation.
Davidson, P.M.
Proc. Roy. Soc. (London), Ser. A 249: 237-47 (1959)
CA 54: 19166d

C 617 Spectrophotometry of the edible oils. II.
DeFrancesco, F.
Olii Minerali, Grassi Saponi, Colori Vernici 36: 73-6 (1959)
CA 53: 22587f

C 618 The use of fluorescence microspectro-
graphy in histochemistry.
DeLerma, B.
Compt. Rend. Assoc. Anatomistes 103:
523-32 (1959)
CA 59: 5977h

C 619 Absorption and fluorescence spectra of
uranium salts and other solids; spectra
of molecules containing tritium. IX.
Dieke, G.H.
U.S. At. Energy Comm. NYO-8090, 12 pp.
(1959)
CA 57: 13308b

C 620 A contribution to the method of fluoro-
photometric determination of adrenaline
and noradrenaline in serum.
Dienstbier, E., and Balik, J.
Casopis Lekaru Ceskych 98: 16-20 (1959)
CA 54: 8974f

C 621 Simple experimental method to determine the intensity of secondary fluorescence.
Dombi, J., Hevesi, J., and Horvai, R.
Acta Univ. Szeged., Acta Phys. Chem. 5:
20-5 (1959)
CA 54: 138853i

C 622 Quenching of excited metal atoms. I.
Excited thallium atoms produced by the
photolysis of thallous iodide vapor,
using an a.c. spark source.
Dowling, D.J., and Warhurst, E.
Trans. Faraday Soc. 55: 532-6 (1959)
CA 54: 2934h

C 623 Quenching of excited metal atoms. II.
Excited thallium and sodium atoms
produced by photodissociation using a
single spark technique.
Dowling, D.J., Jones, G.R.H., and
Warhurst, E.
Trans. Faraday Soc. 55: 537-43 (1959)
CA 54: 2935a

C 624 Fluorescent impurities in liquid para-
ffin and organic solvents.
Druckrey, H., Schmahl, D., and
Preussmann, R.
Arzneimittel-Forsch. 9: 600-4 (1959)
CA 54: 3058g

C 625 Photoluminescence of phthalic and
benzoic acid esters.
Dubinskii, I.B.
Izv. Akad. Nauk SSSR, Ser. Fiz. 23: 116-18
(1959)
CA 53: 11990f

C 626 The sensitized fluorescence of 2-
naphthylamine. A study in transfer of
electronic energy.
Dubois, J.T.
J. Phys. Chem. 63: 8-11 (1959)
CA 53: 7761c

C 627 2,5-bis (Benzimedazol-2-yl)thiophenes.
Duennenberger, M., Siegrist, A.E., and
Maeder, E.
Ger. 1,109,177 (1959)
CA 53: 10157g

C 628 Evolution of the spectra of Nova RS
Ophiuchi after the 1958 outburst.
Dufay, J., Block, M., Bertand, C., and
Dufay, M.
Compt. Rend. 249: 631-3 (1959)
CA 54: 2927e

C 629 Decay of the phosphorescence of ce-
ment and organic oxide phosphors.
Dvorovenko, V.K.
Izv. Akad. Nauk SSSR, Ser. Fiz. 23: 139-41
(1959)
CA 53: 11991e

C 630 Structure of erosinin (Norton and Hans-
berry's "Compound I").
Eisenbeiss, J., and Schmid, H.
Helv. Chim. Acta 42: 61-6 (1959)
CA 53: 20052c

C 631 Monoacetylphenylenediamine.
Ermili, A., and Guiliano, R.
Gazz. Chim. Ital. 89: 517-25 (1959)
CA 54: 12107c

C 632 The depolarization of the sodium
resonance fluorescence.
Ermisch, W., and Seivert, R.
Ann. Physik 2: 393-402 (1959)
CA 53: 18622b

C 633 Dependence of the probability of energy
transfer in sensitized phosphorescence
upon the oscillator strength of triplet-
singlet transition in the molecule of an
energy acceptor.
Ermolaev, V.L.
Opt. i Spektroskopiya 6: 642-7 (1959)
CA 53: 19983f

C 634 Phosphorescence and fluorescence
quantum output of some 1-derivatives
of naphthalene solutions at -196°.
Ermolaev, V.L., and Svitashev, K.K.
Opt. i Spektroskopiya 7: 664-7 (1959)
CA 54: 10507f

C 635 Activation of the fluorescence of chloro-
phyll and its analogs.
Eustigneev, V.B.
Izv. Akad. Nauk SSSR, Ser. Fiz. 23: 74-7
(1959)
CA 53: 12417e

C 636 Response of plastic scintillators to
protons.
Evans, H.C., and Bellamy, E.H.
Proc. Phys. Soc. 74: 483-5 (1959)
CA 54: 20546i

C 637 Phosphorescence spectra of some aromatic hydrocarbons at various temperatures.
Fadeeva, M.S.
Izv. Akad. Nauk SSSR, Ser. Fiz. 23: 147-9
(1959)
CA 53: 11991h

C 638 Milk as an eluant of polycyclic aromatic
hydrocarbons added to wax.
Falk, H.L., Kotin, P., and Miller, A.
Nature 183: 1184 (1959)
CA 53: 22566i

C 639 Ninhydrin and fluorescence of proteins.
Faure, F.
Bull. Soc. Pharm. Bordeaux 98: 187-200
(1959)
CA 54: 14324b

C 640 Some experiments related to the problem of internal conversion of excitation
energy in aromatic molecules and crystals.
Ferguson, J.
J. Mol. Spectry. 3: 177-84 (1959)
CA 54: 23807f

C 641 Unique luminescences of dry chloro-
phylls.
Fernandez, J., and Becker, R.S.
J. Chem. Phys. 31: 467-72 (1959)
CA 54: 2934g

C 642 The in vivo staining of the lining of
mouse forestomach by porphyrins and
other fluorescent substances.
Figge, F.H.J.
J. Histochem. Cytochem. 7: 257-61 (1959)
CA 53: 4803h

C 643 Benzobisthiazoles.
Finzi, G., and Grandolini, G.
Gazz. Chim. Ital. 89: 2543-54 (1959)
CA 55: 5470e

C 644 Isolation and characterization of a yel-
low pteridine from the blue-green alga,
Anacystis nidulans.
Forrest, H.S., Van Baalen, C., and Myers,
J.
Arch. Biochem. Biophys. 83: 508-20 (1959)
CA 54: 1528e

C 645 Transfer mechanisms of electronic
excitation.
Forster, T.
Discussions Faraday Soc. No. 27, 7-17
(1959)
CA 54: 16164b

C 646 Investigation of an organophosphor in
the preexcited state.
Frackowiak, M., and Heldt, J.
Acta Phys. Polon 18: 93-106 (1959)
CA 54: 2935g

C 647 Immunoelectrophoresis analysis by
fluorescence.
Francq, J.C., Eyquem, A., and Grabar, P.
Rev. Franc. Etudes Clin. Biol. 4: 821-2
(1959)
CA 54: 8962c

C 648 Long range energy transfer and self-
absorption in fluorescent solutions.
Freeark, C.W., and Hardwick, E.R.
J. Phys. Chem. 63: 194-8 (1959)
CA 53: 10956f

C 649 Cathodoluminescence of zinc sulfide
and zinc cadmium sulfide activated
with rare-earth elements.
Fridman, S.A., and Shchaenko, V.V.
Materialy VII Soveshch. po Lyuminest.
(Kristallofosfory) Akad. Nauk Est. SSR,
Moscow 1958: 288-97 (1959)
CA 55: 11107g

C 650 The fluorescence spectra of the photo-
reduced forms of chlorophyll and pheo-
phytin.
Gachkovskii, V.F.
Biofizika 4: 16-23 (1959)
CA 57: 14156c

C 651 Fluorescent organosilicon polymers.
George, P.J.
U.S. 2,910,495 (1959)
CA 54: 7225i

C 652 Optical excitation of paramagnetic reso-
nance in an excited state of Cr^{+3} in
aluminum oxide.
Geschwind, S., Collins, R.J., and Schawlow,
A.L.
Phys. Rev. Letters 3: 545-8 (1959)
CA 54: 8297b

C 653 Isoquinoline derivatives. V.
Ghosh, T.N., and Bhattacharya, B.
J. Indian Chem. Soc. 36: 425-8 (1959)
CA 54: 9926f

C 654 The wave length and temperature dependence of the fluorescence efficiency
and the primary photochemical yield in
hexofluoroacetone vapor.
Giacometti, G., Okabe, H., and Stearie,
E.W.R.
Proc. Roy. Soc. (London), Ser. A 250: 287-
300 (1959)
CA 54: 19159f

C 655 Evaluating rate constants in the Jablon-
ski model of excited species in rigid
glasses.
Gilmore, E.H., and Lim, E.C.
J. Phys. Chem. 63: 15-16 (1959)
CA 53: 8825g

C 656 The diffusion of light in solutions of
quinine sulfate.
Giurgea, M., Ghita, C., and Musa, M.
Acad. Rep. Populare Romine, Studii
Cercetari Fiz. 10: 457-64 (1959)
CA 54: 14880d

C 657 Studies in the azole series. VII. Re-
actions of imidazoles with isocyanates.
Gompper, R., Hoyer, E., and Herlinger, H.
Chem. Ber. 92: 550-63 (1959)
CA 53: 13139i

C 658 Spot tests for aromatic compounds with
2,4,7-trinitrofluorenone.
Gordon, H.T., and Huraux, M.J.
Anal. Chem. 31: 302-7 (1959)
CA 53: 12955f

C 659 Isolation and characterization of some
fluorescent substances in the skin of
the frog, Rana nigromaculata.
Goto, T.
Dobutsugaku Zasshi 68: 286-90 (1959)
CA 54: 1758h

C 660 Fluorescence of crystalline substances
and of solutions excited by X-rays:
applications to analysis.
Graulier, M.
Bull. Soc. Chim. France 1959: 1715-21
(1959)
CA 54: 10645h

C 661 Fluorescent emission ascribable to ex-
citon annihilation in pure CdS crystals.
Grillot, E., and Bancie-Grillot, M.
Phys. Chem. Solids 8: 187-90 (1959)
CA 53: 12014f

C 662 Emission near the absorption edge and
other emission effects of GaN.
Grimmeiss, H.G., and Koelmans, H.
Z. Naturforsch. 14a: 264-71 (1959)
CA 53: 12856a

C 663 Paper chromatographic studies for the
differentiation of species and strains.
Taxonomy of red ants.
Groesswald, K., and Schmidt, G.
Umschau 1959, 94 pp. (1959)
CA 54: 15978e

C 664 Primary processes in the photochemis-
try of eosin.
Grosweiner, L.I., and Zwicker, E.F.
J. Chem. Phys. 31: 1141-2 (1959)
CA 54: 8237c

C 665 Influence of a magnetic field on the
lines of blue fluorescence or of lumin-
ous absorption of certain crystals of
CdS cooled to 4.2°K.
Gross, E.F., Grillot, E., Zakhartchenia,
B.P., and Bancie-Grillot, M.
Compt. Rend. 248: 213-16 (1959)
CA 53: 12013d

C 666 Luminescence of dyes of the porphine
series.
Gurinovich, G.P.
Tr. Inst. Fiz. i Mat., Akad. Nauk Belor-
ussk. SSR 1959: 111-30 (1959)
CA 55: 26661d

C 667 The polarization limit of complex
molecule luminescence.
Gurinovich, G.P., Sarzhevskii, A.M., and
Sevchenko, A.N.
Opt. i Spektroskopiya 7: 668-76 (1959)
CA 54: 10506f

C 668 3-Phenyl-7-aminocoumarin derivatives.
Haecisermann, H.
U.S. 2,881,186 (1959)
CA 54: 10105i

C 669 Fluorescence in aerosols.
Harmon, J., and Voldisch, R.
Paint, Oil, Chem. Rev. 112(15): 6-7 (1959)
CA 53: 22985b

C 670 Fluorescence of tetracycline in experimental ulcers and regenerating tissue
injuries.
Hakkinen, J., and Hartiala, K.
Ann. Med. Exptl. Biol. Fenniae 37: 115-20
(1959)
CA 54: 3676a

C 671 Isolation and structure of the fluorescent substances formed in the oxidative
reaction of epinephrine and norepinephrine with ethylenediamine.
Harley-Mason, J., and Laird, A.H.
Tetrahedron 7: 70-6 (1959)
CA 53: 4603i

C 672 Photochemical reduction of thionine.
VI. The photochemical thionine-
hydroquinone system.
Havemann, R., and Pietsch, H.
Z. Physik. Chem. 211: 257-66 (1959)
CA 54: 8237b

C 673 Unusual temperature dependence of
fluorescence of uranyl ions imbedded
in ice.
Hayakawa, S., and Hirata, M.
J. Chem. Phys. 30: 330 (1959)
CA 53: 10994c

C 674 Flavonoids. I. Aluminum salts of
flavonols.
Hayashiya, K.
Nippon Nogeikagaku Kaishi 33: 174-6 (1959)
CA 53: 1427f

C 675 Spectrophotofluorometry of reserpine,
other Rauwolfia alkaloids, and related
compounds.
Haycock, R.P., Sheth, P.B., and Mader,
W.J.
J. Am. Pharm. Assoc. 48: 479-85 (1959)
CA 53: 20700g

C 676 The photolysis and fluorescence of ace-
tone and acetone-biacetyl mixture.
Heicklen, J., and Noyes, W.A.
J. Am. Chem. Soc. 81: 3858-63 (1959)
CA 54: 4146c

C 677 The fluorescence and phosphorescence
of biacetyl vapor and acetone vapor.
Heicklen, J.
J. Am. Chem. Soc. 81: 3863-6 (1959)
CA 53: 4146e

C 677a Absorption and fluorescence spectra
of some mono- and dihydroxy-naphtha-
lenes.
Hercules, D. M., and Rogers, L. B.
Spectrochim. Acta, 1959: 393-408 (1959)
CA 53: 19566i

C 678 Influence of long-wave ultraviolet light
and β-radiation on the fluorescence of
anthracene in various solvent media.
Herforth, L., and Stolz, W.
Monatsber. Deut. Akad. Wiss. Berlin 1:
415-19 (1959)
CA 54: 12774f

C 679 Influence of extractives on eucalypt
pulping and paper-making.
Hillis, W.E., and Carle, A.
Appita 13: 74-81; Discussion, 81-3 (1959)
CA 54: 888f

C 680 (Aluminum with hydroxyazo dyes.)
Holzbecher, Z.
Collection Czech. Chem. Commun. 24:
1457 (1959)

C 681 Fluorescence of the metal salts of sali-
cylaldehyde condensation products.
Holzbecher, Z.
Collection Czech. Chem. Commun. 24:
3915-19 (1959)
CA 54: 10542h

C 682 Fluorescence-microscopy of the potato
tuber. I. Primary fluorescence of
protein crystals.
Holzl, J., and Bancher, E.
Protoplasma 50: 297-302 (1959)
CA 55: 5667a

C 683 The blue-fluorescing substance in the
hair of albino rats.
Hotta, K., Hashimoto, A., Tuboi, S., and
Ishiguro, I.
Seikagaku 31: 218-23 (1959)
CA 54: 3730c

C 684 Effect of some elements on the quality
of luminophors for fluorescent lamps
of the calcium halophosphate type.
Hrabal, L.
Chem. Prumysl 9: 129-31 (1959)
CA 54: 11717c

C 685 Effect of infrared radiation on phos-
phorescence of ZnS·Cu·Sm.
Hsu, J.C., Wang, H.M., Chung, K.I., and
Harang, M.Y.
Wu Li Hsueh Pao 15: 550-8 (1959)
CA 54: 20517g

C 686 Copper and tin activated halophosphate
phosphors.
Hunt, B.E., and McKeag, A.H.
J. Electrochem. Soc. 106: 1032-6 (1959)
CA 54: 2960b

C 687 Fluorescence of coumarin derivatives.
I. Fluorometric analysis of warfarin.
Ichimura, Y.
Yakugaku Zasshi 79: 1079-82 (1959)
CA 53: 22710d

C 688 Fluorescent substances.
Ide, H.
Japan 308 (1959)
CA 53: 19588a

C 689 Arsenate fluorescent substances.
Ide, H., Kuritsa, K., and Matsunaga, K.
Japan 505 (1959)
CA 53: 19588d

C 690 Lifetime of phosphorescence of substi-
tuted naphthalenes.
Iguchi, K.
J. Chem. Phys. 30: 319-20 (1959)
CA 53: 10956i

C 691 New cold cathode using magnesium
oxide.
Imai, T., Mizushima, Y., and Igarashi, Y.
J. Phys. Soc. Japan 14: 979-80 (1959)
CA 54: 11702d

C 692 Fluorometric analysis. IV. Fluorometric analysis. IV. Fluorometric
determination of gallium with 8-
quinolinol.
Ishibashi, M., Shigematsu, T., and
Nishikawa, Y.
Bull. Inst. Chem. Res., Kyoto Univ. 37:
191-7 (1959)
CA 54: 7426g

C 693 Optical bleaching and optically
bleached fabrics investigated with the
photoelectric fluorometer.
Jorder, H.
Melliand Textilber. 40: 1190-4 (1959)
CA 54: 1857i

C 694 Analysis of the fluorescence spectrum
of europium ethyl sulfate.
Judd, B.R.
Mol. Phys. 2: 407-14 (1959)
CA 55: 17250a

C 695 Analysis of the fluorescence spectrum
of neodymium chloride.
Judd, B.R.
Proc. Roy. Soc. (London), Ser. A 251:
134-42 (1959)
CA 54: 17078c

C 696 "Ballistic" method for studying the
decay of phosphorescence.
Kaminskii, M.G.
Opt. i Spektroskopiya 6: 103-6 (1959)
CA 53: 12829i

C 697 The triplet-singlet emission spectra of
phenanthrene and related compounds in
EPA and in petroleum ether at 90°K.
Kanda, Y., and Shimada, R.
Spectrochim. Acta 1959: 211-24 (1959)
CA 53: 19565c

C 698 Fluorescent substance of pterin-like
nature produced by aspergillus fungi.
VII. Conditions for production of the
substance.
Kaneko, Y.
Compt. Rend. Soc. Biol. 153: 887-9 (1959)
CA 54: 4752g

C 699 Luminescence of cerium-containing
       glasses.
       Karapetyan, G.O.
       Izv. Akad. Nauk SSSR, Ser. Fiz. 23: 1382-6
       (1959)
       CA 54: 6333e

C 700 Condensation of halonaphthalic anhy-
       drides with resorcinol.
       Karishin, A.P., and Kustol, D.M.
       Zh. Obshch. Khim. 29: 2241-3 (1959)
       CA 54: 11008d

C 701 Oxygen quenching of the fluorescence
       of heated adsorbates of the series of
       anthraquinone derivatives.
       Karyakin, A.V.
       Izv. Akad. Nauk SSSR, Ser. Fiz. 23: 32-6
       (1959)
       CA 53: 11988i

C 702 Chemiluminescence of lucigenin and
       its derivatives.
       Karayakin, A.V.
       Opt. i Spektroskopiya 7: 122-4 (1959)
       CA 54: 23816e

C 703 Mechanism of the recombination lumin-
       escence of alkali halide crystal phos-
       phors.
       Kats, M.L.
       Materialy VII Soveshch. po Lyuminest.
       (Kristallofosfory), Akad. Nauk Est.
       SSR, Moscow 1958: 130-6 (1959)
       CA 55: 10095d

C 704 Phosphorescence and thermal stimula-
       tion of KBr-In phosphors.
       Kats, M.L.
       Opt. i Spektroskopiya 6: 237 (1959)
       CA 54: 9509e

C 705 Effect of light on the autoxidation of
       fats. I. Bleaching, fluorescence, and
       yellowing.
       Kaufmann, H.P., and Vogelmann, M.
       Farbenchemiker 61: 6-10 (1959)
       CA 53: 23004d

C 706 The effect of light radiation on the aut-
       oxidation of fats. I. Bleaching, fluor-
       escence, and yellowing.
       Kaufmann, H.P., and Vogelmann, M.
       Fette, Seifin, Anstrichmittel 61: 206-10
       (1959)
       CA 54: 927a

C 707 Isolation of D-lactoflavine and isoxan-
       thopterin from the skin of the fire sala-
       mander.
       Kauffmann, T., and Vogt, K.
       Chem. Ber. 92: 2855-61 (1959)
       CA 54: 5684g

C 708 Low-temperature fluorescence spectra
       and crystal-field splittings of rare-
       earth-activated strontium sulfide phos-
       phors.
       Keller, S.P., and Pettit, G.D.
       J. Chem. Phys. 30: 434-41 (1959)
       CA 53: 12013f

C 709 Variation of valence state of europium
       in strontium sulfide phosphors.
       Keller, S.P.
       J. Chem. Phys. 30: 556-60 (1959)
       CA 53: 12013c

C 710 Optical spectra of rare-earth-activated
       $BaTiO_3$.
       Keller, S.P., and Pettit, G.D.
       J. Chem. Phys. 31: 1272-7 (1959)
       CA 53: 5258i

C 711 Phosphor with fluorescence larger than
       the energy gap.
       Keller, S.P., and Pettit, G.D.
       Phys. Rev. 113: 785-6 (1959)
       CA 53: 13795e

C 712 Optical properties of activated and un-
       activated hexagonal zinc sulfide single
       crystals.
       Keller, S.P., and Pettit, G.D.
       Phys. Rev. 115: 526-36 (1959)
       CA 54: 8300f

C 713 Sensitized fluorescence in mixed solu-
       tions.
       Ketskemety, I.
       Acta Phys. Acad. Sci. Hung. 10: 429-39
       (1959)
       CA 54: 12774h

C 714 Role of iron in D-amino acid oxidase.
       I. Fluorescence of flavine adenine di-
       nucleotide.
       Kihara, T.
       Osaka Daigaku Igaku Zasshi 11: 321-3
       (1959)
       CA 53: 10348d

C 715 Effect of reabsorption on the duration of fluorescence for organic materials.
Kilin, S.F., and Rozman, I.M.
Opt. i Spektroskopiya 6: 70-7 (1959)
CA 53: 12856h

C 716 Influence of high pressure on the spectral characteristics of some crystal phosphors.
Kirs, Y.Y., and Laissar, A.I.
Materialy VII Soveshch. po Lyuminest. (Kristallofosfory), Akad. Nauk Est. SSR, Moscow 1958: 59-65 (1959)
CA 55: 8069f

C 717 Decay law for the phosphorescence of trypaflavine in formic acid.
Kislyak, G.M.
Opt. i Spektroskopiya 6: 226-8 (1959)
CA 54: 11697b

C 718 The law of the extinction of the phosphorescence of trypaflavine in formic acid.
Kislyak, G.M.
Opt. i Spektroskopiya 6: 226-8 (1959)
CA 60: 3630a

C 719 Dependence of the lifetime of the excited state of organic luminescent substances on the wave length of the exciting light. II.
Kislyak, G.M.
Izv. Akad. Nauk SSSR, Ser. Fiz. 23: 119-21 (1959)
CA 53: 11990i

C 720 The temperature dependence of the fluorescence of photoconductors.
Klaseno, H.A.
Phys. Chem. Solids 9: 185-97 (1959)
CA 55: 10095e

C 721 Theory of the formation of adsorption mixed crystals and the inclusion of fluorescent organic compounds.
Kleber, W.
Freiberger Forschungsh. B37: 11-28 (1959)
CA 54: 20399a

C 722 Incorporation mechanism in the formation of adsorption mixed crystals.
Kleber, W.
Z. Physik. Chem. 212: 222-32 (1959)
CA 54: 2861b

C 723 The fluorescence of binary and ternary germanates of Group II elements.
Koelmans, H., and Verhagen, C.M.C.
J. Electrochem. Soc. 106: 677-82 (1959)
CA 53: 17682f

C 724 Studies on electrochemically excited molecules by fluorescence spectrum.
Kokubun, H.
Bunko Kenkyu 7(4): 1-14 (1959)
CA 57: 5476b

C 725 Fluorescence of milk and butter.
Konev, S.V., and Kozunin, I.I.
Byul. Nauchn.-Tekhn. Inform. Vses. Nauchn.-Issled. Inst. Zhivotnovodstva 1959: 14-18 (1959)
CA 55: 11693g

C 726 Fluorescence reaction spectra of proteins.
Konev, S.V.
Izv. Akad. Nauk SSSR, Ser. Fiz. 23: 90-3 (1959)
CA 53: 12370g

C 727 Thermochromic effects and constitution of unsymmetrical (hydroxyalkylamino)-p-benzoquinones.
Konig, K.
Chem. Ber. 92: 257-67 (1959)
CA 53: 11287i

C 728 Detection of secondary α-hydroxyethylamines by fluorescence, polarography, paper chromatography, and thermochromy.
Konig, K.H., and Berg, H.
Z. Anal. Chem. 166: 92-100 (1959)
CA 53: 16829c

C 729 Thermochromic effects and constitution of unsymmetrical (hydroxyalkylamino)-p-benzoquinones. II. Variations in the quinone and the amino alcohol component.
Konig, K.H., and Letsch, G.
Chem. Ber. 92: 1789-97 (1959)
CA 53: 4455b

C 730 Photoelectrical zero method for taking fluorescence spectra.
Kortum, G., and Hess, W.
Z. Physik. Chem. 19: 142-55 (1959)
CA 53: 1068i

C 731 The Gudden-Pohl effect of ZnS:Cu.
Kotera, Y., and Naraoka, K.
J. Electrochem. Soc. 106: 1066 (1959)
CA 54: 2960c

C 732 Quantum-mechanical calculation of
adiabatic potentials for the luminescence center in KCl-Tl by the one-
oscillator approximation.
Kristofel, N.N.
Akad. Nauk Est. SSR 1959: 3-36 (1959)
CA 54: 4218b

C 733 A new synthesis of porphine.
Krol, S.
J. Org. Chem. 24: 2065-7 (1959)
CA 54: 11047b

C 734 The action of ultrasonics on the lumin-
escence of phosphors.
Kudryavtsev, B.B., Medvedev, A.N., and
Ponomavev, A.P.
Primenenie Ul'traakustiki k Issled.
Veshchestva 1959: 139-45 (1959)
CA 55: 24266h

C 735 Electron irradiation effects in CdS.
Kulp, B.A., and Kelley, R.H.
Proc. Conf. Nucl. Radiation Effects Semi-
cond. Devices, Mater. Circuits, 2nd,
New York 1959: 131-4 (1959)
CA 59: 14664e

C 736 Effect of the concentration of fluores-
cent substances on the efficacy of
fluorescence inhibitors.
Lavorel, J.
J. Chem. Phys. 55: 911-15 (1959)
CA 53: 11986h

C 737 Effect of rigid media on photochemical
processes in benzene.
Leach, S.
Intern. Symp. Free Radical Stab., Wash-
ington, D.C. 4: 1-12 (1959)
CA 54: 15071c

C 738 Photochemical decomposition of ben-
zene in a rigid medium.
Leach, S., Migiridicyan, E., and Grajcar,
L.
J. Chim. Phys. 56: 749-60 (1959)
CA 54: 7298d

C 739 The application of fluorescence to
problems of glass manufacturing.
Leblanc, J., Taylor, E., and Poole, J.P.
Glastech. Ber., Sonderband 32(1): 29-32
(1959)
CA 54: 14607f

C 740 Polarization of fluorescence in zinc
sulfide and cadmium sulfide single
crystals.
Lempicki, G.
Phys. Rev. Letters 2: 155-7 (1959)
CA 53: 8826g

C 741 Nature of fluorescence of uranium in
fused sodium fluoride.
Le Roux, H.
Nature 183: 1180-1 (1959)
CA 53: 14685d

C 742 Formation of luminescent polymers in
concentrated solutions of Acridine
Orange and an investigation of their
optical properties.
Levshin, V.L., and Klyuev, Y.A.
Izv. Akad. Nauk SSSR, Ser. Fiz. 23: 15-18
(1959)
CA 53: 11987i

C 743 Fluorescence spectra for aromatic hy-
drocarbons of the diphenyl series and
for their oxygen and sulfur analogs.
Levshin, V.L., Mamedov, K.I., Sergienko,
S.R., and Pustil'nikova, S.D.
Izv. Akad. Nauk SSSR, Otd. Khim. Nauk
1959: 1571-8 (1959)
CA 54: 11695f

C 744 Cells and solvents for the measure-
ment of temperature dependence of
electronic spectra.
Lippert, E., Luder, W., and Moll, F.
Spectrochim. Acta 1959: 378-89 (1959)
CA 53: 19571g

C 745 Polarization and relaxation effects in
the temperature dependence of the absorption and fluorescence spectra of
aromatic compounds in polar solvent.
Lippert, E., Luder, W., and Moll, F.
Spectrochim. Acta 1959: 858-69 (1959)
CA 54: 5243e

C 746 Measurement of fluorescence spectra with spectrophotometers and comparison standards.
Lippert, E., Nagele, W., Seibold-Blankenstein, I., Staiger, U., and Voss, W.
Z. Anal. Chem. 170: 1-18 (1959)
CA 54: 2935i

C 747 Investigation of the process of chlorophyll formation and of its state in plant leaves by means of fluorescence spectra.
Litvin, F.F., and Krasnovskii, A.S.
Izv. Akad. Nauk SSSR, Ser. Fiz. 23: 82-5 (1959)
CA 53: 13291d

C 748 Anthracene and its derivatives, sensitizers of photochemical reactions.
Livingston, R.
J. Chim. Phys. 55: 887-91 (1959)
CA 53: 12016f

C 749 Photochemical autoxidation of anthracene.
Livingston, R., and Rao, S.V.
J. Phys. Chem. 63: 794-9 (1959)
CA 53: 21089b

C 750 Action of ultraviolet irradiation on reserpine.
Ljungberg, S.
J. Pharm. Belg. 14: 115-25 (1959)
CA 54: 1803c

C 751 The fluorescence of amino acids in aqueous solutions.
Longin, P.
Compt. Rend. 248: 1971-3 (1959)
CA 53: 14685h

C 752 Mechanism of the photochemical activity of isolated chloroplasts. IV. Fluorescence yield against velocity relations in the Hill reaction of chloroplast fragments.
Lumry, R., Mayne, B., and Spikes, J.D.
Discussions Faraday Soc. No. 27, 149-60 (1959)
CA 54: 14378d

C 753 Recombination luminescence of alkali halide phosphors, activated with mercury-like ions.
Lushchik, C.B., Kyaembre, K.F., and Yaek, I.V.
Materialy VII Soveshch. po Lyuminest. (Kristallofosfory), Akad. Nauk Est. SSR, Moscow 1958: 117-29 (1959)
CA 55: 10095f

C 754 Luminescence spectra of high molecular weight petroleum hydrocarbons.
Mamedov, K.I.
Izv. Akad. Nauk SSSR, Ser. Fiz. 23: 126-30 (1959)
CA 53: 13561g

C 755 Photochemical isomerization of fluorescent whitening agents of stilbene series.
Mashio, F., and Kimura, Y.
Kogyo Kagaku Zasshi 62: 113-18 (1959)
CA 57: 8481i

C 756 New fluorescence colorimetric technique using diaphragms.
Mosser, M.L.
Feingeraetetechnik 8: 505-10 (1959)
CA 54: 9378b

C 757 Imine and imine-d radicals trapped in argon, krypton, and xenon matrixes at 4.2°K.
McCarty, M., and Robinson, G.W.
J. Am. Chem. Soc. 81: 4472-6 (1959)
CA 54: 2932i

C 758 Autoluminophors.
Meckelburg, E.
Chem. Rundschau 12: 688-9 (1959)
CA 54: 10542i

C 759 2- and 3-phenylthianaphthenes.
Middleton, S.
Australian J. Chem. 12: 218-33 (1959)
CA 53: 21871g

C 760 Fluorescence spectra of gasolines.
Mihul, C., Ruscior, C., Pop, V., Schwartz, F.R., and Radvlescu, G.A.
Izv. Akad. Nauk SSSR, Ser. Fiz. 23: 122-5 (1959)
CA 53: 13565d

C 761 Photoreduction of acridine dyes.
Millich, F., and Oster, G.
J. Am. Chem. Soc. 81: 1357-63 (1959)
CA 53: 13799a

C 762 Bile pigment. I. Properties of dihydro-
bilirubin.
Mitsumoto, T.
Okayama Igakkai Zasshi 71: 7185-91 (1959)
CA 54: 24971b

C 763 Mechanism of spherosome fluorescence
with oxazines and other basic dyes.
Mix, M.
Protoplasma 50: 434-70 (1959)
CA 55: 5632f

C 764 Organic analysis. XIII. Estimation of
hexose with 5-hydroxy-1(2H)-naphtha-
lenone.
Momose, T., and Ohkura, Y.
Chem. Pharm. Bull. 7: 31-4 (1959)
CA 54: 18196c

C 765 Electroluminescent phosphors.
Morrison, G.H., Palilla, F.C., and
Zloczower, W.
U.S. 2,999,818 (1959)
CA 54: 2082g

C 766 Fluorimetric microdetection of tetra-
phenylborate.
Mukherji, A.K., and Sant, B.R.
Mikrochim. Acta 1959: 370-1 (1959)
CA 54: 24129f

C 767 Absorption and phosphorescence spectra of mono- and diazanaphthalenes
(π-π-phosphorescence following n-π-
absorption in the diazanaphthalenes).
Muller, R., and Dorr, F.
Z. Elektrochem. 63: 1150-6 (1959)
CA 54: 5243g

C 768 Phenoxazine. I. Oxidation products of
phenoxazine.
Musso, H.
Chem. Ber. 92: 2862-73 (1959)
CA 54: 5657g

C 769 Orcein dyes. IX. Fluorescence of
orcein dyes.
Musso, H., and Matthies, H.G.
Naturwissenschaften 46: 15 (1959)
CA 53: 11842e

C 770 Peculiarities in the luminescence of
orthodisubstituted aromatic hydrocarbons. I. Absorption spectra and fluorescence spectra of the anilides of salicylic and o-methoxybenzoic acids.
Naboikin, Y.V., Paulova, E.N., and
Zadorozhnyi, B.A.
Opt. i Spektroskopiya 6: 366-71 (1959)
CA 55: 10055e

C 771 Peculiarities in the luminescence of
ortho-disubstituted aromatic hydrocarbons. II. Fluorescence of methyl 3-
hydroxy-2-naphthoate and methyl 3-
methoxy-2-naphthoate.
Naboikin, Y.V., Zadorozhnyi, B.H., and
Pavlova, E.N.
Opt. i Spektroskopiya 6: 492-5 (1959)
CA 55: 16140b

C 772 Thermodynamic barrier to micelle
formation and breakdown. I. Hexade-
cyltrimethylammonium salts.
Nash, T.
J. Colloid Sci. 14: 59-73 (1959)
CA 53: 9782h

C 773 Addition of triethylaluminum to tolan.
Nesmeyanov, A.N., Borisov, A.E., and
Savel'eva, I.S.
Izv. Akad. Nauk SSSR, Otd. Khim. Nauk
1959: 1034-6 (1959)
CA 54: 1367a

C 774 Hyperconjugation and fluorescence be-
havior.
Neunhoeffer, O., Alsdorf, G., and Ulrich, H.
Chem. Ber. 92: 252-5 (1959)
CA 53: 10187d

C 775 Acylations with the acid chlorides of
2,5-diphenylfuran-3,4-dicarboxylic acid
and 2,5-dimethylfuran-3,4-dicarboxylic
acid and related compounds.
Nightingale, D.V., and Sukornick, B.
J. Org. Chem. 24: 497-500 (1959)
CA 53: 21870f

C 776 Fluorescence of various dyes, with
special reference to their characteristics as standards for fluorometric determination.
Nihongi, T., and Iwasaki, S.
Tokyo Jikeikai Ika Daigaku Zasshi 74:
949-52 (1959)
CA 54: 17045i

C 777 Paper chromatography of natural deri-
vatives of α- and γ-benzopyrone and
tanning substances.
Nikonov, G.K.
Med. Prom. SSSR 12: 16-21 (1959)
CA 53: 14094b

C 778 The effect of some impurities of after-
glow of $CaWO_4$ phosphor.
Nishikawa, K.
Kogyo Kagaku Zasshi 62: 1635-6 (1959)
CA 57: 15950d

C 779 The synthesis of carbazoles from 3-
vinylindoles with tetracyanoethylene.
Noland, W.E., Kuryla, W.C., and Lange,
R.F.
J. Am. Chem. Soc. 81: 6010-17 (1959)
CA 54: 6683h

C 780 Lactic dehydrogenase. VII. Fluorescence spectra of ternary complexes of
lactic dehydrogenase, reduced diphosphopyridine nucleotide, and carboxylic
acids.
Novoa, W.B., Winer, A.D., Glaid, A.J., and
Schwert, G.W.
J. Biol. Chem. 234: 1155-61 (1959)
CA 53: 14186g

C 781 Spectroscopic studies of molecular
problems – normal coordinates analysis
as a tool for elucidating molecular
structure – iron pentacarbonyl – n→π*
transitions – alkyl nitrites and rotary
dispersions.
O'Dwyer, M.F.
Dissertation, Florida State University
(1959)
CA 54: 6303g

C 782 Changes of fats during cooking.
Ogawa, Y., Saito, N., Suga, K., and Togari,
A.
Eiyo To Shokuryo 12: 83-7 (1959)
CA 59: 12091d

C 783 Investigation of Candida with the fluoro-
microscope. II. Influence of various
chemicals and antibiotics on Candida
albicans.
Ohira, I., Ohashi, S., Iwasaki, F., and
Endo, T.
Chemotherapy (Tokyo) 7: 164-8 (1959)
CA 53: 20269b

C 784 Fluorescence of some metal chelate
compounds of 8-quinolinol. I. Effect of
metallic ions and solvent on spectrum
and quantum yield.
Ohnesorge, W.E., and Rogers, L.B.
Spectrochim. Acta 1959: 27-40 (1959)
CA 53: 13775f

C 785 Intermolecular energy transfer and
concentration depolarization of fluores-
cent light.
Ore, A.
J. Chem. Phys. 31: 442-3 (1959)
CA 54: 2935b

C 786 Phosphorescence of sodium acetate.
Osada, K.
J. Chem. Phys. 30: 1363-4 (1959)
CA 53: 17684a

C 787 Photoreduction of dyes in rigid media.
I. Triphenylmethane dyes.
Oster, G., Joussot-Dubien, J., and Broyde,
B.
J. Am. Chem. Soc. 81: 1869-72 (1959)
CA 53: 18643d

C 788 The (5-phenyl-2-oxazolyl) pyridines as
fluorescent pH indicators.
Oh, D.G.
U.S. At. Energy Comm. LA-2252, 28 pp.
(1959)
CA 53: 11000g

C 789 Quenching of the phosphorescence of
organic dyes by electrolyte ions.
Pankeeva, A.E.
Izv. Akad. Nauk SSSR, Ser. Fiz. 23: 112-
115 (1959)
CA 53: 11992b

C 790 Absorption and fluorescence spectra of
uranyl nitrate solutions at room temp-
erature.
Pant, D.D., and Khandelwal, D.P.
Proc. Indian Acad. Sci. 50A: 323-35 (1959)
CA 54: 14930f

C 791 Temperature dependence of fluores-
cence bands of uranyl nitrate solutions.
Pant, D.D., Khandelwal, D.P., and Bist,
H.D.
Current Sci. 28: 483-4 (1959)
CA 54: 19156a

C 792 Fluorescence spectra of uranyl perchlorate solutions at room temperature.
Pant, D.D., and Khandelwal, D.P.
J. Sci. Ind. Research 18B: 126-7 (1959)
CA 53: 21165c

C 793 Photolysis of thionine in rigid medium.
Parker, C.A., and Rees, W.T.
J. Chim. Phys. 56: 761-70 (1959)
CA 54: 7297g

C 794 Raman spectra in spectrofluorimetry.
Parker, C.A.
Analyst 84: 446-53 (1959)
CA 54: 155f

C 795 Nitrogen mustard analogs of antimalarial drugs.
Peck, R.M., Preston, R.K., and Greech, H.F.
J. Am. Chem. Soc. 81: 3984-9 (1959)
CA 55: 536i and 538e

C 796 Aspects of absorption spectra and decay kinetics of the metastable triplet state.
Pekkarinen, L.
Suomen Kemistilehti 32A: 267-74 (1959)
CA 54: 11697a

C 797 Fluorescence spectra in crystallized solutions at 77°K.
Pesteil, L., and Ciais, A.
Compt. Rend. 249: 528-30 (1959)
CA 54: 2935e

C 798 Effect of the solvent on the electron spectra of phthalimides.
Pikulik, L.G.
Tr. Inst. Fiz. i Mat., Akad. Nauk Belorussk. SSR 1959: 167-75 (1959)
CA 55: 20615g

C 799 The phosphorescence of solid nitrogen.
Pilon, A.M.
Compt. Rend. 249: 1492-3 (1959)
CA 54: 11693e

C 800 Fluorescence in the 8-quinolinol family and the n-π transition.
Popovych, O., and Rogers, L.B.
Spectrochim. Acta 1959: 584-92 (1959)
CA 54: 6303f

C 801 Intramolecular and intermolecular energy conversion involving change of multiplicity.
Porter, G., and Wright, M.R.
Discussions Faraday Soc. No. 27, 18-27 (1959)
CA 54: 14937c

C 802 Some characteristics of large band gap compound semiconductors.
Prener, J.S., and Williams, F.E.
Phys. Chem. Solids 8: 461-4 (1959)
CA 53: 12850g

C 803 Luminescence of stilbene crystals at 20°K.
Prikhot'ko, A.F., and Fugol, I.Y.
Opt. i Spektroskopiya 7: 35-43 (1959)
CA 54: 23815i

C 804 Widely distributed blue fluorescence of organic origin.
Przibram, K.
Oesterr. Akad. Wiss., Math.-Naturw. Kl., Anz. 1959(11): 205-12 (1959)
CA 54: 16193g

C 805 Vibrational structure of phosphorescence spectra of aromatic acids at the temperature of liquid oxygen.
Pyatnitskii, B.A.
Izv. Akad. Nauk SSSR, Ser. Fiz. 23: 135-8 (1959)
CA 53: 11991b

C 806 Filter cigarets.
Pyriki, C., and Moldenhauer, W.
Pharm. Zentralhalle 98: 503-12 (1959)
CA 54: 3864i

C 807 Fluorescent response of cesium iodide crystals to heavy ions.
Quinton, A.R., Anderson, C.E., and Knox, W.J.
Phys. Rev. 115: 886-7 (1959)
CA 54: 9526c

C 808 Fluorimetric determination of boron with resacetophenone as a reagent.
Rao, G.C., and Appalarju, N.
Z. Anal. Chem. 167: 325-9 (1959)
CA 53: 18742i

C 809 Pressure effects in luminescence:
　　　Isobaric experiments on NaI (Tl).
　　　Reiffel, L.
　　　Phys. Rev. 114: 1493-9 (1959)
　　　CA 54: 6333f

C 810 Ethynylation of an *o*-benzoquinone.
　　　Ried, W., Wesselburg, K., and Schmidt,
　　　K.H.
　　　Naturwissenschaften 46: 142-3 (1959)
　　　CA 53: 16018d

C 811 Particle selection in crystals of CsI
　　　(Tl).
　　　Robertson, J.C., and Ward, A.
　　　Proc. Phys. Soc. 73: 523-5 (1959)
　　　CA 55: 147c

C 812 The influence of intermolecular inter-
　　　action on the ultraviolet spectra of aro-
　　　matic compounds.
　　　Romantsova, G.I.
　　　Tr. Nauchn.-Issled. Fiz.-Khim. Inst. 1959:
　　　107-17 (1959)
　　　CA 54: 23836d

C 813 Ultraviolet fluorescence of quinine sul-
　　　fate for detection of phosphate ester
　　　spots on paper.
　　　Rorem, E.S.
　　　Nature 183: 1739-40 (1959)
　　　CA 53: 19706i

C 814 Photoconduction and cis-trans isomer-
　　　ism in β-carotene.
　　　Rosenberg, B.
　　　J. Chem. Phys. 31: 238-46 (1959)
　　　CA 53: 1086f

C 815 Molecular phosphorescence at the tem-
　　　perature of liquid nitrogen: structure
　　　of the vibrations of various aromatic
　　　and heterocyclic derivatives.
　　　Rousset, A., Lochet, R., and Dubarry, J.C.
　　　Compt. Rend. 248: 54-7 (1959)
　　　CA 53: 11986i

C 816 Transfers of activation of molecular
　　　crystals arising from retarded fluores-
　　　cence of impurities in solid solution.
　　　Rousset, A., Lochet, R., Lacueille, R., and
　　　Moyer, Y.
　　　Compt. Rend. 248: 2045-8 (1959)
　　　CA 53: 16689e

C 817 Luminescence spectra of some dicar-
　　　bonic acids at various temperatures.
　　　Ryazanova, E.F.
　　　Izv. Akad. Nauk SSSR, Ser. Fiz. 23: 193-6
　　　(1959)
　　　CA 53: 11991g

C 818 Phosphorescent decay of zinc sulfides.
　　　Soddy, J.
　　　J. Phys. Radium 20: 890-7 (1959)
　　　CA 54: 2959g

C 819 Electron capture by low-lying levels of
　　　attachment in zinc sulfide phosphors
　　　activated by copper and cobalt.
　　　Saichenko, Y.M.
　　　Alma-Ata. Sbornik 1959: 93-6 (1959)
　　　CA 54: 2076a

C 820 Colorimetric and fluorometric deter-
　　　mination of aluminum.
　　　Sandell, E.B.
　　　Colorimetric Determination of Traces of
　　　Metals pp. 219, 304, Interscience, N.Y.
　　　(1959)

C 821 Phase equilibriums and manganese-
　　　activated fluorescence in the system
　　　$Zn_3(PO_4)-Mg_3(PO_4)_2$.
　　　Sarver, J.F., Katnock, F.L., and Hummel,
　　　F.A.
　　　J. Electrochem. Soc. 106: 960-3 (1959)
　　　CA 54: 2020d

C 822 Luminescence method for determining
　　　the volumes of the solvate shells of
　　　molecules in solutions.
　　　Sarzhevskii, A.M., and Sevchenko, A.N.
　　　Zh. Fiz. Khim. 33: 2410-13 (1959)
　　　CA 54: 21998d

C 823 Electronic spectra of exchange-coupled
　　　ion pairs in crystals.
　　　Schawlow, A.L., Wood, D.L., and Clogston,
　　　A.M.
　　　Phys. Rev. Letters 3: 271-3 (1959)
　　　CA 54: 4170b

C 824 The constitution of the red and the color-
　　　less form of the quinolylmethanes.
　　　Scheibe, G., and Riess, W.
　　　Chem. Ber. 92: 2189-98 (1959)
　　　CA 54: 3423h

C 825 Near-ultraviolet spectrum of crystalline durene.
Schnepp, O., and McClure, D.S.
J. Chem. Phys. 30: 874-8 (1959)
CA 53: 15758c

C 826 A measuring method for investigating fluorescent decay processes.
Schutz, H.
Z. Physik 156: 27-37 (1959)
CA 53: 21200d

C 827 Theory of radiation diffusion. II. The nonstationary and stationary fluorescence.
Seiwert, R.
Optik 16: 358-70 (1959)
CA 53: 21165b

C 828 Calculation of the degree of polarization of the resonance fluorescence in the presence of radiation diffusion.
Seiwert, R., and Ermisch, W.
Ann. Physik 5: 4-14 (1959)
CA 53: 4138f

C 829 Quantum yield in the oxidation of firefly luciferin.
Seliger, H.H., and McElroy, W.D.
Biochem. Biophys. Res. Commun. 1: 21-4 (1959)
CA 54: 10006a

C 830 Limiting polarization of fluorescence.
Sevchenko, A.N., Gurinovich, G.P., and Sarzhevskii, A.M.
Dokl. Akad. Nauk SSSR 126: 979-82 (1959)
CA 55: 19473i

C 831 The symmetry of the porphyrin molecules.
Sevchenko, A.N., Gurinovich, G.P., and Solov'ev, K.N.
Dokl. Akad. Nauk SSSR 128: 510-13 (1959)
CA 55: 12030h

C 832 Spectroscopic investigation of uranyl compounds.
Sevchenko, A.N., and Volod'ko, L.V.
Inzh.-Fiz. Zh., Akad. Nauk Belorussk.
SSR 1959: 63-71 (1959)
CA 55: 18293d

C 833 Fluorescent determination of inorganic substances. I. Fluorescent reaction of gallium with Rhodamine S and Rhodamine GG.
Shcherbov, D.P., Solov'yan, I.T., Ivankova, A.I., and Drobachenko, A.V.
Tr. Kazakhsk. Nauchn.-Issled. Inst.
Mineral'n. Syr'ya 1: 188-95 (1959)
CA 55: 7142f

C 834 Fluorescence studies of coenzyme binding to beef-heart lactic dehydrogenase.
Shifria, S., Kaplan, N.O., and Ciotti, M.M.
J. Biol. Chem. 234: 1555-62 (1959)
CA 53: 16255e

C 835 Effect of the position and of the nature of the substitute on the fluorescence spectra of anthraquinone derivatives in frozen solutions.
Shigorin, D.N., Shcheglova, N.A., and Nurmukhametov, R.N.
Izv. Akad. Nauk SSSR, Ser. Fiz. 23: 37-9 (1959)
CA 53: 11989c

C 836 The effect of the position and the nature of substituting groups on the fluorescence spectra of anthraquinone derivatives in frozen solutions.
Shigorin, D.N., Shcheglova, N.A., Nurmukhametov, R.N., and Dokonikhin, N.S.
Dokl. Akad. Nauk SSSR 120: 1242-5 (1959)
CA 53: 15761b

C 837 An artificial comet.
Shklovskii, I.S., Esipov, V.F., Kurt, V.G., Moroz, V.I., and Shcheglov, P.V.
Astron. Zh. 36: 1073-7 (1959)
CA 54: 10499h

C 838 Vibrational analysis of the phosphorescence spectrum of coronene.
Shpol'skii, E.V., and Klimova, L.A.
Izv. Akad. Nauk SSSR, Ser. Fiz. 23: 23-8 (1959)
CA 53: 11988b

C 839 Diimidazoles.
Siegrist, A.E., and Duennenberger, M.
U.S. 2,899,440 (1959)
CA 55: 575b

C 840 Pyrrole derivatives.
Siegrist, A.E., and Duennenberger, M.
U.S. 2,091,480, Aug. 25, (1959)
CA 55: 1659b

C 841 Diimidazole derivatives.
Siegrist, A.E., and Ackermann, F.
Swiss 332,135 (1959)
CA 54: 19716g

C 842 (Fluorometric determination of beryl-
lium.)
Sill, C.W., and Willis, C.P.
Anal. Chem. 31: 598 (1959)

C 843 Nonradiative transitions of 2-naphthyl-
amine.
Simon, Z.
Acad. Rep. Populare Romine, Studii
Cercetari Fiz., Inst. Fiz. At. Inst. Fiz.
10: 317-28 (1959)
CA 54: 7341h

C 844 Relation between quenching of the fluor-
escence of sensitizers by Pinakryptol
Green and photographic activity.
Smirnov, B.R., and Moshkovskii, Y.S.
Zh. Nauchn. i Prikl. Fotogr. i Kinematogr.
4: 234-5 (1959)
CA 53: 21301f

C 845 Some optical measurements on calcium
fluoride.
Sorlich, P., Karras, H., and Kuhne, K.
Sitzber. Deut. Akad. Wiss. Berlin, Kl.
Math., Phys. Tech. 1959, No. 2, 26 pp.
(1959)
CA 54: 21994c

C 846 Chromatography of polycyclic aromatic
hydrocarbons on acetylated paper.
Spotswood, T.M.
J. Chromatog. 2: 90-4 (1959)
CA 53: 21435a

C 847 Measurement of radiative lifetimes. I.
An apparatus for measurement of millimicrosecond radiative lifetimes of gas-
phase molecules. II. The radiative
lifetime of the BO_u^+ state of I_2 by two
absolute absorption methods.
Stafford, F.E.
U.S. At. Energy Comm. UCRL-8854: 1-96
(1959)
CA 54: 11718b

C 848 Furocoumarins.
Stanley, W.L., and Vannier, S.H.
U.S. 2,889,337 (1959)
CA 53: 22021a

C 849 New occurrences of coumarin deriva-
tives in Fraxinus ornus.
Steinegger, E., and Brantschen, A.
Pharm. Acta Helv. 34: 334-44 (1959)
CA 54: 7064i

C 850 Energy transfer in aromatic vapors;
the benzene-sensitized fluorescence of
anthracene vapor at 2652 A.
Stevens, B.
Discussions Faraday Soc. No. 27, 34-9
(1959)
CA 54: 14937e

C 851 Quenching and vibrational-energy trans-
fer of excited iodine molecules.
Stevens, B.
Can. J. Chem. 37: 831-4 (1959)
CA 53: 16681i

C 852 Effect of the electronic structure of the
cation upon fluorescence in metal-8-
quinolinol complexes.
Stevens, H.M.
Anal. Chim. Acta 20: 389-96 (1959)
CA 53: 19661h

C 853 Ultraviolet fluorescence of proteins.
Teale, F.W.J., and Weber, G.
Biochem. J. 72, 15 pp. (1959)
CA 55: 24847h

C 854 Solvent influence on phosphorescence
spectra of sulfobenzoic and bromobenzoic acids at low temperature.
Teplyakov, P.A.
Izv. Vysshikh Uchebn. Zavedenii, Fiz.
1959: 102-6 (1959)
CA 54: 6302e

C 855 Phosphorescence spectra of alcoholic
solutions of aminobenzoic, sulfobenzoic,
and bromobenzoic acids at low temperature.
Teplyakov, P.A.
Izv. Vysshikh Uchebn. Zavedenii, Fiz.
1959: 135-9 (1959)
CA 53: 19567g

C 856 Exciton spectrum of cadmium sulfide.
Thomas, D.G., and Hopfield, J.J.
Phys. Rev. 116: 573-82 (1959)
CA 54: 14971c

C 857 Fluorescence induction phenomena in granular and lamellate chloroplasts.
Thomas, J.B., and Juboer, J.F.W.
J. Phys. Chem. 63: 39-44 (1959)
CA 53: 8319e

C 858 Discharge lamps and phosphors.
Thomas, M.J., and Butler, K.H.
U.S. 2,901,647 (1959)
CA 53: 1132i

C 859 A method of measuring temperature utilizing the thermal sensibility of fluorescent colors.
Thureau, P.
Publ. Sci. Tech. Min. Air No. 349, 131 pp. (1959)
CA 54: 7321d

C 860 Investigation of the spectral distribution of the luminescence decay time of ruby by the method of the pulse taumeter.
Tolstoi, N.A., and Tkachuk, A.M.
Opt. i Spektroskopiya 6: 659-64 (1959)
CA 53: 16712i

C 861 Formal analysis of the theory of two-step excitation of phosphorescence and photoconductivity.
Tolstoi, N.A.
Opt. i Spektroskopiya 6: 665-71 (1959)
CA 53: 16713a

C 862 Usefulness of the thiosulfate method of zinc sulfide production for the synthesis of luminescent materials.
Tombak, M.I., Popova, A.V., Komar, O.F., and Bundel, A.A.
Izv. Akad. Nauk SSSR, Ser. Fiz. 23: 1363-9 (1959)
CA 54: 6331i

C 863 Molecular weights of sugar beet araban fractions.
Tomimatsu, Y., Palmer, K.J., Goodban, A.E., and Ward, W.H.
J. Polymer Sci. 36: 129-39 (1959)
CA 53: 19520h

C 864 The mechanism of photosensitized oxidation-reduction reactions.
Tomita, G., and Takeyama, N.
Kagaku 29: 662 (1959)
CA 54: 20433a

C 865 Fluorescence of deoxyribonucleic acid isolated from tissue of irradiated animals.
Toropova, G.P., and Pozdnyakov, A.L.
Med. Radiol. 4: 57-60 (1959)
CA 53: 19001a

C 866 Phosphoryl chloride enhancement of fluorescence and absorbance of estrogens in sulfuric acid.
Touchstone, J.C., Greene, J.W., and Kukovetz, W.R.
Anal. Chem. 31: 1693-6 (1959)
CA 54: 4739i

C 867 Fluorescence of the azulenium ions.
Treibs, W., and Scholz, M.
Z. Physik. Chem. 212: 118-21 (1959)
CA 54: 2007i

C 868 Copper activated calcium orthophosphate and related phosphors.
Uehara, Y., Kobuke, Y., and Masuda, I.
J. Electrochem. Soc. 106: 200-5 (1959)
CA 54: 8826i

C 869 Capture centers and nonisothermal relaxation processes in ammonium halide crystal phosphors.
Uibo, L.Y.
Materialy VII Soveshch. po Lyuminest. (Kristallofosfory), Akad. Nauk Est. SSR, Moscow 1958: 164-70 (1959)
CA 55: 8070c

C 870 Evaluation of colorants by spectrophotometric methods.
Ulrich, W.F., Kelley, F., and Nelson, D.C.
Paint Ind. Mag. 741: 11-12 (1959)
CA 53: 8657e

C 871 Fluorescence of tryptophan derivatives in trifluoroacetic acid.
Uphaus, R.A., Grossweiner, L.I., Katz, J.J., and Kopple, K.D.
Science 129: 641-2 (1959)
CA 53: 13215i

C 872 Study of photoelectric detection in pulsed operation. Application to spectrometry of the Raman effect.
Valentin, F.
Ann. Phys. 4: 1239-90 (1959)
CA 54: 9497a

C 873 Influence of hydrogen on the red ZnS–Cu
fluorescence.
vanGool, W., and Cleiren, A.P.D.M.
J. Electrochem. Soc. 106: 672-6 (1959)
CA 53: 17683a

C 874 The identification of belladonna leaves
and tincture based on the fluorescence
of chrysatronic acid.
Varady, J.
Gyogyszereszet 3: 296-300 (1959)
CA 59: 10470h

C 875 Absorption spectra and luminescence of
cerium-containing glasses.
Vargin, V.V., and Karapetyan, G.O.
Glastech. Ber. 32: 443-50 (1959)
CA 54: 8278i

C 876 Self-absorption and trapping of sharp-
line resonance radiation in ruby.
Varsanyi, F., Wood, D.L., and Schawlow,
A.L.
Phys. Rev. Letters 3: 544-5 (1959)
CA 54: 8296f

C 877 Monochromatically excited fluores-
cence in rare-earth salts.
Varsanyi, F., and Dieke, G.H.
J. Chem. Phys. 31: 1066-70 (1959)
CA 54: 7336a

C 878 Application of thermography in the in-
vestigation of zinc sulfide.
Vasil'eva, E.G., and Fridman, S.A.
Izv. Akad. Nauk SSSR, Ser. Fiz. 23: 1347-
50 (1959)
CA 54: 6332e

C 879 Instrumental method for measuring
fluorescence and a study of fluorescent
ruthenium polyamine complexes.
Veening, H.
Univ. Microfilms Mic 59-1653 (1959)
CA 53: 18626c

C 880 Fluorometric analysis of coenzyme
binding and thiol interactions of
glyceraldehyde-3-phosphate and lactic
dehydrogenases.
Velick, S.F.
Sulfur Proteins, Proc. Symp. Falmouth,
Mass. 1958: 267-78 (1959)
CA 53: 18116f

C 881 Mutual influence of some 9-monoderi-
vatives of anthracene on the quantum
yields of their photochemical transformations and fluorescence.
Vember, T.M., and Cherkasov, A.S.
Opt. i Spektroskopiya 6: 232-4 (1959)
CA 54: 9491b

C 882 Microscope phase fluorimeter for de-
termining the fluorescence lifetimes of
fluorochromes.
Venetta, B.D.
Rev. Sci. Instr. 30: 450-7 (1959)
CA 54: 13757c

C 883 Estimation of ergot alkaloids in cultures
of claviceps purpurea.
Vining, L.C., and Tober, W.A.
Can. J. Microbiol. 5: 441-51 (1959)
CA 54: 2664f

C 884 Fluorescent substances from Drosophila melanogaster. XIII. Further contri-
butions to the elucidation of the structure of the sepiapterins and the droso-
pterins.
Viscontini, M., and Mohlmann, E.
Helv. Chim. Acta 42: 1679-83 (1959)
CA 54: 4939h

C 885 Fluorescence of aromatic amino acids
in solutions, crystals, and proteins.
Vladimirov, Y.V.
Izv. Akad. Nauk SSSR, Ser. Fiz. 23: 86-9
(1959)
CA 53: 12370e

C 886 Ultraviolet absorption and fluorescence
spectra of radicals produced by high-
frequency discharge in vapors of toluene
and octadeuterriated toluene and trapped
at low temperature.
Vocher, M., and Lortie, Y.
J. Chim. Phys. 56: 732-5 (1959)
CA 54: 6301g

C 887 Additional absorption and fluorescence
in activated alkali halide phosphors and
the lattice energy.
Vorab'ev, A.A.
Nauchn. Dokl. Vysshei Shkoly, Fiz.-Mat.
Nauki 1959: 149-50 (1959)
CA 54: 21986i

C 888 Phthaladehyde as a reagent.
Wachsmuth, H., Denissen, R., and van
Koeckhoven, L.
J. Pharm. Belg. 14: 386-91 (1959)
CA 54: 15834f

C 889 Reactions of ergotamine.
Wachsmuth, H., and van Koeckhoven, L.
J. Pharm. Belg. 14: 461-2 (1959)
CA 54: 15834e

C 890 Uranium-prospecting with the ultra-
violet lamp.
Walenta, K.
Z. Erzbergbau Metallhuettenw. 12: 51-5
(1959)
CA 53: 9920g

C 891 Emission and absorption spectra of
molecules of Acridine Yellow in the
preexcited state.
Walerys, H.
Bull. Acad. Polon. Sci., Ser. Sci., Math.,
Astron. Phys. 7: 47-9 (1959)
CA 53: 18626b

C 892 Luminescence of copper-activated cal-
cium and strontium orthophosphates.
Wanmaker, W.L., and Bakker, C.
J. Electrochem. Soc. 106: 1027-32 (1959)
CA 54: 2960g

C 893 Polarization of the ultraviolet fluores-
cence and electronic energy transfer
in proteins.
Weber, G., and Teale, F.W.J.
Biochem. J. 72: 15-16 (1959)
CA 55: 24847i

C 894 Comparison of concentration measure-
ments of sulfur dioxide and fluorescent
pigment.
Wedin, B., Fressling, N., and Aurivillius,
B.
Advan. Geophys. 6: 425-7 (1959)
CA 54: 3811c

C 895 Fluorescence of some coumarins.
Whellock, E.
J. Am. Chem. Soc. 81: 1348-52 (1959)
CA 54: 14938d

C 896 Effect of pH on fluorescence of tyro-
sine, tryptophan, and related com-
pounds.
White, A.
Biochem. J. 71: 217-20 (1959)
CA 53: 5867b

C 897 Solid-state high-intensity monochroma-
tic light sources.
Wieder, I.
Rev. Sci. Instr. 30: 995-6 (1959)
CA 54: 19028c

C 898 Metalfluorochromic indicators.
Wilkins, D.H.
Talanta 2: 277-8 (1959)
CA 54: 7411c

C 899 Polarization of the phosphorescence of
naphthalene and phenanthrene.
Williams, R.
J. Chem. Phys. 30: 233-7 (1959)
CA 53: 10956g

C 900 Fluorescence of some aromatic com-
pounds in aqueous solution.
Williams, R.T.
J. Roy. Inst. Chem. 83: 611-26 (1959)
CA 54: 2935h

C 901 Absorption and fluorescence polariza-
tion spectra of some mono- and diamin-
oacridines at low temperatures.
Wittwer, A., and Zanker, V.
Z. Physik. Chem. 22: 417-39 (1959)
CA 54: 7338c

C 902 Excitation spectra of the recombination
luminescence of crystalline alkali
halide phosphors.
Yaek, I.V.
Tr. Inst. Fiz. i Astron., Akad. Nauk Est.
SSR 1959: 166-95 (1959)
CA 54: 22038c

C 903 Fluorescence of flavine enzymes.
Yama, M.T.
Osaka Daigaku Igaku Zasshi 11: 4311-17
(1959)
CA 54: 14339a

C 904 Fluorescence of naphthalimide derivatives. Effect of substituents on the radiation transition probability and the
radiationless transition probability.
Yasuda, K., Inukai, K., and Ito, K.
Nippon Kagaku Zasshi 80: 960-2 (1959)
CA 53: 21165h

C 905 Fluorescence spectra of substituted
naphthalimides.
Solvent effect.
Yasuda, K., Okabe, K., Inukai, K., and
Ito, K.
Nippon Kagaku Zasshi 80: 962-5 (1959)
CA 53: 21166a

C 906 Recent results of absorption fluorescence and fluorescence-polarization measurements on the Acridine Orange cation, a further contribution to the problem of metachromasia of the vital stain.
Zanker, V., Held, M., and Rammensee, H.
Z. Naturforsch. 14b: 789-801 (1959)
CA 54: 14938g

C 907 Further spectroscopic data on the deep color of 9-substituted acridines.
Zanker, V., and Reichel, A.
Z. Elektrochem. 63: 1133-40 (1959)
CA 54: 5243c

C 908 Afterglow of zinc sulfide on excitation with an electron beam of small current density.
Zavrazhin, A.G., and Blazhevich, A.I.
Materialy VII Soveshch. po Lyuminest. (Kristallofosfory), Akad. Nauk Est. SSR, Moscow 1958: 316-22 (1959)
CA 55: 10095a

C 909 Relation between the transition probability of complex organic molecules to the metastable state and the spectral composition of the emitted radiation.
Zelinskii, V.V., Kolobkov, V.P., and Reznikova, I.I.
Izv. Akad. Nauk SSSR, Ser. Fiz. 23: 1269-72 (1959)
CA 54: 7336c

C 910 Temperature quenching of fluorescence.
Zelinskii, V.V., Kolobkov, V.P., and Krasnitskaya, N.D.
Opt. i Spektroskopiya 6: 417-19 (1959)
CA 55: 10059c

C910a Connection of the electronic absorption spectra and of the radiation of solutions of organic substances with the chemical nature of the solvent.
Zelinskii, V. V. Kolobkov, V.P., and Reznikova, I.I.
Tr. Soveshch., Moscow 1958: 262-6 (1959)
CA 54: 21998b

C 911 Dependence of the polarization of the fluorescence of molecular crystals on the wave length of emission radiation.
Zhevandrov, N.D., Gribkov, V.I., and Varfolomeeva, V.N.
Izv. Akad. Nauk SSSR, Ser. Fiz. 23: 57-61 (1959)
CA 53: 11989f

C 912 Universal scale for the action of solvents on the electron spectra of organic compounds.
Zhmyreva, I.A., Zelinskii, V.V., Kolobkov, V.P., and Krasnitskaya, N.D.
Dokl. Akad. Nauk SSSR 129: 1089-92 (1959)
CA 55: 26658f

C 913 Absorption and S-S luminescence spectra of aromatic crystals at 20°K: benzene, hexadeuterobenzene, naphthalene, octadeuteronaphthalene.
Zmerli, A.
J. Chim. Phys. 56: 387-404 (1959)
CA 54: 347i

C 914 The T-S luminescence spectra of aromatic crystals at 20°K: benzene, hexadeuterobenzene, naphthalene, octadeuteronaphthalene.
Zmerli, A.
J. Chim. Phys. 56: 405-17 (1959)
CA 54: 347h

C 915 Phosphorescence spectra of some phenols at liquid-oxygen temperature.
Zudin, A.A.
Izv. Akad. Nauk SSSR, Ser. Fiz. 23: 142 (1959)
CA 53: 11989i

(1960-1964)

D 1 Quenching of luminescence by infrared
   radiation.
   Adam, J.
   Zur Physik Chemie Kristallphosphore,
   Tagung Physik. Ges. D.D.R.,1.,
   Greifswald, Ger. 1959: 117-23 (1960)
   CA 60: 3632f

D 2 Ultraviolet flourescence of nucleic acids
   and polyphosphates.
   Agroskin, L.S., Korolev, N.V., Kulaev,
   I.S., Meisel, M.N., and Pomoshchniko-
   va, N.A.
   Dokl. Akad. Nauk SSSR 131: 1440-3 (1960)
   CA 55: 3678d

D 3 Photometric determination of gallium
   and indium with quercetin.
   Alimarin, I.P. , Golovina, A.P., and
   Torgov, V.G.
   Zavodsk. Lab. 26: 709-11 (1960)
   CA 54: 19298f

D 4 Addition of ethylenic compounds to
   tetracyclones.
   Allen, C.F.H., Ryan, R.W., and Van Allan,
   J.A.
   J. Org. Chem. 27: 778-9 (1960)
   CA 57: 4570f

D 5 Effect of metal atom perturbations on the
   luminescent spectra of porphyrins.
   Allison, J.B., and Becker, R.S.
   J. Chem. Phys. 32: 1410-17 (1960)
   CA 54: 21998h

D 6 Naphthalene derivatives in inorganic
   analysis. VIII. Effects of substituents
   in the naphthalene nucleus on the
   fluorimetric detection of tin.
   Anderson, J.R.A., Garnett, J.L., and
   Lock, L.C.
   Anal. Chim. Acta 22: 1-7 (1960)
   CA 54: 6405f

D 7 Effects of added gases on the sensitized
   fluorescence spectrum of a Hg-Tl
   mixture.
   Anderson, R.A., and McFarland, R.H.
   Phys. Rev. 119: 693-700 (1960)
   CA 54: 21992f

D 8 The fluorescence of 2-pyrazoline deriva-
   tives.
   Andreeshchev, E.A., Kouyrzina, K.A., and
   Baron, E.E.
   Stsintillyatory i Stsintillyats. Materialy,
   Vses. Nauchn.-Issled. Inst. Khim.
   Reaktivov, Materialy 2-go Koordinats.
   Soveshch. 1957: 171-81 (1960)
   CA 57: 13274g

D 9 Metallic reflection from molecular
   crystals.
   Anex, B.G., and Simpson, W.T.
   Rev. Mod. Phys. 32: 466-76 (1960)
   CA 54: 22028d

D 10 Polymer vitrification and phosphores-
   cence.
   Anufrieva, E.V., and Zaitseva, A.D.
   Izv. Akad. Nauk SSSR, Ser. Fiz. 24: 755-8
   (1960)
   CA 54: 22002f

D 11 Optical properties of KCl : Tl in the
   extreme ultraviolet range.
   Aoyagi, K., and Kuwabava, G.
   J. Phys. Soc. Japan 15: 2334-42 (1960)
   CA 55: 24264c

D 12 Vanadium-activated zinc and cadmium
   sulfide and selenide phosphors.
   Avinor, M., and Meijer, G.
   Phys. Chem. Solids 12: 211-15 (1960)
   CA 54: 20500b

D 13 Color tests for the detection of sterols
   and estrogens on filter paper.
   Axelrod, L.R., and Pulliam, J.E.
   Arch. Biochem. Biophys. 89: 105-9 (1960)
   CA 54: 21221e

D 14 Transfer of triplet-state energy in
fluid solutions. I. Sensitized phosphorescence and its application to the
determination of triplet-state lifetimes.
Backstrom, H.C.J., and Sandros, K.
Acta Chem. Scand. 14: 48-62 (1960)
CA 56: 8137h

D 15 Instrument for the measurement of
fluorescence of paper chromatographic
spots.
Bailey, G.F.
Anal. Chem. 32: 1726-7 (1960)
CA 55: 6945e

D 16 Dielectric effects and properties of
electronic spectra of multiatomic
organic molecules in solutions.
Bakhshiev, N.G.
Izv. Akad. Nauk SSSR, Ser. Fiz.
24: 587-90 (1960)
CA 54: 22000b

D 17 Interpretation of low-intensity absorp-
tion bands of transition metal cyano
complexes.
Bán, M.I.
Magy. Kem. Folyoirat 66: 325-6 (1960)
CA 55: 7037a

D 18 Fluorescence and light absorption attri-
butable to excitons in CdS semiconduc-
tor crystals.
Bancie-Grillot, M., and Grillot, E.
Proc. Intern. Conf. Color Centers Crystal
Luminescence 1960: 166-86 (1960)
CA 58: 1983e

D 19 Influence of temperature on the two
series of bonds in the green fluores-
cence spectra of pure cadmium sulfide
at low temperature.
Bancie-Grillot, M., Gross, E.F., Grillot,
E., and Razbirin, B.S.
Compt. Rend. 250: 2868-70 (1960)
CA 54: 20515a

D 20 Chemical dosimetry with fluorescent
compounds: the destruction of the
fluorescence of quinine by γ-rays.
Barr, N.F., and Stark, M.B.
Radiation Res. 12: 1-4 (1960)
CA 54: 8325f

D 21 Transfer of excitation energy in solid
solutions of anthracene-polystyrene and
9,10-diphenylanthracene-polystyrene.
Basile, L.J., and Weinreb, A.
J. Chem. Phys. 33: 1028-36 (1960)
CA 55: 10057a

D 22 Time measurements of fluorescence
extinction.
Bauer, R., and Rozwadowski, M.
Postepy Fiz. (Poland) 11: 379-404 (1960)
CA 55: 13035g

D 23 The fluorescence of eight natural estrogens in sulfuric acid.
Bauld, W.S., Givner, M.L., Engel, L.L.,
and Goldzieher, J.W.
Can. J. Biochem. Physiol. 38: 213-32
(1960)
CA 54: 8954a

D 24 New method for the study of glass flow
in flat glass tanks.
Becker, H.
Glastech. Ber. 33: 411-17 (1960)
CA 55: 3943d

D 25 Fluorescence of certain bile acids after
heating with sulfuric acid.
Benard, H., and Broer, Y.
Bull. Soc. Chim. Biol. 42: 99-114 (1960)
CA 55: 3686h

D 26 Instrument to measure fluorescence
lifetimes in the millimicrosecond
region.
Bennett, R.G.
Rev. Sci. Instr. 31: 1275-9 (1960)
CA 55: 20527f

D 27 The ultraviolet absorption of some
$1\alpha,5\alpha$-epidithio steroids.
Bergson, G., Sjoberg, B., Tweit, R.C.,
and Dodson, R.M.
Acta Chem. Scand. 14: 222-3 (1960)
CA 60: 8787b

D 28 Phosphorescence mechanisms. I.
Approach and general analysis.
Billington, C.
Phys. Rev. 120: 697-701 (1960)
CA 55: 4166c

D 29 Liquid scintillators. X. Some aryl-
substituted phenanthrenes and dihyrdro-
phenanthrenes, and related *p* -ter-
phenyls and *p* -quarterphenyls. Determination of Kallmann parameters.
Birkeland, S.P., Daub, G.H., Hayes, F.N.,
and Ott, D.G.
Z. Physik.159: 516-23 (1960)
CA 55: 5122b

D 30 Polarization of fluorescence in CdS and
ZnS single crystals.
Birman, J.
J. Electrochem. Soc. 107: 409-17 (1960)
CA 54: 14970 h

D 31 Relations between light fastness and
dye constitution and their dependence
upon the substrate.
Bitzer, D., and Brielmaier, H.J.
Melliand Textilber. 41: 62-4 (1960)
CA 54: 8090b

D 32 "Last lines" of the spectrum at various
temperatures of 3,4-benzopyrene,
dissolved in normal hydrocarbons.
Bogomolov, G., Pemova, F.D., and
Kolosova, L.P.
Izv. Akad. Nauk SSSR, Ser. Fiz. 24: 725-7
(1960)
CA 54: 22001g

D 33 Quenching photoluminescence of
solutions.
Bojarski, C.
Acta Physiol. Polon. 19: 631-6 (1960)
CA 55: 17209e

D 34 Remarks on the theory of concentration
depolarization of fluorescent solutions.
Bojarski, C.
Ann. Physik 5: 249-51 (1960)
CA 54: 10496f

D 35 Polymerization of flavans. III. The
action of lead tetraacetate of flavans.
Bokadia, M.M., Brown, B.R., and
Cummings, W.
J. Chem. Soc. 1960: 3308-13 (1960)
CA 55: 1601e

D 36 The reaction of pyridoxal 5-phosphate
with cyanide and its analytical use.
Bonavita, V.
Arch. Biochem. Biophys. 88: 366-72 (1960)
CA 54: 19273i

D 37 Basic processes of the deactivation of
excited states in complex organic
molecules.
Borgman, V.A., Zhmyreva, I.A.,
Zelinskii, V.V., and Kolobkov, V.P.
Izv. Akad. Nauk SSSR, Ser. Fiz. 24: 601-6
(1960)
CA 54: 22000e

D 38 The effect of heavy halogens on the
probability of transition to metastable
state, and the chances for a deactiva-
tion of this state.
Borgman, V.A., Zhmyreva, I.A.,
Zelinskii, V.V., and Kolobkov, V.P.
Dokl. Akad. Nauk SSSR 131: 781-4 (1960)
CA 55: 16139e

D 39 Electronic spectra of anthraquinone
vapors.
Borisevich, N.A., and Gruzinskii, V.V.
Izv. Akad. Nauk SSSR, Ser. Fiz. 24: 545-8
(1960)
CA 54: 21999g

D 40 Temperature dependence of the
fluorescence yield in vapors of
complex molecules.
Borisevich, N.A., and Tolkachev, V.A.
Izv. Akad. Nauk SSSR, Ser. Fiz. 24: 521-4
(1960)
CA 54: 21999e

D 41 Use of ultraviolet radiation in paper
chromatography.
Borodin, N.S., Galshin, E.A., Senyakina,
N.A., and Silaeva, V.N.
Metody Lyuminests. Analiza Sb. 1960:
81-2 (1960)
CA 56: 4071i

D 42 Energy transfer in luminescent solu-
tions. I. Solutions containing one
luminescent substance.
Bothe, H.K.
Ann. Physik 5: 339-52 (1960)
CA 54: 10543b

D 43 Energy transfer in luminescent solu-
tions. II. Solutions with two dissolved
luminescent substances.
Bothe, H.K.
Ann. Phys. 6: 156-68 (1960)
CA 55: 5121c

D 44 Fluorescence
Bowen, E.J.
Ciba Rev. 12: 2-12 (1960)
CA 55:5959e

D 45 Causes of removal of inner nonradiating
transitions in organic molecules upon
formation of intercomplex compounds
with cations.
Bozhevol'nov, E.A.
Izv. Akad. Nauk SSSR, Ser. Fiz. 24: 762-6
(1960)
CA 54: 22002i

D 46 Fluorescent properties of fluorescein
isocyanate.
Bozhevol'nov, E.A.
Metody Lyuminests. Analiza Sb. 1960:
65-70 (1960)
CA 56: 2075a

D 47 Determination of aluminum by the
fluorescence method.
Bozhevol'nov, E.A., and Yanishevskaya,
V.M.
Zh. Vses. Khim. Obshchestva im. D. I.
Mendeleeva 5: 356-7 (1960)
CA 54: 19294b

D 48 Triplet-singlet emission spectra of
xylenes in crystalline state of 4.2 and
77 K, and in EPA at 77 K.
Blackwell, L.A., Kanda, Y., and Sponer, H.
J. Chem. Phys. 32: 1465-76 (1960)
CA 54: 21998f

D 49 Measurement of the luminescence decay
by optical excitation in the range of
10^{-8} to 10^{-4} sec.
Blume, H.
Z. Naturforsch. 15a: 743 (1960)
CA 55: 107g

D 50 Delay in intermolecular and intramole-
cular energy transfer and the lifetimes
of photosynthetic pigments.
Brody, S.S.
Z. Elektrochem. 64: 187-94 (1960)
CA 54: 12279c

D 51 Fluorescence properties of the intermediates in the photoreduction of
chlorophyll a and evidence for complex
formation in solution.
Brody, S. S.
J. Am. Chem. Soc. 82: 1570-4 (1960)
CA 54: 16174h

D 52 Phosphorescence of nitrogen and
nitrogen-argon deposited films at 4.2 K.
Broida, H.P., and Nicholls, R.W.
J. Chem. Phys. 32: 623-4 (1960)
CA 54: 14930c

D 53 Formation of vibrationally excited car-
bon monosulfide and sulfur by the flash
photolysis of carbon disulfide.
Callear, A.B., and Norrish, R.G.W.
Nature 188: 53-4 (1960)
CA 55: 6137c

D 54 Optical bleaching agents and paper.
Carr, W.
Papermaker 138: No. 3, 58-60, 62-3;
No. 4, 66, 68, 70-2, 87; No. 5, 52-4,
56, 68 (1960)
CA 50: 11843c

D 55 Ultrazoles based on 2-aryl-1,2,3-
triazoles.
Cepciansky, I., and Vanicek, V.
Sb. Ved. Praci, Vysoka Skola Chem.-
Technol. Pardubice 1960: 107-20 (1960)
CA 55: 22300f

D 56 Depolarization of the fluorescence of
proteins labeled with various fluor-
escent dyes.
Chadwick, C.S., Johnson, P., and
Richards, E.G.
Nature 186: 239-40 (1960)
CA 54: 18618h

D 57 The absorption and fluorescence spec-
tra of naphthalene molecules in
anthracene crystals.
Chandhuri, N.K., and Ganguly, S.C.
Proc. Roy. Soc. A259: 419-23 (1960)
CA 55: 9042f

D 58 Effect of solvent on fluorescence spec-
tra of acetylanthracenes.
Cherkasov, A.S.
Izv. Akad. Nauk SSSR, Ser. Fiz. 24: 591-5
(1960)
CA 54: 23730h

D 59 Fluorescence yields of acetylanthra-
cenes in solvent mixtures.
Cherkasov, A.S.
Opt. i Spektroskopiya 9: 540-2 (1960)
CA 55: 14057b

D 60 The appearance of intermolecular attraction in the fluorescence spectra for solutions of the acetylanthracenes in mixed solvents.
Cherkasov, A.S.
Dokl. Akad. Nauk SSSR 130: 1288-90 (1960)
CA 56: 2090f

D 61 Organophosphors. X. Luminescent systems based on crystalline inorganic substrates and organic activators.
Chomse, H., and Arend, I.
Zur Physik Chemie Kristallphosphore, Tagung Physik. Ges. D.D.R., 1, Geifswald, Ger. 1959: 243-54 (1960)
CA 60: 3633e

D 62 Fluorescence of olive oil.
Ciusa, W., and Nebbia, G.
Boll. Sci. Fac. Chim. Ind. Bologna 18: 113-37 (1960)
CA 55: 9911a

D 63 Polarization of the edge emission in CdS.
Collins, R.J., and Hopfield, J.J.
Phys. Rev. 120: 840-2 (1960)
CA 55: 4166h

D 64 Coherence, narrowing, directionality, and relaxation oscillations in the light emission from ruby.
Collins, R.J., Nelson, D.F., Schawlow, A.L., Bond, W., Garret, C.G.B., and Kaiser, W.
Phys. Rev. Letters 5: 303-5 (1960)
CA 55: 10087f

D 65 Luminescence spectrum of $PmCl_3$.
Conway, J.G., and Gruber, J.B.
J. Chem. Phys. 32: 1586-7 (1960)
CA 54: 21992d

D 66 Immunofluorescence.
Coons, A.H.
Public Health Rept. 75: 937-43 (1960)
CA 55: 760c

D 67 Luminescence of partially deuteriated benzenes in dilute cyclohexane solution at liquid-nitrogen temperatures.
Courpron, C., Lochet, R., Meyer, Y., and Rousset, A.
Compt. Rend. 250: 3549-51 (1960)
CA 55: 1181f

D 68 Fluorimetric microdetermination of selenium in biological material.
Cousins, F.B.
Australian J. Exptl. Biol. Med. Sci. 38: 11-16 (1960)
CA 54: 15512b

D 69 Fluorescence spectra of coordinated holmium and thulium ions.
Crosby, G.A., and Whan, R.E.
Naturwissenschaften 47: 276-7 (1960)
CA 54: 23725a

D 70 Collision processes in mixtures of mercury vapor and foreign gases.
Cunningham, D.E., and Olsen, L.O.
Phys. Rev. 119: 691-3 (1960)
CA 54: 9486b

D 71 Phosphorescence stimulation by infrared.
Curie, G.
Ann. Phys. 5: 365-408 (1960)
CA 54: 14936a

D 72 Electrical fluorescence polarization. The determination of dipole moments of excited molecules from the degree of polarization of fluorescence on strong electric fields.
Czekalla, J.
Z. Elektrochem. 64: 1221-8 (1960)
CA 55: 11072b

D 73 An experimental arrangement for using the temperature jump method with biological systems.
Czerlinski, G.
Z. Elektrochem. 64: 78-9 (1960)
CA 61: 4680a

D 74 Modified titanium dioxide for semiconductors.
Dalton, H.R.
U. S. 2,940,941 (1960)
CA 54: 18094b

D 75 $5\alpha,11\alpha$-Dihydrochromomeno-$(3,4-\beta)$-chromone, the parent compound of the rotenoids.
Dann, O., and Volz, G.
Ann. 631: 111-6 (1960)
CA 54: 22607c

D 76 Luminescent paints.
Deribere, M.
Peintures, Pigments, Vernis 36: 76-82
(1960)
CA 54: 12611b

D 77 Change in the fluorescence of naphthalene derivatives as a function of the concentration of hydrogen ions in solution.
Derkacheva, L.D.
Opt. i Spektroskopiya 9: 209-14 (1960)
CA 55: 25463b

D 78 3,4-Dicyano-6-hydroxy-2-pyridones and the corresponding iminopyrans.
Dickinson, C.L.
U. S. 2,925,422 (1960)
CA 54: P15407a

D 79 Use of fine-structure fluorescent spectra for the detection of carcinogenic substance.
Dikun, P.P.
Vopr. Onkol. 6: 75-83 (1960)
CA 55: 1057f

D 80 Excited states in diketones and quinones and the photochemical behavior of anthraquinone dyes.
Dorr, F.
Z. Elektrochem. 64: 580-2 (1960)
CA 54: 20474h

D 81 The electric semiconductor character of serum albumin.
Douzou, P., and Thuillier, J.M.
J. Chim. Phys. 57: 96-100 (1960)
CA 54: 14967h

D 82 Fluorescence yield of eosin solutions in collodion and in glucose.
Drabent, R.
Bull. Acad. Polon. Sci., Ser. Sci., Math., Astron., Phys. 8: 403-8 (1960)
CA 57: 4195h

D 83 Investigations on stilbenes. XXVI.
Unsymmetrical diarylethylenes.
Drefahl, G., and Ponsold, K.
Chem. Ber. 93: 472-81 (1960)
CA 54: 13132f

D 84 Investigations on stilbenes. XXVII.
Stilbazoles with higher aromatic ring systems.
Drefahl, G., Ponsold, K., and Gerlack, E.
Chem. Ber. 93: 481-5 (1960)
CA 54: 13134c

D 85 Investigations on stilbenes. XXVIII.
Stilbenylimidazoles.
Drefahl, G., and Herma, H.
Chem. Ber. 93: 486-92 (1960)
CA 54: 13135b

D 86 Investigations on stilbenes. XXIX.
Stilbenyloxazoles.
Drefahl, G., and Engelmann, U.
Chem. Ber. 93: 492-7 (1960)
CA 54: 13136b

D 87 Directions of transition moments of absorption bands of polyenes, cyanines, and vitamin B_{12} from dichroism and fluorescence polarization.
Eckert, R., and Kuhn, H.
Z. Elektrochem. 64: 356-64 (1960)
CA 54: 13856d

D 88 Umbellicomplexone and xanthocomplexone. A study of complexometric fluorescent indicators.
Eggers, J.H.
Talanta 4: 38-43 (1960)
CA 54: 17146c

D89 See D843a

D 90 Fluorimetry. III. Fluorimetric determination of naphthionic acid by using short-wave ultraviolet light.
Eisenbrand, J., and Meyer, H.
Z. Anal. Chem. 174: 414-18 (1960)
CA 55: 2943 h

D 91 Fluorimetry. IV. The fluorimetric determination of glycerin following conversion to quinoline by using short-wave ultraviolet light (313 μ).
Eisenbrand, J., and Raisch, M.
Z. Anal. Chem. 177: 1-4 (1960)
CA 55: 4255a

D 92 A method for differentiating Trifolium and Medicago species, based on fluorescence.
Preliminary report
Eifrig, H.
Saatgut-Wirtsch. 12: 194-5 (1960)
CA 55: 26139a

D 93 The 4-pyrones. I. Reactions of some
4-pyrones and 4-thiopyrones involving
the ring oxygen.
Elkaschef, M.A., and Nossier, M.H.
J. Am Chem. Soc. 82: 4344-7 (1960)
CA 55: 2634g

D 94 The occurrence of fluorescent bodies in
the erythrocytes of phenylhydrazine-
poisoned rats.
Eriksen, L., and Jacobsen, F.
Acta Physiol. Scand. 50: 41-2 (1960)
CA 61: 8805h

D 95 Inner conversion from a fluorescent to
a phosphorescent level in naphthalene
derivatives.
Ermolaev, V.L., Kotyar, I.P., and
Svitashev, K.K.
Izv. Akad. Nauk SSSR, Ser. Fiz. 24: 492-5
(1960)
CA 54: 21999c

D 96 The effect of environment on singlet-
triplet transitions of organic mole-
cules.
Evans, D.F.
Proc. Roy. Soc. A 255: 55-61 (1960)
CA 54: 17045a

D 97 The luminescence of thallium-activated
potassium chloride phosphorus.
Ewles, J., and Joshi, R.V.
Proc. Roy. Soc. A 204: 358-71 (1960)
CA 54: 18093c

D 98 The infrared fluorescence of F-centers
and its mechanism in subtractively
colored alkali halide crystals investigated at high temperature.
Ezhik, I.I., and Shaulo, S.T.
Izv. Vysshikh Uchebn. Zavedenii, Fiz.
1960: 190-7 (1960)
CA 55: 16165d

D 99 Luminescence spectra of anthracene
crystals for different methods of their
preparation, excitation, and photography.
Faidish, O.M., and Zima, V.L.
Visn. Kiivs'k. Univ. 1960: No. 3 (1960)
CA 55: 12082d

D 100 Double-beam fluorimeter for quantita-
tive determination of uranium.
Fedorov, V.A., Freibert, S.I.
Metody Lyuminests. Analiza Sb. 1960:
27-31 (1960)
CA 56: 931e

D 101 Nature of the fluorescence of enzyme-
reduced diphosphopyridine nucleotide
(DPNH) complexes.
Fisher, H.F., and McGregor, L.L.
Biochim. Biophys. Acta 38: 562-2 (1960)
CA 54: 15486d

D 102 Effect of formic acid on the absorption
and fluorescence of trypaflavine.
Fokin, V.F.
Zh. Fiz. Khim. 34: 856-9 (1960)
CA 57: 5475h

D 103 Polarized absorption spectrum of ruby
between 14,400 and 16,000 cm^{-1} at
100 K.
Ford, R.A.
Spectrochim. Acta 16: 582-7 (1960)
CA 54: 23757c

D 104 Fluorescence of trivalent dysprosium
in hexagonal zinc sulfide at 77 K.
Ford, R.A., and Williams, M.M.R.
Spectrochim. Acta 16: 721-9 (1960)
CA 54: 23724g

D 105 Pterin Chemistry. III. Reaction of
CN^- with reduced 2-amino-6-hydroxy-
pteridine.
Forrest, H.S., Van Baalen, C., Viscontini,
M., and Piraux, M.
Helv. Chim. Acta 43: 1005-10 (1960)
CA 55: 562d

D 106 Intermolecular transfer of electronic
energy.
Forster, T.
Z. Elektrochem. 64: 157-65 (1960)
CA 54: 11689c

D 107 Yield of the fluorescence and spectra
of chlorophyll in viscous media.
Frackowiak, D., and Marszalek, T.
Bull. Acad. Polon. Sci., Ser. Sci., Math.,
Astron., Phys. 8: 713-7 (1960)
CA 57: 4195c

D 108 Decay of phosphorescence of trypoflav-
ine in gelatin.
Frackowiak, M., and Walerys, H.
Acta Physiol. Polon. 19: 199-215 (1960)
CA 55: 12033b

D 109 Fluorescence enhancement of crystal
phosphors by ultrasonics in the dose
measurement of ultraviolet and
γ-radiation.
Fredrich, K., and Herforth, L.
Zur Physik Chemie Kristallphosphore,
Tagung Physik. Ges., D.D.R., 1,
Greifswald, Ger. 1959: 124-8 (1960)
CA 60: 3632g

D 110 Ultraviolet spectroscopy: aromaticity
of carbonaceous materials: absorption
errors.
Friedel, R.A., and Queiser, J.A.
Am. Soc. Testing Mater., Spec. Tech.
Publ. No. 269: 218-26 (1960)
CA 55: 4146f

D 111 The emission spectra of tryptophan.
Fujimori, E.
Biochim. Biophys. Acta 40: 251-6 (1960)
CA 54: 21998i

D 112 The pigments of the cocoon. VII. The
fluorescent substances of the white
cocoon of the silkworm Bombyx mori.
Fujimoto, N., and Hayashiya, K.
Nippon Sanshigaku Zasshi 29: 495-500
(1960)
CA 61: 9810b

D 113 Rifomycin. VII. Spectrophotometry of
rifomycin B.
Gallo, G.G., Sensi, P., and Rodaelli, P.
Farmoco (Pavia), Ed. Prat. 15: 283-91
(1960)
CA 54: 21649d

D 114 New fluorescent oxidation-reduction
indicators.
Geyer, R., and Steinmetzer, H.
Wiss. Z. Tech. Hochsch. Chem. Leuna-
Merseburg 2: 423-30 (1959/60)
CA 55: 15407f

D 115 The photolysis and the fluorescence of
perfluoro diethyl ketone.
Giacometti, G., Okabe, H., Price, S.J.,
and Steacie, E.W.R.
Can. J. Chem. 38: 104-11 (1960)
CA 54: 11696h

D 116 Metal-ion binding of mononucleotides,
oligonucleotides, and polynucleotides.
Gillchriest, W. C.
U. S. Dept. Comm., Office Tech. Serv.
P. B. Rept. 153, 844 11 pp. (1960)
CA 57: 4138i

D 117 The fluorescence spectra of algae con-
taining bilichromoproteins.
Giraud, G.
Colloq. Intern. Centre Natl. Rech. Sci.
(Paris) 103: 83-90 (1960)
CA 57: 7621i

D 118 Detection of splits and cracks by
ultraviolet fluorescence.
Gobin, F.
Chim. Ind. 84: 532-50 (1960)
CA 55: 6056h

D 119 The physical characteristics and chem-
ical composition of electroluminescent
phosphors.
Goldberg, P., and Faria, S.
J. Electrochem. Soc. 107: 521-6 (1960)
CA 54: 20514i

D 120 Recording spectrofluorometers: two
new designs and their evaluation.
Goldzieher, J.W., Bauld, W.S., Engel,
L.L., and Givner, M.L.
Can. J. Biochem. Physiol. 38: 233-43
(1960)
CA 54: 8170b

D 121 Absorption and luminescence of dia-
mond.
Gomon, G.O.
Materialy Vses. Nauchn.-Issled. Geol.
Inst. 40: 125-46 (1960)
CA 56: 150f

D 122 Use of solvents containing ethyl iodide
in the investigation of phosphorescence
spectra of organic compounds.
Graham-Bryce, I.J., and Corkill, J.M.
Nature 186: 965-6 (1960)
CA 54: 23812h

D 123 Dependence of the polarization of
β-phosphorescence on the intensity of
exciting light.
Gribkovski, V.P.
Dokl. Akad. Nauk Belorusk. SSR 4: No. 5,
199-202 (1960)
CA 57: 10670a

D 124 Fluorescence measurements on mixed
crystals of cyclohexane.
Griessboch, D.
Z. Naturforsch. 15a: 296-301 (1960)
CA 54: 19187d

D 125 Luminescence and photoconductivity of
doped GaN.
Grimmeiss, H.G., Groth, R., and Maak, J.
Z. Naturforsch. 15a: 799-806 (1960)
CA 55: 3205f

D 126 Fungus pigment. XII. Structure and
synthesis of thelephoric acid.
Gripenberg, J.
Tetrahedron 10: 135-43 (1960)
CA 55: 1566e

D 127 Measurement of the contribution of
fluorescence to the brightness of
papers treated with whitening agents.
Grum, F., and Wightman, T.
Tappi 43: 400-5 (1960)
CA 54: 21756f

D 128 Uranium fluorimetry.
Haas, W.E.L.
Rev. Fac. Cien., Univ. Lisboa, 2a Ser.
B7: 77-104 (1959-60)
CA 55: 18446g

D 129 Fluorescent X-ray determination of
symmetry of the green phosphorescence
of heat-pretreated colored KCl
crystals.
Halperin, A., and Lewis, N.
Phys. Rev. 119: 510-15 (1960)
CA 54: 22036g

D 130 Fluorescent X-ray determination of
selenium in plant material.
Handley, R.
Anal. Chem. 32: 1719-20 (1960)
CA 55: 3298a

D 131 Relation of some surface chemical
properties of zinc silicate phosphor to
its behavior in fluorescent lamps.
Harrison, D.E.
J. Electrochem. Soc. 107: 210-17 (1960)
CA 54: 11717h

D 132 The system $BaO-TiO_2-P_2O_5$: phase
relations, fluorescence, and phosphor
preparation.
Harrison, D.E.
J. Electrochem. Soc. 107: 217-21 (1960)
CA 54: 10487e

D 133 Photoelectric spectrophotometer for
measuring absorption and emission of
fluorescent powders.
Hengge, E., Kruger, H.G., and Kubsa, H.
Chem.-Ing.-Tech. 32: 355-9 (1960)
CA 54: 14949i

D 134 Colloidal silica as a standard for
measuring absolute fluorescence
yield.
Hercules, D.M., and Frankel, H.
Science 131: 1611-12 (1960)
CA 54: 18066c

D 135 Luminescence spectra of naphthols
and naphthalenediols: low-temperature
phenomena.
Hercules, D.M., and Rogers, L.B.
J. Phys. Chem. 64: 397-400 (1960)
CA 54: 17045b

D 136 Factors influencing the formation of
phloem and heartwood polyphenols. I.
Examination of a specimen of Acacia
wood containing tension wood.
Hillas, W.E.
Holzforschung 14: 105-10 (1960)
CA 55: 3976b

D 137 Optical bleaching agents in paper
manufacture.
Hinton, A.J.
Textil-Rundschau 15: 654-8 (1960)
CA 55: 8856i

D 138 Luminescence of complex molecules
in relation to the internal conversion
of excitation energy. II. N-Heteroaro-
matics.
Hochstrasser, R.M.
Can. J. Chem. 38: 233-9 (1960)
CA 54: 18067a

D 139 Problem of radiative combinations be-
tween upper singlet states and the
ground state in aromatic molecules.
Hochstrasser, R.M.
Spectrochim. Acta 16: 497-504 (1960)
CA 54: 17044g

D 140 Light sources using resonance fluor-
escence of alkali metals.
Hoffman, K., and Seiwert, R.
Exptl. Tech. Physik (Berlin) 8: 161-77
(1960)
CA 55: 13052c

D 141 Relative quenching cross section in the reaction of Hg (6^3P_1) atoms with isotopic N_2O molecules.
Hoffman, M.Z., and Bernstein, R.E.
J. Chem. Phys. 33: 526-9 (1960)
CA 55: 4144d

D 142 Fluorescence of metal chelates of resorcylaldehyde and its derivatives.
Holzbecher, Z.
Collection Czech. Chem. Commun. 25: 977-82 (1960)
CA 54: 14166g

D 143 Further investigations on hash in fluorescent lamps.
Hoyaux, M., and Gans, P.
Proc. Intern. Conf. Ionization Phenomena Gases, 4th, Uppsala, 1959 1: 502-6 (1960)
CA 54: 20518f

D 144 Effect of deuterium substitution on the lifetime of the phosphorescent triplet state of naphthalene.
Hutchinson, C.A., and Mangum, B.W.
J. Chem. Phys. 32: 1261-2 (1960)
CA 54: 18083f

D 145 Fluorescence of coumarin derivatives. II. Fluorescence spectra of coumarins.
Ichimura, Y.
Yakugaku Zasshi 80: 771-4 (1960)
CA 54: 21646e

D 146 Fluorescence of coumarin derivatives. III. Flourescence of furocoumarins.
Ichimura, Y.
Yakugaku Zasshi 80: 775-8 (1960)
CA 54: 21646g

D 147 Fluorescence of coumarin derivatives. IV. Fluorometric analysis of coumarin derivatives.
Ichimura, Y.
Yakugaku Zasshi 80: 778-81 (1960)
CA 54: 21646h

D 148 Quantum yield of chlorophyll luminescence in various solvents.
Ivanov, N.P.
Izv. Akad. Nauk SSSR, Ser. Fiz. 24: 613-5 (1960)
CA 54: 22001c

D 149 Depolarization of fluorescence of liquid solutions.
Jablonski, A.
Bull. Acad. Polon. Sci., Ser. Sci., Math., Astron., Phys. 8: 655-60 (1960)
CA 57: 4195i

D 150 Triplet state of anthracene in fluid solutions.
Jackson, G., Livingston, R., and Pugh, A.C.
Trans. Faraday Soc. 56: 1635-9 (1960)
CA 55: 15116i

D 151 Sensitized phosphorescence of NaCl-Tl, Mn.
Jaek, I.
Tr. Inst. Fiz. i. Astron., Akad. Nauk Est. SSR 1960: 278-80 (1960)
CA 56: 2075h

D 152 Fluorescence reactions of europium with α-substituted pyridine and quinoline compounds.
Kallistratas, G., Pfau, A., and Ossowski, B.
Naturwissenschaften 47: 468-9 (1960)
CA 55: 7036g

D 153 Shape and nature of colloidal particles of 2,5-dihydroxy-p-benzoquinone copper (II) complex.
Kanda, S.
Nippon Kagaku Zasshi 81: 1347-8 (1960)
CA 55: 22994e

D 154 Sensitization of the oxidation of organic compounds and the quenching of their fluorescence by oxygen.
Karyakin, A.V.
Zh. Fiz. Khim. 34: 144-8 (1960)
CA 55: 15116f

D 155 Theory of fluorescence time constant measurements in liquid and rigid solutions.
Kallmann, H.P.
Phys. Rev. 117: 36-8 (1960)
CA 54: 23729i

D 156 Primary processes in the photobleaching of eosin as revealed by the flash technique.
Kato, S., Watanabe, T., Nagaki, S., and Koizumi, M.
Bull. Chem. Soc. Japan 33: 262-5 (1960)
CA 55: 7038d

D 157 Chlorophyll fluorescence and carbonic acid assimilation. VIII. Fluorescence curve and photochemistry of plants.
Kautsky, H., Appel, W., and Amann, H.
Biochem. Z. 332: 277-92 (1960)
CA 61: 3419c

D 158 Some optical properties of CdSe single crystals.
Keller, S.P., and Pettit, G.D.
Phys. Rev. 120: 1974-7 (1960)
CA 55: 5146g

D 159 The connection of the absorption and fluorescence spectra of solutions.
Ketskemity, I., Dombi, J., and Horvai, R.
Acta Phys. Acad. Sci. Hung. 12: 263-7 (1960)
CA 55: 17209h

D 160 Application of chloramine-T in amperometry.
Khadeev, V.A., Zhdanov, A.K., and Rechkina, L.G.
Uzbek. Khim. Zh. 1960: 28-37 (1960)
CA 56: 1982g

D 161 Effect of temperature on the phosphorescence decay of boron luminophors.
Khalvpovskii, M.D.
Opt. i Spektroskopiya 9: 525-7 (1960)
CA 55: 12062c

D 162 Influence of γ-irradiation on the photoluminescence of molecular crystals.
Khan-Magometova, S.D., Zhevandrov, N.D., and Gribkov, V.I.
Izv. Akad. Nauk SSSR, Ser. Fiz. 24: 561-6 (1960)
CA 54: 21999g

D 163 Spectroscopy of some pyrene derivatives in frozen solutions.
Khesina, A.Y.
Izv. Akad. Nauk SSSR, Ser. Fiz. 24: 623-6 (1960)
CA 54:22001e

D 164 An examination of undistillable Schroeter tar.
Kimber, R.W.L.
Chem. Ind. 1960: 657-8 (1960)
CA 55: 494c

D 165 Phosphorescence of certain solvents and their effect on absorption spectra of organic phosphors.
Kislyak, G.M.
Izv. Akad. Nauk SSSR, Ser. Fiz. 24: 766-8 (1960)
CA 54: 22003b

D 166 The luminescence of cadmium tungstate.
Klikorka, J., Horak, J., and Celikovsky, A.
Collection Czech. Chem. Commun. 25: 388-93 (1960)
CA 54: 14969i

D 167 The measurements of fluorescent materials.
Kling, A., and Kurz, J.
Melliand Textilber. 41: 339-41 (1960)
CA 54: 16836a

D 168 Influence of the surface of the adsorbent on the uranyl ion fluorescence.
Kobyshev, G.I.
Izv. Akad. Nauk SSSR, Ser. Fiz. 24: 752-5 (1960)
CA 54:23725b

D 169 Effect of the solvent on the fluorescence yield.
Kocheminovskii, A.S., and Renzmikova, I.I.
Opt. i Spektroskopiya 8: 399-401 (1960)
CA 57: 218c

D 170 Association and dissociation of centers in luminescent ZnS-In.
Koelmans, H.
Phys. Chem. Solids 17: 69-79 (1960)
CA 55: 17255f

D 171 Color and fluorescence reactions of gallium.
Korenman, I.M., Sheyanova, F.R., and Kunshin, S.D.
Zh. Analit. Khim. 15: 36-42 (1960)
CA 54: 13970c

D 172 Color and fluorescent reactions with quercetin.
Korenman, I.M., Sheyanova, F.K., and Shcherbakova,
Tr. po Khim. i. Khim. Tekhnol. 3: 303-6 (1960)
CA 56: 1423i

D 173 Fluorescence spectra of niobium in compounds NbB_2, NbC, NbN, and in pure niobium.
Korsunkii, M.I., and Genkin, Y.E.
Izv. Akad. Nauk SSSR, Ser. Fiz. 24: 461-4 (1960)
CA 54: 22038i

D 174 Chemical classification of plants. XX. Isolation of hashish substances from Cannabis sativa non indica.
Korte, F., and Sieper, H.
Ann. 630: 71-83 (1960)
CA 54: 24691f

D 175 Indoxyl
Kramer, D.N., and Gelman, C.
U. S. Army, Res. Develop. Lab. Rept. Edgewood Arsenal Maryland p. 541 (1960)

D 176 Optical properties of chlorophyll and pheophytin at low temperatures.
Kravtsov, L.A.
Izv. Akad. Nauk SSSR, Ser. Fiz. 24: 610-12 (1960)
CA 54: 22000g

D 177 Analysis of absorbed X-radiation.
Krokowski, E.
Naturwissenschaften 47: 35-6 (1960)
CA 55: 25455d

D 178 The kinetics of the concentration depolarization of luminescence and of the intermolecular transfer of excitation energy.
Kudryashov, P.I., Sveshnikov, B.Ya., and Shirokov, V.I.
Opt. Spectry. (USSR) 9: 177-81 (1960)
CA 61: 194a

D 179 Depolarization of fluorescence of solutions when excitation energy is transferred by radiation and in radiationless transfer.
Kudryashov, P.I., and Sveshnikov, B.Ya.
Dokl. Akad. Nauk SSSR 134: 792-4 (1960)
CA 56: 15060b

D 180 Phosphorescence concentration depolarization of organoluminophors.
Kudryashov, P.I., and Sveshnikov, B.Ya.
Opt. i Spektroskopiya 8: 651-6 (1960)
CA 55: 1181h

D 181 Displacement of the sulfur atom in CdS by electron bombardment.
Kulp, B.A., and Kelley, R.H.
J. Appl. Phys. 31: 1057-61 (1960)
CA 55: 133g

D 182 Photo- and electroluminescence of a $ZnS \cdot Cu \cdot Nd$ phosphor.
Kynev, K.
Godishnik Sofiiskiya Univ., Fiz.-Mat. 54: 95-112 (1960)
CA 56: 11045f

D 183 Polarization of luminescence in ZnS and CdS single crystals.
Lempicki, A.
J. Electrochem. Soc. 107: 404-9 (1960)
CA 54: 14970i

D 184 Health-physics instrumentation at Brookhaven National Laboratory. Automatic uranium fluorimeter.
Leng, J., Jupe, N.F.L., and Simpson, S.D.
At. Energy Can. Ltd. AECL 802: 109-13 (1960)
CA 55: 8074i

D 185 Isolation of melatonin and 5-methoxyindole-3-acetic acid from lavine pineal glands.
Lerner, A.B., Case, J.D., and Takahashi, Y.
J. Biol. Chem. 235: 1992-7 (1960)
CA 54: 24947h

D 186 2-Pyrrolidino- and 2-piperidinopurine.
Levin, G., and Tamari, M.
J. Chem. Soc. 1960: 2782-3 (1960)
CA 54: 24777e

D 187 Fluorescence spectra of fifty-nine polycyclic aromatic hydrocarbons.
Lijinsky, W., Chestnut, A., and Raha, C.R.
Chicago Med. School Quart. 21: 49-76 (1960)
CA 54: 18066g

D 188 The quenching of triplet states of anthracene and porphyrine by heavy metal ions.
Linschitz, H., and Pekkarinen, L.
J. Am. Chem. Soc. 82: 2411-16 (1960)
CA 54: 19159i

D 189 Physiological and anatomical charac-
teristics of spindling-sprorited
potato tubers (fluorescence).
Lippert, L.F.
Am. Potato J. 37: 313-24 (1960)
CA 56: 5148e

D 190 Absorption and fluorescence spectro-
scopic determination of the dissocia-
tion and tautomeric (equilibrium) con-
stants of lumazine and of its N- and
O-methyl derivatives.
Lippert, E., and Prigge, H.
Z. Elektrochem. 64: 662-71 (1960)
CA 54: 20474i

D 191 Luminescence of various forms of
chlorophyll in plant leaves.
Lituin, F.F., Krasnovskii, A.A., and
Rikhireva, G.T.
Dokl. Akad. Nauk SSSR 135: 1528-31 (1960)
CA 55: 11558a

D 192 Influence of electric field on the
scintillation process in CsI (Tl).
Lemonosov, I.I., and Nemilov, Y.A.
Fiz. Tverd. Tela 2: 1629-31 (1960)
CA 56: 8144 g

D 193 Properties of some crystalline phos-
phors at low temperatures.
Lomonosov, I.I., Nernilov, Yu.A., and
Storozhenko, E.P.
Stsintillyatory i Stsintillyats. Materialy,
Vses. Nauchn.-Issled. Inst. Khim.
Reaktivov, Materialy 2-go Koordinats.
Soveshch. 1957: 74-6 (1960)
CA 57: 19671b

D 194 Paramagnetic and optical spectra of
ytterbium in the cubic field of calcium
fluoride.
Low, W.
Phys. Rev. 118: 1608-9 (1960)
CA 54: 23824f

D 195 Luminescence of uranium-activated
fluorides.
Lys, J.E.A., and Runciman, W.A.
Proc. Phys. Soc. 76: 158-60 (1960)
CA 54: 20475h

D 196 Bicyclo-2,2,1-heptadiene in the
Diels-Alder reaction.
Mackenzie, K.
J. Chem. Soc. 1960: 473-83 (1960)
CA 54: 12084b

D 197 Reduction of current fluctuations in the
Farrand spectrofluorometer.
Mahler, D.J., Humoller, F.L., Beenkery,
H.G., and Loch, R.D.
Anal. Chem. 32: 1374 (1960)
CA 55: 22943b

D 198 Evaluation of cigaret filter efficiency
by photofluorimetry.
McConnell, W.V., Mumpower, R.C., and
Touey, G.P.
Tobacco Sci. 4: 56-61 (1960)
CA 60: 16229e

D 199 Energy transfer in molecular com-
plexes. II. The anthracene-sym-
trinitrobenzene complex.
McGlynn, S.P., Boggus, J.D., Elder, E.
J. Chem. Phys. 32: 357-61 (1960)
CA 54: 13856h

D 200 Valence-bond resonance structure for
a triplet state.
McLachlan, A.D.
J. Chem. Phys. 33: 663-4 (1960)
CA 55: 6132a

D 201 A vibrational effect on the polarization
of molecular crystal fluorescence.
McRae, E.G.
J. Chem. Phys. 33: 932-3 (1960)
CA 55: 7066i

D 202 Excitation spectra of vanadium-
activated zinc and cadmium sulfide and
selenide phosphors.
Meijer, G., and Avinor, M.
Philips Res. Rept. 15: 225-37 (1960)
CA 54: 20516i

D 203 A standard fluorescence spectrum for
calibrating spectrofluorophotometers.
Melhuisch, W.H.
J. Phys. Chem. 64: 762-4 (1960)
CA 54: 23847b

D 204 Decay of fluorescence.
Metcalf, W.S.
J. Chem. Soc. 1960: 3726-9 (1960)
CA 55: 18295i

D 205 Fluorescence and ionization rate of an
acid.
Metcalf, W.S.
J. Chem. Soc. 1960: 3729-33 (1960)
CA 56: 18296b

D 206 Displacement of the fluorescence
spectra of anthracene and 9,10-
dibromoanthracene solutions as a
function of the solvent.
Mihul, C., Pop, V., and Haba, M.
Acad. Rep. Populare Romine, Filiala
Iasi, Studii Cercetari Stiint., Fiz.
Stiinte Tehnice 11, 2: 175-81 (1960)
CA 56: 1063a

D 207 Spectroscopic investigation of vibrational energy transformation at the
collision of complex molecules. II.
Action of foreign gases on the fluorescence yield of 3-di-methylamino-6-
aminophohalimide vapors.
Mirumyants, S.O., and Neporent, B.S.
Opt. i Spektroskopiya 8: 787-98 (1960)
CA 55: 13052f

D 208 Action of foreign gases on the electron
absorption intensity of 3-dimethylamino-
6-aminophthalimide vapors.
Mirumyants, S.O., and Neporent, B.S.
Opt. i Spektroskopiya 9: 215 (1960)
CA 54: 23846g

D 209 Apatite from the Morefield permatite,
Amelia County, Va.
Mitchell, R.S., and McGauock, E.H.
Rock Minerals 35: 553-5 (1960)
CA 59: 13408e

D 210 Quenching of luminescence owing to
concentration (of the activator) in
organoluminophors.
Mokeeva, G.A., and Sveshnikov, B.Ya.
Opt. i Spektroskopiya 9: 601-7 (1960)
CA 55: 12032h

D 211 Organic analysis. XIX. Micro fluor-
metric estimation of isoniazid.
Momose, T., Veda, Y., Mukai, Y., and
Watanabe, K.
Yakugaku Zasshi 80: 225-8 (1960)
CA 54: 11861e

D 212 Identification of polycyclic aromatic
hydrocarbons by ultraviolet fluorescence.
Monkman, J.L., and Porro, T.J.
Proc. Symp. Instr. Methods Anal. 6: D4
10 pp (1960)
CA 60: 7468d

D 213 Fluorescence and phosphorescence of
cadmium iodide.
Monod-Herzen, G., Chung, N.T., and
Roodenbeck, A.
Compt. Rend. 250: 3618-19 (1960)
CA 54: 20518e

D 214 Photochemical evidence for triplet-
state quenching by paramagnetic
species.
Moore, W.M., Hammond, G.S., and Foss,
R.P.
J. Chem. Phys. 32: 1594-5 (1960)
CA 54: 23657a

D 215 Use of fluorescent dye to whiten wool.
Moore, M.A., and Mitchell, B.W.
Hilgardia 30: 237-45 (1960)
CA 55: 5959f

D 216 An improved spectrofluorimeter for
biochemical analysis.
Moss, D.W.
Clin. Chim. Acta 5: 283-8 (1960)
CA 54: 15500e

D 217 "Photographic" effects of an intense
bombardment by charged heavy parti-
cles.
Muhlestein, E.
Ann. Guebhard 35-36: 90-118 (1959-60)
CA 57: 320d

D 218 Synthesis of 5'-iodo-2'-
hydroxychalcones and 6-iodo-
flavonones, -flavones, and -flavonols.
Mulchandani, N.B., and Shah, N.M.
Chem. Ber. 93: 1913-18 (1960)
CA 54: 24688h

D 219 The fluorescence of acetaldehyde
vapor.
Murad, E.
J. Phys. Chem. 64: 942-5 (1960)
CA 55: 4146b

D 220 Thermochromism of dixanthylenes.
Reactions with substituted xanthones.
III.
Mustafa, A., Asker, W., and El-Din Sobhy,
M.E.
J. Org. Chem. 25: 1519-25 (1960)
CA 55: 7401c

D 221 Luminescence and scintillation proper-
ties of some disubstituted hydrocar-
bons.
Naboikin, Yu.V., Zadorozhnyi, B.A., and
Pavlova, E.N.
Stsintillyatory i Stsintillyats. Materialy,
Vses. Nauchn.-Issled. Inst. Khim.
Reaktivov, Materialy 2-go Koordinats.
Soveshch. 1957: 130-4 (1960)
CA 57: 13275d

D 222 Luminescence peculiarities of ortho-
disubstituted aromatic hydrocarbons.
III. Fluorescence and absorption
spectra of some carbonic acids.
Naboikin, Yu.V., Zadorozhnyi, B.A., and
Pavlova, E.N.
Opt. i Spektroskopiya 8: 657-62 (1960)
CA 55: 2271i

D 223 Infrared spectra of uranyl phosphate,
oxalate, and salicylate in the solid
state.
Narasimham, K.V.
Indian J. Phys. 34: 321-30 (1960)
CA 55: 1183g

D 224 The determination of the fluorescence
spectra with an accessory of the Beckman DU spectrophotometer.
Nebbia, G.
Boll. Sci. Fac. Chim. Ind. Bologna 18:
35-46 (1960)
CA 55: 16150a

D 225 Fluorescence as an indicator of skin
deposition on electrophoresis of
thiocaine.
Nechaev, A.V.
Vopr. Kurortol., Fizioterapii i Lecheb.
Fiz. Kul't. 25: 404-8 (1960)
CA 55: 9672c

D 226 Effect of the internal field on the
spectral characteristics of polyatomic
organic molecules in solution.
Neporent, B.S., and Bakhashiev, N.G.
Gosudarst Univ. im. A. A. Zhdanova
Sb. Statei 1960: 35-51 (1960)
CA 56: 4256h

D 227 Fluorescence of some complexes of
manganese halides with pyridene.
Nikolic, K., de la Garanderie, H., and
Schlivitch, S.
Compt. Rend. 250: 4143-5 (1960)
CA 54: 20476a

D 228 The inconstancy of the fluorescence of
ryegrass.
Nitzsche, W.
Z. Acker- u. Pflanzenblau 110: 267-88
(1960)
CA 55: 20101i

D 229 Luminescence of thioindigo solutions
at low temperatures.
Nurmukhametov, R.N., Shigorin, D.N.,
and Dokunikhim, N.S.
Zh. Fiz. Khim. 34: 2055-9 (1960)
CA 55: 12061h

D 230 Luminescence of chrysene solutions
and powders at 77°K.
Nurmukhametov, R.N., Popova, E.G., and
Dokunikhin, N.S.
Opt. i Spektroskopiya 9: 593-600 (1960)
CA 55: 12063f

D 231 Fluorometric method in chemical
prospecting of U deposits. I. Design
of a transmission-type fluorometer
and a doubly rotary oven.
Ohashi, S.
Nippon Kogyo Kaishi 76: 530-6 (1960)
CA 55: 21683h

D 232 Spectrophotometric studies of some
red algal constituents.
O'hEocha, C.
Colloq. Intern. Centre Natl. Rech. Sci.
(Paris) 103: 121-34 (1960)
CA 57: 7622b

D 233 Action spectra for fluorescence excita-
tion of pyridine nucleotide in photosyn-
thetic bacteria and algae.
Olson, J.M., and Amesz, J.
Biochem. Biophys. Acta 37: 14-24 (1960)
CA 54: 11145e

D 234 The primary processes of radiation-
initiated polymerization.
Okomura, S., Manabe, T., Hignshimura,
T., Oishi, Y., and Futami, S.
Large Radiation Sources Ind., Proc. Conf.,
Warsaw, 1959, 1: 391-405 (1960)
CA 55: 10949d

D 235 Mineral acid zinc sulfide precipitation.
Ortmann, H., and Piwanka, R.
Monatsber. Deut. Akad. Wiss. Berlin 2:
549-51 (1960)
CA 55: 8138d

D 236 Rapid Scanning Microspectrofluori-
meter.
Olson, R.A.
Rev. Sci. Instr. 31: 844-9 (1960)
CA 55: 24144h

D 237 Synthesis of oxazolophenoxazines.
Osman, A.M., and Bassiouni, I.
J. Am. Chem. Soc. 82: 1607-9 (1960)
CA 54: 15389g

D 238 Fluorescence of Sb- and Mn-activated
Ca halophosphates.
Ostaszewicz, E.
Acta Phys. Polon. 19: 421-42 (1960)
CA 55: 17253h

D 239 Liquid Scintillators. XII. Absorption
and fluorescence spectra of
2,5-dianyl-1,3,4-oxadiazoles.
Ott, D.G., Kerr, V.N., Hayes, F.N., and
Hansbury, E.
J. Org. Chem. 25: 872-3 (1960)
CA 55: 107i

D 240 Emission spectrum and structure of
the crystal phosphor NH_4I-Tl.
Pae, A.Ya.
Tr. Inst. Fiz. i Astron., Akad. Nauk Est.
SSR 1960: 49-61 (1960)
CA 55: 26710c

D 241 Aromatic diazo and azo compounds.
Panchartek, J., Allan, Z.J., and Muzik, F.
Collection Czech. Chem. Commun. 25:
2783-99 (1960)
CA 55: 5500e

D 242 Fluorescence of various modifications
of uranyl sulfate.
Pande, D.N.
J. Sci. Ind. Res. (India) 19B: 71-2 (1960)
CA 54: 19187h

D 243 Fluorescence spectra of solutions of
uranyl nitrate at liquid-air tempera-
ture. I. Rapidly cooled solutions.
Pant, D.D., and Khandelwal, D.P.
Proc. Indian Acad. Sci. 51A: 60-74 (1960)
CA 55: 103b

D 244 Spectrophotofluorimetry of lubricating
oils: determination of oil mist in air.
Parker, C.A., and Barnes, W.J.
Analyst 85: 3-8 (1960)
CA 54: 18947g

D 245 Correction of fluorescence spectra and
measurement of fluorescence quantum
efficiency.
Parker, C.A., and Rees, W.T.
Analyst 85: 587-600 (1960)
CA 55: 5121a

D 246 New base-catalyzed aromatization
reaction.
Pawellek, D., and Bradsher, C.K.
J. Org. Chem. 25: 281-2 (1960)
CA 54: 13130h

D 247 Change in structure of blue and green
fluorescence in cadmium sulfide at low
temperatures.
Pedrotti, L.S., and Reynolds, D.C.
Phys. Rev. 119: 1897-8 (1960)
CA 55: 133c

D 248 Energy model for edge emission in
cadmium sulfide.
Pedrotti, L.S., and Reynolds, D.C.
Phys. Rev. 120: 1664-9 (1960)
CA 55: 5148i

D 249 Quenching of fluorescence in liquid
scintillation counting of labeled
organic compounds.
Perry, C.T.
Anal. Chem. 32: 1292-6 (1960)
CA 55: 148c

D 250 New methods for determination of very
small quantities of ozone.
Peregud, E.A., and Stepanenko, E.M.
Zh. Analit. Khim. 15: 96-8 (1960)
CA 54: 13959i

D 251 Fluorescence measurements on thin
films of aromatic hydrocarbons.
Z. Physik. Chem. 24: 1-10 (1960)
CA 54: 16174d

D 252 Impurity fluorescence in crystals.
Perlin, Yu.E.
Fiz. Tverd. Tela 2: 1915-27 (1960)
CA 55: 14077e

D 253 Consideration of nonradiative transi-
tions in the theory of impurity fluor-
escence.
Perlin, Yu.E.
Fiz. Tverd. Tela 2: 1928-35 (1960)
CA 55: 14077h

D 254 Absorption and fluorescence spectra
of perylene at low temperatures.
Personov, R.I.
Izv. Akad. Nauk SSSR, Ser. Fiz. 24: 620-2
(1960)
CA 54: 22000i

D 255 Fluorescence spectra of pure and
octadeuteriated naphthalene.
Pesteil, L.
Compt. Rend. 250: 497-9 (1960)
CA 54: 14934a

D 256 Fluorescence spectra of naphthalene.
Pesteil, P., and Clais, A.
Compt. Rend. 250: 494-6 (1960)
CA 54: 14937g

D 257 Phosphorescence and dosimetry with
samarium-activated calcium sulfate.
Peter, H.
Atomkernenergie 5: 453-5 (1960)
CA 55: 10094h

D 258 Temperature dependence of the
quantum yield of the fluorescence of
some phthalimides in various solvents.
Pikulik, L.G., and Sevchenko, A.N.
Izv. Akad. Nauk SSSR, Ser. Fiz. 24:
729-33 (1960)
CA 54: 22001i

D 259 Effect of temperature on luminescence
and absorption spectra of complex
molecules in solutions.
Pikulik, L.G., and Solomakho, M.A.
Inzh.-Fiz. Zh., Akad. Nauk Belorussk.
SSR 3: 53-60 (1960)
CA 55: 13079h

D 260 Effect of temperature on the electronic
spectra of complex molecules.
Pikulik, L.G., and Solomakho, M.A.
Opt. i Spektroskopiya 8: 338-41 (1960)
CA 57: 236f

D 261 The anti-Stokes phosphorescence of
organic phosphors.
Pilipovich, V.A.
Opt. i Spektroskopiya 9: 754-8 (1960)
CA 55: 17251e

D 262 Structure of the F band in KCl crys-
tals.
Politkov, N.G.
Tr. Inst. Fiz. Akad. Nauk Gruz. SSR 7:
221-30 (1960)
CA 55: 11043c

D 263 Fluorescence of certain metal
8-quinolinolates as a function of sol-
vent and substituents.
Popovych, O., and Rogers, L.B.
Spectrochim. Acta 16: 49-57 (1960)
CA 54: 13855a

D 264 Photooxidation of biacetyl.
Porter, G.B.
J. Chem. Phys. 32: 1587-8 (1960)
CA 54: 23656i

D 265 Elimination of contraction in Nuclear
Emulsions used in the study of
α-radioactivity of the atmosphere.
Potsyus, V.Yu., and Nedvetskaite, T.N.
Nauchn. Soobshch. Inst. Geol. i Geogr.
Akad. Nauk Lit. SSR 11: 42-7 (1960)
CA 57: 360i

D 266 Light absorption of bivalent rare
earths in calcium fluoride.
Przibram, K.
Oesterr. Akad. Wiss., Math.-Naturw. Kl.,
Anz. 97: 299-303 (1960)
CA 55: 4252b

D 267 Fluorescence of organic traces in
inorganic substances.
Przibram, K.
Nature 188: 657 (1960)
CA 55: 10080h

D 268 Widely distributed blue fluorescence
of organic origin. II.
Przibram, K.
Oesterr. Akad. Wiss., Math.-Naturw. Kl.,
Anz. 4: 65-8 (1960)
CA 55: 4144b

D 269 Widely distributed blue fluorescence
of organic origin. III.
Przibram, K.
Oesterr. Akad. Wiss., Math.-Naturw. Kl.,
Anz. 4: 165-70 (1960)
CA 55: 16185d

D 270 The connection between light absorption, decay time, and absolute quantum
yield of fluorescence of organic dye
molecules.
Prammensee, H., and Zanker, V.
Z. Angew. Phys. 12: 237-40 (1960)
CA 54: 17045h

D 271 Kinetics of the processes in crystal
   phosphors.
   Rebane, K.K., and Sil'd, O.
   Tr. Inst. Fiz. i Astron., Akad. Nauk Est.
   SSR 1960: 130-49 (1960)
   CA 55: 26709f

D 272 Potential anticancer agents. XXXIV.
   Nonredox analogs of riboflavine 1.
   Model studies.
   Reist, E.J., Hamlow, H.P., Junga, I.G.,
   Silverstein, R.M., and Baker, B.R.
   J. Org. Chem. 25: 1368-78 (1960)
   CA 54: 24738b

D 273 Temperature dependence of edge
   emission in cadmium sulfide.
   Reynolds, D.C.
   Phys. Rev. 118: 478-9 (1960)
   CA 54: 23773a

D 274 Fluorescence of bone after quercetin
   ingestion.
   Ribelin, W.E., Masri, M.S., and De Eds,
   F.
   Proc. Soc. Exptl. Biol. Med. 103: 271-2
   (1960)
   CA 54: 11269a

D 275 The Mannich reaction. V. Mannich
   reactions with bifunctional isocyclic
   ethynyl hydrocarbons.
   Ried, W., and Wesselborg, K.
   Ann. 635: 97-108 (1960)
   CA 55: 1620f

D 276 Magnesium-calcium fluorophosphate
   phosphors.
   Rinbach, H.W.
   U. S. 291,912 (1960)
   CA 54: P7352d

D 277 The phosphorescence of adsorbed
   acriflavine.
   Rosenberg, J.L., and Shombert, P.J.
   J. Am. Chem. Soc. 82: 3252-7 (1960)
   CA 54: 20516h

D 278 Effect of temperature and viscosity on
   the fluorescence of solutions of some
   organic compounds by means of ultraviolet excitation.
   Sahu, J.
   J. Indian Chem. Soc. 37: 411-14 (1960)
   CA 55: 15116b

D 279 Procedure for measuring the light sum
   of crystallophosphors.
   Saichenko, Yu.M.
   Sb. Nauchn. Rabot Kafedry Opt. Kafedry
   Eksperim. Fiz. Kazakhsk. Univ. 1960:
   2: 141-3
   CA 57: 13276d

D 280 Spectroscopic study of the effect of
   detergent micelles on dyes. Fluorescence quenching and energy transfer.
   Sardisco, J.B.
   Univ. Microfilms, L.C. Card No. Mic
   60-2983, 91 pp (1960)
   CA 55: 2107h

D 281 Ultraviolet, visible and fluorescence
   spectral analysis of polynuclear
   hydrocarbons.
   Sawicki, E., Hauser, T.R., and Stanley,
   T.W.
   Intern. J. Air Pollution 2: 253-72 (1960)
   CA 54: 15779e

D 282 Absorption spectra of solid xenon,
   krypton, and argon in the vacuum
   ultraviolet.
   Schnepp, O., and Dressler, K.
   J. Chem. Phys. 33: 49-55 (1960)
   CA 55: 101g

D 283 The photochemistry of bis(4-methyl-
   2-naphthalene) indigo. I.
   Schulte-Frohlinde, D., and Erhardt, F.
   Chem. Ber. 93: 2880-4 (1960)
   CA 55: 5451d

D 284 Sensitized fluorescence of Mg-Na
   vapor mixture.
   Seiwert, R., and Tilch, J.
   Monatsber. Deut. Akad. Wiss. Berlin 2:
   689-90 (1960)
   CA 57: 16002b

D 285 Quantitative luminescence determina-
   tion of magnesium.
   Serebryakova, G.V., Lukin, A.M., and
   Bozhevol'nov, E.A.
   U.S.S.R. 129,273, June 15, 1960
   CA 55: 2365b

D 286 Limiting polarization of the fluor-
   escence of polyatomic molecules.
   Sevehenko, A.N.
   Analele Stiint. Univ. "A.I. Cuza," Iasi 6:
   809-18 (1960)
   CA 59: 8276g

D 287 Determination of boron with benzoin
by using an objective fluorimeter for
liquids.
Shcherbov, D.P., Korzheva, R.N., and
Ponomarenko, A.I.
Metody Lyuminests. Analiza Sb. 1960:
37-42 (1960)
CA 55: 4084d

D 288 Increasing the sensitivity and reproducibility of the fluorescent analysis
of solutions.
Shcherbov, D.P., and Korzheva, R.N.
Metody Lyuminests. Analiza Sb. 1960:
43-49 (1960)
CA 55: 4i

D 289 Fluorescent detection of boron in solu-
tions using phenylfluorone.
Shcherbov, D.P., and Korzheva, R.N.
Tr. Kazakhsk. Nauchn.-Issled. Inst.
Mineral'n. Syr'yu. 1960: 217-22 (1960)
CA 60: 1097g

D 290 Simplified fluorometer with an FEU-19
photomultiplier.
Shcherbov, D.P., and Ponomarenko, A.I.
Zavodsk. Lab. 26: 1143-56 (1960)
CA 57: 609a

D 291 Fluorescent liquid based on kerosine
for flow detection.
Shuikin, N.I., Bekauri, N.G., and Klimov,
A.I.
U.S.S.R. 138,084 (1960)
CA 59: 6252b

D 292 Luminescence of crystalline naphthalene containing small amounts of
impurities.
Shpak, M.T., and Sheka, E.F.
Izv. Akad. Nauk SSSR, Ser. Fiz. 24: 553-5
(1960)
CA 54: 21999h

D 293 Line spectra for the fluorescence of
organic compounds and their use.
Shpol'skii, E.V.
Usp. Fiz. Nauk 71: 243-87 (1960)
CA 54: 23838g

D 294 Attenuation of the phosphorescence
intensity of petroleum products
(analysis).
Sidorov, N.K., and Rodomakina, G.M.
Uch. Zap. Saratovsk. Gos. Univ. 69: 161-9
(1960)
CA 57: 12782d

D 295 $\pi-\pi*$ phosphorescence of chloro-
phylls a and b.
Singh, I.S., and Becker, R.S.
J. Am. Chem. Soc. 82: 2083-4 (1960)
CA 54: 14937a

D 296 Fluorescence radiation in contact
microradiography, its application in
mineralogy and petrography.
Soldsztaub, S., and Schmitt, J.
X-Ray Microscopy X-Ray Microanal.,
Proc. Intern. Symp., 2nd, Stockholm,
1959: 149-52 (1960)
CA 55: 15001c

D 297 Temperature relation for the lumin-
escence of NaI (Tl) crystals at 0-270°.
Startsev, V.I., Baturicheva, Z.B., and
Tsirlin, Ya.A.
Opt. i Spektroskopiya 8: 541-4 (1960)
CA 57: 1688c

D 298 Solubility of 3,4-benzopyrene in
aqueous protein solutions.
Stauff, J., and Reske, G.
Z. Naturforsch. 15B: 578-84 (1960)
CA 55: 7013e

D 299 Fluorescence and excitation spectra of
anthracene vapor at low pressures.
Stevens, B., and Hutton, E.
Mol. Phys. 3: 71-8 (1960)
CA 54: 14936e

D 300 Spectrum of delayed fluorescence of
phenanthrene vapor: a criterion of
purity.
Stevens, B., Hutton, E., and Porter, G.
Nature 185: 917-8 (1960)
CA 54: 19159f

D 301 Fluorescence self-quenching in aro-
matic vapors; the role of excited
dimers.
Stevens, B., and McCartin, P.J.
Mol. Phys. 3: 425-33 (1960)
CA 55: 7037f

D 302 Purine N-oxides. VII. Reaction of
aminopurine 1-N-oxides with acetic
anhydride.
Stevens, M.A., Smith, H.W., and Brown,
G.B.
J. Am. Chem. Soc. 82: 1148-52 (1960)
CA 54: 12147e

D 303 Gudden-Pohl effect in single crystals of zinc sulfide.
Steinberger, I.T., and Bar, V.
Proc. Intern. Conf. Color Centers Crystal Luminescence 1960: 222-7 (1960)
CA 54: 6296f

D 304 Nonlinear optical phenomena in a system of particles with three energy levels.
Stepanov, B.I., and Gribkovskii, V.P.
Izv. Akad. Nauk SSSR, Ser. Fiz. 24: 534-8 (1960)
CA 55: 7066b

D 305 Preparation of some surface active alcohols containing the anthracene nucleus.
Stewart, F.H.C.
Australian J. Chem. 13: 478-87 (1960)
CA 55: 7375f

D 306 Effect of deuterium substitution on the lifetime of the phosphorescent triplet state of naphthalene.
Sternlicht, H., and McConnell, H.M.
J. Chem. Phys. 33: 302-3 (1960)
CA 55: 2276g

D 307 Sensitized fluorescence in solutions.
Sveshnikov, B. Ya., Kudryashov, P.I., and Limareva, L.A.
Opt. i Spektroskopiya 9: 203-8 (1960)
CA 55: 26659c

D 308 Zeeman effect of the purely cubic field fluorescence line of MgO:Cr^{+++} crystals.
Sugano, S., Schawlow, A.L., and Varsanyi, F.
Phys. Rev. 120: 2045-53 (1960)
CA 55: 5147g

D 309 The fluorescent substances in the bile. I. Identification of the fluorescent substances.
Sugiura, M., Tanaka, E., Yoshida, O., and Hotta, K.
Bitamin 20: 181-3 (1960)
CA 60: 8422f

D 310 Tetrahydro- and octahydro-4,7-phenanthroline and octahydro-1,5-diaza-anthracene.
Sykes, W.O.
J. Chem. Soc. 1960: 4583-87 (1960)
CA 55: 5494a

D 311 Ultraviolet fluorescence of proteins in neutral solution.
Teale, F.W.J.
Biochem. J. 76: 381-8 (1960)
CA 54: 23729f

D 312 Influence of concentration on phosphorescence spectra of sulfobenzoic acid solutions at low temperature.
Teplyakov, P.A.
Izv. Vysshikh Uchebn. Zavedenii, Fiz. 1960: No. 3, 87-92 (1960)
CA 54: 23730e

D 313 The excitation spectrum of zinc oxide.
Thomas, D.G.
Phys. Chem. Solids 15: 86-96 (1960)
CA 55: 12050d

D 314 Fluorescent spectra of aromatic hydrocarbons found in polluted atmosphere.
Thomas, J.F., Tebbens, B.D., Sanborn, E.N., and Cripps, J.M.
Intern. J. Air Pollution 2: 210-20 (1960)
CA 54: 15782b

D 315 Pyrophosphate luminescent substances.
Thorm Electrical Industries Ltd.
Ger. 1,079,766, Apr. 14, 1960
CA 55: 21852f

D 316 Some X-ray and fast neutron response characteristics of silver metaphosphate glass dosimeters.
Thornton, W.T., and Auxier, J.A.
U.S. At. Energy Comm. ORNL-2912 57 pp. (1960)
CA 55: 3218c

D 317 Radiation damage to solids and catalysts.
Turkevich, J.
Large Radiation Sources Ind., Proc. Conf., Warsaw, 1959 2: 111-17 (1960)
CA 55: 20649d

D 318 The quantitative interpretation of luminescent properties of Cu-activated calcium orthophosphate.
Uehara, Y.
J. Phys. Soc. Japan 15: 612-29 (1960)
CA 55: 17254b

D 319 Luminescent properties of silver-activated magnesium borate phosphors.
Uehara, Y., Kofuya, Y., and Masuda, I.
J. Electrochem. Soc. 107: 104-7 (1960)
CA 54: 9509a

D 320 Chemical behavior of the intermediates produced during the reductive photobleaching of cosin in evacuated alcoholic solution.
Uchida, K., Kato, S., and Koizumi, M.
Bull. Chem. Soc. Japan 33: 169-73 (1960)
CA 54: 20431c

D 321 Investigations on mematocides. III. Polythienyls and related compounds.
Uhlenbroek, J.H., and Bijloo, J.D.
Rec. Trav. Chim. 79: 1181-96 (1960)
CA 55: 7382i

D 322 Removal of fluorescence from mineral oils.
VEB Lemma-Werke "Walter Ulbricht"
Ger. 1,086,376, Aug. 4, 1960
CA 55: 14900g

D 323 Fluorescence spectra of aromatic hydrocarbons and heterocyclic aromatic compounds.
Van Duuren, B.L.
Anal. Chem. 32: 1436-42 (1960)
CA 55: 4147e

D 324 Types of foreign element insertions in solid compounds.
Van Gool, W.
Mededel. Vlaam. Chem. Ver. 22: 38-44 (1960)
CA 54: 21915a

D 325 Self-activated and copper-activated fluorescence of zinc sulfide.
Van Gool, W., and Cleiren, A.P.D.M.
Philips Res. Rept. 15: 238-53 (1960)
CA 54: 20516 g

D 326 Fluorescence of some activated zinc sulfide phosphors.
Van Gool, W., Cleiren, A.P.D.M., and Heijligers, H.J.M.
Philips Res. Rept. 15: 254-74 (1960)
CA 54: 20517a

D 327 Fluorescence and reflection spectra of NaI single crystals.
Van Sciver, W.J.
Phys. Rev. 120: 1193-1205 (1960)
CA 55: 5148a

D 328 Emission spectrum of trivalent holmium in the scheelite structure.
Van Uitert, L.G., and Soden, R.R.
J. Chem. Phys. 33: 1532-4 (1960)
CA 55: 14052a

D 329 Fluorescent determination of gallium in ores with rhodamine S.
Vasil'ev, P.I., Podual'naya, R.L., and Voronkova, M.A.
Mineral. Syr'ye Moscow, Sb. 1960: 302-6 (1960)
CA 55: 26841c

D 330 Fluorescent histiocytes in sputum related to smoking.
Vassar, P.S., Culling, C.F.A., and Saunders, A.M.
A.M.A. Arch. Pathol. 70: 649-52 (1960)
CA 55: 13636d

D 331 Fluorometry. V. Quenching effect of keto compounds on the fluorescence of some naphthalene derivatives.
Vecerek, B., Hynie, I., and Kael, K.
Collection Czech. Chem. Commun. 25: 2221-4 (1960)
CA 55: 4147h

D 332 Influence of some fluorescence quenchers on the quantum yields of photochemical transformations of 9-methylanthracene and 9-methyl-10-methoxymethylanthracene.
Vember, T.M., and Cherkasov, A.S.
Izv. Akad. Nauk SSSR, Ser. Fiz. 24: 577-81 (1960)
CA 54: 22000c

D 333 Aging of ZnS luminophors.
Vereshehagin, I.K.
Opt. i Spektroskopiya 9: 519-21 (1960)
CA 55: 12062i

D 334 Synthesis of 2,5-substituted 1,3.4-thiadiazoles.
Vereshchagina, N.N., and Postovskii, I.Ya.
Zh. Obshch. Khim. 30: 4024-6 (1960)
CA 55: 22300h

D 335 Duration of phosphorescence of solutions of organic compounds at -196°.
Viktorova, E. N.,Zhmyreva, I.A., Kolobkov, U.P., and Sagamenko, A.A.
Opt. i Spektroskopiya 9: 349-52 (1960)
CA 54: 18391f

D 336 New examples of pronounced depend-
ence of the fluorescence yield on the
position of the emission spectra.
Viktorova, E.N., Kochemirovskii, A.S.,
Krasnitskaya, N.D., and Reznikova, I.I.
Opt. i Spektroskopiya 9: 544-6 (1960)
CA 55: 12031g

D 337 Interpretation of electronic and vibra-
tion absorption spectra of uranyl
nitrates.
Volod'ko, L.V., Sevchenko, A.N., Umreiko,
D.S.
Dokl. Akad. Nauk SSSR 135: 560-3 (1960)
CA 54: 1068i

D 338 Correspondence between absorption
and luminescence spectra in solutions
of uranyl compounds.
Volod'ko, L.V., Sevchenko, A.N., and
Umreiko, D.S.
Izv. Akad. Nauk SSSR, Ser. Fiz. 24:
749-51 (1960)
CA 54: 22002 d

D 339 Universal double-dish phosphoroscope.
Volod'ko, L.V., and Umreiko, D.S.
Inzh.-Fiz. Zh., Akad. Nauk Belorussk.
SSR 3: 120-4 (1960)
CA 55: 7931i

D 340 Differences in the fluorescence pro-
ducts of ZnS (Ag) for particles of
different ionizing powers.
Warncke, J., and Neuert, H.
Naturwissenschaften 47: No. 1, 11 (1960)
CA 54: 16194c

D 341 Fluorescence-polarization spectrum
and electron-energy transfer in tyrosine, tryptophan, and related compounds. II. In proteins.
Weber, G.
Biochem. J. 75: 335-45 (1960)
CA 55: 1180g

D 342 Condensation of catechols with
ethylenediamine.
Weil-Malherbe, H.
Biochim. Biophys. Acta 40: 351-3 (1960)
CA 54: 24494 g

D 343 Fluorescent studies of ox-brain lactic
and malic dehydrogenases.
Winer, A.D.
Biochem. J. 76: 5p-6p (1960)
CA 60: 12291b

D 344 Influence of Brownian motion on the
transfer of energy in solutions.
Weinreb, A.
U.S. At. Energy Comm. TID-7612: 59-76
(1960)
CA 54: 9439i

D 345 Investigation of rapid reactions of ex-
cited molecules by fluorescence
measurements.
Weller, A.
Z. Elektrochem. 64: 55-9 (1960)
CA 54: 11655h

D 346 Electroluminescence of copper-
activated zinc sulfides.
Wendel, G., and Richter, G.
Z. Physik. Chem. 214: 253-60 (1960)
CA 54: 20517c

D 347 Intersystem crossing from higher ex-
cited states in complex molecules.
Whan, R.E., and Crosby, G.A.
U.S. At. Energy Comm. SCTM 234-60 (51):
1-24 (1960)
CA 55: 7037e

D 348 Fluorescence spectra of metal
chelates.
White, C.E., Ho, M., and Weimer, E.Q.
Spectrochim. Acta 16: 236-7 (1960)
CA 54: 13854c

D 349 Methods for obtaining correction
factors for fluorescence spectra as
determined with the Aminco-Bowman
spectrophotofluorometer.
White, C.E., Ho, M., and Weimer, E.Q.
Anal. Chem. 32: 438-40 (1960)
CA 54: 12878b

D 350 Fluorometric analysis.
White, C.E.
Anal. Chem. 32: 47R-53R (1960)
CA 54: 11804g

D 351 9,10-Dihydroanthracene derivatives
containing hetero bridge atoms.
Wittig, G., Krauss, E., and Niethammer,
K.
Ann. 630: 10-18 (1960)
CA 54: 24650d

D 352 Emission of thallium-activated borates of metals from the second group of the periodic system.
Witzmann, H., and Malur, L.
Zur Physik Chemie Kristallphosphore, Tagung Physik. Ges. D.D.R., 1, Greifswald, Ger. 1959: 150-5 (1960)
CA 60: 3632h

D 353 Selected studies in papermaking.
Wurz, O., and Mossopust, W.
Wochbl. Papierfabrik. 88: 478-84 (1960)
CA 54: 21754g

D 354 Fluorescent bleaching agents. IX. Relation between fluorescent intensity and concentration of fluorescent bleaching agents on the substrate. The constant of Merritt's equation.
Yabe, A., and Hayashi, M.
Kogyo Kagaku Zasshi 63: 987-90 (1960)
CA 60: 10860d

D 355 Condensation reaction of Dopa with ethylenediamine.
Yagi, K., Nagatsu, T., and Nagatsu, I.
J. Biochem. (Tokyo) 48: 617-20 (1960)
CA 55: 6431b

D 356 Luminescence spectra.
Yaroshenko, V.F.
Tr. Tbilissk. Gos. Ped. Inst. 15: 23-8 (1960)
CA 54: 12081b

D 357 Fluorescence spectra of monosubstituted naphthalenes. Effect of substituents on the quantum yield and the probabilities of radiation transition and radiationless transition.
Yasuda, K., Okabe, K., and Ito, K.
Nippon Kagaku Zasshi 81: 1361-5 (1960)
CA 55: 3192b

D 358 Heterocycle formation in the reaction of α-diazoacetophenone with bases.
Yates, P., and Farnum, D.G.
Tetrahedron Letters 1960: No. 17, 22-6 (1960)
CA 55: 1644f

D 359 Absolute fluorescence and phosphorescence quantum yields for various acridine and fluorescein dyes.
Zaaker, U., and Rammensee, H.
Z. Physik. Chem. 26: 168-86 (1960)
CA 55: 6138e

D 360 High-temperature phosphorescence with coronene.
Zander, M.
Naturwissenschaften 47: 443 (1960)
CA 55: 5121g

D 361 Low-temperature absorption and fluorescence-polarization spectra of the monohydroxy-acridines; contribution to the tautomery and to the bond problem of acridone and 9-aminoacridine.
Zanker, V., and Wittwer, A.
Z. Physik. Chem. 24: 183-205 (1960)
CA 54: 16172e

D 362 Fluorometric analysis. XII. Fluorometric determination of aluminum by extraction of its Pontachrome Blue Black complex and the application to the analysis of pure magnesium.
Nippon Kagaku Zasshi 81: 259-62 (1960)
CA 55: 13173h

D 363 Effect of the solvent on the electron spectra of organic molecules.
Zhmyreva, I.A., Zelinskii, V.V., Kolobkov, V.P., Kockmirovskii, A.S., and Reznikova, I.I.
Opt. i Spektroskopiya 8: 412-14 (1960)
CA 57: 285e

D 364 Polarization of electronic (absorption) bands of aromatics. I. Naphthalene, anthracene, tetracene.
Zimmermann, H., and Joop, N.
Z. Elektrochem. 64: 1215-19 (1960)
CA 55: 11072a

D 365 Polarized absorption and fluorescence spectra of 2,6-dimethylnaphthalene at different temperatures.
Zmerli, A., and Poulet, H.
J. Chem. Phys. 33: 1177-83 (1960)
CA 55: 11076d

D 366 Luminescence investigation of recrystallization processes. I. Fading away of Tl-activated NaCl recrystallized phosphors.
Zaltan, M.
Magy. Fiz. Folyoirat 8: 293-305 (1960)
CA 54: 13659f

D 367 The deep color of 9-substituted acridines. III. The planarity of these compounds and the constitution of 9-hydroxyphenyl- and 9-hydroxystyrylacridine.
Zanker, V., and Reichel, A.
Z. Elektrochem. 64: 431-7 (1960)
CA 54: 13855i

D 368 Influence of the structure on the luminescent characteristics of complex organic molecules.
Zelinskii, V.V., and Reznikova, I.I.
Izv. Akad. Nauk SSSR, Ser. Fiz. 24: 607-9 (1960)

D 369 Ore prospecting with ultraviolet lamps.
Zeschke, G.
Z. Erzbergbau Metallhuettenw. 13: 228-32 (1960)
CA 54: 16296i

D 370 The spectrum of Nova Herculis 1960, from 3100 to 5100 A, and the spectrum of Nova Herculis in the near infrared.
Compt. Rend. 251: 1969-71 and 2289-91 (1960)
CA 55: 11066c

D 371 (Fluorometric Determination of Uranium.)
U.S. At. Energy Comm., Rept. No. TID-7015: Suppl. 3, Meth. No. 1,219,240 (Jan. 1960)

D 372 Excitation spectra and brightness of specific cell fluorescence.
Agroskin, L.S., and Barskii, I.Ya.
Dokl. Akad. Nauk SSSR 139: 987-90 (1961)
CA 56: 1856d

D 373 Microscope — spectrofluorimeter for the ultraviolet region.
Agroskin, L.S., and Korolev, N.V.
Biofizika 6: 478-85 (1961)
CA 56: 5782i

D 374 Fluorescence of nucleic acids in solution.
Agroskin, L.S., Korolev, N.V., Kulaev, I.S., and Pomoshchnikova, N.A.
Dokl. Akad. Nauk SSSR 136: 226-9 (1961)
CA 55: 18297d

D 375 The kinetics of electroluminescence of zinc sulfide single crystals activated by copper and chloride.
Andreev, I.S., and Kisin, V.I.
Izv. Akad. Nauk Uz. SSR 1961: 78-82 (1961)
CA 56: 11044g

D 376 Luminescence of alkali halide phosphors activated with antimony trichloride.
Andrianov, A.S., and Kats, M.L.
Izv. Akad. Nauk SSSR, Ser. Fiz. 25: 390-2 (1961)
CA 55: 19505h

D 377 Fluorescence curves of phosphors with comparable residence periods of electrons in traps of various kinds.
Antonov-Romanovskii, V.V.
Opt. i Spektroskopiya 10: 644-8 (1961)
CA 56: 2993b

D 378 Microdetermination of uranium by a transmission type fluorometer. I. Uranium in natural water.
Aoyama, Y.
Nippon Kagaku Zasshi 82: 336-9 (1961)
CA 57: 595i

D 379 Fluorescence, phosphorescence, and quenching of a willemite sample activated with manganese.
Ardvino, H.A., Guerci, J.C., Levialdi, A., and Majilis, N.
J. Phys. Radium 22: 220-4 (1961)
CA 55: 19509c

D 380 Luminescence phenomena as a means for the solution of biological and chemical problems.
Arend, I., Chomse, H., and Floegel, G.
Arch. Exptl. Veterinaermed. 15: 966-9 (1961)
CA 61: 12310h

D 381 Existence of pigments fluorescing red under ultraviolet light in tissues of some mollusks.
Arvy, L., and Lerma, B.
Compt. Rend. Soc. Biol. 155: 715-9 (1961)
CA 56: 3929e

D 382 The temperature dependence of the energy transport in anthracene-tetracene mixed crystals.
Avakian, P., and Wolf, H.
Z. Physik 165: 439-44 (1961)
CA 56: 9440e

D 383 Electronic spectra of substituted an-
thracenes. Evidences for appearance
of the ^1L_b transition.
Baba, H., and Suzuki, S.
J. Chem. Phys. 35: 1501-2 (1961)
CA 56: 6799f

D 384 Apatite from veins of Alpine type in the
Arctic Ural.
Bakanov, V.V.
Zap. Vses. Mineralog. Obshchestva 90:
591-8 (1961)
CA 56: 12564e

D 385 Microdetermination of codeine and
codethyline by their fluorescence.
Balatre, P., Traismel, M., and Delcambre,
J.P.
Ann. Pharm. Franc. 19: 171-4 (1961)
CA 55: 27779e

D 386 Irradiation of diethylstilbestrol.
Banes, D.
J. Assoc. Offic. Agr. Chemists 44: 323-8
(1961)
CA 55: 23464d

D 387 Direction of transition moments of ab-
sorption bands of phthalocyanimes and
porphyrins from fluorescence polari-
zation measurements.
Bar, F., Zand, H., Schnabel, E., and
Kuhn, H.
Z. Elektrochem. 65: 346-54 (1961)
CA 55: 21793e

D 388 Infrared phosphorescence detection,
using pulsed excitation.
Barnett, R.H., and Moxham, R.M.
Rev. Sci. Instr. 32: 740-1 (1961)
CA 56: 8180c

D 389 Fluorescence and vibrational relaxation
of nitric oxide studied by kinetic spec-
troscopy.
Basco, N., Callaer, A.B., and Norrish,
R.G.W.
Proc. Roy. Soc. A 260: 459-74 (1961)
CA 55: 17171h

D 390 Possibility of an early detection of blue
mold infection on Nicotiana tabacum by
optical fluorescence.
Beck, W., and Diskus, A.
Fachliche Mitt. Oesterr. Tabakregie 2:
25-32 (1961)
CA 56: 1202b

D 391 Photo- and thermostimulated electron
emission from alkali-halide crystal
phosphors excited with ultraviolet
radiation.
Belkind, A.I., and Kyaembre, Kh.F.
Izv. Akad. Nauk SSSR, Ser. Fiz. 25: 381-3
(1961)
CA 55: 19507g

D 392 Pyrene dimers.
Berlman, I.B.
Nature 191: 594-5 (1961)
CA 56: 8138f

D 393 The use of photoelectric fluorimetry for
investigation of the luminescence of
bitumens and oils.
Bilalova, B.N.
Tr. Vses. Neft. Nauchn.-Issled.
Geologorazved. Inst. No. 186: 432-40
(1961)
CA 57: 15417a

D 394 Fluorescence lifetime studies of
pyrene solutions.
Birks, J.B., and Munro, I.H.
Luminescence Org. Inorg. Mater., Intern.
Conf., New York 1961: 230-4 (1961)
CA 56: 2951e

D 395 Reflection and emission of polarized
light in ZnS and CdS.
Birman, J.L., Samelson, H., and
Lempicki, A.
G T E (Gen. Telephone Electron.) Res.
Develop. J. 1: 2-15 (1961)
CA 55: 26711a

D 396 A reaction-kinetic model with five
transitions for photoconductors and
phosphors.
Boer, K.W., and Voigt, J.
Z. Naturforsch. 16a: 873-79 (1961)
CA 56: 12417a

D 397 Effect of self-extinction on the concen-
tration depolarization of photolumin-
escence of stable solutions.
Bojarski, C.
Ann. Physik 7: 402-11 (1961)
CA 57: 2967h

D 398 Fluorescence lifetime of terbium and
europium salts.
Bonrath, H., Heber, H., Hellwege, K.H.,
Huefner, S., and Laemmermann, H.
Naturwissenschaften 48: 713 (1961)
CA 57: 11977a

D 399 Carcinogenic substances in water and
soil. VI. Detection of polycyclic
hydrocarbons by means of fluorescence
spectral analysis.
Borneff, J., and Fischer, R.
Arch. Hyg. Bakteriol. 145: 241-55 (1961)
CA 55: 25107d

D 400 Fluorescence measurement.
Bowen, E.J.
Photoelec. Spectrometry Group Bull. 13:
331-33 (1961)
CA 57: 6754d

D 401 Molecular biophysics of dye-polymer
complexes.
Bradley, D.F.
Trans. N.Y. Acad. Sci. 24: No. 1, 64-74
(1961)
CA 57: 7413e

D 402 Fluorescence of photosynthetic organ-
isms at room and liquid nitrogen
temperatures.
Brody, M., and Linschitz, H.
Science 133: 705-6 (1961)
CA 55: 14546a

D 403 Study of energy transfer by quenching
experiments.
Brown, F.H., Furst, M., Kallmann, H.P.
U.S. At. Energy Comm. TID-7612: 37-58
(1961)
CA 56: 3309h

D 404 Ultraviolet fluorescence of cells during
mitosis.
Brumberg, E.M., Meisel, M.N., Barshii,
I.Ya., Zelenin, A.V., and Lyapunova,
E.A.
Dokl. Akad. Nauk SSSR 141: 723-5 (1961)
CA 56: 12106a

D 405 Photoelectric fluorimeter FM-1.
Bruskin, V.Ya.
Zavodsk. Lab. 27: 1151-6 (1961)
CA 56: 9383g

D 406 Generalized temperature function for
the fluorescence intensity of inorganic
crystalline luminophors.
Buhrow, J.
Z. Physik. Chem. 218: 55-63 (1961)
CA 56: 11045e

D 407 The state of the Nd^{+++} ions as derived
from the absorption and fluorescence
spectra of $NdCl_3$ and their Zeeman
effects.
Carlson, E.H., and Dieke, G.H.
J. Chem. Phys. 34: 1602-9 (1961)
CA 55: 20609f

D 408 Fluorescence analyses for polycyclic
aromatic hydrocarbons.
Chaudet, J.H., and Kaye, W.I.
Anal. Chem. 33: 113-17 (1961)
CA 55: 8181b

D 409 Effect of fluorescence quenchers on the
fluorescence spectra of solutions of
certain anthracene and phthalimide
derivatives in solvent mixtures.
Cherkasov, A.S.
Dokl. Akad. Nauk SSSR 139: 658-61 (1961)
CA 56: 15059g

D 410 Effect of viscosity of solvent on the
fluorescence spectra of certain organic
compounds.
Cherkasov, A.S., and Gragneva, G.I.
Opt. i Spektroskopiya 10: 466-72 (1961)
CA 55: 18296e

D 411 Spectral investigation of the products
formed during copolymerization of
styrene with 9-vinylanthracene.
Cherkasov, A.S., and Voldaikina, K.G.
Vysokomolekul. Soedin. 3: 570-6 (1961)
CA 56: 2566h

D 412 (Fluorometric detection of aluminum
and beryllium with 2-hydroxy-3-
naphthoic acid.)
Cherkasov, A.S., and Zhigalkina, T.S.
Zavodsk. Lab. 27: 658 (1961)

D 413 Development of fluorescent compounds
and redox indicator dyes.
Cheronis, N.D.
U.S. Dept. Comm. Office Tech. Serv.,
AD 278,359, 52 pp. (1961)
CA 60: 6846f

D 414 The luminescence of some substituted
naphthalenes.
Corkill, J.M., and Graham-Bryce, I.J.
J. Chem. Soc. 1961: 3893-7 (1961)
CA 56: 4216d

D 415 Crystal spectra of very weal interactions. I. Measurements of the naphthalene 3200-A system at 4 K.
Craig, D.P., Lyons, L.E., and Walsh, J.R.
Mol. Phys. 4: 97-112 (1961)
CA 56: 3035c

D 416 The soft X-ray emission spectra of sodium, beryllium, boron, silicon, and lithium.
Crisp, R.S., and Williams, S.E.
Phil. Mag. 6: 365-9 (1961)
CA 55: 18288a

D 417 Spectrum and magnetic properties of hexagonal PyCl$_3$.
Crosswhite, H.M., and Dieke, G.H.
J. Chem. Phys. 35: 1535-48 (1961)
CA 56: 12425g

D 418 Anomalous phosphorescence of naphthalene in poly(methyl methacrylate).
Czarnecki, S.
Bull. Acad. Polon. Sci., Ser. Sci., Math., Astron., Phys. 9: 561-3 (1961)
CA 59: 9439f

D 419 Two electro-optical methods for the determination of dipole moments of excited molecules.
Czekalla, J.
Chimia (Aarau) 15: 26-31 (1961)
CA 61: 120e

D 420 Measurements of the dipole moment in a donor-acceptor complex in the first excited singlet state.
Czekalla, J., and Meyer, K.O.
Z. Physik. Chem. (Frankfurt) 27: 185-98 (1961)
CA 55: 13992c

D 421 Bose statistics and Y* production and decay in K$^-$-p collisions.
Dalitz, R.H., and Miller, D.H.
Phys. Rev. Letters 6: 652-7 (1961)
CA 55: 25518i

D 422 Quantification of the antigen-antibody reaction by the polarization of fluorescence.
Dandliken, W.B., and Feigen, G.A.
Biochem. Biophys. Res. Commun. 5: 299-304 (1961)
CA 55: 23782i

D 423 Natural and induced fluorescence in microscopic organisms.
Darken, M.A.
Appl. Microbiol. 9: 354-60 (1961)
CA 55: 23678d

D 424 Electrical properties of organic insulating liquids containing fluorescent solutes.
Darveniza, M., and Tropper, H.
Proc. Phys. Soc. 78: 854-68 (1961)
CA 60: 4903d

D 425 Fluorescence of the alkali halides observed with the scanning microscope.
Davoine, F., Bernard, R., and Pinard, P.
Proc. European Regional Conf. Electron Microscopy, Delft 1: 165-8 (1960) (Pub. 1961)
CA 56: 12417f

D 426 Determination of uranium by means of luminescence in phosphate solutions.
Davydov, A.V., Dobrolyubskaya, T.S., and Nemodruk, A.A.
Zh. Analit. Khim. 16: 68-72 (1961)
CA 56: 1997h

D 427 Fluorescence of polyacrylonitrile solutions.
Dechant, J.
Faserforsch. Textiltech. 12: 471-4 (1961)
CA 56: 7532e

D 428 See D 429

D 429 Effect of deuterium and chlorine substitution on triplet → singlet transition probabilities in naphthalene.
de Groot, M.S., and Van der Waals, J.H.
Mol. Phys. 4: 189-90 (1961)
CA 56: 6800g

D 430 The fluorescence of concentrated porphyrins in minerals of biologic origin.
Déribéré, M.
Bull. Soc. Franc. Mineral. Crist. 84: 94-5 (1961)
CA 55: 19619a

D 431 Absorption and fluorescence spectra with magnetic properties of ErCl$_3$.
Dieke, G.H., and Singh, S.
J. Chem. Phys. 35: 555-63 (1961)
CA 56: 6780e

D 432 Fine-structure fluorescent spectra of monomethyl derivatives of 1,2-benzanthracene.
Dikun, P.P.
Vopr. Onkol. 7: 64-73 (1961)
CA 56: 8177e

D 433 Fluorescence of porphyrins.
Djuric, D.
Arhiv Farm. (Belgrade) 11: 1-6 (1961)
CA 56: 3750e

D 434 Mean lifetime of the lowest excited singlet state of benzene.
Donovan, J.W., and Duncan, A.B.F.
J. Chem. Phys. 35: 1389-91 (1961)
CA 56: 9561h

D 435 Investigations on stilbenes. XL. Conjugated stilbazoles.
Drefahl, G., Plotner, G., and Buchner, G.
Chem. Ber. 94: 1824-33 (1961)
CA 55: 23518d

D 436 Quenching solution of fluorescent photographic paper used in making halftone negatives.
Duffy, J.E., and Sottysiak, J.
U.S. 2,983,604 May 9, 1961
CA 55: 17323c

D 437 Phosphorescence of cement luminophors prepared without heat treatment.
Dvorovenko, V.K.
Nauchn. Zap., Odessk. Gos. Ped. Inst. Kafedry Mate., Fiz. i Estestvozn. 25: 83-5 (1961)
CA 60: 4925d

D 438 Certain peculiarities of the phosphorescence of cement luminophors.
Dvorovenko, V.K.
Ukr. Fiz. Zh. 6: 768-70 (1961)
CA 56: 2953d

D 439 Phosphorescence of magnesium cement.
Dvorovenko, V.K.
Nauchn. Zap., Odessk. Gos. Ped. Inst. Kafedry Mate., Fiz. i Estestvozn. 25: 81-2 (1961)
CA 60: 3590d

D 440 An investigation of the benzoin method for the fluorometric determination of boron.
Elliot, G., and Radley, J.A.
Analyst 86: 62-9 (1961)
CA 55: 21981a

D 441 Fluorimetry. V. Quinine sulfate as a fluorescence standard. Measurements of its fluorescence intensity in aqueous solutions and acetic acid solutions with the addition of sulfuric or perchloric acid. Influence of buffer mixtures.
Eisenbrand, J.
Z. Anal. Chem. 179: 170-5 (1961) (Ger.)
CA 55: 13162f

D 442 Fluorimetry. VI. Relation of acid concentration to the effect of chloride ion on the fluorescence of quinine sulfate in aqueous solution.
Eisenbrand, J., and Raisch, M.
Z. Anal. Chem. 179: 352-5 (1961) (Ger.)
CA 55: 13162g

D 443 Fluorimetry. VII. Influence of different concentrations of halides on the fluorescence of quinine sulfate in aqueous solution.
Eisenbrand, J., and Raisch, M.
Z. Anal. Chem. 179: 406-9 (1961) (Ger.)
CA 55: 18426g

D 444 Fluorimetric method for the determination of hippuric acid.
Ellman, G.L., Burkhalter, A., and LaDou, J.
J. Lab. Clin. Med. 57: 813-18 (1961)
CA 55: 15601e

D 445 Energy transfer in H-bonded N-heterocyclic complexes and their possible role as energy sinks.
El-Bayoumi, M.A., and Kosha, M.
J. Chem. Phys. 34: 2181-2 (1961)
CA 56: 2993i

D 446 Intramolecular excitation transfer. The lowest $n \rightarrow \pi^*$ transitions in pyrazine.
El-Sayed, M.F.A., and Robinson, G.W.
Mol. Phys. 4: 273-86 (1961)
CA 56: 12442b

D 447 Luminescence of simple benzene derivatives. I. Aromatic amines.
Ermolaev, V.L.
Opt. i Spektroskopiya 11: 492-7 (1961)
CA 56: 6764g

D 448 A fluorimeter for determining catechol amines in urine.
Esikov, A.D., and Men'shikov, V.V.
Lab. Delo 7: 22-5 (1961)
CA 55: 22467b

D 449 Improved technique for the fluorimetric estimation of catechol amines.
Euler, U.S.V., and Lishajko, F.
Acta Physiol. Scand. 51: 348-56 (1961)
CA 55: 23655f

D 450 Action of M, R, and F color centers on the mechanism of infrared fluorescence of F centers in alkali halide crystals.
Ezhik, I.I., and Shavlo, S.T.
Izv. Vysshikh Uchebn. Zavedenii, Fiz. 1961: 46-53 (1961)
CA 55: 19505a

D 451 1,2-Naphthoquinone-4-sulfonate and fluorescence of proteins, of substances with a guanidine radical, and of streptomycin.
Faure, F., and Blanquet, P.
Bull. Soc. Pharm. Bordeaux 100: 105-19 (1961)
CA 56: 1731b

D 452 Substances giving the reaction of Strecker and fluorescence of proteins.
Faure, F., and Blanquet, P.
Bull. Soc. Chim. Biol. 43: 953-67 (1961)
CA 56: 7681d

D 453 Fluorescence effects in ion-exchange resins.
Flint, T.R., and Eicholz, G.G.
Dept. Mines Tech. Surv., Mines Br. Res. Rept. R91: 1-35 (1961)
CA 57: 16867c

D 454 Luminescence parameters of oils.
Florovskaya, U.N., and Klyuev, Yu.A.
Dokl. Akad. Nauk Belorussk. SSR 5: 438-41 (1961)
CA 59: 7277c

D 455 Phosphorescent materials.
Forster, C.F.
Brit. 870,504 June 19, 1961
CA 55: 26714e

D 456 Coenzyme binding, observed by fluorescence enhancement, apparently unrelated to the enzymic activity of glutamic dehydrogenase.
Frieden, C.
Biochim. Biophys. Acta 47: 428-30 (1961)
CA 55: 21196f

D 456a Physicochemical studies on Actimonycins. V. Spectrophotometric measurements of hydrolytic and photolytic reactions.
Fritzsche, H., Loeber, G., and Berg, H.
Z. Physic. Chem. 218:291-309 (1961)
CA 57:104i

D 457 Polyacrylonitrile solution. I. A few problems on the light-scattering measurement.
Fujisaki, N.
Kobunshi Kagaku 18: 581-8 (1961)
CA 56: 14452c

D 458 Fluorescence microscopic studies on the differentiation of living and dead staphylococci by means of acridine orange vital staining. I. Investigation of the acridine orange soluble blue double staining method.
Fukuskima, K.
Tokyo Jikeikai Ika Daigaku Zasshi 76: 915-19 (1961)
CA 56: 7146f

D 458a Fluorescence microscopic studies on the differentiation of living and dead staphylococci by means of acridine orange vital staining. II. Effect of pretreatment of the bacteria with trypsin.
Fukuskima, K.
Tokyo Jikeikai Ika Daigaku Zasshi 76: 935-9 (1961)
CA 56: 7146f

D 459 Stimulated emission into optical whispering modes of spheres.
Garrett, C.G.B., Kaiser, W., and Bond, W.L.
Phys. Rev. 124: 1807-9 (1961)
CA 56: 8122b

D 460 Cross section for the decomposition of C^{12} into three α-particles induced by 90-m.e.v. protons, and the probability for a transitory α-structure.
Gauvin, H., Chastel, R., and Vigneron, L.
Compt. Rend. 253: 257-9 (1961)
CA 56: p1107i

D 461 The role of grain boundaries and dislocations in the luminescence of activated alkali halide crystals.
Gindina, R.I.
Tr. Inst. Fiz. i Anstron., Akad. Nauk Est. SSR 1961: 168-89 (1961)
CA 56: 12417a

D 462 Luminosity behind shock waves in xenon.
Gloersen, P.
Phys. Fluids 4: 790 (1961)
CA 56: 8136e

D 463 Determination of the K fluorescence yield of argon by proportional-counter spectrometry.
Godeau, C.
U.S. Dept. Comm., Office Tech. Serv., P.B. Rept. 161,592 8 pp. (1961)
CA 57: 10668a

D 464 Phosphine oxide complexes. V. Tetrahedral complexes of Mn(III) containing triphenylphosphine oxide and triphenylarsine oxide as ligands.
Goodgame, D.M.L., and Cotton, F.A.
J. Chem. Soc. 1961: 3735-41 (1961)
CA 56: 2145g

D 465 N → π* transitions in the azines.
Goodman, L.
J. Mol. Spectroscopy 6: 109-37 (1961)
CA 55: 20614a

D 466 Spectra of porphyrins.
Gouterman, M.
J. Mol. Spectroscopy 6: 138-63 (1961)
CA 55: 20615e

D 467 Condensed pyrrole compounds. I. 3H-Pyrrolo-(2,3-c)-quinolines.
Govindachari, T.R., Rajappa, S., and Sudarsanam, V.
Tetrahedron 16: 1-4 (1961)
CA 57: 791h

D 468 Photochemical reactions of some aromatic halogen compounds.
Grabowski, Z.R.
Z. Physik. Chem. (Frankfurt) 27: 239-52 (1961)
CA 55: 13017g

D 469 Emission and absorption spectra of the benzl radical at 77 K near 4600 A.
Grajcar, L., and Leach, S.
Compt. Rend. 252: 1014-16 (1961)
CA 56: 11082i

D 470 Fluorescence of sulfur dioxide.
Greenough, K.F., and Duncan, A.B.F.
J. Am. Chem. Soc. 83: 555-60 (1961)
CA 55: 13046b

D 471 Anomalies in the conductance of Ag^+ ions in crystalline quartz.
Gross, F.
Acta Phys. Austriaca 14: 75-85 (1961)
CA 55: 24171d

D 471a The freezing-out of the energy transfer in organic mixed crystals.
Gschwendtner, K., and Wolf, H.C.
Naturwissenschaften 48 : 42-3 (1961)
CA 55: 15117a

D 472 X-ray fluorescent intensity of elements evaporated from solution onto thin film.
Gunn, E.L.
Anal. Chem. 33: 921-7 (1961)
CA 55: 23162f

D 473 Luminescence of rare earth phthalocyamines.
Gurevich, M.G., and Solov'ev, K.N.
Dokl. Akad. Nauk Belorussk. SSR 5: 291-4 (1961)
CA 57: 15948e

D 474 Limiting polarization of the fluorescence of porphyrins.
Gurinovich, G.P., Sevchenko, A.N., and Solov'ev, K.N.
Opt. i Spektroskopiya 10: 750-8 (1961)
CA 56: 4048d

D 475 The significance of the temperature dependence of fluorescence intensity.
Haake, C.H.
J. Electrochem. Soc. 108: 78-82 (1961)
CA 55: 9069b

D 476 Band edge emission properties of CdTe.
Halsted, R.E., Lorenz, M.R., and Segall, B.
Phys. Chem. Solids 22: 109-16 (1961)
CA 57: 5476e

D 477 Photofluorescence decay times of organic phosphors.
Hamilton, T.D.S.
Proc. Phys. Soc. 78: 743-52 (1961)
CA 56: 8136f

D 478 Improved fluorimeter for uranium analysis.
Haran, E.N.
J. Sci. Instr. 38: 273-7 (1961)
CA 55: 26554g

D 479 Crystal growth and structural study of the barium titanium phosphate phosphor.
Harrison, D.E., and Shirane, G.
J. Electrochem. Soc. 108: 788-90 (1961)
CA 55: 25410i

D 480 See D456a

D 481 Optical bleaching agents for wool and synthetic fibers.
Hayashi, M., Baba, T., and Yabe, A.
Kaseigaku Zasshi 12: 136-9 (1961)
CA 55: 22834h

D 482 Investigation of the applicability of solid state fluorescence to pharmaceutical analysis.
Head, W.F.
J. Pharm. Sci. 50: 1041-4 (1961)
CA 56: 7427e

D 483 Experiences with a fluorometric method for determining corticasteroids in man and rat.
Hedner, P.
Acta Pharmacol. Toxicol. 18: 65-74 (1961)
CA 55: 21222f

D 484 Fluorescence of DL-aldosterone in an acid medium.
Hedner, P.
Acta Pharmacol. Toxicol. 18: 407-13 (1961)
CA 57: 1482f

D 485 Theory of the pulsation of fluorescent light from ruby.
Hellworth, R.W.
Phys. Rev. Letters 6: 9-12 (1961)
CA 55: 25510b

D 486 The fluorescence of polarization of lignified cell walls.
Hengartner, H.
Holz Roh-Werkstoff 19: 303-9 (1961)
CA 59: 15475g

D 487 Fluorometric determination of zirconium with quercetin. Separation of interferences by extraction with TTA (thenoyltrifluoroacetone).
Hercules, D.M.
Talanta 8: 485-91 (1961)
CA 55: 20787i

D 488 The effect of intramolecular twisting on the emission spectra of hindered aromatic molecules. I. 1,1'-Binaphthyl.
Hochstrasser, R.M.
Can. J. Chem. 39: 459-70 (1961)
CA 55: 13057h

D 489 Luminescence of complex molecules in relation to the internal conversion of excitation energy. III. The total emission spectra of 1-naphthoic acid.
Hochstrasser, R.M.
Can. J. Chem. 39: 1776-82 (1961)
CA 56: 4257i

D 490 The polarization of the fluorescence of 1- and 2-naphthoic acid.
Hochstrasser, R.M.
Can. J. Chem. 39: 1853-5 (1961)
CA 56: 3034c

D 491 Inelastic collisions between excited and nonexcited potassium atoms.
Hoffman, K., and Seiwert, R.
Ann. Physik 7: 71-6 (1961)
CA 55: 15109h

D 492 Cordierite-indialite —a new manganese-activated phosphor.
Hummel, F.A.
J. Electrochem. Soc. 108: 809-10 (1961)
CA 56: 5508a

D 493 Paramagnetic resonance absorption in naphthalene in its phosphorescent state.
Hutchinson, C.A., and Mangum, B.W.
J. Chem. Phys. 34: 908-22 (1961)
CA 56: 1084f

D 494 Luminescence of alcoholic solutions of benzene at -196°.
Ivanova, T.V., and Sveshnikov, B.Ya.
Opt. i Spektroskopiya 11: 598-605 (1961)
CA 56: 5541i

D 495 The lifetime of ultraviolet fluorescence of some aromatic compounds.
Ivanova, T.V., Kudryashov, P.I., and Sveshnikov, B.Ya.
Dokl. Akad. Nauk SSSR 138: 572-4 (1961)
CA 55: 26652f

D 496 Decay phenomena of polarized photo-
luminescence.
Jabionski, A.
Z. Naturforsch. 16a: 1-4 (1961)
CA 55: 14057a

D 497 Acidity constants in the triplet state.
Jackson, G., and Porter, G.
Proc. Roy. Soc. A260: 13-30 (1961)
CA 55: 10030f

D 498 Unimolecular decay of the triplet
state of anthracene in fluid and viscous
media.
Jackson, G., and Livingston, R.
J. Chem. Phys. 35: 2182-6 (1961)
CA 56: 14968i

D 499 Temperature dependence of the ultra-
violet and visible phosphorescence.
Jashi, R.V.
Physica 27: 1119-28 (1961)
CA 56: 1062g

D 500 Induction kinetics of the photosynthesis
in the chlorella pyrenoidosa. II. Kinetics of oxygen emission and fluorescence during the initial phase of illumination.
Joliot, P.
J. Chim. Phys. 58: 584-95 (1961)
CA 55: 26111b

D 501 Photochemical activity and chalking of
zinc oxide (pigment).
Jonas, K.
Veszpremi Vegyip. Egyet. Kozlemen. 5:
99-108 (1961)
CA 57: 4222d

D 502 Spectrum and radius of OH^- in solution.
Jortner, J., Aaz, B., and Stein, G.
J. Chem. Phys. 34: 1455-6 (1961)
CA 55: 20612c

D 503 Splitting of the emission lines of ruby
by an external electric field.
Kaiser, W., Sugano, S., and Wood, D.L.
Phys. Rev. Letters 6: 605-7 (1961)
CA 55: 25510c

D 504 Two-photon excitation in $CaF_2:E_u^{++}$
Kaiser, W., and Garrett, C.G.B.
Phys. Rev. Letters 7: 229-31 (1961)
CA 56: 4211b

D 505 Fluorescence and optical maser effects
in $CaF_2:Sm^{++}$.
Kaiser, W., Garrett, C.G.B., and Wood,
D.L.
Phys. Rev. 123: 766-76 (1961)
CA 55: 21847b

D 506 Solid-state radiation-induced pheno-
mena.
Kallmann, H.P.
U.S. Dept. Comm., Office Tech. Serv.,
AD 264,879, 29 pp. (1961)
CA 56: 1990b

D 507 Phosphorescence spectrum of biphenyl
at 90°K.
Kanda, Y., Shimada, R., and Sakai, Y.
Spectrochim. Acta 17: 1-6 (1961)
CA 55: 18296d

D 508 Triplet-singlet emission spectra of
benzene in carbon tetrachloride and
dioxane matrixes at 90°K.
Kanda, Y., and Shimada, R.
Spectrochim. Acta 17: 7-13 (1961)
CA 55: 18296c

D 509 Triplet-singlet emission spectra of
toluene, o-, m-, and p-xylenes and
mesitylene in cyclohexane matrixes at
90°K.
Kanda, Y., and Shimada, R.
Spectrochim. Acta 17: 279-85 (1961)
CA 55: 18296h

D 510 Absorption spectrum of benzene in
carbon tetrachloride at 77°K.
Kanda, Y., Gondo, Y., and Shimada, R.
Spectrochim. Acta 17: 424-35 (1961)
CA 55: 23040d

D 511 Phosphorescence spectra of polycyclic
compounds.
Kanda, Y., Shimada, R., Hanada, K., and
Kajigaeshi, S.
Spectrochim. Acta 17: 1268-74 (1961)
CA 57: 5477c

D 512 Pigments fluorescing under daylight
and ultraviolet-light excitation.
Kacprzak, F., Olszewski, Z., and Sekula,
W.
Pol. 44,970, 1959 (1961)
CA 56: 11580e

D 513 Phosphorescent enamels.
Kerstan, W.
Sprechsaal 94: 210-13 (1961)
CA 55: 18047d

D 514 Fluorescence emission, absorption,
and temperature radiation of solutions.
Ketskemetry, I., Dombi, J., and Horvai, R.
Ann. Physik 7: 342-52 (1961)
CA 57: 288a

D 515 A complex investigation on the effect of
activator anions on the optical proper-
ties of crystals of alkali halide phos-
phors.
Khalilov, A.Kh., and Isaev, F.K.
Izv. Akad. Nauk Azerb. SSR, Ser. Fiz.-
Mat. i Tekhn. Nauk 1961: 61-71 (1961)
CA 55: 26709g

D 516 Complex investigation of optical and
thermo-optical properties of polyacti-
vated alkali-halide crystal phosphors.
Khalilov, A.Kh., Salaev, E.Yu., Mamedov,
A.P., Alieva, T.D., and Isaev, F.K.
Izv. Akad. Nauk SSSR, Ser. Fiz. 25: 335-40
(1961)
CA 55: 21846d

D 517 Emission and absorption spectra of
frozen crystalline solutions of certain
pyrene derivatives.
Khesina, A.Ya.
Opt. i Spektroskopiya 10: 607-16 (1961)
CA 55: 26660e

D 518 Phosphorescence of organic molecules
having two metastable levels.
Khalupovskii, M.D.
Opt. i Spektroskopiya 11: 617-22 (1961)
CA 56: 9589a

D 519 Line shapes in the method of inter-
secting energy levels.
Kibble, B.P., and Serics, G.W.
Proc. Phys. Soc. 78: 70-4 (1961)
CA 55: 24224f

D 520 Spectral study of centers formed in
silver halide emulsions at various
stages of photolysis.
Kirillov, E.A., and Nesterovskaya, E.A.
Nauchn. Ezhegodnik Odessk. Gos. Univ.,
Fiz.-Mat. Fak. i Nauchn.-Issled. Inst.
Fiz. 1961 (2): 151-7 (1961)
CA 57: 321i

D 521 Concentration phosphorescence ex-
tinction.
Kislyak, G.M.
Ukr. Fiz. Zh. 6: 774-6 (1961)
CA 57: 1692e

D 522 Quenching of fluorescence in solutions
by alien substances in the presence of
high concentrations of fluorescent
material.
Kiyanskaya, L.A., Kudryashov, P.I., and
Sveshnikov, B.Ya.
Dokl. Akad. Nauk SSSR 143: 563-6 (1961)
CA 57: 4152g

D 523 Quenching of fluorescence induced by
iodides in solutions at liquid air
temperature.
Kiyanskaya, L.A., and Sveshnikov, B.Ya.
Opt. i Spektroskopiya 11: 613-16 (1961)
CA 56: 6800d

D 524 Basic processes causing the appear-
ance of fluorescence.
Kling, A., and Kurz, J.
Melliand Textilber. 42: 698-701 (1961)
CA 55: 18116c

D 525 Phase fluorometer.
Kloss, H.G., and Wendel, G.
Z. Naturforsch. 16a: 61-6 (1961)
CA 55: 14056i

D 526 Fluorescence of undoped silver halide
crystals.
Koswig, H.D.
Z. Naturforsch. 16a: 1103-6 (1961)
CA 57: 4152e

D 527 Activation by anions in the oxy-acid
phosphors.
Kotera, Y., Yonemova, M., and Sekine, T.
J. Electrochem. Soc. 108: 540-5 (1961)
CA 55: 23082c

D 528 Simple fluorometer for determination
of B vitamins.
Koziol, J.
Chem. Anal. (Warsaw) 6: 251-60 (1961)
CA 55: 21485c

D 529 Production of fluorescence in packaged
chicken.
Kraft, A.A., and Ayres, J.C.
Appl. Microbiol. 9: 549-53 (1961)
CA 56: 6426e

D 530 Preparation of thin single-crystal
leaflets of anthracene.
Krajsovszky, J., and Ujhelyi, S.
Magy. Fiz. Folyoirat 9: 81-4 (1961)
CA 61: 13957a

D 531 Glycinin absorption spectra in relation
to its association-dissociation.
Kretovich, V.L., Smirnova, T.I., and
Karyakin, A.V.
Biokhimiya 26: 800-6 (1961)
CA 56: 1062g

D 532 Polarization of triplet → singlet
emission spectra.
Krishna, V.G., and Goodman, L.
Nature 191: 800-1 (1961)
CA 56: 4257h

D 533 Semiconducting properties of type
α-IIb diamonds. Luminescence in
semiconducting diamond.
Krumme, J.B., and Leivo, W.J.
U.S. Dept. Comm., Off. Tech. Serv., PB
Rept. 155,100, 37 pp. (1961)
CA 56: 1016d

D 534 A method of studying polymerization
rates at high conversion.
Kryszewski, M., and Grosmanowa, B.
J. Polymer Sci. 52: 85-90 (1961)
CA 56: 11779i

D 535 Spectrophotometric studies of organic
substances. IX. Ultraviolet absorption spectra of derivatives of heterocyclic N-oxides. General properties
on the solvent effect and the substituting effect.
Kubota, T., and Miyazaki, H.
Chem. Pharm. Bull. (Tokyo) 9: 948-61
(1961)
CA 57: 2991g

D 536 Effect of electron bombardment on the
near-infrared fluorescence of single-
crystal CdS.
Kulp, B.A.
J. Appl. Phys. 32: 1966-9 (1961)
CA 56: 4217h

D 537 The fluorescence of 2-hydroxyquinoxa-
line derivatives.
Kumashiro, I.
Nippon Kagaku Zasshi 82: 1224-35 (1961)
CA 56: 4047g

D 538 Fluorescence of 2-hydroxyquinoxaline
derivatives.
Kumashiro, I.
Nippon Kagaku Zasshi 82: 1386-7 (1961)
CA 59: 2607h

D 539 Analysis of fluorescence in elastin,
collagen, and procollagen.
LaBella, F.S.
Can. Conf. Res. Rheumat. Diseases, 2nd,
Toronto 1960: 221-4 (1960)
CA 56: 15994i

D 540 Variation of the ratio of polarization
with wavelength in the fluorescence
spectra of some coloring materials in
solid solution: concentration effect.
Laffitte, E., and Pujols, C.
Compt. Rend. 252: 1008-10 (1961)
CA 56: 15059e

D 541 Fluorescence polarization and the
oscillator form for triphenylmethane
dyes.
Laffitte, E., Mace, N., and Pujols, C.
Compt. Rend. 253: 2911-13 (1961)
CA 57: 8079i

D 542 Mechanism of concentration-dependent
depolarization of fluorescence from
solid solutions in Plexiglas.
Laffitte, E., Mace, N., and Pujols, C.
Compt. Rend. 253: No. 23, 2665-70 (1961)
CA 57: 14550a

D 543 Fluorescent and scintillation spectra
of a CsI(Tl) crystal.
Lagu, R.G., and Thosar, B.U.
Proc. Indian Acad. Sci., Sect. A53: 219-26
(1961)
CA 55: 26710i

D 544 Optical and paramagnetic resonance
spectra of trivalent Pu^{239} in the tri-
chloride and ethyl sulfate.
Lammermann, H., and Stapleton, H.J.
J. Chem. Phys. 35: 1514-16 (1961)
CA 56: 8193b

D 545 Fluorimetry of gallium and indium 8-
quinolinolates.
Landi, M.F.
Metallurgia Ital. 53: 247-50 (1961)
CA 55: 25595d

D 546 The electronic spectra of the Group IV
tetraphenyls.
La Paglia, S.R.
J. Mol. Spectroscopy 7: 427-34 (1961)
CA 56: 11081e

D 547 Edge emission in ZnSe single crystals
at low temperatures.
Larson, O.W.
U.S. Dept. Comm., Office Tech. Serv.,
AD 259,968, 33 pp. (1961)
CA 56: 1011d

D 548 The $^3E_{1u}$ - $^1A_{1g}$ transition of benzene
and the attribution of weak bands near
2600 A to the $^1B_{2u}$ - $^1A_{1g}$ transition.
Leach, S., Lopez-Delgado, R., and Delmos,
F.
J. Mol. Spectroscopy 7: 304-5 (1961)
CA 56: 6799c

D 549 Automatic uranium fluorimeter.
Leng, J.
At. Energy Can. Ltd. AECL 1178, 42 pp.
(1961)
CA 55: 13930i

D 550 Detection of short-life intermediate
products in the fluorescence extinction.
Lemhardt, H., and Weller, A.
Z. Physik. Chem. 29: 277-80 (1961)
CA 57: 8079e

D 551 A barium-calcium carbonate from Hot
Spring County, Arkansas.
Lippmann, F.
Fortschr. Mineral. 39: 81 (1961)
CA 55: 19619g

D 552 Four-specimen liquid helium cryostat
for fluorescence.
Lipsett, F.R.
Rev. Sci. Instr. 32: 840-1 (1961)
CA 56: 11382h

D 553 Spectral photometric and fluorescent
investigation on the complex formation
of chromotropic acid with molybdates
and tungstates.
Liska, M., and Plsko, E.
Sb. Prac. Chem. Fak. SUST 1961: 87-92
(1961)
CA 57: 6869b

D 554 Tyrosine phosphorescence of proteins.
Longworth, J.W.
Biochem. J. 81: 23p-24p (1961)
CA 56: 5087c

D 555 Dual atmospheric tracer techniques
for diffusion studies using phosphores-
cence-fluorescence analysis.
Ludwick, J.D.
U.S. At. Energy Comm. HW-70892, 17 pp.
(1961)
CA 59: 2143e

D 556 Liquid scintillation techniques applied
to counting phosphorescence emission.
Measurement of trace quantities of
zinc sulfide.
Ludwick, J.D., and Perkins, R.W.
Anal. Chem. 33: 1230-5 (1961)
CA 56: 22d

D 557 Qualitative test of the fluorescence
changes of organic solutions after the
action of ultraviolet light.
Ludwig, W., and Herforth, L.
Z. Naturforsch. 16b: 638-44 (1961)
CA 57: 218e

D 558 Stimulated optical emission in fluores-
cent solids. I. Theoretical considera-
tions. II. Spectroscopy and stimulated
emission in ruby.
Maiman, T.H., Hoskins, R.H., and
D'Haenens, I.J.
Phys. Rev. 123: 1145-50 (1961)
CA 55: 24237c

D 559 Quenching of fluorescence of dyes by
neutral salts.
Majumdar, D.Z.
Z. Physik. Chem. 217: 200-6 (1961)
CA 56: 1063b

D 560 Polarization of low-temperature lumin-
escence (fluorescence) from thallium
containing sodium nitrate single crys-
tals.
Makishima, S., Tomotsu, T., Hirata, M.,
Kambe, R., and Shionoya, S.
Phys. Chem. Solids 18: 262-4 (1961)
CA 55: 25511c

D 561 Preparation and investigation of sub-
limed zinc sulfide phosphors.
Malysheva, A.F., and Jogi, H.
Tr. Inst. Fiz. i Astron., Akad. Nauk Est.
SSR 1961: 105-19 (1961)
CA 56: 2953h

D 562 Direct fluorescence from single crystals of naphthalene doped with anthracene.
Martin-Bouyer, M., and Meinnel, J.
J. Phys. Radium 22: 126-8 (1961)
CA 55: 15145e

D 563 Wave-length dependence of quantum efficiency and absorption coefficient of $ZnSiO_3$/Mn powder phosphor.
Masuda, I.
J. Phys. Soc. Japan 16: 106-7 (1961)
CA 55: 12061g

D 564 Flavines in some types of invertebrate muscles.
Mattisson, A.G.M.
Arkiv Zool. 13: 545-52 (1961)
CA 57: 8882c

D 565 The electronic structure, spectra, and magnetic properties of actinyl ions. I. The uranyl ion.
McGlynn, S.P., and Smith, J.K.
J. Mol. Spectroscopy 6: 164-187 (1961)
CA 55: 20610b

D 566 Molecular vibrations in the excitation theory for molecular aggregates.
McRae, E.G.
Australian J. Chem. 14: 329-43 (1961)
CA 56: 2988g

D 567 Molecular vibrations in the excitation theory for molecular aggregates. III. Polymeric systems.
McRae, E.G.
Australian J. Chem. 14: 354-71 (1961)
CA 56: 2988b

D 568 Zero-field splitting of molecular Zeeman levels (phos. naphthalene).
McWeeny, R.
J. Chem. Phys. 34: 399-401 (1961)
CA 56: 1057a

D 569 Decay of phosphorescence in $CaCO_3$, $MgCO_3$, $CaMg(CO_3)_2$, and $CaSO_4$.
Medlin, W.L.
Phys. Rev. 122: 837-42 (1961)
CA 55: 20657a

D 570 Decay of phosphorescence from a distribution of trapping levels.
Medlin, W.L.
Phys. Rev. 123: 502-9 (1961)
CA 56: 4217c

D 571 Fluorescence studies of the changes undergone by nucleoproteins and their derivatives in irradiated cells.
Meisel, M.N., Brumberg, E.M., Kondrat'eva, T.M., and Barskii, I.Ya.
Initial Effects Ionizing Radiations Cells 1961: 107-29
CA 57: 6275e

D 572 Quantum efficiencies of fluorescence of organic substances: effect of solvent and concentration of the fluorescent solute.
Melhuish, W.H.
J. Phys. Chem. 65: 229-35 (1961)
CA 55: 24228d

D 573 Distribution of fluorescence from a dish-shaped cuvette.
Melhuish, W.H.
J. Opt. Soc. Am. 51: 278-9 (1961)
CA 55: 13052e

D 574 Theory of the fluorescent polarization for nonspherical symmetric molecules.
Memming, R.
Z. Physik. Chem. 28: 168-89 (1961)
CA 55: 18285h

D 575 Fluorescence in mineral oils.
Mihul, C., Ruscior, C., Pop, V., Suciu, M., and Radulescu, Gh.
Acad. Rep. Populare Romine, Filiala Iasi, Studii Cercetari Stiint., Fiz. Stiinte Technice 12: 7-14 (1961)
CA 57: 1163h

D 576 Infrared spectroscopic study of the photolysis of chlorine azide in solid argon at 4.2 K (phos. observ.).
Milligan, D.E.
J. Chem. Phys. 35: 372-3 (1961)
CA 56: 1066d

D 577 The sulfuric acid-induced fluorescence of aldosterone.
Miras, C., and Contaxis, C.
Chim. Chronika (Athens, Greece) 26: 124-6 (1961)
CA 55: 27499g

D 578 Chemical and chemotheropeutical studies on furan derivatives. XX. Increased antibacterial activity of 1,5-bis(5-nitro-2-furyl)-3-pentadienone hydrochloride by heat treatment.
Miura, K., Ikeda, M., Ohashi, T., Igarashi, Y., and Ichimura, K.
Yakugaku Zasshi 81: 1372-4 (1961)
CA 56: 7321d

D 579 The mechanism of the transition of
excited organic molecules into the
metastable state.
Mokeeva, G.A., and Sveshnikov, B.Ya.
Opt. i Spektroskopiya 10: 86-90 (1961)
CA 55: 10059a

D 580 Fluorescence of crystals of magnesium
and calcium platinum tetracyanides.
Moncuit, C.
Compt. Rend. 252: 397-9 (1961)
CA 55: 15112a

D 581 Fluorescence of soap attached to the
skin and to cloth.
Mori, I.
Kyoto Furitsu Ika Daigaku Zasshi 70:
453-6 (1961)
CA 56: 10409c

D 582 Effect of ozone on the fluorescence and
ultraviolet absorption spectra of
3,4-benzopyrene.
Morlin, Z., and Saringer, K.M.
Nature 191: 907-8 (1961)
CA 56: 4256f

D 583 The effect of ozone on the absorption
and fluorescence spectra of 3,4-benzo-
pyrene.
Morlin, Z., and Saringer, M.
Egeszsegtudomany 5: 383-9 (1961)
CA 56: 11082f

D 584 Fluorimetric determination of gold
with kojic acid.
Murata, A., and Ujihara, T.
Bunseki Kagaku 10: 497-501 (1961)
CA 56: 6180d

D 585 Orcein pigments. XII. Synthesis of
α-hydroxyorcein.
Musso, H., and Beecken, H.
Chem. Ber. 94: 585-600 (1961)
CA 55: 14463h

D 586 Fluorescence spectra of some uranyl
salts.
Narasimhan, K.V.
Indian J. Phys. 35: 282-98 (1961)
CA 56: 1059i

D 587 Fluorescence and phosphorescence of
cadmium iodide at low temperatures.
Nguyen, Chung-Tu, and Mink, Pho Duc
Ann. Fac. Sci., Univ. Saigon 1961: 29-34
(1961)
CA 56: 13303f

D 588 Luminescence of anthraquinone and its
β-derivatives in frozen solutions.
Nurmukhametov, R.N., and Shigorin, D.N.
Zh. Fiz. Khim. 35: 72-9 (1961)
CA 55: 12032b

D 589 Effect of hydrogen bond on the lumin-
escence of hydroxy and aminoazo
compounds.
Nurmukhametov, R.N., Shigorin, D.N.,
Kozlov, Yu. I., and Puchkov, V.A.
Opt. i Spektroskopiya 11: 606-12 (1961)
CA 56: 5542e

D 590 Structure of tanshinone-II.
Okumura, Y., Kakisawa, H., Kato, M., and
Hirata, Y.
Bull. Chem. Soc. Japan 34: 895-7 (1961)
CA 56: 3434e

D 591 Polarized phosphorescence in crystal-
line hexachlorobenzene at 77 K.
Olds, D.W.
J. Chem. Phys. 35: 2248-9 (1961)
CA 56: 15030i

D 592 Lifetime studies of phosphorescence in
aromatic compounds under various
conditions at low temperature.
Olness, D.V.
Dissertation, Duke Univ. (1961)
CA 56: 359f

D 593 Orientation of chlorophyll molecules in
vivo—evidence from polarized fluores-
cence.
Olson, R.A., Butler, W.L., and Jennings,
W.H.
Biochim. Biophys. Acta 54: 615-7 (1961)
CA 56: 13258g

D 594 See D 471a

D 595 Fluorescence of tetracyclines in bone
tumors, normal bones, and teeth.
Owen, L.N.
Nature 190: 500-2 (1961)
CA 61: 12481f

D 596 Influence of the temperature upon the
phosphorescence of organic dyes.
Pankeeva, A.Y.
Ukr. Fiz. Zh. 6: 800-3 (1961)
CA 57: 288b

D 597 Lifetime of the pyrene dimer.
　　Parker, C.A., and Hatchard, C.G.
　　Nature 190: 165-6 (1961)
　　CA 55: 19879f

D 598 The fluorescence and phosphorescence
　　phenomena in the photoluminous pig-
　　ments.
　　Pasquarelli, O.
　　Pitture Vernici 37: 3-6 (1961)
　　CA 55: 13869i

D 599 Photosensitization and the effect of
　　ultraviolet radiation on the production
　　of unpaired electrons in the presence
　　of furocoumarins (psoralens).
　　Pathak, M.A., Allen, B., Ingram, D.J.E.,
　　and Fellman, J.H.
　　Biochim. Biophys. Acta 54: 506-15 (1961)
　　CA 56: 11988c

D 600 Cyclopentadienes, fulvenes, and fulva-
　　lenes. I. A heraphenyl fulvalene.
　　Pauson, P.L., and Williams, B.J.
　　J. Chem. Soc. 1961: 4153-7 (1961)
　　CA 56: 4671d

D 601 Electronic spectra of acenaphthene and
　　denterioacenaphthene at a low tempera-
　　ture.
　　Pesteil, L.
　　J. Chim. Phys. 58: 204-7 (1961)
　　CA 55: 13050i

D 602 Shift of the electronic spectra of
　　phthalimides in solution under the
　　effect of temperature.
　　Pikulik, L.G., and Dreitser, F.F.
　　Dokl. Akad. Nauk Belorussk. SSR 5: 57-60
　　(1961)
　　CA 57: 13306e

D 603 Polarization of phosphorescence in
　　organophosphors.
　　Pilipovich, V.A.
　　Opt. i Spektroskopiya 10: 209-13 (1961)
　　CA 55: 12060e

D 604 A study of the fluorescence of cellulosic
　　polymers.
　　Plitt, K.F., and Toner, S.D.
　　J. Appl. Polymer Sci. 5: 534-8 (1961)
　　CA 56: 3694f

D 605 The fluorescence of essential oils and
　　aromatic alcohols (flavors for liqueur
　　vodkas).
　　Podlubnaya, E.T., and Babkova, A.N.
　　Tr., Tsentr. Nauchn.-Issled. Inst. Spirt. i
　　Likero-Vodochn. Prom. 1961: 262-71
　　(1961)
　　CA 56: 14679i

D 606 Complex nature of the F band in KCl
　　crystals.
　　Politov, N.G.
　　Opt. i Spektroskopiya 10: 173-6 (1961)
　　CA 55: 12061c

D 607 Phosphorescence of polystyrene.
　　Pozzi-Escot Herold, L.
　　Dissertation, Fordham Univ. (1961)
　　CA 56: 581e

D 608 Observed phosphorescence and singlet-
　　triplet absorption in s -triazine and
　　trimethyl-s -triazine.
　　Paris, J.P., Hirt, R.C., and Schmitt, R.G.
　　J. Chem. Phys. 34: 1851-2 (1961)
　　CA 55: 23041h

D 609 Fluorometric determination of sub-
　　microgram amounts of selenium.
　　Parker, C.A., and Harvey, L.G.
　　Analyst 86: 54-62 (1961)
　　CA 55: 20782b

D 610 Triplet-singlet emission in fluid solu-
　　tions. Phosphorescence of cosin.
　　Parker, C.A., and Hatchard, C.G.
　　Trans. Faraday Soc. 51: 1894-1904 (1961)
　　CA 57: 289a

D 611 Temperature effect in the phosphores-
　　cence of solid nitrogen.
　　Pilon, A.M.
　　Preprints Papers Intern. Symp. Free
　　Radicals, 5th, Uppsala 1961: 50/1-50/4
　　(1961)
　　CA 59: 8278h

D 612 Fluorescence excitation by the absorp-
　　tion of two consecutive photons.
　　Porter, J.F., Jr.
　　Phys. Rev. Letters 7: 414-15 (1961)
　　CA 61: 192a

D 613 Temperature changes in the luminescence spectrum of samarium in certain perovskite lattices.
Rabkin, L.M.
Materialy Nauchn. Kong. Aspirantov Sb. 1961: 124-30
CA 60: 7585e

D 614 Fluorometric determination of phenothiazines in biological samples.
Rogland, J.B., and Kinross-Wright, V.J.
Federation Proc. 20: 397 (1961)

D 615 Differential fluorometric determinations of metadrenaline and normetadrenaline.
Randrup, A.
Clin. Chim. Acta 6: 584-6 (1961)
CA 55: 26089f

D 616 A versatile high-resolution spectrofluorometer.
Rehwoldt, R.E., King, R.M., and Hercules, D.M.
Anal. Chem. 33: 1362-5 (1961)
CA 57: 1513c

D 617 Temperature shift of edge emission in cadmium sulfide crystals.
Reynolds, D.C., and Pedrotti, L.S.
Proc. Intern. Conf. Semicond. Phys., Prague 1960: 1049-52 (Pub. 1961)
CA 56: 2993f

D 618 Edge emission in zinc selenide single crystals.
Reynolds, D.C., Pedrotti, L.S., and Larson, D.W.
J. Appl. Phys. 32: 2250-4 (1961)
CA 56: 4228b

D 619 Substituted benzidines and related compounds as reagents in analytical chemistry. XVII. The N,N,N',N'-tetra-carbormethyl derivatives of some 3,3'-disubstituted benzidines.
Rees, D.I., and Stephen, W.I.
J. Chem. Soc. 1961: 5101-5 (1961)
CA 56: 12776e

D 620 The separation of mine species of Iridaceae by paper chromatography.
Riley, H.P., and Bryant, T.R.
Am. J. Botany 48: 133-7 (1961)
CA 60: 9593d

D 621 Spectra and energy-transfer phenomena in crystalline rare-gas solvents.
Robinson, G.W.
J. Mol. Spectroscopy 6: 58-83 (1961)
CA 55: 20611f

D 622 Fluorescence lifetimes of fluorescein anion.
Rohatgi, K.K.
Z. Physik. Chem. 217: 353-6 (1961)
CA 57: 215f

D 623 Average extinction coefficient for reabsorption of fluorescence radiation through uranin solution.
Rohatgi, K.K., and Singhal, G.S.
Nature 191: 799-800 (1961)
CA 56: 4258b

D 624 Optical absorption and fluorescence of oxygen in alkali halide crystals.
Rolfe, J., Lipsett, F.R., and King, W.J.
Phys. Rev. 123: 447-54 (1961)
CA 55: 19505b

D 625 Effect of substitution on the fluorescence of phenal and aniline.
Rosen, A., and Williams, R.T.
Photoelec. Spectrometry Group Bull. 13: 339-45 (1961)
CA 57: 6754a

D 626 Photo- and semiconduction in crystalline chlorophylls a and b.
Rosenberg, B., and Couniscoli, J.F.
J. Chem. Phys. 35: 982-92 (1961)
CA 56: 9563b

D 627 A new method for studying pore sizes by the use of dye luminescence.
Rosenberg, J.L., and Shambert, D.J.
J. Phys. Chem. 65: 2103-5 (1961)
CA 56: 12415i

D 628 pH dependence on the fluorescence yield of tyrosine copolymers.
Rosenheck, K., and Weber, G.
Biochem. J. 79, 29 pp. (1961)
CA 60: 8253g

D 629 Transfer of excitation energy in rigid solutions of organic compounds.
Rossman, I.M.
Opt. i Spektroskopiya 10: 354-61 (1961)
CA 55: 15114g

D 630 Estimation of copper by a luminescence
activation method.
Rapp, R.C., and Shearer, N.W.
Anal. Chem. 33: 1240-2 (1961)
CA 56: 1989g

D 631 Singlet →triplet absorption in a few
polysubstituted benzenes in the vapor
state.
Roy, J.K.
Indian J. Phys. 35: 628-36 (1961)
CA 57: 1756b

D 632 Effect of pH on the fluorescence of
fluorescein solutions.
Rozwadowski, M.
Acta Physiol. Polon. 20: 1005-17 (1961)
CA 56: 1062f

D 633 Fluorometric determination of small
amounts of uranium — an improved
furnace and examination of procedure.
Sakanone, M., and Ichikawa, M.
Bunseki Kagaku 10: 645-51 (1961)
CA 56: 6658h

D 634 The preparation of fluorescent german -
ia.
Sarver, J.F., and Hummel, F.A.
J. Electrochem. Soc. 108: 195-6 (1961)
CA 55: 9067i

D 635 Polarization of fluorescence and dichroism of anisotropic [poly(vinyl alcohol)] films activated with phthalimides.
Sarzhevskii, A.M.
Dokl. Akad. Nauk Belorussk. SSR 5:
249-52 (1961)
CA 57: 14594g

D 636 Effect of temperature and viscosity on
the polarization of fluorescence in
phthalimides.
Sarzhevskii, A.M.
Opt. i Spektroskopiya 10: 621-6 (1961)
CA 56: 1058d

D 637 The fluorescence spectra of carbazole.
Sawicki, E., Hauser, T.R., Stanley, T.W.,
and Elbert, W.C.
Anal. Chem. 33: 1574 (1961)

D 638 The acidity of some 4-nitro- and
2,4-dinitrophenylhydrazones.
Schaal, R., and Gadet, C.
Bull. Soc. Chim. France 1961: 2154-61
CA 57: 128c

D 639 Strain-induced effects on the degener-
ate spectral lines of chromium in MgO
crystals.
Schanlow, A.L., Piksis, A.H., and Sugano,
S.
Phys. Rev. 122: 1469-76 (1961)
CA 55: 18342g

D 640 Reaction of blood and muscle proteins
with fluorescent curarelike substances
derived from stilbene.
Schell, H.D.
Acad. Rep. Populare Romine, Studii
Cercetari Biochim. 4: 365-9 (1961)
CA 57: 6539d

D 641 Effect of molecular structure on
scintillation and fluorescence.
Schmid, W.F.
L. C. Card No. MIC 61-808, 262 pp.,
Dissertation Univ. of Colorado, Diss.
Abs. 21: 3285 (1961)
CA 55: 17250d

D 642 Fluorescence of pyrene in solid hydro-
carbons.
Schmillen, A.
Z. Naturforsch. 16a: 5-10 (1961)
CA 55: 14079d

D 643 Fluorescence decay of some aromatic
hydrocarbons by alpha-ray excitation.
Schmillen, A., and Kramer, K.
Z. Naturforsch. 16a: 1192-9 (1961)
CA 56: 12456h

D 644 Fluorescence induced by vacuum
ultraviolet radiation.
Schoen, R.I., Judge, D.L., and Weissler,
G.L.
U.S. Dept. Comm., Office Tech. Serv.,
AD 268,005, 9 pp. (1961)
CA 56: 3002g

D 645 Kinetics of photoconduction and phos-
phorescence.
Schon, M.
U.S. At. Energy Comm. AEC-tr-4339
90 pp. (1961)
CA 55: 17246g

D 646 Electronic spectroscopy and theoretical
treatments of ferrocene and nickelo-
cene.
Scott, D.R., and Becker, R.S.
J. Chem. Phys. 35: 516-31 (1961)
CA 56: 6797i

D 647 The secondary fluorescence as a tool in
cytochemistry.
Sebruyns, M.
Biol. Jaarboek Konink. Naturrw. Genoot.
Dodonaea, Gent 29: 27-41 (1961)
CA 56: 6297i

D 648 Radioluminescence from β-radiation.
Serra, M., and Del Monaco, A.
Energia Nucl. (Milan) 8: 566-70 (1961)
CA 56: 11148i

D 649 High-energy radiation damage to
fluorescent organic solids.
Sharn, C.F.
J. Chem. Phys. 34: 240-6 (1961)
CA 55: 19433a

D 650 Fluorimetric determination of gallium
with rhodamine S.
Shcherbov, D.P., and Kogarlitskaya, N.V.
Tr. Kazakhsk. Nauchn.-Issled. Inst.
Mineral'n. Syr'ya 1961: 225-9 (1961)
CA 56: 11946h

D 651 Fluorescence of some oxinates. Fluor-
ometry of scandium in the presence of
rare earth elements.
Shcherbov, D.P., and Lovchi, A.K.
Tr. Kazakhsk. Nauchn.-Issled. Inst.
Mineral'n. Syr'ya 1961: No. 6, 183-91
(1961)
CA 56: 13304g

D 652 Unit based on spectrophotometer SF-4
for recording excitation and fluores-
cence spectra.
Shcherbov, D.P., and Ponomarenko, A.I.
Zavodsk. Lab. 27: 1156-8 (1961)
CA 56: 8152c

D 653 Lowest multiplicity forbidden transi-
tions in diazines. I. The phosphores-
cence spectrum of pyrazine at 90°K. II.
The phosphorescence spectrum of
pyrimidine at 90°K.
Shimada, R.
Spectrochim. Acta 17: 14-29 (1961)
CA 55: 17208f

D 654 Behavior of excited electrons and holes
in zinc sulfide phosphors.
Shionoya, S., Kallmann, H.P., and Kramer,
B.
Phys. Rev. 121: 1607-19 (1961)
CA 55: 18341e

D 655 Determination of excitation spectra
with a recording spectrophotometer.
Sill, C.W.
Anal. Chem. 33: 1579-84 (1961)
CA 56: 1981i

D 656 (Fluorometric determination of
beryllium.)
Sill, C.W., Willis, C.P., and Flygare, J.K.
Anal. Chem. 33: 1671 (1961)

D 657 Inner-molecular radiationless transi-
tions.
Simon, Z.
Rev. Phys., Acad. Rep. Populaire
Roumaine 6: 105-18 (1961)
CA 56: 12434b

D 658 Nonradiative transitions and lumines-
cence of molecules. Derivatives of the
diphenylmethyl cation and other torsion-
able aromatic systems.
Simon, Z.
Acad. Rep. Populare Romine, Studii
Cercetari Chim. 9: 667-72 (1961)
CA 57: 216e

D 659 Energy-recording spectrofluorimeter.
Slavin, W., Mooney, R.W., and Palumbo,
D.T.
J. Opt. Soc. Am. 51: 93-7 (1961)
CA 55: 10984a

D 660 Effects of rare-earth substitutions on
the fluorescence of terbium hexa-
antipyrine triiodide.
Soden, R.R.
J. Appl. Phys. 32: 750-1 (1961)
CA 55: 21789f

D 661 X-ray excitation of fluorescence of
dilute aqueous solutions of aromatic
compounds.
Sommermeyer, K., Birkwald, V., and
Pruetz, W.
Naturwissenschaften 48: 666-7 (1961)
CA 56: 12435c

D 662 The fluorescence of aqueous solutions
of aromatic compounds excited by
X-rays.
Sommermeyer, K., Birkwald, V.B., and
Pruetz, W.
Strahlentherapie 116: 354-63 (1961)
CA 56: 9588b

D 663 Excitation of fluorescence by X-rays in highly dilute aqueous solutions of aromatic compounds.
Sommermeyer, K., Pruetz, W., and Birkwald, V.
Atomkernenergie 6: 445-51 (1961)
CA 56: 11044h

D 664 Solid-state optical maser using bivalent samarium in calcium fluoride.
Sorokin, P.P., and Stevenson, M.J.
IBM J. Res. Develop. 5: 56-8 (1961)
CA 55: 15126f

D 665 Properties of tertiary phosphines. I. Triphenylphosphine oxide and its adducts with bromine and dinitrogen tetraoxide.
Stachlewska-Wroblowa, A., and Okon, K.
Biul. Wojskowej Akad. Tech. 10: 3-13 (1961)
CA 56: 14322e

D 666 The electrorefining effect on phosphorescence and thermoluminescence of crystalline quartz provoked by γ-irradiation.
Starodubtsev. S.V., and Vakhidov, Sh.A.
Izv. Akad. Nauk Uz. SSR, Ser. Fiz.-Mat. Nauk 1961: 49-53 (1961)
CA 56: 2074h

D 667 Influence of pH and urea on the ultraviolet fluorescence of several globular proteins.
Steiner, R.F., and Edelhock, H.
Nature 192: 873-4 (1961)
CA 36: 9068i

D 668 Experimental proof for the radiation of plasma oscillations in thin silver films.
Steinmann, W.
Z. Physik 163: 92-107 (1961)
CA 55: 17235h

D 669 Fluorescence and absorption studies of reversible aggregation in chlorophyll.
Stensby, P.S., and Rosenberg, J.L.
J. Phys. Chem. 65: 906-9 (1961)
CA 55: 19474b

D 670 Lifetime of the pyrene dimer.
Stevens, B., and Hutton, E.
Nature 190: 166-7 (1961)
CA 55: 19879g

D 671 The use of rivanol for obtaining fluorescent γ-globulins brucellosis serums.
Stolbikov, E.P.
Veterinariya 38: 76-9 (1961)
CA 56: 1888d

D 672 Arenesulfonamide-aminotriazine-formaldehyde resins containing fluorescent pigments.
Switzes, K.
Brit. 869,801 (1961)
CA 56: 673h

D 673 Optical bleaches. I.
Szekely, R., and Szekely, A.
Melliand Textilber. 42: 923-5 (1961)
CA 55: 22834i

D 674 Modification of a spectrophotofluorimeter for scanning paper strips.
Taketomo, Y.
Nature 190: 1094-5 (1961)
CA 56: 3752b

D 675 Structural dependence of protein ultraviolet fluorescence.
Teale, F.W.J.
Biochem. J. 80: 14 pp. (1961)
CA 55: 21183c

D 676 Ultraviolet fluorescence of proteins.
Teale, W.F.J.
Photoelect. Spectrometry Group Bull. 13: 346-8 (1961)
CA 57: 8868d

D 677 Effect of solvent and of temperature on phosphorescence spectra of phenanthrene.
Teplyakov, P.A., and Pyatnitskii, B.A.
Izv. Vysshikh Uchebn. Zavedenii, Fiz. 1961: 84-9 (1961)
CA 56: 8177i

D 678 Phosphorescence spectra of benzoic and sulfobenzoic acids in nonane.
Teplyakov, P.A., and Grosul, V.P.
Ukr. Fiz. Zh. 6: 816-19 (1961)
CA 57: 287d

D 679 Excitons and band splitting produced by uniaxial stress in CdTe.
Thomas, D.G.
J. Appl. Phys. 32: 2298-304 (1961)
CA 56: 4211g

D 680 Bound exciton complexes.
Thomas, D.G., and Hopfield, J.J.
Phys. Rev. Letters 7: 316-19 (1961)
CA 60: 8740e

D 681 Spedrofluorometric determination of
anthracene, fluorene, and phenanthrene
in mixtures.
Thommes, G.A., and Leininger, E.
Talanta 7: 181-6 (1961)
CA 56: 8011b

D 682 Pseudoaromatics from 2-indanones.
III. 1,2: 5,6-Dibenzoxalenes.
Treibs, W., and Schroth, W.
Ann. 642: 82-96 (1961)
CA 55: 27289c

D 683 Effect of solvent on the extinction of
phosphorescence in some organic
acids at low temperature.
Trusov, V.V.
Nauchn. Zap., Odessk. Gos. Ped. Inst.
Kafedry Mate., Fiz. i Estestvozn. 25:
89-92 (1961)
CA 61: 10200g

D 684 Phosphorescence of crystalline quartz
under the effect of γ-irradiation.
Vakhidov, Sh.A. and Starodubtsev, S.V.
Tr. Tashkentsk. Konf. po Mirnomu
Ispol'z. At. Energii, Akad. Nauk Uz.
SSR 1: 171-4 (1961)
CA 57: 1685f

D 685 Ion-pair resonance mechanism of
energy transfer in rare earth crystal
fluorescence.
Vaasanyi, F., and Dieke, G.H.
Phys. Rev. Letters 7: 442-3 (1961)
CA 56: 12418i

D 686 Solvent effects in the fluorescence of
indole and substituted indoles.
VanDuren, B.L.
J. Org. Chem. 26: 2954-60 (1961)
CA 56: 441i

D 687 Fluorescence centers in zinc sulfide.
Van Gool, W.
Philips Res. Rept., Suppl. 3: 1-119 (1961)
CA 56: 5506b

D 688 Emission spectra of trivalent thulium.
Van Uitert, L.G., and Soden, R.R.
J. Chem. Phys. 34: 276-9 (1961)
CA 55: 20609h

D 689 Comparison of the thermal activation
energies of electrical conduction with
absorption and phosphorescence spectra in a series of organic compounds.
Vartanyan, A.T., and Rozenshtein, L.D.
Izv. Akad. Nauk SSSR, Ser. Fiz. 25:
428-30 (1961)
CA 55: 17251f

D 690 Single-crystal cathodoluminescent
screens.
Vasil'eva, M.A., Kuprevich, V.V.,
Stepanov, I.V., and Feofilov, P.P.
Izv. Akad. Nauk SSSR, Ser. Fiz. 25: 321-3
(1961)
CA 55: 20658a

D 691 Effect of p-toluidine and potassium
iodide on the fluorescence and photo-
oxidation of 9,10-di-n-propylanthracene.
Vember, T.M., and Cherkasov, A.S.
Opt. i Spektroskopiya 10: 544-6 (1961)
CA 55: 17209f

D 692 Certain regularities in the nature of
yield variation as related to the position of the fluorescence band in a number of organic compounds.
Viktorova, E.N.
Opt. i Spektroskopiya 10: 279-81 (1961)
CA 55: 12031i

D 693 Isolation, structure, and synthesis of
pterins from Ephestia kuhnellia Zeller.
Viscontini, M., and Stierlin, H.
Helv. Chim. Acta 44: 1783-5 (1961)
CA 56: 7319d

D 694 ZnS:Cu, Si phosphors.
Wachtel, A.
J. Electrochem. Soc. 108: 534-40 (1961)
CA 55: 23082d

D 695 Experimental study of energy transfer
between unlike molecules in solution.
Ware, W.R.
J. Am.Chem. Soc. 83: 4374-7 (1961)
CA 57: 5311e

D 696 Fluorescence in gemstone identifica-
tion.
Webster, R.
Lapidary J. 14: 494-6, 500-9 (1961)
CA 61: 13966c

D 697 Energy-transfer and quenching pro-
cesses in the system cyclohexane-
benzene-terphenyl-oxygen.
Weinreb, A.
J. Chem. Phys. 34: 1316-19 (1961)
CA 56: 1062b

D 698 Some effects of temperature and vis-
cosity on fluorescence and energy
transfer in solutions.
Weinreb, A.
J. Chem. Phys. 35: 91-102 (1961)
CA 55: 21793a

D 699 The photolysis and fluorescence of
diethyl ketone and diethyl ketone-
biacetyl mixtures at 3130 and 2537 A.
Weir, D.S.
J. Am. Chem. Soc. 83: 2629-33 (1961)
CA 55: 23041f

D 700 Luminescent properties of the mixed
phenyl- and p-biphenylyl-substituted
silanes under ultraviolet γ-, and
β-excitation.
Weis, C.G.
U.S. Dept. Comm., Office Tech. Serv.
AD 259,662 72 pp. (1961)
CA 56: 7577e

D 701 Spectrofluorometric analysis of cardio-
tonic steroids.
Wells, D., Katzung, B., and Meyers, F.H.
J. Pharm. Pharmacol. 13: 389-95 (1961)
CA 56: 543g

D 702 Phosphate phosphors.
Westinghouse Electric Corp.
Brit. 882,347 (1959) (1961)
CA 56: 12431c

D 703 Stimulated optical emission from
exchange-coupled ions of Cr^{+3} in Al_2O_3.
Wieder, I., and Sarles, L.R.
Phys. Rev. Letters 6: 95-6 (1961)
CA 56: 110e

D 704 Pressure dependence of fluorescence
spectra.
Wilson, D.J., and Noble, B.
J. Chem. Phys. 34: 1392-6 (1961)
CA 55: 19459h

D 705 The true fluorescence of naphthalene
at 4.2°K.
Wolf, H.C.
Naturwissenschaften 48: 43-4 (1961)
CA 55: 14079c

D 706 Analysis of the absorption and fluor-
escence spectrum of $NdCl_3$ diluted with
$LaCl_2$.
Wong, E.Y.
J. Chem. Phys. 34: 1989-93 (1961)
CA 55: 18290g

D 707 Mode of binding in D-amino acid oxidase studied by the fluorescence of
flavine adenine dinucleotide.
Yagi, K., Ozawa, T., and Harada, M.
Koso Kagaku Shimpoziumu 14: 87-91, 91-3
(1961)
CA 55: 16622i

D 708 Decay and buildup of luminescence of
(ZnS:Cu, Cl) phosphors
Yoshida, T.
Oyo Butsuri 30: 764-9 (1961)
CA 57: 8045e

D 709 Diaminostilbenedisulfonic acid deriva-
tives.
Yoshida, T.
Japan 14,921 (1959) (1961)
CA 56: 10171e

D 710 Fluorescence microscopic studies on
the differentiation of living and dead
coli bacilli by means of acridine
orange vital staining.
Yoshiba, S., and Fukushima, K.
Tokyo Jikeikai Ika Daigaku Zasshi 76:
905-14 (1961)
CA 56: 7146e

D 711 Fluorescence of organic substances in
an adsorbed state.
Zhmyreva, I.A., and Kochemirovskii, A.S.
Zh. Fiz. Khim. 35: 1163-5 (1961)
CA 56: 4258i

D 712 Effect of the size of a solute molecule
on the susceptibility of its electronic
spectra to the action of the solvent.
Zhmyreva, I.A., and Reznikova, I.I.
Opt. i Spektroskopiya 10: 281-4 (1961)
CA 55: 12033a

D 713 Dependence of phosphorescence quenching on electron-trap filling and on temperature.
Zhukova, N.V., Eudokimova, G.K., and Levshin, V.L.
Izv. Akad. Nauk SSSR, Ser. Fiz. 25: 476-8 (1961)
CA 55: 20655f

D 714 Polarization of electronic (absorption) bands of aromatics. III. Quinoline, isoquinoline, indole. IV. Phenanthrene, chrysene, tetraphene.
Zimmermann, H., and Joop, N.
Z. Elektrochem. 65: 61-5 (1961) (Ger.)
CA 55: 14055a

D 715 Polarization of electronic (absorption) bands of aromatics. V. Benzene, coronene, triphenylene, pyrene, perylene.
Zimmermann, H., and Joop, N.
Z. Elektrochem. 65: 138-42 (1961) (Ger.)
CA 55: 16136f

D 716 Phosphorescence spectra of acenaphthene at low temperatures.
Zmerli, A.
J. Chem. Phys. 34: 2130-5 (1961)
CA 55: 18296i

D 717 Fluorescent coating compositions.
Sterling Drug Ltd.
Brit. 859,891 (1961)
CA 60: 13451d

D 718 Optical double-photon absorption in cesium vapor.
Abella, I.D.
Phys. Rev. Letters 9: 453-9 (1962)
CA 56: 4031c

D 719 Luminescence in Al nitride and Al_2O_3.
Adams, I., and AuCoin, T.R.
Luminescence Org. Inorg. Mater. Intern. Conf., New York 1961: 638-40 (1962)
CA 56: 2952f

D 720 Electrofluorescence of ruby powder.
Adams, I., Aucoin, T.R., and Mellichamp, J.W.
J. Appl. Phys. 33: 245 (1962)
CA 57: 14547b

D 721 Electrofluorescence of rare-earth-activated Al_2O_3.
Adams, I., and Mellichamp, J.W.
J. Chem. Phys. 36: 2456-9 (1962)
CA 57: 10636d

D 722 Effect of reabsorption on the quenching of phosphorescence of molecules in an infinite plane-parallel layer of finite thickness.
Adamov, V.S., and Kantardzhyan, L.T.
Opt. i Spektroskopiya 13: No. 1, 100-6 (1962)
CA 57: 13276i

D 723 Growth of single crystals of zinc sulfide.
Addamiano, A.
U.S. 3,022,144 (1958) (1962)
CA 58: 11071i

D 724 Fluorescence properties of mono- and polyazaindoles.
Adler, T.K.
Anal. Chem. 34: 685-9 (1962)
CA 57: 4195b

D 725 Fluorescence spectrum and the energy levels of the Sm^{++} ion.
Aiche, G.H., and Sarup, R.
J. Chem. Phys. 36: 371-7 (1962)
CA 57: 5426c

D 726 Electron transitions in cesium and rubidium under action of pressure.
Alekseev, E.S., and Arkhipov, R.G.
Fiz. Tverd. Tela 4: 1077-81 (1962)
CA 57: 10635b

D 727 Determination of zirconium in iron- and titanium-containing ores using quercetin.
Alimarin, I.P., Golovina, A.P., and Tenyakova, L.A.
Metody Analiza Khim. Reaktivov i Preparatov. Gos. Kom. Sov. Min. SSSR po Khim. 4: 128-30 (1962)
CA 61: 2480f

D 728 The lowest excited states of carbonyl compounds.
Amako, Y., Akagi, M., and Azumi, H.
Proc. Intern. Symp. Mol. Struct. Spectry., Tokyo 1962: (1962)
CA 61: 5105b

D 729 The general method of investigation of the excited phosphor glow and thermo-decoloration curves.
Antonov-Romanovskii, V.V.
Zur Physik. Chemie Kristallphosphore, Tagung Unterkomm. Leuchstoffe Sekt. Physik, 2, Berlin 1961: 32-52 (1962)
CA 59: 10867e

D 730 Optically bleached polymers.
Bayrische Anilin und Soda-Fabrik, A.G.
Brit. 913,735 (1962)
CA 56: 5842g

D 731 Proposal for standardization of methods of reporting fluorescence emission spectra.
Anon.
Photoelec. Spectrometry Group Bull. 14: 378-9 (1962)
CA 56: 5157h

D 732 Effect of temperature on the kinetics of α-phosphorescence in organic substances.
Aristov, A.V., and Sveshnikov, B.Ya.
Opt. i Spektroskopiya 13: 222-8 (1962)
CA 56: 2955a

D 733 Temperature quenching of the β-phosphorescence of organic luminophors.
Aristov, A.V., and Sveshnikov, B.Ya.
Opt. i Spektroskopiya 13: 383-5 (1962)
CA 56: 112d

D 734 Emission mechanisms of the radical NH_2 in comets.
Arpigny, C., and Woszczyk, A.
Bull. Soc. Roy. Sci. Liege 31: 390-5 (1962)
CA 57: 6755a

D 735 An ellipsoid fluorimeter.
Arrhenius, S.
Arkiv Kemi 18: 165-72 (1962)
CA 57: 2011g

D 736 Distortion of the complexes Cr^{+++} $(H_2O)_6$ and Ni^{++} $(H_2O)_6$ and the splitting of the spin levels of Cr^{+++} and Ni^{++} due to the Jahn-Teller phenomenon.
Avvakumov, V.I.
Opt. i Spektroskopiya 13: 588-91 (1962)
CA 56: 5158a

D 737 Radiation effects in solids.
Ashkin, J.
U.S. At. Energy Comm. TID-77027, 14 pp. (1962)
CA 61: 8965d

D 738 Fluorescence of pyridine nucleotides in mitochondria.
Aui-Dor, Y., Olson, J.M., Doherty, M.D., and Kaplan, N.O.
J. Biol. Chem. 237: 2377-83 (1962)
CA 57: 8886a

D 739 Phosphorescence spectra. I. Phosphorescence of 1-naphthol and 2-naphthol.
Azumi, T.
Bull. Chem. Soc. Japan 35: No. 5, 788-90 (1962)
CA 57: 5475e

D 740 Polarization study of emission spectra.
Azumi, T., and McGlynn, S.P.
Proc. Intern. Symp. Mol. Struct. Spectry., Tokyo 1962: 4 pp. (1962)
CA 61: 9063c

D 741 Polarization of the luminescence of phenanthrene.
Azumi, T., and McGlynn, S.P.
J. Chem. Phys. 37: 2413-20 (1962)
CA 56: 4044h

D 742 Electronic spectra of substituted aromatic hydrocarbons. III. Anthrols.
Baba, H., and Suzuki, S.
Bull. Chem. Soc. Japan 35: 683-7 (1962)
CA 56: 2017h

D 743 Determination of trace amounts of metals as complexes of metal thiocyanates (halides) and a basic dye by a fluorescence method. II. Extraction-fluorescence determination of zinc as a zinc-thiocyanate-rhodamine compound.
Babko, A.K., and Chalaya, Z.I.
Zh. Analit. Khim. 17: 286-90 (1962)
CA 57: 9211a

D 744 A new series of synthetic borates isostructural with the carbonate mineral huntite.
Ballman, A.A.
Am. Mineralogist 47: 1380-3 (1962)
CA 56: 5114f

D 745 Rare earth aluminum or chromium
borates.
Ballman, A.A.
U.S. 3,057,636 Oct. 9, 1962; Appl. Oct. 6,
1960, 4 pp.
CA 58: 6478d

D 746 Universal molecular interactions and
their effect on the position of the electronic spectra of molecules in two-
component solutions. III. Derivatives
of naphthalene, stilbene, biphenyl,
aniline, fluorene, and pyridine (liquid
solutions).
Bakhshiev, N.G.
Opt. i Spektroskopiya 12: 473-8 (1962)
CA 57: 5446e

D 747 Absorption and fluorescence of the
vapor of anthracene and its derivatives.
Bakhshiev, N.G., Klochkov, V.P., Neporent,
B.S., and Cherkasov, A.S.
Opt. i Spektroskopiya 12: 582-5 (1962)
CA 57: 10667g

D 748 Light emission by recombination of
free carries created by electron
bombardment.
Balkanski, M., and Gans, F.
Luminescence Org. Inorg. Mater., Intern.
Conf., New York 1961: 318-33 (1962)
CA 56: 5130b

D 749 Fluorescence of 2,6-dimethylnaphtha-
lene in pure liquid helium.
Banarroche, M.
Compt. Rend. 254: 3836-8 (1962)
CA 57: 8080a

D 750 Fluorescence spectra of pure CdS.
Bancie-Grillot, M.
Proc. Intern. Meeting Mol. Spectry., 4th,
Bologna, 1952 2: 618-26 (1962)
CA 59: 82766e

D 751 Two fluorescent emission bands of ZnS
activated by copper.
Bancie-Grillot, M.
Compt. Rend. 254: 1247-9 (1962)
CA 57: 4159h

D 752 Short-lived phosphorescence of
DL-trytophane in solutions.
Barenboim, G.M.
Biofizika 7: 227-32 (1962)
CA 57: 5426i

D 753 Transfer of excitation energy in rigid
solutions of organic scintillators.
Basile, L.J.
U.S. At. Energy Comm. TID-18167, 40 pp.
(1962)
CA 61: 263c

D 754 Effect of styrene monomer on the
fluorescence properties of polystyrene.
Basile, L.J.
J. Chem. Phys. 36: 2204-10 (1962)
CA 57: 8077g

D 755 Luminescence of Iceland spar.
Barsanov, G.P., and Sarsembaeva, Kh.K.
Tr. Mineralog. Muzeya, Akad. Nauk SSSR
1962: No. 13, 147-52 (1962)
CA 58: 7475c

D 756 Ultraviolet fluorescence of muscles.
Barskii, I.Ya., Brumberg, E.M., and
Brumberg, V.A.
Dokl. Akad. Nauk SSSR 147: 474-6 (1962)
CA 58: 7179g

D 757 Protolytic dissociation of electronically
excited organic acids.
Bartok, W., Lucchesi, P.J., and Suider,
N.S.
J. Am. Chem. Soc. 84: 1842-4 (1962)
CA 57: 6756h

D 758 Absorption and emission spectra of
nucleic acids at long wavelengths.
Basu, S., and Loh, L.
J. Chim. Phys. 59: 1031-2 (1962)
CA 56: 5156g

D 759 Bifluorescence of bone tissues treated
with tetracyclines.
Baud, C.A., and Dupont, D.H.
Compt. Rend. 254: 3129-30 (1962)
CA 61: 16347b

D 760 Depolarization of fluorescence of dye
solutions by thermal motion of the
molecules.
Bauer, R., and Szczurek, T.
Acta Physiol. Polon. 22: 29-36 (1962)
CA 59: 2301c

D 761 The problem of fluorescence and polar-
ization — microscopic examinations in
the human epiphysis.
Bayerova, G., Bayer, A., and Obrucnik, M.
Acta Histochem. 14: 276-83 (1962)
CA 59: 946a

D 762 Activation of zinc sulfide with element-
al phosphorus by introducing it from
the vapor phase.
Belyanova, I.M., Guretskaya, Z.I., and
Bundel, A.A.
Tr. Mosk. Khim.-Tekhnol. Inst. 39: 44-9
(1962)
CA 60: 12760e

D 763 The electronic spectra of crystalline
durene at low temperatures.
Benarroche, M.
Compt. Rend. 254: 459-61 (1962)
CA 58: 15040c

D 764 Absorption and fluorescence spectra of
pure crystalline fluorine at 4°K.
Benarroche, M.
Compt. Rend. 254: 3520-2 (1962)
CA 57: 8079b

D 765 The measurement of radiative lifetimes.
Berg, R.
U.S. At. Energy Comm. UCRL-9954,
108 pp. (1962)
CA 57: 6753g

D 766 Dimethylchlorotetracycline-induced
fluorescence of gastric sediment. Use
to differentiate benign and malignant
gastric lesions.
Berk, J.E., and Kantor, S.M.
J. Am. Med. Assoc. 179: 997-1000 (1962)
CA 57: 2751e

D 767 Fluorescence quenching of a scintilla-
tion solution by oxygen.
Berlman, I.B., and Walter, T.A.
J. Chem. Phys. 37: 1888-9 (1962)
CA 56: 204c

D 768 The fluorescence spectrum and decay
time of naphthalene.
Berlman, I.B., and Weinreb, A.
Mol. Phys. 5: 313-19 (1962)
CA 57: 16035d

D 769 Thermoluminescence due to water
desorption.
Bettinali, C., Ferraresso, G., and
Stampacchia, G.
Atti Accad. Nazl. i Lincei, Rend.,
Classe Sci. Fis., Mat. Nat. 32: 948-50
(1962)
CA 58: 12081h

D 770 The effect of dodecyl sulfate on the
ultraviolet spectra of proteins.
Bigelow, C.C., and Sonenberg, M.
Biochemistry 1: 197-204 (1962)
CA 57: 287h

D 771 Spectroscopic determination of dipole
moments of excited molecules.
Bilot, L., and Kawski, A.
Acta Phys. Polon. 22: 289-91 (1962)
CA 61: 1185f

D 772 The fluorescence and scintillation
decay times of crystalline anthracene.
Briks, J.B.
Proc. Phys. Soc. 79: 494-6 (1962)
CA 57: 215i

D 773 Excimer fluorescence of aromatic
hydrocarbons in solution.
Birks, J.B., and Christophorou, L.G.
Nature 194: 442-4 (1962)
CA 56: 6334h

D 774 Resonance interactions of fluorescent
organic molecules in solution.
Birks, J.B., and Christophorou, L.G.
Nature 196: 33-5 (1962)
CA 57: 16001d

D 775 The influence of impurities on the
thermoluminescent properties of ZnO.
Blanchard, M.L.
Compt. Rend. 254: 249-51 (1962)
CA 58: 15031d

D 776 Kinetics of the fast component of scin-
tillation in a pure organic medium.
Application to the case of anthracene.
Blanc, D., Cambou, F., and deLafond, Y.G.
Compt. Rend. 254: 3187-9 (1962)
CA 57: 6735d

D 777 Flash excitation of acridine orange in
acidic and basic solvents.
Blauer, G., and Zinschitz, H.
J. Phys. Chem. 66: 453-5 (1962)
CA 57: 288f

D 778 Paper-chromatographic differentiation
of some species and varieties of the
genus populus.
Boertitz, S.
Zuechter 32: 24-33 (1962)
CA 61: 12332g

D 779 The efficiency of solution fluorescence.
Bowen, E.J., and Seaman, O.
Luminescence Org. Inorg. Mater., Intern.
   Conf., New York 1961: 153-60 (1962)
   CA 58: 7513g

D 780 Hydrogen bonding of excited states.
Bowen, E.J., and Woodger, G.B.
J. Phys. Chem. 66: 2491-2 (1962)
CA 56: 4041f

D 781 Determination of aluminum with
   salicylal-o-aminophenol in sodium
   acetate.
Bozhevol'nov, E.A.
Metody Analiza Khim. Reaktivov i
   Preparatov, Gos. Kom. Sov. Min. SSSR
   po Khim. 4: 46-9 (1962)
   CA 61: 2461f

D 782 Determination of Ga in Se with Lumo-
   gallion (2,2',4'-trihydroxy-5-chloroazo-
   benzenesulfonic acid).
Bozhevol'nov, E.A.
Metody Analiza Khim. Reaktivov i
   Preparatov, Gos. Kom. Sov. Min. SSSR
   po Khim. 4: 72-5 (1962)
   CA 61: 3679c

D 783 Determination of copper in water and
   acids by lumocupferron.
Bozhevol'nov, E.A., and Kreingol'd, S.U.
Metody Analiza Khim. Reaktivov i
   Preparatov, Gos. Kom. Sov. Min. SSSR
   po Khim. 4: 96-9 (1962)
   CA 61: 2467g

D 784 Determination of zinc in acids and
   potassium sodium tartrate with
   8-(α-toluenesulfonylamino)quinoline.
Bozhevol'nov, E.A., and Serebryakova,
   G.V.
Metody Analiza Khim. Reaktivov i
   Preparatov, Gos. Kom. Sov. Min. SSSR
   po Khim. 4: 120-5 (1962)
   CA 61: 3690c

D 785 Fluorescence of intracomplex com-
   pounds of cations.
Bozhevol'nov, E.A., and Serebryakova,
   G.V.
Opt. i Spektroskopiya 13: 390-5 (1962)
CA 57: 16002e

D 786 Fluorescent properties of fluorescein
   complexon.
Bozhevol'nov, E.A., and Kreingol'd, S.U.
Zh. Analit. Khim. 17: 291-4 (1962)
CA 57: 6617e

D 787 Fluorescence peculiarities of vapors
   of complex molecules on excitation
   with high-energy quanta.
Borisevich, N.A., Tolkachev, V.A., and
   Gruzinskii, V.V.
Fiz. Probl. Spektroskopii, Akad. Nauk
   SSSR, Materialy 13-go Soveshch.,
   Leningrad, 1960 1: 225-9 (1962)
   CA 59: 8279b

D 788 The vapor phase fluorescence and its
   relationship to photolysis of the
   butyraldehydes.
Borkowski, R.P., and Ausloos, P.
Bull. Soc. Chim. Belges 71: 660 (1962)
CA 58: 7537g

D 789 Vapor-phase fluorescence and its rela-
   tion to the photolysis of propionalde-
   hyde and the butyraldehydes.
Borkowski, R.P., and Ausloos, P.
J. Am. Chem. Soc. 84: 4044-8 (1962)
CA 56: 1079f

D 790 Energy transport phenomena in the
   case of molecular fluorescence.
Budo, A., and Kelskemety, I.
Acta Phys. Acad. Sci. Hung. 14: 167-76
   (1962)
   CA 61: 15548f

D 791 Isolation and properties of a yellow-
   green fluorescent peptide from
   azotobacter medium.
Bulen, W.A., and LeComte, J.R.
Biochem. Biophys. Res. Commun. 9: 523-8
   (1962)
   CA 58: 9348c

D 792 Effects of red and far-red light on the
   fluorescence yield of chlorophyll in
   vivo.
Butler, W.L.
Biochim. Biophys. Acta 64: 309-17 (1962)
CA 56: 2654e

D 793 New analogs of lucizenin. I.
   N'N'-Bis(p-chlorophenyl) diacridylium
   nitrate.
Braun, A.
Roczniki Chem. 36: 151-6 (1962)
CA 57: 15071h

D 794 Pressure dependence on fluorescence
    spectra. III. Effect of infinite pulse
    length.
    Brauner, J.W., and Wilson, D.J.
    J. Chem. Phys. 36: 2547-8 (1962)
    CA 57: 11977g

D 795 Fluorescence of aromatic aldehydes.
    Bredereck, K., Foerster, T., and
    Oesterlin, H.G.
    Luminescence Org. Inorg. Mater., Intern.
    Conf., New York 1961: 161-75 (1962)
    CA 58: 7514b

D 796 Phase fluorometer to measure radia-
    tive lifetimes of 10^{-5} to 10^{-9} sec.
    Brewer, L., James, C.G., Brewer, R.G.,
    Stafford, F.E., Berg, R.A., and
    Rosenblatt, G.M.
    Rev. Sci. Instr. 33: 1450-5 (1962)
    CA 58: 10869a

D 797 Fluorescence of some substituted
    benzenes.
    Bridges, J.W., and Williams, R.T.
    Nature 196: 59-61 (1962)
    CA 57: 16000i

D 798 Absolute efficiency measurements of
    infrared fluorescent zinc and cadmium
    sulfide activated with V-Ag and V-Cu.
    Bril, A., and Meurs-Hoekstra, W.
    Philips Res. Rept. 17: 280-2 (1962)
    CA 56: 2951d

D 799 Fluorescence properties of aggregated
    chlorophyll in vivo and in vitro.
    Brody, S.S., and Brody, M.
    Trans. Faraday Soc. 58: 416-28 (1962)
    CA 57: 12868h

D 800 Fluorescence of aldosterone in sulfuric
    acid.
    Bruivels, J., and Van Noordwijk, J.
    Nature 193: 1260-2 (1962)
    CA 57: 10196c

D 801 Photoluminescence of some dibasic
    organic acids at liquid-oxygen temper-
    atures.
    Bruns, S.A.
    Nauchn. Soobshch. Inst. Gorn. Dela, Akad.
    Nauk SSSR 16: 114-24 (1962)
    CA 60: 7583c

D 802 Emission spectrum of ionized nitrous
    oxide, N_2O^+.
    Callomon, J.H.
    Proc. Intern. Meeting Mol. Spectry., 4th,
    Bologna, 1959 1: 375-6 (1962)
    CA 59: 5950b

D 803 Fluorescence in comets as a Markov
    process.
    Carrington, T.
    Astrophys. J. 135: 883-91 (1962)
    CA 60: 2444f

D 804 Fluorescence and rotational relaxation
    of OH radicals in flames.
    Carrington, T.
    Symp. Combust., 8th, Pasadena, Calif.
    1960: 257-62 (1962)
    CA 57: 6752c

D 805 Energy of interaction of ground- and
    excited-state hydrogen atoms.
    Cade, P.E.
    Dissertation Abstr. 22: 3003-4 (1962)
    CA 57: 2989f

D 806 Interpretation of rate experiments
    with resolved quantum levels.
    Carrington, T.
    Discussions Faraday Soc. 33: 44-51 (1962)
    CA 56: 4047d

D 807 Intracellular oxidation-reduction
    states in vivo.
    Chance, B., Cohen, P., Jobsis, F., and
    Schoener, B.
    Science 137: 499-508 (1962)
    CA 57: 14323e

D 808 Metabolically linked changes in fluor-
    escence emission spectra of cortex of
    rat brain, kidney, and adrenal gland.
    Chance, B., Legallais, V., and Schoener, B.
    Nature 195: No. 4846, 1073-5 (1962)
    CA 57: 17034h

D 809 Spectral detection of s-cis- and
    s-trans-isomers of 2-vinylanthracene.
    Cherkasov, A.S.
    Dokl. Akad. Nauk SSSR 146: 852-5 (1962)
    CA 56: 2017c

D 810 The effect of the solvent on the position of fluorescence spectra of some anthracene derivatives.
Cherkasov, A.S.
Fiz. Probl. Spektroskopii, Akad. Nauk SSSR, Materialy 13-go Soveshch., Leningrad, 1960 1: 248-51 (1962)
CA 59: 10901g

D 811 3-Hydroxy-2-naphthoic acid as colorimetric and fluorescent reagent (for Fe (III) and Be).
Cherkesov, A.I., and Zhigalkina, T.S.
Tr. Astrakhansk. Tekhn. Inst. Rybn. Prom. i Khoz. 8: 25-49 (1962)
CA 60: 4756b

D 812 Ultraviolet fluorescent of perinuclear bodies of spermatids among some locusts.
Chernogryadskaya, N.A., and Shudel, M.S.
Dokl. Akad. Nauk SSSR 145: 917-19 (1962)
CA 57: 15632g

D 813 Phosphorescence traps of "fluoromeite" substances.
Chevalier, N., Gaume-Mahn, F., Janin, J., and Oriol, J.
Compt. Rend. 255: 1096-8 (1962)
CA 59: 4642e

D 814 Fluorescence of s -tetrazine.
Chowdhory, M., and Goodman, L.
J. Chem. Phys. 36: 548-9 (1962)
CA 57: 5431h

D 815 Polarization of fluorescence of reoxidized muramidase.
Churckich, J.E.
Biochim. Biophys. Acta 65: 349-50 (1962)
CA 58: 7089g

D 816 The mode of displacement of electronic spectra of aromatic hydrocarbons in solution.
Ciais, A., and Pesteil, P.
J. Chim. Phys. 59: 811 (1962)
CA 56: 2018b

D 817 1,3,4-Thiadiazoles.
Ciba Ltd.
Brit. 900,815, July 11, 1962
CA 60: 4287d

D 818 Polarization of luminescence.
Clark, C.D., Maycraft, G.W., and Mitchell, E.W.J.
J. Appl. Phys. 33: 378-82 (1962)
CA 57: 13275b

D 819 Quantum theory of the optical pumping cycle. I. Prediction of new effects based on theory.
Cohen-Tannoudji, C.
Ann. Phys. 7: 423-60 (1962)
CA 58: 9754a

D 820 Phosphorescence of fused quartz and sapphire.
Coop, W.H., and Hammond, J.A.
J. Opt. Soc. Am. 52: 835 (1962)
CA 57: 10667h

D 821 A light-sensitive fluorescent substance in bovine and rabbit lenses.
Cremer-Bartels, G.
Exptl. Eye Res. 1: 443-8 (1962)
CA 58: 7192g

D 822 Fluorescence spectra of some simple coumarins.
Crosby, D.G., and Berthold, R.V.
Anal. Biochem. 4: 349-57 (1962)
CA 56: 5926b

D 823 Effect of the change transfer complex formation on the spectrum and the fading time of the phosphorescent aromatic hydrocarbons.
Czekalla, J.
Proc. Intern. Meeting Mol. Spectry, 4th, Bologna, 1959 2: 627-32 (1962)
CA 59: 5948h

D 824 Variations in the singlet-triplet transitions of aromatic hydrocarbons owing to formation of complexes.
Czekalla, J., and Mager, K.J.
Z. Elektrochem. 66: 65-73 (1962)
CA 57: 289c

D 825 Investigations on hydrolysis products from the proteins of skin of mice painted with 3,4-benzopyrene.
Davdel, P., Muel, B., Lacroix, G., and Prodi, G.
J. Chim. Phys. 59: 263-6 (1962)
CA 57: 2751h

D 826 Fluorimetry and chemical analysis.
Davey, S.C.B.
Hilger J. 7: 46-51 (1962)
CA 58: 11933h

D 827 Theory of temporal growth of ionization
between parallel plates in the inert
gases.
Davidson, P.M.
Proc. Phys. Soc. 80: 143-50 (1962)
CA 57: 11951f

D 828 Pressure dependence of fluorescence
spectra. IV. Effects of vibrational
energy transfer between fluorescing
molecules.
Davis, R.C., and Wilson, D.J.
J. Chem. Phys. 37: 848-53 (1962)
CA 57: 16002f

D 829 An investigative immunological method
of general interest: immunofluorescence.
deRosnay, D.C., and Boineau, J.
J. Med. Bordeaux Sud-Ouest 139: 1453-70
(1962)
CA 59: 4420g

D 830 Strontium and other minor elements in
marble-bearing rocks of the Apuan
Alps and in other Triassic rocks from
Central Italy.
Dessau, G.
Boll. Soc. Geol. Ital. 81: 365-84 (1962)
CA 61: 9316a

D 831 The displacement of the selenium atom
in single-crystal zinc selenide by
electron bombardment.
Detweiler, R.M.
U.S. Dept. Comm., Office Tech. Serv.
AD 284,018, 67 pp. (1962)
CA 60: 4885d

D 832 Minor absorption and fluorescence in
ZnTe.
Dietz, R.E., and Thomas, D.G.
Phys. Rev. Letters 8: 391-3 (1962)
CA 57: 5430g

D 833 Quenching of porphyrins in solution and
adsorbate state. I. Theories of
quenching of fluorescence.
Diuric, D.
Arhiv Farm. (Belgrade) 12: 19-82 (1962)
CA 58: 12081g

D 834 Fluorescence quenching of porphyrins
in solution and in adsorbed state.
Djuric, D.
Arhiv Farm. (Belgrade) 12: 263-71 (1962)
CA 58: 8524a

D 835 Physical and chemical methods of
measurement application.
Dlouhy, J.
Jaderna Energie 8: 432-3 (1962)
CA 58: 7578a

D 836 The energy levels of Pr^{+3} in $CaWO_4$.
Dodd, D.M., and Wood, D.L.
Proc. Intern. Symp. Mol. Struct. Spectry.,
Tokyo 1962: A406, 4 pp. (1962)
CA 61: 6527h

D 837 Concentration reversal of the fluores-
cence in pyrene derivatives.
Doeller, E.
Z. Physik. Chem. 34: 151-62 (1962)
CA 56: 3014a

D 838 The concentration reversal of the
fluorescence of pyrene.
Doeller, E., and Foerster, Th.
Z. Physik. Chem. 34: 132-50 (1962)
CA 56: 3013h

D 839 Anisotropy of the triplet-singlet
phosphorescence of aromatic com-
pounds.
Doerr, F., and Gropper, H.
Angew. Chem. 74: 354 (1962)
CA 57: 5475i

D 840 The polarization of the $\pi-\pi$ phosphorescence of N-heterocyclics on excitation in the $n-\pi$ band: quinoxaline,
2-3-dimethyl-quinoxaline, quinazoline,
benzoquinoxaline.
Doerr, F., Gropper, H., and Mika, N.
Ber. Bunsenges. Physik. Chem. 67: 202-5
(1962)
CA 58: 13310b

D 841 Luminescence of long duration from
serum albumen and its principal con-
stituents.
Douzou, P., and Francq, J.C.
J. Chim. Phys. 59: 578-83 (1962)
CA 57: 10669c

D 841a Fluorescence of optical bleaches
(also describes fluorometer).
Dresner, H., and Weber, K.
Kem. Ind. (Zagreb) 11: 485-9 (1962)
CA 62: 4795g

D 842 The sensitization of biacetyl fluores-
cence in fluid solutions.
Dubois, J.T., and Stevens, B.
Luminescence Org. Inorg. Mater., Intern.
Conf., New York 1961: 115-31 (1962)
CA 56: 1060e

D 843 Ultraviolet spectra of stilbene, *p*-
mono-halogen stilbenes, and azobenzene and the trans-to-cis photocrom-
erization process.
Dyck, R.H., and McClure, D.S.
J. Chem. Phys. 36: 2326-45 (1962)
CA 57: 6757e

D 843a Spectrophotometry of the comet
Bernem 1959K.
Egibekov, P.
Byul. Inst. Astrofiz., Akad. Nauk Tadzh.
SSR 1962: 42-5 (1962)
CA 60: 6339g

D 844 A.G.C. amplifier to correct for effects
of light source variations in spectro-
fluorometric measurements.
Eisenberg, L., Rosen, P., and Edelman,
G.M.
Rev. Sci. Instr. 33: 1435-40 (1962)
CA 58: 9754e

D 845 Spectrophotometric study of low-
intensity fluorescence. II. Hard paraffin.
Eisenbrand, J.
Deut. Lebensm.-Rundschau 58: 319-21
(1962)
CA 56: 6335c

D 846 Spectrophotometric study of low-
intensity fluorescence. I. Liquid
paraffin.
Eisenbrand, J.
Deut. Lebensm.-Rundschau 58: 230-3
(1962)
CA 57: 16001h

D 847 Extinction of halogen salts on the
fluorescence of quinoline and various
quinoline derivatives in acid solution.
Eisenbrand, J., and Raisch, M.
Nahrung 6: 157-65 (1962)
CA 57: 5475f

D 848 Recombination luminescence of
oxygen- and fluorine — containing
compounds activated by mercury-like
ions.
Eksina, T.I.
Tr. Inst. Fiz. i Astron., Akad. Nauk Est.
SSR 1962: 117-38 (1962)
CA 59: 2263c

D 849 Comparison of localized and delocal-
ized models for n →n* transitions: a
possible interpretation of the observed
sym-tetrazine fluorescence.
El-Bayoumi, M.A., and Kearns, D.R.
J. Chem. Phys. 36: 2516-17 (1962)
CA 57: 9330i

D 850 Proposed effect of high pressures on
the radiationless processes.
El-Sayed, M.A.
J. Chem. Phys. 37: 1568-9 (1962)
CA 56: 2954g

D 851 Retardation of singlet and triplet excit-
ation migration in organic crystals by
isotopic dilution.
El-Sayed, M.A., Wauk, M.T., and Robinson,
G.W.
Mol. Phys. 5: 205-7 (1962)
CA 57: 1682e

D 852 Measurement of the quantum yields of
sensitized phosphorescence as a
means of studying the quenching processes at the triplet level of organic
molecules.
Ermolaev, V.L.
Opt. i Spektroskopiya 13: 90-5 (1962)
CA 56: 1059a

D 853 Nonradiative energy transfer between
the triplet and singlet states of organic
molecules.
Ermolaev, V.L., and Sveshnikova, E.B.
Izv. Akad. Nauk SSSR, Ser. Fiz. 26: 29-31
(1962)
CA 57: 1688a

D 854 Geochemical survey of the Nottingham-
shire oil fields and related sediments.
Evans, W.D., Cooper, B.S., Corbett, D.W.,
and Gough, K.
Quart. J. Geol. Soc. London 118: 23-38
(1962)
CA 59: 15052a

D 855 Spectral characteristics of the deoxy-
ribonucleic acid (D.N.A.)-Acridine
Orange complex.
Faddeeva, M.D.
Tsitologiya 4: 231 (1962)
CA 57: 8078a

D 856 Observation on the possibilities of the
cellular localization of monoamines by
a fluorescence method.
Falck, B.
Acta Physiol Scand., Suppl. 197: 25 pp.
(1962)
CA 58: 12923g

D 857 Fluorescence of catechol amines and
related compounds condensed with
formaldehyde.
Falck, B., Hillarp, N.A., Thieme, G., and
Torp, A.
J. Histochem. Cytochem. 10: 348-54 (1962)
CA 57: 8044f

D 858 Fluorescence decay time measure-
ments of organic scintillators.
Falk, W.R., and Katz, L.
Can. J. Phys. 40: 978-91 (1962)
CA 57: 10734e

D 859 Pyrrolines as optical bluing agents.
Farbwerke Hoechst, A.-G.
Belg. 610,232, 1960 (1962)
CA 58: 11521f

D 860 Effect of electric field on the lumines-
cence of ZnS single crystals.
Fedyushin, B.T.
Opt. i Spektroskopiya 13: 558-63 (1962)
CA 56: 2951f

D 861 Luminescence of KI under X-ray ir-
radiation.
Fieschi, R., and Spinalo, G.
Nuovo Cimento 23: 738-50 (1962)
CA 57: 4150h

D 862 Fluorescent properties of rare-earth
chelates in vinylic hosts.
Filipescu, N., Kagan, M.R., McAvoy, N.,
and Serafin, F.A.
Nature 196: 467-8 (1962)
CA 56: 1062a

D 863 Development of host cell reaction
during chemical skin carcinogenesis
of mouse.
Fiore-Donati, L., DeBenedictis, G., and
Chieco-Bianchi, L.
Nature 193: 287-8 (1962)
CA 58: 13432g

D 864 Optical glass in nuclear physics and
engineering-interactions between high-
energy radiation and glasses.
Fischer, R.
Isotopen Tech. 2: 360-3 (1962)
CA 58: 11077c

D 865 Are the neurohormones of arthropods
identical with fluorescent substances
from the nervous system?
Fischer, F., Kapilza, W., Gersch, M.,
and Unger, H.
Z. Naturforsch. 176: 834-6 (1962)
CA 58: 4843c

D 866 Luminescence of fluorescein dyes.
Forster, L.S., and Dudley, D.
J. Phys. Chem. 66: 838-40 (1962)
CA 60: 14025g

D 867 Fluorescence.
Forziati, A.F.
High Polymers, Anal. Chem. of Polymer,
II, Mol. Structure and Chem. Groups
12: 335-57 (1962)
CA 57: 8713e

D 868 Spectral studies of the aqueous solution
of pyronine G.
Fujiki, K., Iwanaga, C., and Koizumi, M.
Bull. Chem. Soc. Japan 35: 185-93 (1962)
CA 57: 1758a

D 869 Fluorescence of α centers in alkali
halides.
Fujita, I.
Sci. Light (Tokyo) 11: 142-56 (1962)
CA 61: 190c

D 870 Quenching and energy transfer by the
same substance.
Furst, M., and Kallmann, H.P.
J. Chem. Phys. 37: 2159-61 (1962)
CA 56: 3012f

D 871 Spectra of universal fluorescence of
polymers.
Gachkovskii, V.F.
Dokl. Akad. Nauk SSSR 143: 150-2 (1962)
CA 57: 3613b

D 872 Effect of temperature on the fluores-
cence of some aromatic amino acids
and proteins.
Gally, J.A., and Edelman, G.M.
Biochim. Biophys. Acta 60: 499-509 (1962)
CA 57: 10668b

D 873 A relation between the lifetime and in-
tensity of fluorescence in the region of
thermal extinction.
Garanderie, H.
Compt. Rend. 255: 2585-7 (1962)
CA 56: 6296h

D 874 Sharp-line fluorescence, electron para-
magnetic resonance and thermolumin-
escence of Mn^{4+} in α-Al_2O_3.
Geschwind, S., Kisliuk, P., Klein, M.P.,
Remeika, J.P., and Wood, D.L.
Phys. Rev. 126: 1684-6 (1962)
CA 57: 8042d

D 875 The scintillation phenomenon in anthra-
cene. II. Scintillation pulse shape.
Gibbons, P.E., Northrop, D.C., and
Simpson, O.
Proc. Phys. Soc. 79: 373-82 (1962)
CA 57: 219e

D 876 Polarization of the quasilinear lumin-
escence spectra of perylene in the
electric field at 77°K.
Gobov, G.V., Kalimbet, A.Z., Fedotov,
A.P., and Sheremet'ev, G.D.
Opt. i Spektroskopiya 13: 879 (1962)
CA 56: 5157g

D 877 The amplification of X-ray fluorescence
by electric fields and infrared radiation.
Gobrecht, H., Gumlich, H.E., Nelkowski,
H., and Lacmann, K.
Z. Physik 168: 273-82 (1962)
CA 57: 15947f

D 878 Datiscin — a new fluorimetric reagent
for zirconium.
Golovina, A.P., Alimarin, I.P., Bozhevol-
nov, E.A., and Agasyan, L.B.
Zh. Analit. Khim. 17: 591-4 (1962)
CA 56: 931a

D 879 Vibrational spectra of alkali metal
uranyl tricarbonates and uranyl tri-
nitrates in crystals.
Gorbenko-Germanov, D.S., and Zenkova,
R.A.
Fiz. Probl. Spektroskopii, Akad. Nauk
SSSR, Materialy 13-go Soveshch.,
Leningrad, 1960 1: 427-9 (1962)
CA 59: 13452a

D 880 Fluorescence of the fluorinated
phenylalimines.
Guroff, G., and Chirigas, M.A.
Anal. Biochem. 3: 330-6 (1962)
CA 57: 9364e

D 881 Fluorescence polarization of some
porphyrins.
Gouterman, M., and Stryer, L.
J. Chem. Phys. 37: 2260-6 (1962)
CA 56: 3012g

D 882 Porphyrin charge-transfer complexes
with symtrinitrobenzene.
Gouterman, M., and Stevenson, P.E.
J. Chem. Phys. 37: 2266-9 (1962)
CA 56: 150c

D 883 The fluorescence of olive oils in the
ultraviolet. I. Origin of the compounds
responsible for the blue fluorescence
observed with a Wood filter.
Gracian, J., and Martel, J.
Grasas Aceites (Seville, Spain) 13: 128-33
(1962)
CA 58: 7050e

D 884 Fluorescence and absorption attributed
to the exciton in pure CdS.
Grillot, E.
Proc. Intern. Meeting Mol. Spectry., 4th,
Bologna, 1959 2: 634-40 (1962)
CA 59: 4694e

D 885 Dependence of fluorescence polariza-
tion on the dye concentration in
(plexiglas) solid solutions.
Grzywacz, J.
Fiz. Chem. Wyzsza Szkota Pedagog.
Gdansk. 2: 81-4 (1962)
CA 59: 12982e

D 886 Electronic spectrum of decaborane.
Haaland, A., and Eberhardt, W.H.
J. Chem. Phys. 36: 2386-92 (1962)
CA 57: 6752e

D 887 Absorption and fluorescence properties of the 8-quinolinol compounds of some elements of the 2nd, 3rd, and 4th groups of the periodic system.
Haar, R., and Umland, R.
Z. Anal. Chem. 191: 81-94 (1962)

D 888 Polarization of fluorescence as a measure of antigen-antibody interaction.
Haber, E., and Bennett, J.C.
Proc. Natl. Acad. Sci. U.S. 48: 1935-42 (1962)

D 889 Angular distribution of intensity and polarization in the triplet-singlet emission of benzene.
Hameka, H.F.
J. Chem. Phys. 37: 1085-7 (1962)
CA 57: 14596h

D 890 X-ray photoconduction, ultraviolet photoconduction, and fluorescence in anthracene.
Hartmann, H.K.
Z. Angew. Phys. 14: 727-34 (1962)
CA 58: 10839g

D 891 Organic liquid scintillators. III. Quantum yield of fluorescence and the quenching of fluorescence by oxygen.
Heller, A.
J. Chem. Phys. 36: 2858-60 (1962)
CA 57: 13308c

D 892 Fluorescent lamps and their phosphor compositions.
Henderson, S.T.
Brit. 888,337, 1959 (1962)
CA 58: 13675a

D 893 Color and fluorescence of cyclic Si compounds. I. The preparation of a siloxene for optical studies. II. Fluorescence and color of siloxene derivatives.
Hengge, E.
Chem. Ber. 95: 645-7 (1962); 648-57 (1962)
CA 57: 4196a

D 894 Universal fluorescence spectrophotometer.
Hengge, E.
Proc. Intern. Meeting Mol. Spectry., 4th, Bologna, 1959, 3: 1313−18 (1962)
CA 59: 3457b

D 895 Continuous absorption and phosphorescence spectra associated with the dissociation of SO_2 and the recombination of SO and O.
Herman, L., Akruche, J., and Grenat, H.
J. Quant. Spectry. Radiative Transfer 2: 215-24 (1962)
CA 56: 4037e

D 896 Relative yield and the degree of polarization of viscous fluorescent solutions quenched by potassium iodide.
Hevesi, J.
Acta Univ. Szeged., Acta Phys. Chem. 8: 8(1-2): 16-24 (1962)
CA 56: 5130g

D 897 Proton transfer of 3-hydroxy-2-naphthoic acid in the excited state.
Hirota, K.
Z. Physik. Chem. 35: 222-33 (1962)
CA 58: 10866d

D 898 Mixed dimer emission from pyrene crystals containing perylene.
Hochstrasser, R.M.
J. Chem. Phys. 36: 1099-100 (1962)
CA 57: 2966g

D 899 The luminescence of organic molecular crystals.
Hochstrasser, R.M.
Rev. Mod. Phys. 34: 531-50 (1962)
CA 56: 111d

D 900 The influence of temperature on the emission spectra of stilbene monocrystals and mixed crystals with tetracene.
Hochstrasser, R.M.
J. Mol. Spectroscopy 8: 485-506 (1962)
CA 57: 1365a

D 901 Elimination of the fluorescence of mineral oils.
Hofling, W., and Linke, A.
Brit. 913,518 (1962)
CA 58: 6636h

D 902 The fluorescence spectrum of naphthalene vapor in the 3100-A region.
Hollas, J.M.
J. Mol. Spectroscopy 9: 138-69 (1962)
CA 57: 13312e

D 903 Fluorometric determination of aluminum with salicylaldehyde formaylhydrazone.
Holzbecher, Z., and Pulkrab, P.
Collection Czech. Chem. Commun. 27: 1142-9 (1962)
CA 57: 9195g

D 904 Electron-phonon coupling in impurity states.
Hopfield, J.J.
Proc. Intern. Conf. Phys. Semicond., Exeter, Eng. 1962: 75-80 (1962)
CA 60: 3586e

D 905 Electroluminescence in SiC single crystals.
Hori, J.
Japan. J. Appl. Phys. 1: 262-9 (1962)
CA 56: 2951b

D 906 Effect of foreign dusts on the luminescence of quartz dusts.
Holzapfel, L., Lehmann, J., Herr, H.D., and Weiglin, W.
Staub 22: 265-70 (1962)
CA 56: 1062g

D 907 The extraction and constitution of peat wax. Chromatographic fractionation of wax.
Howard, A.J., and Hamer, D.
J. Am. Oil Chemists' Soc. 39: 250-5 (1962)
CA 57: 4784f

D 908 Pyrene-sensitized delayed fluorescence of naphthalene vapor.
Hutton, E., and Stevens, B.
Spectrochim. Acta 18: 425-6 (1962)
CA 57: 9364f

D 909 Arsenate phosphors. VI. Some new arsenate phosphors with Mn activator.
Ide, H.
Kogyo Kagaku Zasshi 65: 1743-8 (1962)
CA 58: 10834g

D 910 Exciton processes in molecular crystals. I. Vibrationally perturbed excitons.
Iguchi, K.
J. Phys. Radium 23: 433-45 (1962)
CA 56: 1007e

D 911 Fluorescent screen.
Intezet, T.K.
Hung. 150,563 (1962)
CA 60: 2426g

D 912 Photosensitization by benzene vapor: biacetyl. The triplet state of benzene.
Ishikawa, H., and Noyes, W.A., Jr.
J. Chem. Phys. 37: 583-91 (1962)
CA 57: 16038e

D 913 The cause of fluorescence in sea water.
Ivanoff, A.
Compt. Rend. 254: 4190-2 (1962)
CA 57: 9595e

D 914 Dependence of the fluorescence of benzene, toluene, and p-xylene solutions on the concentration of the fluorescent substance.
Ivanova, T.V., Mokeeva, G.A., and Sveshnikov, B.Ya.
Opt. i Spektroskopiya 12: 586-92 (1962)
CA 57: 10666i

D 915 Decay and polarization of fluorescence of solutions.
Jablonski, A.
Luminescence Org. Inorg. Mater., Intern. Conf., New York 1961: 110-14 (1962)
CA 56: 2951c

D 916 Two unusual effects in hydrogen-fired CaO:Mn,Li phosphors.
Jaffe, P.M.
J. Electrochem. Soc. 109: 872-4 (1962)
CA 57: 9339b

D 917 The orientation of the transition moments of absorption bands of acridinephenazine, phenoxazine, and xanthene dyes from dichroism and fluorescence-polarization studies.
Jakobi, H., and Kuhn, H.
Z. Elektrochem. 66(1): 48-53 (1962)
CA 57: 1757c

D 918 Optical maser characteristics of Nd^{+++} in $SnMoO_4$.
Johnson, L.F., and Soden, R.R.
J. Appl. Phys. 33: 757 (1962)
CA 56: 2979e

D 919 A simple instrument for studying the polarization of fluorescence.
Johnson, P., and Richards, E.G.
Arch. Biochem. Biophys. 97: 250-9 (1962)
CA 57: 6283d

D 920 Luminescence analysis of lubricating oils.
Khalupovskii, M.D.
Zavodsk. Lab. 28: 206-7 (1962)
CA 57: 2494f

D 921 Temperature effects in boron luminophors.
Khalupovskii, M.D.
Opt. i Spektroskopiya 12(1): 81-5 (1962)
CA 57: 2967b

D 922 Spectroscopy and luminescence properties of CdF_2:Eu.
Kingsley, J.D., and Prener, J.S.
Phys. Rev. 126: 458-65 (1962)
CA 57: 4150i

D 923 The effect of temperature on the phosphorescence of phthalic acid in aluminum sulfate.
Kislyak, G.M., and Lisenko, G.M.
Ukr. Fiz. Zh. 7: 1314-17 (1962)
CA 58: 13310c

D 924 Concerning the relation between fluorescent processes and the properties of optical brighteners. I.
Kling, A., and Kurz, J.
Textil-Praxis 17: 143-6 (1962)
CA 57: 1114f

D 925 Displacement of the cadmium atom in single-crystal cadmium sulfide by electron bombardment.
Kulp, B.A.
Phys. Rev. 125: 1865-9 (1962)
CA 57: 1704c

D 926 The primary absorption act in a one-electron photooxidation in a rigid medium.
Kalantar, A.H., and Albrecht, A.C.
J. Phys. Chem. 66: 2279-80 (1962)
CA 56: 2054g

D 927 Chemical reactions of gibberellic acid.
Kallistratos, G., Padral, D., and Pfau, A.
Eigenschaften Wirkungen Gibberelline, Symp., Giessen, Ger. 1960: 20-4 (1962)
CA 60: 6147h

D 928 Energy transfer phenomena in pure and mixed crystals of durene.
Kallmann, H.P., Hayakawa, S., and Magnante, P.
Luminescence Org. Inorg. Mater., Intern. Conf., New York 1961: 244-51 (1962)
CA 56: 6295h

D 929 Fluorescence spectra of organic compounds.
Kanda, Y.
Kagaku No Ryoiki 16: 580-8 (1962)
CA 58: 9751f

D 930 Anomalous vibrational structure in the phosphorescence spectra of several aromatic compounds.
Kanda, Y., Shimada, R., Gondo, Y., Nakamizo, M., Hanada, K., Koyanagi, M., and Takenoshita, Y.
Proc. Intern. Symp. Mol. Struct. Spectry., Tokyo 1962 (1962)
CA 61: 1406c

D 931 Singlet-triplet absorption spectra of several carbonyl and oxalyl compounds.
Kanda, Y., Shimada, R., Shimada, H., Kaseda, H., and Matsumuru, T.
Proc. Intern. Symp. Mol. Struct. Spectry., Tokyo 1962 (1962)
CA 61: 5101c

D 932 Spectral investigation of water state in chlorophyll.
Karyakin, A.V., and Chibisov, A.K.
Biofizika 7: 561-7 (1962)
CA 56: 2024h

D 933 Investigation of the keto-enol tautomerism of chlorophyll with the aid of infrared absorption spectroscopy.
Karyakin, A.V., and Chibisov, A.K.
Opt. i Spektroskopiya 13: 379-82 (1962)
CA 57: 16009i

D 934 Raman spectra of adsorbed molecules: quenching of fluorescence by adsorption.
Karagoonis, G., and Issa, R.
Nature 195: 1196-7 (1962)
CA 57: 14599c

D 935 Variation with exciting wavelength of the fluorescence efficiencies of some alkyl benzenes.
Kato, S., Lipsky, S., and Braun, C.L.
J. Chem. Phys. 37: 190-1 (1962)
CA 57: 14596b

D 936 Influence of a photosensitizing dye on
the arthus phenomenon investigated
with fluorescein-labeled protein and
microfluorophotometry.
Kawamura, M., Shimoda, S., and Suzuc, M.
Kanko Shikiso 66: 13-21 (1962)
CA 56: 834b

D 937 The fluorescence-polarization spectra
of some dyes in polymethyl methacryl-
ate.
Kawski, A., and Polacka, P.
Z. Naturforsch. 17a: 1119-20 (1962)
CA 56: 6336g

D 938 Concentration quenching of fluores-
cence of anthracene in Plexiglas.
Kawski, A., and Polacki, Z.
Bull. Acad. Polon. Sci., Ser. Sci., Math.,
Astron. Phys. 8: 817-19 (1960) (pub.
1962)
CA 57: 15947h

D 939 Delayed fluorescence in cyanine dyes.
Kern, J. Doerr, F., and Scheibe, G.
Z. Elektrochem. 66: 462-6 (1962)
CA 57: 11976h

D 940 Intermolecular energy transition in
fluorescent solutions.
Ketskeméty, I.
Z. Naturforsch. 17a: 666-70 (1962)
CA 56: 150h

D 941 Fluorescence, absorption, and heat
radiation of solutions.
Ketskeméty, I., Dombi, J., and Horvai, R.
Acta Physiol. Acad. Sci. Hung. 14: 165-6
(1962)
CA 61: 14054c

D 942 Effect of microdefects on the spectral
properties of luminescence centers in
some potassium chloride and sodium
chloride phospors.
Khalilov, A.K., Salaev, E.Y., Mamedov,
A.P., Aliev, T.D., and Isaev, F.K.
Fiz. Shchelochnogalordn. Kristallov, Riga,
Sb. 1962: 168-71 (1962)
CA 60: 7586d

D 943 Interaction of paramagnetic ions with
lattice vibrations.
Kiel, A.
U.S. Dept. Comm., Office Tech. Serv.
AD 275,763, 166 pp. (1962)
CA 59: 14775c

D 944 Nonlinear effect in KI-Tl and NaI-Tl
luminescence.
Kink, R., and Liidja, G.
Tr. Inst. Fiz. i Astron., Akad. Nauk Est.
SSR 1962, No. 18, 79-92 (1962)
CA 56: 6336c

D 945 Intensity of fluorescence of dyes in
solution.
Kishore, J., Machive, M.K., Gopala, K.,
and Chaudhuri, K.D.
Indian J. Phys. 36: 415-8 (1962)
CA 56: 3995h

D 946 Some luminescent properties of
acridine dyes.
Kislyak, G.M., and Lisenko, G.M.
Fiz. Probl. Spektroskopii, Akad. Nauk
SSSR, Materialy 13-go Soveshch.,
Leningrad, 1960 1: 336-9 (1962)
CA 59:12318b

D 947 Phosphorescence of borophthalic
phosphors.
Kislyak, G.M., and Lisenko, G.M.
Ukr. Fiz. Zh. 7: 1309-13 (1962)
CA 58: 13262c

D 948 Photosynthesis and respiration of
potato plants under conditions of the
far north.
Kislyakova, T.E.
Tr. Lab. Evolyutsionnoi i Ekolog. Fiziol.,
Akad. Nauk SSSR, Inst. Fiziol. Rast. 4:
39-76 (1962)
CA 59: 1964e

D 949 Energy levels of bivalent thulium in
CaF_2.
Kiss, Z.J.
Phys. Rev. 127: 718-24 (1962)
CA 57: 13272f

D 950 Photoconductive and emission-
spectroscopic properties of organic
molecular materials.
Kleinerman, M., Azarraga, L., and
McGlynn, S.P.
Luminescence Org. Inorg. Mater., Intern.
Conf., New York 1961: 196-225 (1962)
CA 56: 1995g

D 951 Emissivity and photoconductivity of
organic molecular crystals.
Kleinerman, M., Azarraga, L., and
McGlynn, S.P.
J. Chem. Phys. 37: 1825-34 (1962)
CA 59: 2268h

D 952 The relation between fluorescent pro-
cesses and the properties of optical
brighteners. II.
Kling, A., and Kurz, J.
Textil-Praxis 17: 250-3 (1962)
CA 57: 11414g

D 953 Polarization of fluorescence and the
spectral classification of complex
molecules.
Klochkov, V.P., and Neporent, B.S.
Opt. i Spektroskopiya 12: 233-8 (1962)
CA 57: 10669i

D 954 Measurement of extremely short decay
times of organic phosphors, excited by
an electron beam.
Kloss, H.G.
Cesk. Casopis Fys. 12: 628-33 (1962)
CA 58: 10834h

D 955 Fine structure of emission spectra of
carboxyamido compounds.
Kobyshev, G.I., and Terenin, A.N.
Fiz. Probl. Spektroskopii, Akad. Nauk
SSSR, Materialy 13-go Soveshch.,
Leningrad, 1960 1: 206-8 (1962)
CA 59: 12316h

D 956 Scintillation solution enhancers.
Kollman, H.P., Furst, M., and Brown, F.H.
U.S. 3,068,178, 1959, 18 pp. (1962)
CA 58: 13281b

D 957 Temperature dependence of the fluor-
escence of AgBr crystals containing S.
Koswig, H.D.
Zur Physik Chemie Kristallphosphore,
Tagung Unterkomm. Leuchtstoffe Sekt.
Physik, 2., Berlin 1961: 166-72 (1962)
CA 59: 12278b

D 958 Luminescence of certain solvents.
Kouyrzine, K.A., and Rozman, I.M.
Opt. i Spektroskopiya 12: 248-53 (1962)
CA 57: 9333g

D 959 Metachromasia of fluorescence of
acridine dyes.
Koyper, Ch. M.A.
Histochemie 3: 46-53 (1962)
CA 57: 11977e

D 960 Fluorescence of woody plant cells in
the frozen state.
Krasautsev, O.A.
Fiziol. Rast. 9: 359-67 (1962)
CA 57: 12899d

D 961 Determination of copper in water and
acids with fluorexon.
Kreingol'd, S.U., and Bozhevol'nov, E.A.
Metody Analiza Khim. Reaktivov i
Preparatov, Gos. Kom. Sov. Min. SSSR
po Khim. 4: 100-7 (1962)
CA 61: 2468b

D 962 Polarization of T →S emission spectra
of azines.
Krishna, V.G., and Goodman, L.
J. Chem. Phys. 36: 2217-22 (1962)
CA 57: 5473h

D 963 Effect of 1-methylnaphthalene on the
optical characteristics of certain sub-
stituted oxazoles and oxadiazoles.
Kutsyna, L.M., and Verkhovtseva, E.T.
Opt. i Spektroskopiya 12: 785-7 (1962)
CA 57: 15960i

D 964 Effect of structure on the optical char-
acteristics of the derivatives of some
five-membered heterocycles.
Kutsyna, L.M., Sidorova, R.P., Voevoda,
L.V., Ishchenko, I.K., and Demchevko,
N.P.
Izv. Akad. Nauk SSSR, Ser. Fiz. 26: 1304-5
(1962)
CA 58: 7512c

D 965 Luminescence of cyclopentanone.
La Paglia, S.R., and Roquitte, B.C.
J. Phys. Chem. 66: 1739-40 (1962)
CA 57: 15999b

D 966 Heterogeneity of chlorophyll in vivo. I.
Fluorescence spectra.
Lavorel, J.
Biochim. Biophys. Acta 60: 510-23 (1962)
CA 57: 8901h

D 967 Effect of temperature on the spectro-
scopic properties of phycocyanine.
Lavorel, J., and Moniot, C.
J. Chim. Phys. 59: 1007-12 (1962)
CA 56: 6329e

D 968 The effect of rigid matrixes on the
electronic spectrum of benzene.
Leach, S., and Lopez-Delgado, R.
Proc. Intern. Meeting Mol. Spectry., 4th,
Bologna, 1959, 1: 419-35 (1962)
CA 59: 7082a

D 969 Fluorescence quenching studied by
flash spectroscopy.
Leonhardt, H., and Weller, A.
Luminescence Org. Inorg. Mater., Intern.
Conf., New York 1961: 74-82 (1962)
CA 56: 1059g

D 970 The luminescence of solids and one of
its most recent applications, the determination of surface temperatures.
Leroux, J.P.
Nature (Paris) 3324: 179-85; 187-8
(1962)
CA 57: 4154b

D 971 Fluorescence and absorption spectra
of stilbene in octane at low tempera-
tures.
Leushin, V.L., and Mamelov, Kh.I.
Opt. i Spektroskopiya 12: 593-8 (1962)
CA 57: 10667b

D 972 Cu doping in thermal treatment of
some alkali halide single crystals.
Levialdi, A., and Spinolo, G.
Nuovo Cimento 26: 1153-63 (1962)
CA 58: 10861h

D 973 Association nature of the concentration
quenching of the luminescence of sodium fluorescein in aqueous and glycerol
solutions.
Levshin, V.L., and Krotova, L.V.
Opt. i Spektroskopiya 13: 809-18 (1962)
CA 56: 5158b

D 974 Radiationless transitions and radiative
lifetime of $^3B_{1u}$ state of benzene.
Lim, E.C.
J. Chem. Phys. 36: 3497-8 (1962)
CA 57: 15993a

D 975 Delayed fluorescence of acriflavine in
rigid media.
Lim, E.C., and Swenson, G.W.
J. Chem. Phys. 36: 118-22 (1962)
CA 57: 5430i

D 976 Fluorescence of quinine in an alkaline
medium and in absolute EtOH.
Linnewiel, H.A., and Visser, B.J.
Nature 195: 699 (1962)
CA 57: 10669d

D 977 The Franck-Condon principle applied
to solutions of aromatic compounds.
Lippert, E.
Luminescence Org. Inorg. Mater., Intern.
Conf., New York 1961: 271-3 (1962)
CA 56: 2020c

D 978 Photochemical cis-trans isomerization
of p-dimethylaminocinnamic acid
nitrile.
Lippert, E., and Lueder, W.
J. Phys. Chem. 66: 2430-3 (1962)
CA 56: 4082b

D 979 Electronic spectra of 3,4-benzocinno-
line. Observation of a $\pi * \to n$ fluores-
cence.
Lippert, E., and Voss, W.
Z. Physik. Chem. 31: 321-7 (1962)
CA 57: 1756d

D 980 Spectroscopic investigations and dipole
measurements on the electron structure of cis- and trans-p-dimethylamin-
ocinnamic acid nitrile and related
nitriles.
Lippert, E., and Lueder, W.
Z. Physik. Chem. 33: 60-81 (1962)
CA 57: 15999g

D 981 The fluorescence spectrum and
Franck-Condon principle in solutions
of aromatic compounds.
Lippert, E., Lueder, W., and Boos, H.
Proc. Intern. Meeting Mol. Spectry., 4th,
Bologna, 1959, 1: 443-57 (1962)
CA 59: 7088a

D 982 Quenching of electronic energy trans-
fer in organic liquids.
Lipsky, S., Helman, W.P., and Merklin,
J.F.
Luminescence Org. Inorg. Mater., Intern.
Conf., New York 1961: 83-99 (1962)
CA 56: 1060a

D 983 Edge emission in CdS crystals that
show mechanically excited emission.
Litton, C.W., and Reynolds, D.C.
Phys. Rev. 125: 516-23 (1962)
CA 58: 11050e

D 984 Fine structure of absorption and fluorescence spectra of phthalocyanine and protoperphyrin.
Litvin, F.F., and Personov, R.I.
Fiz. Probl. Spektroskopii, Akad. Nauk SSSR Materialy 13-go Soveshch., Leningrad 1: 229-30 (1962)
CA 59: 12296h

D 985 Phosphorescence in light scintillation counting.
Lloyd, R.A., Ellis, S.C., and Hallowes, K.H.
Tritium Phys. Biol. Sci., Proc. Symp. Detection Use, Vienna, Austria 1961, 1: 263-79 (1962)
CA 57: 10739b

D 986 Extinction of the fluorescence of solutions by surface activity.
Lucatu, E.
J. Chim. Phys. 59: 128-31 (1962)
CA 58: 14955c

D 987 Change in fluorescence capacity of organic solutions after the action of ultraviolet light.
Ludwig, W., and Herforth, L.
Zur Physik Chemie Kristallphosphore, Tagung Unterkomm. Leuchtstoffe Sekt. Physik 2., Berlin 1961: 200-10 (1962)
CA 59: 13495h

D 988 Change in fluorescence capacity of organic solutions after the action of ultraviolet light.
Ludwig, W., and Herforth, L.
Wiss. Z. Tech. Univ. Dresden 11: 905-9 (1962)
CA 59: 4695c

D 989 Fluorescence of europium tungstate.
MacDonald, R.E., Vogel, M.J., and Brookman, J.W.
IBM J. Res. Develop. 6: 363-64 (1962)
CA 57: 14547h

D 990 Absorption and fluorescence spectra of hexagonal $SmCl_3$ and their Zeeman effects.
Magno, M.S., and Dieke, G.H.
J. Chem. Phys. 37: 2354-63 (1962)
CA 56: 151a

D 991 Processes involved in excited states of the NO_2- group in solids.
Makishima, S., Shionoya, S., Tomotsu, T., and Hirata, M.
Proc. Intern. Symp. Mol. Struct. Spectry., Tokyo 1962: (A405) 4 pp. (1962)
CA 61: 1406f

D 992 Further studies of viral inclusions in experimental influenza using fluorescence microscopy.
Maksimovich, N.A., and Petrovskaya, O.G.
Acta Virol. (Prague) 6: 127-31 (1962)
CA 57: 5212a

D 993 Application of the additive statistical method for the study of fluorescent and absorption spectra.
Maslov, P.G.
Proc. Intern. Meeting Mol. Spectry., 4th, Bologna, 1959, 1: 266-87 (1962)
CA 59: 7087h

D 994 Fluorescence of oils. II. Fluorescent spectra of lubricating oils.
Masuko, K.
Kagaku Keisatsu Kenkyusho Hokoku 15: 325-7 (1962)
CA 58: 10019d

D 995 Hydrogen bonding of excited carbazole and 2-naphthylamine with various proton acceptors in nonpolar solvents.
Mataga, N., Torihashi, Y., and Kaifu, Y.
Z. Physik. Chem. 34: 379-92 (1962)
CA 56: 5156c

D 996 Ionic dissociation of excited carbazole in aqueous solution.
Mataga, N., and Torihashi, Y.
Z. Physik. Chem. 34: 401-3 (1962)
CA 56: 5156d

D 997 Hydrogen bonding of electronically excited molecules in solution.
Mataga, N., and Kaifu, Y.
Proc. Intern. Symp. Mol. Struct. Spectry., Tokyo 1962: (D113) 4 pp. (1962)
CA 61: 3746g

D 998 Intermolecular proton transfer in the excited hydrogen-bonded complex in nonpolar solvent and fluorescence quenching due to hydrogen bonding.
Mataga, N., and Kaifu, Y.
J. Chem. Phys. 36: 2804-5 (1962)
CA 57: 10668d

D 999 Separation of active and inactive absorption in the long wavelength region of excitation of luminescence.
Mazurenko, Y.T.
Opt. i Spektroskopiya 13: 854-5 (1962)
CA 56: 5157f

D 1000 Certain fluorescence characteristics of intact bright-leaf tobacco.
McClure, W.E., and Hassler, F.J.
Tobacco Sci. 6: 10-15 (1962)
Tobacco 154: 68-73 (1962)
CA 58: 13249h

D 1001 The external heavy-atom spin-orbital coupling effect. III. Phosphorescence spectra and lifetimes of extremely perturbed naphthalenes.
McGlynn, S.P., Reynolds, M.J., Gaigre, G.W., and Christodouleas, N.D.
J. Phys. Chem. 66: 2499-505 (1962)
CA 58: 8524b

D 1002 The external heavy-atom spin-orbital coupling effect.
McGlynn, S.P., Smith, F., and Christodouleas, N.
Proc. Intern. Symp. Mol. Struct. Spectry., Tokyo 1962
CA 61: 6528e

D 1003 Thermoluminescent emission spectra of X-ray-irradiated alkali halides.
McLachlan, L.A.
Phys. Chem. Solids 23: 1344-6 (1962)
CA 56: 4029g

D 1004 Relation of tetracycline to carcinoma.
McLeary, J.F., and Walshe, B.R.
J. Obstet. Gynaecol. Brit. Commonwealth 69: 313-7 (1962)
CA 56: 861g

D 1005 Activation of luminescent substances with monovalent manganese.
McKeag, A.H.
Acta Physiol. Acad. Sci. Hung. 14: 301-10 (1962)
CA 61: 12808a

D 1006 Lifetime of the triplet state of anthracene in Lucite.
Melhuish, W.H., and Hardwick, R.
Trans. Faraday Soc. 58: 1908-11 (1962)
CA 58: 7513c

D 1007 Fluorescence of organic compounds.
Melikadze, L.D.
Tr. Inst. Khim., Akad. Nauk Gruz. SSR 16: 31-50 (1962)
CA 59: 14764e

D 1008 Events leading to the proposed publication on methods of reporting fluorescence spectra.
Menzies, A.C.
Photoelec. Spectrometry Group Bull. No. 14: 377 (1962)
CA 56: 5157h

D 1009 Quantitative study of a one-electron photooxidation in a rigid medium.
Meyer, W.C., and Albrecht, A.C.
J. Phys. Chem. 66: 1168-77 (1962)
CA 57: 5506i

D 1010 Enhancement of fluorescence of a conjugate of soybean inhibitor accompanying interaction with trypsin.
Miller, D., Minzghor, K., and Steiner, R.
Biochim. Biophys. Acta 65: 153-5 (1962)
CA 58: 7089h

D 1011 Infrared resonance fluorescence in the fundamental vibration-rotation band of carbon monoxide.
Millikan, R.C.
Phys. Rev. Letters 8: 253-4 (1962)
CA 57: 6627d

D 1012 Fluorimetric determination of traces of beryllium with morin in silicate minerals. I. Reaction and measurement conditions.
Minczewski, J., and Rutkowski, W.
Chem. Anal. 7: 1107-18 (1962)
CA 58: 13116d

D 1013 Optical constants of the crystalline alkaline earth platino cyanides.
Moncuit, C., and Poulet, H.
J. Phys. Radium 23: 353-62 (1962)
CA 57: 13286i

D 1014 Elettrofluorescenza elettroluminescenza. 2nd ed.
Morati, L.
Milan, 429 pp. (1962) (Book)
CA 58: 9748h

D 1015 Phosphorescence and thermolumines-
cence of recrystallized NaCl·Tl and
KCl·Tl phosphors.
Morlin, Z.
Zur Physik Chemie Kristallphosphore,
Tagung Unterkomm. Leuchtstoffe Sekt.
Physik, 2., Berlin 1961: 152-61 (1962)
CA 59: 12278f

D 1016 Extinction of the fluorescence of
3,4-benzopyrene.
Morlin, Z., and Saringer, M.
Acta Physiol. Acad. Sci. Hung. 14: 211-6
(1962)
CA 61: 5103d

D 1017 Effect of crystallization of the solvent
on the luminescence of dyes.
Morozov, Y.V., Naberukhim, Y.I., and
Gorskii, G.U.
Opt. i Spektroskopiya 12: 598-605 (1962)
CA 57: 80431h

D 1018 Excitation of a retarded fluorescence
of photons with energies lower than
those of the emitted photons. Interpretation.
Muel, B.
Compt. Rend. 255: 3149-51 (1962)
CA 56: 6335b

D 1019 First triplet of polycyclic aromatic
hydrocarbons. I. Phosphorescent
spectra in the red and the near infrared
of solutions congealed at -180°. II.
First triplet and cancerigenic activity.
Muel, B., and Hubert-Habart, M.
Proc. Intern. Meeting Mol. Spectry., 4th,
Bologna, 1959, 2:647, 56 (1962)
CA 59: 4695a

D 1020 Optical study of certain 1,2-diaryl
derivatives of ethylene in polystyrene.
Nagornaya, L.L., Mulkes, L.Y., and
Shubina, L.V.
Opt. i Spektroskopiya 12: 117-21 (1962)
CA 57: 5447c

D 1021 Fluorescence spectra of uranyl phos-
phate and oxalate.
Narasimham, K.V., and Rao, V.
Spectrochim. Acta 18: 1055-8 (1962)
CA 56: 5158d

D 1022 Interactions between excited mole-
cules and metal ions; the catalysis of
radiationless transitions in oxazine
dyes by heavy metal ions and com-
plexes.
Nelson, R.L., Bell, J.A., Norland, K., and
Linschitz, H.
Proc. Intern. Symp. Mol. Struct. Spectry.,
Tokyo (1962)
CA 61: 6544h

D 1023 Objective indicators for qualitative
alterations in oils after vegetable
frying.
Nemets, S.M., and Nesterova, I.M.
Konserv. i Ovoshchesushil'. Prom.
17(4): 4-7 (1962)
CA 57: 12968g

D 1024 Decay of fluorescence of some inorganic scintillators on excitation by
α-particles, protons, or γ-rays (electrons).
Neuert, H.
Halbleiterprobleme 7: 239-59 (1962)
CA 61: 3831e

D 1025 Rapid triplet excitation migration in
organic crystals.
Nieman, G.C., and Robinson, G.W.
J. Chem. Phys. 37: 2150-1 (1962)
CA 56: 1056e

D 1026 Plasticization of polyethylene in low-
temperature radiolysis.
Nikal'skii, V.G., and Boben, N.Ya.
Dokl. Akad. Nauk SSSR 147: 1406-8 (1962)
CA 58: 9244c

D 1027 Internal structure of polymer solu-
tions as studied by a fluorescence
polarization method.
Nishijima, Y.
Luminescence Org. Inorg. Mater., Intern.
Conf., New York 1961: 235-43 (1962)
Internal Structure
CA 56: 1547b

D 1028 Effect on the crystal state on the
structure of electron spectra and the
character of the diffraction pattern of
chrysene and fluorene.
Nurmukhametov, R.N., and Popova, E.G.
Fiz. Probl. Spektroskopii, Akad. Nauk
SSSR, Materialy 13-go Soveshch., Leningrad, 1960, 1: 216-18 (1962)
CA 59: 12317b

D 1029 Luminescence and structure of azo
   compounds.
   Nurmukhametov, R.N., Shigorin, D.N.,
   Kozlov, Yu.I., and Puchkov, V. A.
   Fiz. Probl. Spektroskopii, Akad. Nauk
   SSSR, Materialy 13-go Soveshch.,
   Leningrad, 1961, 1: 283-7 (1962)
   CA 59: 12318d

D 1030 Luminescence spectra of fluorene.
   Nurmukhametov, R.N., and Gobov, G.V.
   Opt. i Spektroskopiya 13: 676-82 (1962)
   CA 56: 4046a

D 1031 Observation of anomalous phosphores-
   cence-fluorescence intensity ratio in
   excitation of upper electronic states of
   certain aromatic hydrocarbons.
   O'Dwyer, M.F., El-Bayoumi, M.A., and
   Strickler, S.J.
   J. Chem. Phys. 36: 1395-6 (1962)
   CA 57: 2966g

D 1032 Intensities of crystal spectra of
   rare-earth ions.
   Ofelt, G.S.
   J. Chem. Phys. 37: 511-20 (1962)
   CA 58: 147f

D 1033 ESR (electron spin resonance) study
   of the radiation chemical process in
   polyethylene.
   Ohnishi, S.
   Bull. Chem. Soc. Japan 35: 254-9 (1962)
   CA 57: 1038e

D 1034 Greek bands in the alkali halide.
   Onaka, R., and Fujita, J.
   J. Quant. Spectry. Radiative Transfer 2:
   599-611 (1962)
   CA 58: 10862f

D 1035 Correlation between Stokes shift and
   the stability constant of 6,7-dihydroxy-
   2-naphthalenesulfonic acid metal
   chelates.
   Ono, N., and Hirota, K.
   J. Chem. Phys. 39: 1907-8 (1963)
   CA 59: 14642e

D 1036 2-Arylnaphthoxazoles and some other
   condensed oxazoles.
   Osman, A.M., and Bassiouni, I.
   J. Org. Chem. 27: 558-61 (1962)
   CA 57: 796g

D 1037 Interdependence of excited states of
   dichroic molecules.
   Oster, G., and Oster, G.K.
   Luminescence Org. Inorg. Mater., Intern.
   Conf., New York 1961: 186-95 (1962)
   CA 56: 1061b

D 1038 Luminescence in plastics.
   Oster, G., Geocintov, N., and Khan, A.
   Nature 196: 1089-90 (1962)
   CA 56: 4047b

D 1039 Phosphorescent mixture.
   Palumo, D.T., Shaffer, F.N., and Rulon,
   R.M.
   Belg. 614,288, 9 pp. (1962)
   CA 57: 14579c

D 1040 Prophyrin fluorescence quenching as
   a criterion of analytical evaluation.
   Panconesi, E., Nannelli, M., and
   Gabrielli, G.
   Panminerva Med. 4: 387-8 (1962)
   CA 61: 2173a

D 1041 Determination of gallium in cadmium
   sulfide.
   Pantaler, R.P., and Timofeeva, N.B.
   Metody Analiza Veshchestv, Osoboi
   Chistoty i Monokristallov Gos. Kom.
   Sov. Min. SSSR po Khim. 1962: 96-8
   (1962)

D 1042 Spectrofluorometer calibration in the
   ultraviolet region.
   Parker, C.A.
   Anal. Chem. 34: 502-5 (1962)
   CA 57: 4i

D 1043 Luminescence of some piazselenols.
   A new fluorimetric reagent for sele-
   nium.
   Parker, C.A., and Harvey, L.G.
   Analyst 87: 558-65 (1962)
   CA 57: 7900a

D 1044 Delayed fluorescence from solutions
   of anthracene and phenanthrene.
   Parker, C.A., and Hatchard, C.G.
   Proc. Chem. Soc. 1962: 147
   CA 57: 2992h

D 1045 Sensitized anti-Stokes delayed fluor-
   escence.
   Parker, C.A., and Hatchard, C.G.
   Proc. Chem. Soc. 1962: 386-7
   CA 56: 6336h

D 1046 Phosphorescence measurement in
chemical analysis: Tests with a new
instrument.
Parker, C.A., and Hatchard, C.G.
Analyst 87: 644-76 (1962)
CA 56: 1894c

D 1047 Triplet-singlet emission in fluid
solution.
Parker, C.A., and Hatchard, C.G.
J. Phys. Chem. 66: 2506-11 (1962)
CA 56: 4041d

D 1048 Luminescence of NaCl-Cu phosphor.
Parfianovich, I.A., and Shuraleva, E.I.
Fiz. Shchelochnogaloidnykh Kristallov,
Latv. Gos. Univ., Tr. 2-go Vses.
Soveshch., Riga 1961: 172-8 (1962)
CA 60: 12762d

D 1049 Particulars of the luminescence and
structure of some alkali halide phosphors.
Parfianovich, I.A., and Shuraleva, E.I.
Izv. Akad. Nauk SSSR, Ser. Fiz. 26:
497-505 (1962)
CA 57: 5432c

D 1050 Electronic spectra of the phenan-
threnes, $C_{14}H_{10} + C_{14}D_{10}$ at low temper-
ature.
Pesteil, L., and Rabaud, M.
J. Chim. Phys. 59: 167-76 (1962)
CA 57: 1757g

D 1051 Direct observation of rise of fluores-
cence in Tb^{+++} following a short
burst of ultraviolet excitation.
Peterson, G.E., and Bridenbaugh, P.M.
J. Opt. Soc. Am. 52: 1079-80 (1962)
CA 56: 1061g

D 1052 Paramagnetic resonance of some
benzophenone derivatives in their
phosphorescent state.
Piette, L.H., Sharp, J.H., Kuwana, T., and
Pitts, J.N., Jr.
J. Chem. Phys. 36: 3094-5 (1962)
CA 57: 11989b

D 1053 Fluorescence polarization of complex
molecules.
Pikulik, L.G., and Snopko, V.N.
Dokl. Akad. Nauk Belorussk. SSR 6(3):
155-8 (1962)
CA 57: 15947i

D 1054 Temperature effects of the dislocation
of electronic spectra of complex mole-
cules in solutions.
Pikulik, L.G.
Fiz. Probl. Spektroskopii, Akad. Nauk
SSSR, Materialy 13-go Soveshch.,
Leningrad, 1960, 1: 297-300 (1962)
CA 59: 12315e

D 1055 Relations between temperature and
phosphorescence of organic phosphors.
Pilipovich, V.A.
Dokl. Akad. Nauk Belorussk. SSR 6: 90-3
(1962)
CA 57: 11978d

D 1056 Structural effects in the photochemi-
cal processes of ketones in solution.
Pitts, J.N., Johnson, H.W., and Kuwana, T.
J. Phys. Chem. 66: 2456-61 (1962)
CA 56: 4080f

D 1057 Simple method of obtaining the true
absorption spectrum of a fluorescent
solution.
Popovych, O., and Rogers, L.B.
Spectrochim. Acta 18: 1279 (1962)
CA 58: 5158b

D 1058 Energy transfer from molecules in
the triplet state.
Porter, G.
Pure Appl. Chem. 4: 141-2 (1962)
CA 57: 6625d

D 1059 Fluorescence spectrum of anisole
vapor.
Prakash, S.
Nature 193: 268 (1962)
CA 57: 1691b

D 1060 Influence of scattering on fluorescence
spectra of dilute solutions obtained
with the Aminco-Bowman Spectrophoto-
fluorometer.
Price, J.M., Kaihara, M., and Howerton,
H.K.
Appl. Opt. 1: 521-33 (1962)
CA 57: 9364h

D 1061 Fluorescence of organic traces in inorganic materials and their distribution
in nature.
Przibram, K.
Geochim. Cosmochim. Acta 26: 1045-54
(1962)
CA 56: 4329f

D 1062 The phosphorescence of naphthalene
and some of its derivatives at liquid-
oxygen temperatures.
Pyratmitskii, B.A., Grossman, A.Ya.,
Krasnova, V.V., and Vlasenko, A.I.
Izv. Vysshikh Uchebn. Zavedenii, Fiz.
1962 (3): 41-4 (1962)
CA 57: 15950b

D 1063 Stationary luminescence in ZnS-type
phosphors. Excitation valence band-
electron traps.
Rebane, K.
Tr. Inst. Fiz. i Astron., Akad. Nauk Est.
SSR 1962: 247-56 (1962)
CA 59: 3919f

D 1064 Microfluorimetric studies of microorganisms. III. Secondary fluorescence conferred by fluorescent antibodies.
Repertigny, J.D., and Sonea, S.
Ann. Inst. Pasteur 102: 182-91 (1962)
CA 57: 1216g

D 1065 Geologic interpretations of hydrocar-
bon fluorescence.
Riecker, R.E.
Dissertation Abstr. 22: 3609-10 (1962)
CA 57: 6918a

D 1066 Hydrocarbon fluorescence and migra-
tion of petroleum.
Riecker, R.E.
Bull. Am. Assoc. Petrol. Geologists 46:
60-75 (1962)
CA 58: 11878f

D 1067 Fluorescence spectra and lifetimes of
some rare-earth compounds.
Rieke, F.F., and Allison, R.
J. Chem. Phys. 37: 3011-12 (1962)
CA 58: 7513f

D 1068 Application of direct and indirect
immuno-fluorescence for identification
of entero-viruses and titrating their
antibodies.
Riggs, J.L., and Brown, G.C.
Proc. Soc. Exptl. Biol. Med. 110: 833-7
(1962)
CA 57: 17261h

D 1069 Determination of average molar ab-
sorptivity for self-absorption of fluor-
escent radiation in fluorescein solution.
Rohatgi, K.K., and Singhal, G.S.
Anal. Chem. 34: 1702-6 (1962)
CA 56: 3036h

D 1070 Biochemical changes in seedlings of
bean infected with colletotrichum linde-
muthianum.
Romanowski, R.D., Kuc, J., and
Quackenbash, F.W.
Phytopathology 52: 1259-64 (1962)
CA 58: 7142f

D 1071 Electronic spectrum of 2,2'-paracy-
clophane.
Ron, A., and Schnepp, O.
J. Chem. Phys. 37: 2540-6 (1962)
CA 56: 1057a

D 1072 Term analysis of trivalent erlium in
vacious oxide modifications.
Rosenberger, D.
Z. Physik 167: 349-59 (1962)
CA 56: 1667b

D 1073 Fluorescent pigment standardization.
Roux, M.
Couleurs No. 46, 16-19 (1962)
CA 57: 13312d

D 1074 Migration and transfer of energy in a
two-component mixture.
Rubanov, V.S.
Opt. i Spektroskopiya 13: 454-7 (1962)
CA 57: 15875d

D 1075 The average duration of fluorescence
of protochlorophyll in the process of
greening of etiolated leaves.
Rubin, A.B., Minchenkova, L.E.,
Krasnovskii, A.A., and Tumerman, L.A.
Biofizika 7: 571-7 (1962)
CA 56: 1717e

D 1076 Effect of uniaxial stress on the spec-
trum of CaF_2 (Sm^{++}).
Runciman, W.A., and Stager, C.V.
J. Chem. Phys. 37: 196-7 (1962)
CA 56: 2015e

D 1077 Effect of temperature on phosphores-
cence of azelaic acid.
Ryazanova, E.F.
Uch. Zap., Gor'kovsk. Gos. Ped. Inst.
1962: 70-6 (1962)
CA 61: 14008a

D 1078 Fluorescence analysis of anthranilic
       acid.
       Sanders, P.P., and Parks, L.W.
       Anal. Biochem. 3: 354-6 (1962)
       CA 57: 9364d

D 1079 Transfer of triplet-state energy in
       fluid solutions. II. Quenching of
       biacetyl phosphorescence in solution.
       Sandros, K., and Backstrom, H.L.J.
       Acta Chem. Scand. 16: 958-68 (1962)
       CA 57: 16028d

D 1080 Metastable state of amino acids, pro-
       teins, nucleic acids, and other biologi-
       cal material.
       Sapezhinskii, I.I., and Emanuel, N.M.
       Fiz. Probl. Spektroskopii, Akad. Nauk
       SSSR, Materialy 13-go Soveshch.,
       Leningrad, 1960 1: 296-7 (1962)
       CA 59: 1177e

D 1081 The polarization of the fluorescence
       of crystals of phthalimide derivatives.
       Sarzhevskii, A.M.
       Dokl. Akad. Nauk Belorussk. SSR 6: 556-9
       (1962)
       CA 58: 10869g

D 1082 Spectrophotofluorometric test for
       formaldehyde and its precursors.
       Sawicki, E., Hauser, T.R., and McPherson,
       S.
       Anal. Chem. 34: 1460 (1962)

D 1083 The application of fluorescence polar-
       ization to the antigen-antibody reaction.
       Schapiro, H.
       Dissertation, University of Miami (1962)
       CA 58: 9514e

D 1084 Improved phosphorescent screen in
       electron tubes.
       Scharrer, E.
       Ger. 1,127,499, 1959 (1962)
       CA 61: 1375a

D 1085 Deactivation of Hg (G^3P_1) by CO and
       N_2.
       Scheer, M.D., and Fine, J.
       J. Chem. Phys. 36: 1264-7 (1962)
       CA 57: 2958e

D 1086 Intramolecular energy transfer in a
       naphthalene-anthracene system.
       Schnepp, O., and Levy, M.
       J. Am. Chem. Soc. 84: 172-7 (1962)
       CA 58: 14183h

D 1087 Cracking products of substances
       accompanying fats.
       Schmid, L.
       Mitt. Gebiete Lebensm. Hyg. 53: 507-10
       (1962)
       CA 60: 1038c

D 1088 Fluorescence decay times of organic
       crystals.
       Schmillen, A.
       Luminescence Org. Inorg. Mater., Intern.
       Conf., New York 1961: 30-43 (1962)
       CA 56: 6297a

D 1089 Fluorescence of a reaction product of
       6-chloro-4-hydroxy-2-ketopyrane-3-
       carboxylic acid with pyridine in sol-
       vents.
       Schulte, H., and Chomse, H.
       Zur Physik Chemie Kristallphosphore,
       Tagung Unterkomm. Leuchtstoffe Sekt.
       Physik, 2, Berlin 1961: 196-9 (1962)
       CA 59: 13493e

D 1090 Determination of the quantum yield of
       ruby fluorescence on irradiation with
       one of the blue absorption lines.
       Schulz, G.U.
       Z. Physik 167: 446-51 (1962)
       CA 57: 2966d

D 1091 Precise measurements of the distance
       between two spectral lines by use of
       two photomultipliers.
       Semenov, R.I.
       Opt. i Spektroskopiya 13: 134-6 (1962)
       CA 57: 11966a

D 1092 The luminescence of TlCl and KCl
       phosphor.
       Sen, S.C., and Bose, H.N.
       Z. Physik. Chem. 167: 20-5 (1962)
       CA 57: 214c

D 1093 Lifetimes of excited F centers.
       Swank, R.K., and Brown, F.C.
       Phys. Rev. Letters 8: 10-12 (1962)
       CA 58: 8134f

D 1094 Anisotropy of electron vibration
       transitions.
       Sevchenko, A.N.
       Fiz. Probl. Spektroskopii, Akad. Nauk
       SSSR, Materialy 13-go Soveshch.,
       Leningrad, 1960 1: 223-5 (1962)
       CA 59: 8263c

D 1095 Mechanism of recombination lumines-
cence in alkali halide phosphors.
Shamovskii, L.M.
Fiz. Shchelochnogaloidnykh Kristallov,
Riga, Sb. 1962: 221-9 (1962)
CA 60: 4974b

D 1096 Dependence of the luminescence
properties of crystal phosphors on
their structural properties.
Shamovskii, L.M., and Shibanov, A.S.
Pribory dlya Geofiz. Issled. Skvazhin
Radioakt. Metodami 1962: 49-58 (1962)
CA 60: 94e

D 1097 Portable fluorimeter built from mass-
production parts.
Shcherbov, D.P., Voinov, S.A., and
Kaptil'nyi, M.A.
Tr. Kazakhsk. Nauchn.-Issled. Inst.
Mineral'n. Syr'ya 1962: 209-13 (1962)
CA 61: 4e

D 1098 Determination of zinc in iron ores and
iron-containing minerals with 8-(p-
toluene-sulfonamido)quinoline.
Shcherbov, D.P., and Kolmogorova, V.V.
Metody Analiza Khim. Reaktivov i
Preparatov, Gos. Kom. Sov. Min. SSSR
po Khim. No. 4: 125-8 (1962)
CA 61: 2480d

D 1099 Line spectra of polycyclic aromatic
hydrocarbons in frozen crystalline
solutions. II. Singlet-singlet and
triplet-singlet spectra of 1,2-benzo-
pyrene at 77° and 40°K.
Shpol'skii, E.V., Klimova, L.A., and
Personov, R.I.
Opt. i Spektroskopiya 13: 34-52 (1962)
CA 56: 3009g

D 1100 Spectroscopy of some pyrene deriva-
tives at 20° and 4°K.
Shpol'skii, E.V., and Klimova, L.A.
Fiz. Probl. Spektroskopii, Akad. Nauk
SSSR, Materialy 13-go
Soveshch., Leningrad, 1960 1: 209-16
(1962)
CA 59: 1474b

D 1101 The fluorescence of organic scintil-
lant materials.
Shortt, B.A.
Dissertation Abstr. 22: 4049-50 (1962)
CA 57: 5476g

D 1102 Study of the triplet states of molecules
by the luminescence and electron para-
magnetic resonance methods.
Shigorin, D.N., Volkoua, N.V., Piskurov,
A.K., and Gurevich, A.I.
Opt. i Spektroskopiya 12: 657-9 (1962)
CA 57: 9331e

D 1103 Luminescence of indigo solutions at
77°K.
Shigorin, D.N., Nurmukhametov, R.N., and
Kozlov, Yu.I.
Opt. i Spektroskopiya 12: 659-61 (1962)
CA 57: 10637b

D 1104 The luminescence spectra of anthra-
quinone, thiondigo, and their deriva-
tives in frozen solutions.
Shigorin, D.N., Nurmukhametov, R.N.,
Sheheglova, N.A., and Dokunikhin, N.S.
Proc. Intern. Meeting Mol. Spectry., 4th,
Bologna, 1959 2: 672-87 (1962)
CA 59: 8277e

D 1105 Fluorescence of some quinoline salts.
Sljivic, S., and Nikolic, K.
Archiv Farm. (Belgrade) 12: 61-3 (1962)
CA 57: 4513c

D 1106 Hepatotoxic action of chromatographi-
cally separated fractions of aspergillus
flavas extracts.
Smith, R.H., and McKernan, W.
Nature 195: 1301-3 (1962)
CA 56: 763f

D 1107 Immunofluorescence as applied to
pathology.
Smith, C.W., Metzger, J.F., and Hoggan,
M.D.
Am. J. Clin. Pathol. 38: 26-42 (1962)
CA 57: 11761i

D 1108 Spectroscopy and optical maser action
in SrF_2:Sm^{++}.
Sorokin, P.P., Stevenson, M.J., Lankard,
J.R., and Pettit, G.D.
Phys. Rev. 127: 503-8 (1962)
CA 57: 13290g

D 1109 Delayed fluorescence and phosphores-
cence in crystals of aromatic molecules
at 4.2°K.
Sponer, H.
Luminescence Org. Inorg. Mater., Intern.
Conf., New York 1961: 143-52 (1962)
CA 56: 1060c

D 1110 Effects of molecular orientation on fluorescence emission and energy transfer in crystalline aromatic hydrocarbons.
Stevens, B.
Spectrochim. Acta 18: 439-48 (1962)
CA 57: 11977f

D 1111 Photo-oxidation of biacetyl in solution.
Stevens, B., and Dubois, J.T.
J. Chem. Soc. 1962: 2813-15 (1962)
CA 57: 5505a

D 1112 The sensitivity of the universal relation of absorption and luminescence spectra of complex molecules to the presence of an impurity.
Stepanov, B.I., Krautsov, L.A., and Rubinov, A.N.
Dokl. Akad. Nauk Belorussk. SSR 6: No. 1, 14-18 (1962)
CA 57: 289f

D 1113 Synthesis and properties of 4-methyl-2-oxo-1,2-benzopyran-7-yl β-O-galactoside (galactoside of 4-methylambelliferone).
Strachan, R., Wood, J., and Hirschmann,R.
J. Org. Chem. 27: 1074-75 (1962)
CA 57: 7370h

D 1114 Relation between absorption intensity and fluorescence lifetime of molecules.
Strickler, S.J., and Berg, R.A.
J. Chem. Phys. 37: 814-22 (1962)
CA 57: 16001f

D 1115 Relation between the n → π* absorption spectrum and the emission spectrum of benzaldehyde.
Stockburger, M.
Z. Physik. Chem. 31: 350-62 (1962)
CA 57: 1756b

D 1116 The fluorescence of multicomponent organic solutions.
Stolz, W.
Wiss. Z. Tech. Univ. Dresden 11: 687-91 (1962)
CA 58: 12082c

D 1116a Fluorescence of multicomponent organic solutions.
Stolz, W., and Hereforth, L.
Zur Physik Chemie Kristallphosphore, Tagung Unterkomm. Leuchtstoffe Sekt. Physik, 2, Berlin 1961: 211-17 (1962)
CA 59: 13496a

D 1117 Infrared spectra of perfluorovinyl derivatives of a series of elements.
Sterlin, R.N., and Dobov, S.S.
Zh. Vses. Khim. Obshchestva im. D. I. Mendeleeva 7: No. 1, 117-18 (1962)
CA 57: 294e

D 1118 Fluorescence of α-chymotrypsin in the presence of substrates and inhibitors.
Sturtevant, J.M.
Biochem. Biophys. Res. Commun. 8: 321-5 (1962)
CA 57: 12890g

D 1119 Fluorescent crayons.
Switzer, J.L.
U.S. 3,057,806, 1959 (1962)
CA 58: 8145e

D 1120 Structure and reactivity of the oxy anions of transition metals. XIII. The effect of environment and certain complex-forming cations on the visible and ultraviolet absorption spectra of permanganate and chromate.
Symons, M.C.R., and Trevalion, P.A.
J. Chem. Soc. 1962: 3503-9 (1962)
CA 57: 13306i

D 1121 The fundamental polarization of the fluorescence of viscous solutions.
Szalay, L., Gati, L., and Sarkany, B.
Acta Phys. Acad. Sci. Hung. 14: 217-24 (1962)
CA 61: 5081f

D 1122 The sphere of influence in theories which refer to the concentration-depolarization of fluorescence light.
Szalay, L., and Sarkany, B.
Acta Univ. Szeged., Acta Phys. Chem. 8: 25-9 (1962)
CA 56: 3996e

D 1123 Phosphorescence spectra of benzonitrile and related compounds.
Takei, K., and Kanda, Y.
Spectrochim. Acta 18: 1201-16 (1962)
CA 56: 5158c

D 1124 The formation of colloidal silver in photosensitive glass examined by the use of a high-temperature spectrophotometer.
Tashiro, M., and Soga, N.
Kogyo Kagaku Zasshi 65: 337-41 (1962)
CA 57: 10689g

D 1125 Pigment interaction in chloroplast
fluorescence.
Teale, F.W.J.
Biochem. J. 85, 14 pp. (1962)
CA 56: 5981b

D 1126 Bound excitons in CdS and ZnTe.
Thomas, D.G.
Proc. Intern. Conf. Phys. Semicond.,
Exeter, Eng. 1962: 403-8 (1962)
CA 60: 4913d

D 1127 Fluorescence in CdS and its possible
use for an optical maser.
Thomas, D.G., and Hopfield, J.J.
J. Appl. Phys. 33: 3243-9 (1962)
CA 56: 1028d

D 1128 Optical properties of bound exciton
complexes in cadmium sulfide.
Thomas, D.G., and Hopfield, J.J.
Phys. Rev. 128: 2135-48 (1962)
CA 56: 104f

D 1129 A geometrical aspect of fluorescent
spectroscopy.
Thomas, J.F., Mukai, M., and Tebbens,
B.D.
Natl. Cancer Inst. Monograph No. 9: 127-33
(1962)
CA 56: 1060f

D 1130 Fluorescence-induction phenomena in
isolated chloroplasts.
Thomas, J.B., Voskuil, W., Olsman, H.,
and deBoois, H.M.
Biochim. Biophys. Acta 59: 224-6 (1962)
CA 57: 7617e

D 1131 Measurement of surface temperatures
by photoluminescence.
Thureau, P.
Journees Intern. Transmission Chaleur,
Paris, 1961 2: 865-9, 869-72 (1962)
CA 59: 4557b

D 1132 Electronic spectroscopy of rigid
polymers.
Tinoco, I.
Am. Chem. Soc., Div. Org. Coatings,
Plastics Chem., Preprints 22: 28-31
(1962)
CA 61: 5796d

D 1133 Quantitative fluorimetric determina-
tion of methandrostenolone.
Tishler, F., Sheth, P.B., Giaimo, M.B.,
and Mader, W.J.
J. Pharm. Sci. 51: 1175-80 (1962)
CA 58: 9380h

D 1134 Low temperature spectra of plant
leaves and the chlorophyll state.
Litvin, F.F., Rikhireva, G.T., and Kras-
novskii, A.A.
Biofizika 7: 578-91 (1962)
CA 56: 4805g

D 1135 Excitation energy transfer between
pigments in photosynthetic cells.
Tomita, G., and Rabinowitch, E.
Biophys. J. 2: 483-99 (1962)
CA 59: 5991a

D 1136 Spectrofluorometric studies of de-
graded cotton cellulose.
Toner, S.D., and Plitt, K.F.
Tappi 45: 681-8 (1962)
CA 57: 12764i

D 1137 Luminescence of purines and pyrimi-
dines.
Longworth, J.W.
Biochem. J. 84: 104p-105p (1962)
CA 56: 3635e

D 1138 Quenching fluorescence by certain
organophosphorus insecticides.
Tropenko, M.A.
Novoe v Oblasti Sanit.-Khim. Analiza, Sb.
1962: 175-8 (1962)
CA 59: 14512g

D 1139 Fluorescence analysis in pharmacy.
Ischudi-Steiner, I., and Leupin, K.
Pharm. Acta Helv. 37: 274-82 (1962)
CA 57: 11311f

D 1140 Effect of radioprotective chemicals on
phosphorescent emission of riboflavine
in oxygenated and deoxygenated solutions.
Van den Brenk, H.A.S., and Jamieson, D.
Experientia 18: 231-3 (1962)
CA 57: 6275d

D 1141 Separation and identification of aliphatic and aromatic carcinogens from en-
vironmental sources.
Van Duuren, B.L.
Natl. Cancer Inst. Monograph No. 9,
135-51 (1962)
CA 58: 8304f

D 1142 Association of centers in zinc sulfide.
Van Gool, W., and Diemer, G.
Luminescence Org. Inorg. Mater., Intern.
Conf., New York 1961: 391-401 (1962)
CA 56: 1983g

D 1143 Quenching interactions between rare-
earth ions.
Van Uitert, L.G., and Iida, S.
J. Chem. Phys. 37: 986-92 (1962)
CA 57: 14590a

D 1144 Role of f-orbital electron wave func-
tion mixing in the concentration
quenching of Eu^{+++}
Van Uitert, L.G., Linares, R.C., Soden,
R.R., and Ballman, A.A.
J. Chem. Phys. 36: 702-5 (1962)
CA 57: 212a

D 1145 Influence of ion size upon the intensity
of Eu^{+++} fluorescence in the tungstates.
Van Uitert, L.G., and Soden, R.R.
J. Chem. Phys. 36: 517-19 (1962)
CA 57: 8079h

D 1146 Effects of rare-earth ion substitution
upon the fluorescence of terbium hexa-
antipyrene tri-iodide and sodium europium tungstate.
Van Uitert, L.G., and Soden, R.R.
J. Chem. Phys. 36: 1797-1800 (1962)
CA 57: 5429b

D 1147 Enhancement of rare-earth ion fluor-
escence by lattice processes in oxides.
Van Uitert, L.G., Soden, R.R., and
Linares, R.C.
J. Chem. Phys. 36: 1793-6 (1962)
CA 57: 5429g

D 1148 Stimulation and quenching of chemi-
luminescence by oxygen.
Vasil'ev, R.F., Vichutinskii, A.A.,
Karpukhin, O.N., and Shlyapintokh, V.Ya.
Fiz. Probl. Spektroskopii, Akad. Nauk
SSSR, Materialy 13-go Soveshch.,
Leningrad, 1960 1: 320-1 (1962)
CA 59: 12317h

D 1149 Mechanism of the effect of aromatic
amines on the fluorescence and photochemical oxidation of anthracene compounds.
Vember, T.M.
Dokl. Akad. Nauk SSSR 147: 123-6 (1962)
CA 58: 7474h

D 1150 Luminous spectra of liquids and gases
produced by radiation of fast electrons.
Vereshchinskii, I.V., Glazunov, P.Ya.,
Kustanovich, I.M., and Polak, L.S.
Tr. 2-go Vses. Soveshch. po Radiats.
Khim., Akad. Nauk SSSR, Otd. Khim.
Nauk, Moscow 1960: 83-6 (1962)
CA 56: 5157e

D 1151 Luminescent derivatives of borazone
and their application to the counting of
thermal neutrons.
Videla, G.J., Molinari, M.A., Rojo, E.A.,
Lines, O.A., and Caras, L.H.
Arg., Rep., Com. Nacl. Energia At.,
Informe No. 77, 6 pp. (1962)
CA 59: 8329c

D 1152 Fluorescent substances from anagasta
kuehniella. III. Isolation and structures of erythropterin, ekapterion, and
lepidopterin.
Viscontini, M., and Stierlin, H.
Helv. Chim. Acta 45: 2479-87 (1962)
CA 58: 9062d

D 1153 Determination of thallium with Rhoda-
mine 6Zh in metallic zinc and cadmium.
Vladimirova, V.M., and Davidovich, N.K.
Metody Analiza Khim. Reaktivov i
Preparatov, Gos. Kom. Sov. Min. SSSR
po Khim. No. 4: 116-19 (1962)
CA 61: 3688a

D 1154 Luminescence spectra of proteins and
aromatic amino acids at different pH.
Vladimirov, Y.A., and Li, Chin-Kuo
Biofizika 7: 270-80 (1962)
CA 57: 11978c

D 1155 The line and band spectrum of
radiation-discolored lithium fluoride.
von der Osten, W., and Waidelich, W.
Naturwissenschaften 49: 342-3 (1962)
CA 57: 11972i

D 1156 Additional optical absorption, fluorescence spectra, and band energy in ionic
crystals.
Vorob'ev, A.A.
Izv. Vysshikh Uchebn. Zavedenii, Fiz.
1962: No. 2, 165-70 (1962)
CA 57: 9365c

D 1157 Infrared absorption of some hexagonal
rare-earth chlorides.
Varsanyi, F., and Dieke, G.H.
J. Chem. Phys. 36: 835-40 (1962)
CA 57: 1763a

D 1158 Energy levels of hexagonal ErCl$_3$.
Varsanyi, F., and Dieke, G.H.
J. Chem. Phys. 36: 2951-61 (1962)
CA 57: 11972c

D 1159 Spin waves in complex lattices.
Wallace, D.C.
Phys. Rev. 128: 1614-18 (1962)
CA 56: 2979h

D 1160 Oxygen quenching of fluorescence in
solution: an experimental study of the
diffusion process.
Ware, W.R.
J. Phys. Chem. 66: 455-8 (1962)
CA 57: 1688g

D 1161 Location of the lowest triplet level in
azulene.
Ware, W.R.
J. Chem. Phys. 37: 923-4 (1962)
CA 56: 3010g

D 1162 Fluorescent decay of anthracene in
the region 0.1-10 μsec.
Wasson, M.N.
At. Energy Res. Estab. (Gt. Brit.) Memo.
AERE-M 1153, 3 pp. (1962)
CA 59: 2301e

D 1163 The deposition of tetracycline drugs
in bones and scales of fish and its possible use for marking.
Weber, D.D., and Ridgeway, G.J.
Progressive Fish Culturist 24: 150-5
(1962)
CA 56: 3828h

D 1164 Fluorescence energy transfer and
oxygen quenching in solutions of diphen-
yloxazole in cyclohexane.
Weinreb, A.
J. Chem. Phys. 36: 890-4 (1962)
CA 57: 4195f

D 1165 Quenching of the phosphorescence and
primary dissociation of biacetyl vapor
at 4358 A, and high temperature.
Weir, D.S.
J. Chem. Phys. 36: 1113-14 (1962)
CA 57: 2992i

D 1166 Energy exchange in the 3-methyl-2-
butanone-biacetyl system at 3130 A.
Weir, D.S.
J. Am. Chem. Soc. 84: 4039-40 (1962)
CA 56: 1061d

D 1167 Spectrophotofluorimetric identification
of chromatographically separated
plastic pigments.
Weiss, J.B., and Smith, I.
Chromatog. Symp., 2nd, Brussels 1962:
293-7 (1962)
CA 60: 7141h

D 1168 Cross-section measurements for
some photon-gas interaction processes.
Weissler, G.L.
J. Quant. Spectry. Radiative Transfer 2:
383-92 (1962)
CA 58: 9760g

D 1169 Relation between absorption and
fluorescence spectra of triphenyl-
methyl.
Weissman, S.J.
J. Chem. Phys. 37: 1886-7 (1962)
CA 56: 2021d

D 1170 Fluorescent aromatic azines. III. In-
vestigation and significance of the elec-
tronic spectra of azines.
Weller, A., and Wolf, H.
Ann. 657: 64-79 (1962)
CA 58: 1382f

D 1171 Spectroscopy of carbon vapor con-
densed in inert matrixes at 4° and 20°K.
Weltner, W., and Walsh, P.N.
J. Chem. Phys. 37: 1153-4 (1962)
CA 56: 6330a

D 1172 Effect of dimerization on the α-phos-
phorescence of acriflavine in glucose
glass.
Wen, W.Y., and Hsu, R.
J. Phys. Chem. 66: 1353-4 (1962)
CA 57: 10668f

D 1173 Energy-transport mechanism investi-
gations by measuring the fade-out time
of organic luminophors.
Wendel, G., and Haertig, G.
Z. Physik. Chem. 221: 17-28 (1962)
CA 56: 1991a

D 1174 Significance of the triplet state of sensitizing dyes in optical sensitization.
West, W.
Sci. Phot., Proc. Intern. Colloq., Liege 1959: 557-68 (1962)
CA 59: 9478f

D 1175 Review of fundamental developments in analysis. Fluorometric analysis.
White, C.E., and Weissler, A.
Anal. Chem. 34: 81R-91R (1962)
CA 58: 13520f

D 1176 Pressure dependence of fluorescence spectra. II. Transient effects.
Wilson, D.J.
J. Chem. Phys. 36: 1293-7 (1962)
CA 57: 4196d

D 1177 Nature of fluorescence of an enzyme-reduced coenzyme-reduced substrate complex.
Winer, A.D.
Biochim. Biophys. Acta, 59: 219-21 (1962)
CA 57: 7612c

D 1178 Effects of elevated temperatures on the fluorescence and optical maser action of ruby.
Wittke, J.P.
J. Appl. Phys. 33: 2333-5 (1962)
CA 56: 2979f

D 1179 Luminescence of the system strontium oxide-cerium oxide.
Witzmann, H., and Herzog, G.
Z. Physik. Chem. 219: 369-86 (1962)
CA 57: 9332d

D 1180 Fluorescence of rare earths.
Witzmann, H., and Moller-Buschbaum, Hk.
Naturwissenschaften 49: 180 (1962)
CA 57: 13276e

D 1181 The effect of ultraviolet and X rays on 3,4-benzopyrene.
Woenckhaus, J.W., Woenckhaus, Ch.W., and Koch, R.
Z. Naturforsch. 176(5): 295-9 (1962)
CA 57: 6777e

D 1182 Temperature dependence of energy transfer in organic molecular crystals.
Wolf, H.C.
Proc. Intern. Symp. Mol. Struct. Spectry., Tokyo 1962, 4 pp. (1962)
CA 61: 5102c

D 1183 Fluorescent aromatic azines. I. Preparation of o-hydroxylated aldazines and ketazines capable of fluorescence.
Wolf, H., and Westphal, O.
Ann. 657: 39-51 (1962)
CA 58: 1380b

D 1184 Absorption and fluorescence spectra of Pr^{+++} in $LaBr_3$.
Wong, E.Y., and Richman, I.
J. Chem. Phys. 36: 1889-92 (1962)
CA 57: 8044g

D 1185 Absorption and fluorescence of Sm^{++} in CaF_2, SrF_2, and BaF_2.
Wood, D.L., and Kaiser, W.
Phys. Rev. 126: 2079-88 (1962)
CA 57: 10634e

D 1186 Ultraviolet absorption and fluorescence spectroscopy in protein research.
Yanari, S.S.
Am. Chem. Soc., Div. Org. Coatings, Plastics Chem., Preprints 22: 32-43 (1962)
CA 61: 718f

D 1187 Fluorescence and polymers. I. Polystyrene and styrene — methyl methacrylate copolymers.
Yanari, S.S., and Bovey, F.A.
Am. Chem. Soc., Div. Org. Coatings, Plastics Chem., Preprints 22: 44-55 (1962)
CA 61: 1959g

D 1188 Optical maser emission from trivalent praseodymium in calcium tungstate.
Yariv, A., Porto, S.P.S., and Nassau, K.
J. Appl. Phys. 33: 2519-21 (1962)
CA 56: 1027h

D 1189 Luminescence decay times: concentration effects.
Yguerabide, J., and Burton, M.
J. Chem. Phys. 37: 1757-74 (1962)
CA 56: 2021b

D 1190 Radiometer glass for use of γ-ray.
Toshiba Electric Co. (by R. Yokota)
Japan 5024 (1962), June 18, 1962
CA 59: 6114c

D 1191 Phosphorescence of solid triphenylene.
Zander, M.
Naturwissenschaften 49: 7-8 (1962)
CA 57: 214d

D 1192 Measurement of fluorescence and
phosphorescence yields and lifetimes
in dihalogenated acridines and fluor-
esceins.
Zanker, U., and Koerber, W.
Z. Angew. Phys. 14: 43-8 (1962)
CA 57: 1755e

D 1193 Preparation and properties of subli-
mated layers of luminescent zinc oxide
and sulfide.
Zelikin, Ya.
Pribory i Tekhn. Eksperim. 7: 130-2
(1962)
CA 57: 5432f

D 1194 Transitions and states of π electrons
in 9-substituted acridines.
Zanker, V., and Reichel, A.
Proc. Intern. Meeting Mol. Spectry. 4th,
Bologna, 1959 2: 596-610 (1962)
CA 59: 5997b

D 1195 Migration of energy between impurity
molecules in molecular crystals.
Zhevandrov, N.D., Gribkov, V.I., and
Khan-magometova, Sh.D.
Opt. i Spektroskopiya 13(1): 96-99
(1962)
CA 56: 1990a

D 1196 The relation between the luminescence
and absorption spectrum of some aro-
matic compounds.
Zurauskiene, E., and Vaicumas, S.
Lietuvos Fiz. Rinkinys, Lietuvos TSR
Mokslu Akad., Lietuvos TSR Aukstosios,
Mokyklos 1: 179-86 (1961) (1962)
CA 58: 12082e

D 1197 Conversion of a grating monochroma-
tor to a spectrograph.
Adelman, A.H., and Verber, C.M.
Rev. Sci. Instr. 34: 591-2 (1963)
CA 60: 11514f

D 1198 Two-photon excitation of phosphores-
cence and fluorescence.
Adelman, A.H., and Verber, C.M.
J. Chem. Phys. 39: 931-3 (1963)
CA 59: 5951d

D 1199 Vibronic-spin-orbit perturbations and
the assignment of the lowest triplet
state of benzene.
Albrecht, A.C.
J. Chem. Phys. 38: 354-65 (1963)
CA 58: 4060c

D 1200 Luminescence of activated LiF crys-
tals.
Alekeeva, E.P.
Izv. Vysshikh. Uchebn. Zavedenii, Fiz.
1963: 110-16 (1963)
CA 61: 192c

D 1201 The synthesis and study of the absorp-
tion of some naphthysilanes in the
ultraviolet region.
Allen, C.S.
U.S. Dept. Comm., Office Tech. Serv.,
AD 419,040, 120 pp. (1963)
CA 60: 14361g

D 1202 Immunofluorescence applied to pro-
tein solutions.
Allen, J.C.
J. Lab. Clin. Med. 62: 517-24 (1963)
CA 59: 15763e

D 1203 Relation between the absorption and
dispersion of light in concentrated
fluorescent solutions of dyes.
Al'perovich, L.I.
Opt. i Spektroskopiya 14: 756-9 (1963)
CA 60: 3612c

D 1204 Excitation energy transfer in rigid
solution of organic substances. II. III.
Andreeshehev, E.A., Baroni, E.E.,
Viktorova, V.S., Kovyrzina, K.A.,
Rozman, I.M., and Shoniya, V.M.
Opt. i Spektroskopiya, Akad. Nauk SSSR,
Otd. Fiz.-Mat. Nauk, Sb. Statei 1:
128-35 (1963)
CA 59: 9435e

D 1205 Kinetics of phosphorescence decay
upon α-excitation.
Antonov-Romanovskii, V.V.
Opt. i Spektroskopiya, Akad. Nauk SSSR,
Otd. Fiz.-Mat. Nauk, Sb. Statei 1:
274-80 (1963)
CA 59: 8235b

D 1206 Phosphorescence induced by anti-
Stokes excitation.
Aristov, A.V.
Opt. i Spektroskopiya 15: 284-6 (1963)
CA 59: 13495g

D 1207 Intramolecular quenching of fluores-
cence in solutions of organic compounds.
Aristov, A.V., and Sveshnikov, B.Ya.
Opt. i Spektroskopiya Akad. Nauk SSSR,
Otd. Fiz.-Mat. Nauk, Sb. Statei 1: 58-
60 (1963)
CA 59: 5955c

D 1208 Effect of temperature on the frequency of transition of the activator molecule in organoluminophors into the triplet state.
Aristov, A.V., and Sveshnikov, B.Ya.
Opt. i Spektroskopiya, Akad. Nauk SSSR, Otd. Fiz.-Mat. Nauk, Sb. Statei 1: 94-7 (1963)
CA 59: 5905c

D 1209 Existence of several metastable states of the activator molecule in organoluminophors.
Aristov, A.V., and Sveshnikov, B.Ya.
Opt. i Spektroskopiya 14: 732-4 (1963)
CA 59: 3413h

D 1210 Temperature influence on the transition probability of the molecule in the phosphorescent state.
Aristov, A.V., and Sveshnikov, B.Ya.
Izv. Akad. Nauk SSSR, Ser. Fiz. 27: 638-40 (1963)
CA 59: 8235a

D 1211 A fluorometric procedure for the determination of cerous ion.
Armstrong, W.A., Grant, D.W., and Humphreys, W.G.
Anal. Chem. 35: 1300 (1963)

D 1212 Symmetry considerations in the spectrum of ruby.
Artman, J.O., and Murphy, J.C.
J. Chem. Phys. 38: 1544-7 (1963)
CA 58: 10859h

D 1213 Fluorescence spectrum of di-cesium uranyl nitrate.
Asundi, R.K., and Dixit, R.M.
Nature 200: 668 (1963)
CA 60: 6360b

D 1214 Indirect observation of singlet-triplet absorption in anthracene.
Avakiau, P., Abramson, E., Kepler, R.G., and Caris, J.C.
J. Chem. Phys. 39: 1127-8 (1963)
CA 60: 132h

D 1215 Radiative transition probabilities within $4f^n$ configurations; the fluorescence spectrum of europium ethylsulfate.
Axe, J.D.
J. Chem. Phys. 39: 1154-60 (1963)
CA 59: 7088f

D 1216 Bivalent rare earth spectra selection rules and spectroscopy of $SrCl_2:Sm^{++}$.
Axe, J.D., and Sorokin, P.P.
Phys. Rev. 130: 945-52 (1963)
CA 58: 12087c

D 1217 Delayed fluorescence of solid solutions of polyacenes.
Azumi, T., and McGlynn, S.P.
J. Chem. Phys. 38: 2773-4 (1963)
CA 59: 10900d

D 1218 Delayed fluorescence of solid solutions of polyacenes. II. Kinetic considerations.
Azumi, T., and McGlynn, S.P.
J. Chem. Phys. 39: 1186-94 (1963)
CA 59: 9480e

D 1219 Delayed fluorescence of solid solutions of polyacenes. IV. The origin of excimer fluorescence.
Azumi, T., and McGlynn, S.P.
J. Chem. Phys. 39: 3533-4 (1963)
CA 60: 8749b

D 1220 Determination of the estrogenic activity of extracts from calendula officimalis flowers.
Banaszkiewiez, W., Kowalska, M., and Mrozikiewicz, A.
Poznan. Towarz. Przyjaciol Nauk, Wydzial Lekar. Prace Komisji Farm. 1: 53-63 (1963)
CA 61: 2365a

D 1221 Physical and photochemical properties of a fluorescent chlorophyll colloid.
Bannister, T.T., and Bernadini, J.E.
Photochem. Photobiol. 2: 535-49 (1963)
CA 60: 7068d

D 1222 Concentration quenching of Rhodamine 6 Zh solutions in alcohols.
Baranova, E.G., and Leushin, V.L.
Izv. Akad. Nauk SSSR, Ser. Fiz. 27: 554-7 (1963)
CA 59: 8263a

D 1223 Interaction of excited bromolecules with oxygen. II. Extinguishing of tryptophan and tyrosine X-ray fluorescence with oxygen and nitric acid.
Barenboim, G.M., and Domanskii, A.N.
Biofizika 8: 321-30 (1963)
CA 59: 7767c

D 1224 On the phosphorescence of deoxyribo-
nucleic acid (DNA).
Barsohn, R.
Biochem. Biophys. Res. Commun. 13:
205-8 (1963)
CA 59: 15530c

D 1225 Absorption and emission spectra of
nucleosides, purines, and pyrimidines.
Basu, S., and Greist, J.H.
J. Chim. Phys. 60: 407-11 (1963)
CA 59: 14757g

D 1226 Polarization and decay fluorescence
of solutions.
Bauer, R.K.
Z. Naturforsch. 18a: 718-24 (1963)
CA 59: 8277b

D 1227 Determination of number-average
molecular weights of polymers by
using fluorescence spectroscopy.
Baumbauch, D.O.
J. Polymer Sci. 1: 669-70 (1963)
CA 60: 8146e

D 1228 Fluorescence of NH radicals during
the photodissociation of NH_3 in the
vacuum ultraviolet.
Becker, K.A., and Welge, K.H.
Z. Naturforsch. 18a: 600-3 (1963)
CA 59: 7105e

D 1229 Metalloporphyrins. Electronic spec-
tra and nature of perturbations. I.
Transition metal ion derivatives.
Becker, R.S., and Allison, J.B.
J. Phys. Chem. 67: 2662-9 (1963)
CA 60: 6358a

D 1230 Photoluminescence of the minerals of
Madagascar.
Behier, J.
Bull. Acad. Malgache 39: 84-96 (1961)
(1963)
CA 61: 6779h

D 1231 S-S luminescence of benzene crystal-
lized pure at 4°K.
Benarroche, M.
Compt. Rend. 257: 1249-51 (1963)
CA 60: 132b

D 1232 Fluorescence and energy transfer in
phenanthrene crystals.
Benz, K.W., and Wolf, H.C.
Z. Naturforsch. 19a: 181-90 (1963)
CA 60: 15313g

D 1233 Localization, physicochemical proper-
ties, and action of phycocyanim in
anacystis nidulans.
Bergeron, J.A.
Natl. Acad. Sci. — Natl. Res. Council,
Misc. Publ. No. 1145, 527-36 (1963)
CA 60: 13576h

D 1234 Fluorescence decay times of rare-
earth chelates.
Bhaumik, M.L., Lyons, H.L., and
Fletcher, P.C.
J. Chem. Phys. 38: 568-9 (1963)
CA 58: 9769a

D 1235 Lattice work performed by excited
molecules.
Bhaumik, M.L., and Hardwick, R.
J. Chem. Phys. 39: 1595-8 (1963)
CA 59: 8279d

D 1236 Effect of the solvent on the electronic
spectrum of luminescent molecules.
Bilot, L., and Kawski, A.
Z. Naturforsch. 18a: 10-15 (1963)
CA 58: 13309e

D 1237 The delayed fluorescence of pyrene
solutions.
Birks, J.B.
J. Phys. Chem. 67: 2199-2200 (1963)
CA 59: 13499e

D 1238 The photodimerization and excimer
fluorescence of 9-methyl anthracene.
Birks, J.B., and Aladekomo, J.B.
Photochem. Photobiol. 2: 415-18 (1963)
CA 60: 8814a

D 1239 Excimer fluorescence. I. Solution
spectra of 1,2-benzanthracene deriva-
tives.
Birks, J.B., and Christophorou, L.G.
Proc. Roy. Soc., Ser. A 274: 552-64 (1963)
CA 59: 4698b

D 1240 Excimer formation in polycyclic
hydrocarbons and their derivatives.
Birks, J.B., and Christophorou, L.G.
Nature 197: 1064-65 (1963)
CA 59: 2620e

D 1241 The relations between the fluorescence
and absorption properties of organic
molecules.
Birks, J.B., and Dyson, D.J.
Proc. Roy. Soc., Ser. A 275: 135-48 (1963)
CA 59: 5952a

D 1242 Excimer fluorescence. II. Lifetime
studies of pyrene solutions.
Birks, J.B., Dyson, D.J., and Munro, I.H.
Proc. Roy. Soc., Ser. A 257: 135-48 (1963)
CA 59: 10900f

D 1243 Emission spectra of organic liquid
scintillators.
Birks, J.B., Geake, J.E., and Lumb, M.D.
Brit. J. Appl. Phys. 14: 141-3 (1963)
CA 58: 10868g

D 1244 Complex formation and fluorescence.
I. Complexes of 8-hydroxyquinoline-
5-sulfonic acid. II. The use of 8-
hydroxyquinoline-5-sulfonic acid as an
indicator. III. Salicylate complexes.
Bishop, J.A.
Anal. Chim. Acta 29: 172-84 (1963)
CA 59: 13347e

D 1245 Florida: better coagulation processes
for better water.
Black, A.P.
Water Works Eng. 116: 375-6, 414-5 (1963)
CA 59: 3649a

D 1246 Enhancement by copper of the visible
region fluorescence spectrum of zinc
oxide.
Blanchard, M.L., and Monod-Herzen, G.
Compt. Rend. 256: 4189-91 (1963)
CA 59: 4694h

D 1247 Fluorimetric determination of small
amounts of thallium.
Bock, R., and Zimmer, E.
Z. Anal. Chem. 198: 170-3 (1963)
CA 60: 4795f

D 1248 Fluorescence of vapors of complex
molecules.
Borisevich, N.A.
Izv. Akad. Nauk SSSR, Ser. Fiz. 27: 562-9
(1963)
CA 59: 9476d

D 1249 The effect of temperature, magnitude
of exciting quanta, and foreign gases on
the structural electronic spectra of
molecular vapors.
Borisevich, N.A., and Gruzinskii, V.V.
Dokl. Akad. Nauk Belorussk. SSR 7: 309-12
(1963)
CA 59: 13470f

D 1250 Application of the universal relationship between fluorescence and absorption spectra to the study of the excited
state of complex molecules in vapor
state.
Boriesevich, N.A., and Gruzinskii, V.V.
Opt. i Spektroskopiya 14: 39-44 (1963)
CA 58: 8524a

D 1251 Quantum yield of fluorescence of
molecular vapors.
Borisevich, N.A., and Tolkachev, V.A.
Dokl. Akad. Nauk Belorussk. SSR 7(2):
87-91 (1963)
CA 58: 13305h

D 1252 Determination of the degree of spiral-
ling in transport ribonucleic acids
from fluorescent properties of their
complexes with acridine dyes.
Borisova, O.F., Kiseler, L.L., and
Tumerman, L.A.
Dokl. Akad. Nauk SSSR 152: 1001-4 (1963)
CA 60: 5790f

D 1253 Effects of protanation, alkylation, and
cooling on the fluorescence of purine
nucleosides in liquid solution.
Borresen, H.C.
Acta Chem. Scand. 17: 2359-60 (1963)
CA 60: 7061d

D 1254 Fluorometric determination of tung-
sten with flavanol.
Bottei, R.S., and Trusk, B.A.
Anal. Chem. 35: 1910-12 (1963)
CA 60: 1109g

D 1256 Use of luminescent reagents in the
kinetic method of analysis.
Bozhevol'nov, E.A., Kreingol'd, S.V.,
Lastovskii, R.P., and Sidorenko, V.V.
Dokl. Akad. Nauk SSSR 153: 97-100 (1963)
CA 60: 7427g

D 1257 Increasing the sensitivity of lumines-
cent reactions for cations with organic
reagents by freezing the solutions.
Bozhevol'nov, E.A., and Solov'ev, E.A.
Dokl. Akad. Nauk SSSR 148: 335-7 (1963)
CA 58: 13109b

D 1258 Apparent concentration quenching of
morphine fluorescence.
Brandt, R., Olsen, M.J., and Cheronis,
N.D.
Science 139: 1063-4 (1963)
CA 58: 13304a

D 1259 Appearance of fluorescence on treatment of histidine residues with N-bromosuccinimide.
Brand, L., and Shaltiel, S.
Biochim. Biophys. Acta 75: 145-8 (1963)
CA 59: 7769c

D 1260 Internal conversion from upper electronic states to the first excited singlet state of benzene, toluene, p-xylene, and mesitylene.
Braun, C.L., Kato, S., and Lipsky, S.
J. Chem. Phys. 39: 1645-52 (1963)
CA 59: 9456d

D 1261 A new approach to unimolecular reaction rate theory and the effect of finite pulse lengths on transient fluorescence spectra and nuclear magnetic resonance studies of the triethylamine-iodine, pyridine-iodine, and benzene-iodine charge-transfer complexes.
Brauner, J.W.
Dissertation, University of Rochester (1963)
CA 60: 11509d

D 1262 Radiative lifetime of I_2 fluorescence.
Brewer, L., Berg, R.A., and Rosenblatt, G.M.
J. Chem. Phys. 38: 1381-8 (1963)
CA 58: 8524h

D 1263 Excited state ionization of pyridoxine and other compounds.
Bridges, J.W., Creaven, P.T., Davies, D.S., and Williams, R.T.
Biochim. J. 88, 65 pp. (1963)
CA 60: 10975e

D 1264 Anion and solvent dependence of an interionic electron exchange in ion pairs in N-(2,6-dichlorobenzyl) quinoline salts.
Briegleb, G., Jung, W., and Herre, W.
Z. Physik. Chem. 38: 253-63 (1963)
CA 60: 2381d

D 1265 Phosphorescence of o-phenanthroline.
Brinen, J.S., Rosebrook, D.D., and Hirt, R.C.
J. Phys. Chem. 67: 2651-5 (1963)
CA 60: 1246b

D 1266 Aggregated chlorophyll in vivo.
Brody, S.S., and Brody, M.
Natl. Acad. Sci. — Natl. Res. Council, Misc. Publ. No. 1145: 455-78 (1963)
CA 60: 13575e

D 1267 Rotational, vibrational, and electronic energy transfer in the fluorescence of nitric oxide.
Broida, H.P., and Carrington, T.
J. Chem. Phys. 38: 136-47 (1963)
CA 58: 3014b

D 1268 Simplified system for the measurement of fluorescence lifetimes by using the stroboscopic method.
Brown, G.C., Jr.
Rev. Sci. Instr. 34: 414-15 (1963)
CA 59: 5953c

D 1269 Use of PAN film EFKE 25 in professional and scientific work.
Broz, I., and Weber, K.
Kem. Ind. (Zagreb) 12: 591-6 (1963)
CA 60: 4991e

D 1270 The nature of ultraviolet fluorescence of cells.
Brumberg, E.M., Barskii, I.Ya., Chernogryadskaya, N.A., and Shudel, M.S.
Dokl. Akad. Nauk SSSR 150: 1356-8 (1963)
CA 59: 10548a

D 1271 Properties of metallic photocathodes and of fluorescent films in the far ultraviolet.
Brunet, M., Cantin, M., Julliot, C., and Vasseur, J.
J. Phys. Radium, Suppl. 24: 53A-59A (1963)
CA 59: 5953g

D 1272 Concentrations extinguishing in phosphorescence of dibasic organic acids.
Bruns, S.A.
Nauchn. Soobshch., Inst. Gorn. Dela., Akad. Nauk SSSR 19: 66-75 (1963)
CA 61: 7847g

D 1273 The true fluorescence decay time of solutions.
Budo, A., and Szalay, L.
Z. Naturforsch. 18a: 90-1 (1963)
CA 58: 13306d

D 1274 Parameter computation for the concentration dependence of the fluorescence of non-photoconductive crystalline luminophors.
Buhrow, J.
Czech. J. Phys. 13: 161-4 (1963)
CA 59: 4641h

D 1275 Effect of light intensity on the far-red
inhibition of chlorophyll a fluorescence
in vivo.
Butler, W.L.
Biochim. Biophys. Acta 66: 275-6 (1963)
CA 58: 14358f

D 1276 Low-temperature spectra of chloro-
plast fragments.
Butler, W.L., and Baker, J.E.
Biochim. Biophys. Acta 66: 206-11 (1963)
CA 58: 10865f

D 1277 Action of two-pigment system on
fluorescence yield of chlorophyll a.
Butler, W.L., and Bishop, N.I.
Natl. Acad. Sci. — Natl. Res. Council,
Misc. Publ. No. 1145: 91-100 (1963)
CA 60: 16216d

D 1278 Lifetime of the long-wavelength
chlorophyll fluorescence.
Butler, W.L., and Norris, K.H.
Biochim. Biophys. Acta 66: 72-7 (1963)
CA 58: 7073d

D 1279 Action of protamine sulfate on reserve
leukocytes.
Capri, A., Mastroeni, P., and Orecchio, F.
Boll. Soc. Ital. Biol. Sper. 39: 1178-81
(1963)
CA 60: 9797a

D 1280 Fluorescence and stimulated emission
from trivalent equopium in yttrium
oxide.
Chang, N.C.
J. Appl. Phys. 34: 3500-4 (1963)
CA 60: 133h

D 1281 Proposal for standardization of meth-
ods for reporting fluorescence emis-
sion spectra.
Chapman, J.H., Kortom, G., Lippert, E.,
Melhuish, W.H., Nebbia, G., and
Parker, C.A.
Z. Anal. Chem. 197: 431-3 (1963)
CA 59: 14762g

D 1282 Spectral study of the interaction be-
tween anthracene derivatives and mono-
mers in the process of polymerization.
II. Interaction between anthracene and
styrene.
Cherkasov, A.S., and Voldaikina, K.G.
Vysokomolekul. Soedin 5: 79-86 (1963)
CA 59: 8875g

D 1283 Spectral study of the interaction between anthracene derivatives with
monomers in the process of polymerization. III. Interaction of 1-vinyl- and
2-vinylanthracene with styrene.
Cherkasov, A.S., and Voldaikina, K.G.
Vysokomolekul. Soedin. Karbotseknye
Vysokomolekul. Soedin., Sb. Statei
1963: 170-8 (1963)
CA 61: 5768e

D 1284 Absorption and fluorescence of vinyl-
anthracenes and the change of molecule
configuration in the excited state.
Cherkasov, A.S., and Voldaikina, K.G.
Izv. Akad. Nauk SSSR, Ser. Fiz. 27: 628-33
(1963)
CA 59: 7057d

D 1285 Polarization spectra of phycoerythrin
fluorescence.
Chernitskii, E.A., and Konev, S.V.
Biofizika 8: 561-8 (1963)
CA 60: 795b

D 1286 Polarization of fluorescence and phos-
phorescence of tryptophan and indole.
Chernitskii, E.A., Konev, S.V., and
Bobrovich, V.P.
Dokl. Akad. Nauk Belorussk. SSR 7: 628-32
(1963)
CA 60: 3612g

D 1287 Nature of s-tetrazine emission
spectra.
Chowdhury, M., and Goodman, L.
J. Chem. Phys. 38: 2979-85 (1963)
CA 59: 2300c

D 1288 Fluorescence studies on soluble ribo-
nucleic acid (s-RNA) labeled with
acriflavine.
Churchich, J.E.
Biochim. Biophys. Acta 75: 274-6 (1963)
CA 60: 789f

D 1289 The fluorescent response of NaI (Tl)
and CsI (Tl) to X-rays and γ-rays.
Collinson, A.J.L., and Hill, R.
Proc. Phys. Soc. 81: 883-92 (1963)
CA 58: 12084a

D 1290 The fluorescence of tyrosine in alka-
line solution.
Cornog, J.L., Jr., and Adams, W.R.
Biochim. Biophys. Acta 66: 356-65 (1963)
CA 59: 1861h

D 1291 Fluorescence of a chromium-doped
single crystal of $AlLaO_3$ perovskite.
Couture, L., Brunetiere, F., Forrat, F.,
and Trevoux, P.
Compt. Rend. 256: 3046-9 (1963)
CA 59: 128b

D 1292 Fluorescence and the structure of
proteins. I. Effects of substituents on
the fluorescence of indole and phenol
compounds.
Cowgill, R.W.
Arch. Biochem. Biophys. 100: 36-44 (1963)
CA 58: 8162b

D 1293 Fluorescence and the structure of
proteins. II. Fluorescence of peptides
containing tryptophan or tyrosine.
Cowgill, R.W.
Biochim. Biophys. Acta 75: 272-3 (1963)
CA 60: 132f

D 1294 Spectrophotofluorometric studies of
some aromatic aldehydes and their
acetals.
Crowell, E.P., and Varsel, C.J.
Anal. Chem. 35: 189-92 (1963)
CA 58: 8409c

D 1295 Effect of an external electric field on
the fluorescence of molecules. II. Determination of dipole moments of excited molecules.
Czekalla, J., Liptay, W., and Meyer, K.O.
Ber. Bunsenges. Physik. Chem. 67: 465-70
(1963)
CA 59: 4640g

D 1296 Daylight fluorescent pigments.
D'Alelio, G.F., and Voedisch, R.W.
U.S. 3,116,256, 1961 (1963)
CA 60: 8235e

D 1297 Daylight fluorescent coatings.
Dane, C.D.
Prod. Finishing (London) 16: 46, 77-8
(1963)
CA 60: 7027b

D 1298 Bleaching of viscose staple fibers
with optical brightening agents.
Darvina, V.V., and Makarova, T.P.
Khim. Volokna 1963: 38-9 (1963)
CA 59: 11715c

D 1299 Additional fluorimetric studies of
olive and olive-husk oils.
DeGori, R., Grandi, F., and Cantagalli, P.
Boll. Lab. Chim. Provinciali (Bologna) 14:
455-9 (1963)
CA 60: 14743e

D 1300 A study of M-center formation in add-
itively colored KCl.
Delbecq, C.J.
Z. Physik 171: 560-81 (1963)
CA 58: 13252c

D 1301 Spectra and energy levels of Eu^{+3} in
$LaCl_3$.
DeShazer, L.G., and Dieke, G.H.
J. Chem. Phys. 38: 2190-9 (1963)
CA 58: 12075h

D 1302 Fluorometric and spectrophotometric
determination of magnesium with
O,O'-dihydroxyazobenzene.
Diehl, H., Olsen, R., Spielholtz, G.I., and
Jensen, R.
Anal. Chem. 35: 1144-54 (1963)
CA 59: 9290g

D 1303 Detection of cancerogenic hydrocarbons from fine structure fluorescent
spectra.
Dickun, P.P.
Acta, Unio Intern. Contra Cancrum 19:
484-5 (1963)
CA 61: 8726e

D 1304 Nature of quenching fluorescence of
chlorophyll by oxidizing and reducing
reagents.
Dilung, I.I., and Chernyuk, I.N.
Zh. Fiz. Khim. 37: 1100-5 (1963)
CA 59: 4694b

D 1305 Fluorescence decay in solutions of optical bleaching agents and its influence on the brightening effect.
Divatia, A.S., and Vaidya, B.K.
Indian J. Technol. 1: 109-11 (1963)
CA 59: 797b

D 1306 Luminescence from the H-O_2 reaction.
Dixon, R.N., and Mason, B.F.
Nature 197: 1198 (1963)
CA 59: 130f

D 1307 Processes involved in intermolecular electron transfer under pulsed illumination.
Dmitrievskii, O.D., and Terenin, A.N.
Dokl. Akad. Nauk SSSR 151: 122-4 (1963)
CA 59: 12298c

D 1308 The variation within a definite time of the electron density in the phosphorescence of N plasma.
Doebler, F.
Z. Naturforsch. 18a: 431 (1963)
CA 59: 4626e

D 1309 The polarization of the triplet-singlet-phosphorescence of some aryl and heterocyclic compounds. II. Quinoline, isoquinoline, fluorene, chrysene, triphenylene, dibenzoquinoxaline, 1,2-3,4-dibenzophenazine, coronene.
Doerr, F., and Gropper, H.
Ber. Bunsenges. Physik. Chem. 67: 193-201 (1963)
CA 58: 13310a

D 1310 Effect of heavy substituents on the polarization of the triplet-singlet emission. Monohalide derivatives of naphthalene.
Doerr, F., Gropper, H., and Mika, N.
Z. Naturforsch. 18a: 1025-6 (1963)
CA 59: 14729h

D 1311 Liquid and solid sodium-silicon solutions (waterglass) as solvents of luminescent dyes.
Drabent, R., and Drabent, Z.
Bull. Acad. Polon. Sci., Ser. Sci., Math., Astron. Phys. 11: 415-19 (1963)
CA 60: 2461c

D 1312 Sensitized fluorescence after light-releases energy transfer through thin layers.
Drexhage, K.H., Zwick, M.M., and Kuhn, H.
Ber. Bunsenges. Physik. Chem. 67(1): 67-70 (1963)
CA 58: 13307a

D 1313 Correction of luminescence spectra and calculation of quantum efficiencies using computer techniques.
Drushel, H.V., Sommers, A.L., and Cox, R.C.
Anal. Chem. 35: 2166 (1963)

D 1314 Radiative lifetime of triplet biacetyl.
Dubois, J.T.
J. Chem. Phys. 39: 899-901 (1963)
CA 59: 5951g

D 1315 Singlet-singlet energy transfer in fluid solutions.
Dubois, J.T., and Cox, M.
J. Chem. Phys. 38: 2536-41 (1963)
CA 58: 13310g

D 1316 Triplet state of benzene.
Dubois, J.T., and Wilkinson, F.
J. Chem. Phys. 38: 2541-7 (1963)
CA 58: 13311a

D 1317 Leukoconcentration (staining of leukocytes) by the fluorescence produced with acridine orange in leukemias.
Duheille, J., Herbeuval, R., and Herbeuval, H.
Ann. Med. Nancy 2: 353-60 (1963)
CA 59: 9180c

D 1318 Phosphorescence spectra of cement and organo-oxide luminophors.
Duorovenko, V.K.
Opt. i Spektroskopiya Akad. Nauk SSSR, Otd. Fiz.-Mat. Nauk Sb Statei 1: 151-5 (1963)
CA 59: 10867b

D 1319 The fluorescence of an excited triplet state of dilute 9,10-benzophenanthrene in hexane at 77°K.
Dupoy, F., Nouchi, G., and Rousset, A.
Compt. Rend. 256: 2976-9 (1963)
CA 59: 131c

D 1320 Role of two photosynthetic pigment
systems in cytochrome oxidation, pyridine nucleotide reduction, and fluorescence.
Duysens, L.N.M.
Proc. Roy. Soc., Ser. B 157: 301-13 (1963)
CA 58: 14444b

D 1321 Mechanism of two photochemical re-
actions in algae as studied by means of
fluorescence.
Duysens, L.N.M., and Sweers, H.E.
Studies Microalgae Photosyn. Bacteria,
Collection Papers 1963: 353-72 (1963)
CA 61: 7367g

D 1322 Fluorometric determination of selenium in plants and animals with
3,3'-diaminobenzidine.
Dye, W.B., Bretthauer, E., Scim, H.J.,
and Blincoe, C.
Anal. Chem. 35: 1687-93 (1963)

D 1323 Observation of anticrossings in optical
resonance fluorescence.
Eck, T.G., Toldy, L.L., and Wieder, H.
Phys. Rev. Letters 10: 239-42 (1963)
CA 58: 13306e

D 1324 Identification and determination of
polycyclic aromatic hydrocarbons by
fluorescence spectra of solid solutions
at low temperatures.
Eichhoff, H.J., and Koehler, M.
Z. Anal. Chem. 197: 271-83 (1963)
CA 60: 29d

D 1325 Directions of development of lumines-
cence analysis in the last 30 years.
Eisenbrand, J.
Z. Anal. Chem. 192: 83-91 (1963)
CA 58: 8390g

D 1326 Spectrophotometric investigation of
the fluorescence intensities of white
and yellow petrolatum and ceresin.
Eisenbrand, J.
Deut. Apotheker-Ztg. 103: 623-5 (1963)
CA 59: 4696a

D 1327 Triplet-triplet transfer in plastic at
298° and 77°K.
Eisenthal, K.B., and Murashige, R.
J. Chem. Phys. 39: 2108-9 (1963)
CA 59: 14715d

D 1328 Physicochemical changes in pyrimidine and purine bases in an ultrasonic
field with formation of a series of
fluorescing substances.
Elpiner, I.E., and Sokolovskaya, A.V.
Dokl. Akad. Nauk SSSR 153: 200-3 (1963)
CA 60: 7060e

D 1329 Origin of the phosphorescence radia-
tion in aromatic hydrocarbons.
El-Sayed, M.A.
Nature 197: 481-2 (1963)
CA 58: 12082g

D 1330 Polarization of molecular lumines-
cence in plastic media by the method of
photoselection.
El-Sayed, M.A.
J. Opt. Soc. Am. 53: 797-800 (1963)
CA 59: 5902f

D 1331 Spin-orbit coupling and the radiation-
less processes in nitrogen heterocyc-
lics.
El-Sayed, M.A.
J. Chem. Phys. 38: 2834-8 (1963)
CA 59: 1206d

D 1332 Polarization of the $\pi^* \to \pi$ and $\pi^* \to n$
phosphorescence spectra of N-hetero-
cyclics.
El-Sayed, M.A., and Brewer, R.G.
J. Chem. Phys. 39: 1623-8 (1963)
CA 59: 12316b

D 1333 Intramolecular heavy-atom effect on
the polarization of naphthalene phos-
phorescence.
El-Sayed, M.A., and Pavlopoulas, T.
J. Chem. Phys. 39: 1899-1900 (1963)
CA 59: 14730b

D 1334 Histochemical demonstration of
fluorogenic amines in the cytoplasm of
sympathetic ganglion cells of the rat.
Eranko, O., and Harkonen, M.
Acta Physiol. Scand. 58: 285-6 (1963)
CA 60: 2126e

D 1335 Inductive-resonance energy transfer
from aromatic molecules in the triplet
state.
Ermolaev, V.L., and Sveshnikova, E.V.
Dokl. Akad. Nauk SSSR 149: 1295-8 (1963)
CA 59: 4665h

D 1336 The effect of pressure on the
quenching of fluorescence.
Ewald, A.H.
J. Phys. Chem. 67: 1727-8 (1963)
CA 59: 9435d

D 1337 Infrared natural phosphorescence and
X-ray luminescence of NaCl, KCl, and
KBr crystals at room temperature.
Ezhik, I.I.
Opt. i Spektroskopiya, Akad. Nauk SSSR,
Otd. Fiz.-Mat. Nauk, Sb. Statei 1: 175-7
(1963)
CA 59: 7061a

D 1338 Ultraviolet absorption of fluorescent
mixtures of sodium β-naphthoquinone-
4-sulfonate and several substances
with a quanidine function (streptomycin
and dihydrostreptomycin).
Faure, F., and Blanquet, M.P.
Bull. Soc. Pharm. Bordeaux 102: 79-88
(1963)
CA 59: 13484h

D 1339 Magnetic dipole character of the
$^3A_{2g} \leftarrow ^3T_{2g}$ transition in octahedral
nickel (II) compounds.
Ferguson, J., Guggenheim, H.J., Johnson,
L.F., and Kamimora, H.
J. Chem. Phys. 38: 2579-80 (1963)
CA 59: 5932g

D 1340 Determination of micro quantities of
aluminum.
Flavia, J.
Anales Asoc. Quim. Arg. 51: 96-122 (1963)
CA 60: 1096h

D 1341 Study of multicomponent mixtures in
solution with a vertical-axis trans-
mission-type filter-fluorometer.
Fletcher, M.H.
Anal. Chem. 35: 278-88 (1963)
CA 58: 11931e

D 1342 A vertical-axis transmission-type
filter-fluorometer for solutions.
Fletcher, M.H.
Anal. Chem. 35: 288 (1963)

D 1343 Fluorescence effects in ion-exchange
resins.
Flint, T.R., and Eichholz, G.G.
Can. J. Chem. Eng. 41: 33-7 (1963)
CA 58: 14209f

D 1344 Triplet phosphorescence and electron
spin resonance (E.S.R.) absorption of
some organic molecules in glassy
solutions.
Foerster, G.V.
Z. Naturforsch. 18a: 620-6 (1963)
CA 59: 5969g

D 1345 Concentration inversion of the fluor-
escence in solution at higher pressures.
Foerster, T., Leiber, C.O., Seidel, H.P.,
and Weller, A.
Z. Physik. Chem. 39: 265-9 (1963)
CA 60: 7584c

D 1346 Crystallographic and optical proper-
ties of LaClO$_3$.
Forrat, F., Jansen, R., and Trevoux, P.
Compt. Rend. 256: 1271-4 (1963)
CA 58: 9693c

D 1347 Phosphorescence in plastics.
Forster, C.F., and Rickard, E.F.
Nature 197: 1199-1200 (1963)
CA 59: 808h

D 1348 The yield of fluorescence of chloro-
phyll a.
Frackowiak, D.
Bull. Acad. Polon. Sci., Ser. Sci., Math.,
Astron. Phys. 11: 561-6 (1963)
CA 60: 10078h

D 1349 Yield of anti-Stokes fluorescence of
chlorophyll.
Frackowiak, D., and Marszalek, T.
Bull. Acad. Polon. Sci., Ser. Sci., Math.,
Astron. Phys. 9: 53-5 (1963)
CA 59: 131e

D 1350 Fluorescent screens for electron-
optical image transformers.
Franz, K.
Ger. 1,159,568, (1960) (1963)
CA 60: 7563h

D 1351 Energy migration in rare earth
complexes.
Freeman, J.J.
U.S. At. Energy Comm. SC-DC-3564,
219 pp. (1963)
CA 61: 11436c

D 1352 Color reactions and fluorescence
reactions of digitalis compounds.
Frerejacquet, M., and De Graeve, P.
Ann. Pharm. Franc. 21: 209-28 (1963)
CA 59: 12591b

D 1353 Fluorescence, energy transfer, and
SH groups in photosynthetic pigments
of red and blue-green algae.
Fujimori, E.F., and Quinlan, K.
Natl. Acad. Sci. — Natl. Res. Council,
Misc. Publ. No. 1145: 519-26 (1963)
CA 60: 13576e

D 1354 Fluorescence. Intermediate transfer
of excitation in xylene-base scintilla-
tors.
Furst, M., Koechlin, Y., and Raviart, A.
Compt. Rend. 256: 3639-42 (1963)
CA 59: 3429h

D 1355 The universal fluorescence of poly-
mers. I. Qualitative results.
Gachkovskii, V.F.
Zh. Strukt. Khim. 4: 424-32 (1963)
CA 59: 8892e

D 1356 Labeling by sublimable fluorescent
compounds.
National Cash Register Co.
Ger. 1,150,540 (1963)
CA 60: 1222e

D 1357 Mechanisms involved in the production
of red fluorescence of human and ex-
perimental tumors.
Ghadially, F.N., Neish, W.J.P., and
Dawkins, H.C.
J. Pathol. Bacteriol. 85: 77-92 (1963)
CA 58: 14546e

D 1358 Structure and semiconductor proper-
ties of thin anthracene layers. I.
Structure of anthracene layers obtained
by thermal evaporation in vacuum.
Gheorghita-Oancea, C.
Rev. Phys., Acad. Rep. Populaire
Roumaine 8: 361-75 (1963)
CA 60: 13980b

D 1359 The study of electronic spectra of
organic molecules in diffusing media.
Gladchenko, L.F., and Pikulik, L.G.
Vestsi Akad.Navuk Belarusk. SSR, Ser.
Fiz.-Tekhn. Navuk 1963: 53-8 (1963)
CA 59: 4695e

D 1360 Fluorometer cuvette adapters for
measurements of 0.05- or 0.01-milli-
liter volumes.
Glick, D., and Von Redlick, D.
Anal. Biochem. 6: 471-4 (1963)
CA 60: 7118h

D 1361 Influence of neutral salts on the photo-
luminescence of various ionic forms of
fluorescein.
Glowacki, J., and Kaminsko, V.
Acta Phys. Polon. 23: 43-51 (1963)
CA 60: 8796g

D 1362 Dependence of the quenching of fluor-
escence of dyes on the position of the
maxima of fluorescence and absorption.
Glowacki, J.
Acta Phys. Polon. 24: 555-6 (1963)
CA 61: 6544f

D 1363 Dependence of the quenching constant
of fluorescence on the polarizability
and diamagnetic susceptibility of
quenchers.
Glowacki, J.
Bull. Acad. Polon. Sci., Ser. Sci., Math.,
Astron. Phys. 11: 487-92 (1963)
CA 60: 2461e

D 1364 Spectroscopy of frozen crystalline
solutions of diphenyl polyenes and
polyphenols. II.
Gobov, G.U.
Opt. i Spektroskopiya 15: 362-70 (1963)
CA 60: 129h

D 1365 Fluorescence of synthetic LiF crys-
tals.
Georlich, P., Karras, H., and Koetitz, G.
Phys. Status Solidi 3: 1803-18 (1963)
CA 60: 10078e

D 1366 Excitation of the spontaneous fluores-
cence of synthetic CaF_2/U^{+3} crystals.
Goerlich, P., Karras, H., and Koetitz, G.
Phys. Status Solidi 3: 1935-40 (1963)
CA 60: 7535f

D 1367 Determination of the spectral compo-
sition of the Gudden-Pohl flash in phos-
phors based on zinc sulfide.
Gol'dman, A.G., and Proskuva, A.I.
Dokl. Akad. Nauk SSSR 149: 567-70 (1963)
CA 59: 4642d

D 1368 Theory of the polarization of fluores-
cence of macromolecules with a mobile
emitter group around a rotation axis.
Gotlib, Yu.Ya., and Wahl, P.
J. Chim. Phys. 60: 849-56 (1963)
CA 59: 13497b

D 1369 Quenching cross sections for the extinction of the potassium fluorescence
radiation as a function of the relative
velocity of the colliding particles.
Satzke, J.
Z. Physik. Chem. 223: 321-6 (1963)
CA 60: 2475a

D 1370 The tissue affinity of one hundred
forty-eight fluorescent organic com-
pounds.
Gouaze, A., and Caslaing, J.
Compt. Rend. 257: 4230-3 (1963)
CA 60: 11229f

D 1371 Emerson enhancement effect and two
light reactions in photosynthesis.
Govindjee, R.
Natl. Acad. Sci. — Natl. Res. Council,
Misc. Publ. No. 1145, 318-34 (1963)
CA 60: 14834e

D 1372 Fluorescence of pyrazoles under
ultraviolet light.
Grandberg, I.I., Tabak, S.V., and Kost,
A.N.
Zh. Obshch. Khim. 33: 525-33 (1963)
CA 59: 1616c

D 1373 Fluorescent end-group reagent for
proteins and peptides.
Gray, W.R., and Hartley, B.S.
Biochem. J. 89, 59 pp. (1963)
CA 60: 9507b

D 1374 Fluorescence and light absorption by
the exciton in pure CdS.
Grillot, E., and Bracie-Grillot, M.
Phys. Status Solidi 3: 229-49 (1963)
CA 59: 13497e

D 1375 Excitation and ionization of nitrogen
by degrees.
Gromova, I.I., and Yakavleva, A.V.
Izv. Akad. Nauk SSSR, Ser. Fiz. 27:
1097-101 (1963)
CA 60: 7584d

D 1376 The orientation of the optical transition moments in phenanthrene and its
derivatives.
Gropper, H., and Doerr, F.
Ber. Bunsenges. Physik. Chem. 67: 46-54
(1963)
CA 58: 10869e

D 1377 Long-wavelength excitation of the
multiband edge of luminescence in GaP
crystals.
Gross, E.F., and Medzvetskii, D.S.
Dokl. Akad. Nauk SSSR 152: 1335-8 (1963)
CA 60: 7586h

D 1378 The study of edge luminescence ex-
citation spectra in copper halides.
Gross, E.F., and Shekhamet'ev, R.I.
Fiz. Tverd. Tela 5: 502-5 (1963)
CA 58: 12082a

D 1379 Spectrum of tripositive neptumium in
single crystals of lanthanun trichloride.
Gruber, J.B.
J. Inorg. Nucl. Chem. 25: 1093-5 (1963)
CA 59: 13465b

D 1380 Application of a universal relation to
structured fluorescence and absorption
spectra of vapors of aromatic molecules.
Gruzinskii, V.V.
Izv. Akad. Nauk SSSR, Ser. Fiz. 27: 580-3
(1963)
CA 59: 7088e

D 1381 Application of the universal relation
between fluorescence and absorption
spectra to the study of the excited
state of complex molecules in the vapor
state. II. Structural spectra.
Gruzinskii, V.V., and Borisevich, N.A.
Opt. i Spektroskopiya 15: 457-63 (1963)
CA 61: 2615a

D 1382 Spectral investigation of easine in
methyl polymethacrylate.
Grzywacz, J.
Bull. Acad. Polon. Sci., Ser. Sci., Math.,
Astron. Phys. 11: 347-50 (1963)
CA 60: 129b

D 1383 Emission and polarization spectra of
eosin.
Grzywacz, J., and Pohoski, R.
Bull. Acad. Polon. Sci., Ser. Sci., Math.,
Astron. Phys. 11: 573-6 (1963)
CA 60: 8776c

D 1384 New data on the relation between the degree of polarization and the wave-length of fluorescence.
Gurinovich, G.P., Sarzhevskii, A.M., and Sevchonko, A.N.
Opt. i Spektroskopiya 14: 809-12 (1963)
CA 59: 7087c

D 1385 Radiationless triplet-singlet transi-tions in naphthalene.
Hadley, S.G., Rast, H.E., Jr., and Keller, R.A.
J. Chem. Phys. 39: 705-11 (1963)
CA 59: 4697f

D 1386 Study of anthracene fluorescence ex-cited by the ruby giant pulse laser.
Hall, J.L., Jennings, D.A., and McClintock, R.M.
Phys. Rev. Letters 11: 364-6 (1963)
CA 60: 2471a

D 1387 Double-acceptor fluorescence in II-VI compounds.
Halsted, R.E., and Segall, B.
Phys. Rev. Letters 10: 392-5 (1963)
CA 59: 5900d

D 1388 A new family of self-activated phos-phors.
Harrison, D.E., Melamed, T., and Subbarao, E.C.
J. Electrochem. Soc. 110: 23-8 (1963)
CA 58: 3997g

D 1389 Ascorbic acid-induced fluorescence of a noradrenaline oxidation product.
Harrison, W.H.
Biochim. Biophys. Acta 78: 705-10 (1963)
CA 60: 5836h

D 1390 Detection of intermediate oxidation states of adrenaline and noradrenaline by fluorescence spectrometric analysis.
Harrison, W.H.
Arch. Biochem. Biophys. 101: 116-30 (1963)
CA 59: 881d

D 1391 Phosphorescence of naphthalene-doped crystals of durene.
Hayakawa, S., and Nakamura, T.
J. Phys. Soc. Japan 18: 531-5 (1963)
CA 59: 105d

D 1392 Prototropic equilibria of electronically excited molecules. I. 3-hydroxyquino-line.
Haylock, J.C., Mason, S.F., and Smith, B.E.
J. Chem. Soc. 1963: 4897-4901 (1963)
CA 59: 12611e

D 1393 The fluorescence of carbon disulfide vapor.
Heicklen, J.
J. Am. Chem. Soc. 85: 3562-5 (1963)
CA 60: 133h

D 1394 Organic liquid scintillators. V. Scin-tillation and fluorescent properties of several anthracene derivatives.
Heller, A., and Rio, G.
Bull. Soc. Chim. France 1963: 1707-9 (1963)
CA 60: 3631g

D 1395 Measurement of fluorescence spectra of gaseous scintillators excited by low-intensity α-radiation.
Henck, R., and Coche, A.
J. Phys. 24: 166-7 (1963)
CA 59: 7987a

D 1396 Color and fluorescence of cyclic sili-con compounds. III. The band system of siloxene.
Hengge, E., and Pretzer, K.
Ber. 96: 470-7 (1963)
CA 59: 1264a

D 1397 Mechanisms of photoreactions in solu-tions. XIV. Stereoisomeric triplet states of an α-diketone.
Herkstroeter, W.G., Saltiel, J., and Hammond, G.S.
J. Am. Chem. Soc. 85: 482-3 (1963)
CA 58: 12105b

D 1398 Phosphorescence in alkali halides containing lattice defects.
Hersh, H.N., and Hadley, W.B.
Phys. Rev. Letters 10: 437-8 (1963)
CA 59: 13496c

D 1399 Steady-state unimolecular processes in multilevel systems.
Hoare, M.
J. Chem. Phys. 38: 1630-5 (1963)
CA 58: 10768g

D 1400 Fluorescence techniques in the microdetermination of metals in biological
materials. Utility of 2,4-bis-[N,N'-di-
(carboxymethyl)aminomethyl]fluores-
cein in the fluorometric estimation of
Al^{+3}, alkaline earths, Co^{+2}, Cu^{+2}, Ni^{+2},
and Zn^{+2} in micromolar concentrations.
Hoelzl Wallach, D., and Steck, T.L.
Anal. Chem. 35: 1035 (1963)

D 1401 The systems $BaO-MgO-P_2O_5$ and BaO-
$ZnO-P_2O_5$ compounds and fluorescence.
Hoffman, M.V.
J. Electrochem. Soc. 110: 1223-7 (1963)
CA 60: 49h

D 1402 Measurement of the fluorescence
yields of the L shell.
Hohmuth, K., Mueller, G., and
Schintlmeister, J.
Nucl. Phys. 48: 209-24 (1963)
CA 59: 10901b

D 1403 Direct evidence for energy transfer
between rare earth ions in terbium-
europium tungstates.
Holloway, W.W.
Phys. Rev. Letters 11: 458-60 (1963)
CA 59: 3629d

D 1404 Anomalous shifts in the fluorescence
of MnF_2 and $KMnF_3$.
Holloway, W.W., Kestigian, M., Newman,
R., and Prohofsky, E.W.
Phys. Rev. Letters 11: 82-4 (1963)
CA 59: 7057g

D 1405 Spectra and apparent association constants of the fluorescent complexes of
some salicylaldehyde derivatives.
Holzbecker, Z.
Collection Czech. Chem. Commun. 28:
716-27 (1963)
CA 58: 13309g

D 1406 Pair-spectra in GaP.
Hopfield, J.J., Thomas, D.G., and
Gershenzon, M.
Phys. Rev. Letters 10: 162-4 (1963)
CA 58: 10868d

D 1407 The system $CaO-SiO_2-MnO$.
Hueniger, M., and Ruffler, H.
Tech.-Wiss. Abhandl. Osram-Ges. 8:
41-55 (1963)
CA 61: 5183c

D 1408 Fluorescence and pre-ionization in
nitrogen excited by vacuum ultraviolet
radiation.
Huffman, R.E., Tanaka, Y., and Larrabee,
J.C.
J. Chem. Phys. 38: 1920-6 (1963)

D 1409 The photo-initiation of the polymeriza-
tion of vinyl compounds by means of
cerous ions.
Hussain, F., and Norrish, R.G.W.
Proc. Roy. Soc. 275: 161-74 (1963)
CA 59: 8873d

D 1410 Paramagnetic resonance of phosphor-
escent organic molecules.
Hutchinson, C.A.
Record Chem. Progr. (Kresge-Hooker Sci.
Lib.) 24: 105-27 (1963)
CA 60: 101h

D 1411 The application of electron magnetic
resonance techniques to the study of the
phosphorescence of organic molecules.
Hutchinson, C.A.
Proc. Colloq. Spectros. Intern., 10th,
Univ. Maryland 1962: 681-705 (1963)
CA 61: 10206b

D 1412 Exciton processes in molecular crys-
tals. II. Fluorescence quenching in
mixed crystals.
Iguchi, K.
J. Phys. 24: 209-15 (1963)
CA 59: 8229a

D 1413 Resins from formaldehyde. LXXVII.
Synthesis of some anthracene deriva-
tives as a wavelength shifter for plastic
scintillators.
Imoto, M., and Nakaya, T.
Bull. Chem. Soc. Japan 36: 785-8 (1963)
CA 59: 13901a

D 1414 Nutritional studies on royal jelly. III.
Fluorescent substances and kynurenine
contents in royal jelly.
Ishiguro, I., Naito, J., Takatori, K., and
Haranda, J.
Gifu Yakka Daigaku Kiyo 13: 10-14 (1963)
CA 60: 14883g

D 1415 New fluorometric micromethod for the
simultaneous determination of digitoxin
and digoxin.
Jakovljevic, I.M.
Anal. Chem. 35: 1513-17 (1963)

D 1416 Optical maser characteristics of rare-earth ions in crystals.
Johnson, L.F.
J. Appl. Phys. 34: 897-909 (1963)
CA 58: 8507b

D 1417 Optical maser oscillation from Ni^{+2} in MgF_2 involving simultaneous emission of phonons.
Johnson, L.F., Dietz, R.E., and Guggenheim, H.J.
Phys. Rev. Letters 11: 318-20 (1963)
CA 60: 149e

D 1418 L-shell fluorescence yields in heavy elements.
Jopson, R.C., Mark, H., Swift, C.D., and Williamson, M.A.
Phys. Rev. 131: 1165-9 (1963)
CA 59: 7089g

D 1419 Triplet-energy transfer and triplet-triplet interaction in aromatic crystals.
Jortner, J., Choi, S.I., Kotz, J.L., and Rice, S.A.
Phys. Rev. Letters 11: 323-6 (1963)
CA 60: 10078c

D 1420 Wavelength analysis of fluorescence from gases excited by vacuum ultraviolet radiation.
Judge, D.L., Morse, A.L., and Weissler, G.L.
Compt. Rend. Conf. Intern. Phenomenes Ionisation Gaz, 6^e, Paris, 1963, 3: 373-9 (1963)
CA 61: 9064f

D 1421 Presence of pyrene, 1,2-benzopyrene, and 3,4-benzopyrene in different vegetable oils.
Jung, L., and Morand, P.
Compt. Rend. 257: 1638-40 (1963)
CA 60: 771b

D 1422 Effects of an electric field on the laser emission of ruby.
Kaiser, W., and Lessing, H.
Appl. Phys. Letters 2: 206-8 (1963)
CA 59: 3456b

D 1423 Fluorescence of laser ruby rods.
Kamiyama, M., Tokyama, Y., Namba, S., and Kim, P.H.
Sci. Papers Inst. Phys. Chem. Res. (Tokyo) 57: 47-50 (1963)
CA 60: 2472b

D 1424 Phosphorescence of organic compounds.
Kanda, K.
Kagaku (Kyoto) 18: 713-9 (1963)
CA 60: 10036f

D 1425 Scheelite from the ore veins of Banka Stiavnica, Hodrusa, and Vyhne.
Kantor, J., and Elias, K.
Geol. Prace, Zpravy 27: 5-34 (1963)
CA 61: 461c

D 1426 Investigation of the photo-transformations of chlorophyll by differential spectrophotometry.
Karapetyan, N.V., Litvin, F.F., and Krasnovskii, A. A.
Biofizika 8: 191-200 (1963)
CA 58: 14443f

D 1427 Changes in luminescence in a study of differential spectra of photosynthesizing organisms.
Karapetyan, N.V., Litvin, F.F., and Krasnovskii, A.A.
Dokl. Akad. Nauk SSSR 149: 1428-31 (1963)
CA 59: 5500g

D 1428 Luminescence of glycinine.
Karyakin, A.V., and Chmutina, L.A.
Izv. Akad. Nauk SSSR, Ser. Fiz. 27: 791-5 (1963)
CA 59: 19762a

D 1429 Fluorescence quenching by intermolecular energy crossing.
Kawski, A.
Acta Physiol. Polon. 24: 641-9 (1963)
CA 61: 12808g

D 1430 Fluorescence-polarization (F.P.) spectra of dyes.
Kawski, A.
Bull. Acad. Polon. Sci., Ser. Sci., Math., Astron. Phys. 11: 37-8 (1963)
CA 59: 1193b

D 1431 The lack of dependence of fluorescence spectra on the wavelength of irradiation.
Kawski, A.
Bull. Acad. Polon. Sci., Ser. Sci., Math., Astron. Phys. 11: 567-72 (1963)
CA 60: 8796g

D 1432 Intermolecular energy transport in fluorescent solutions.
Kawski, A.
Z. Naturforsch. 18: 961-6 (1963)
CA 60: 131h

D 1433 Solvent effects on the electronic spectrum of POPOP.
Kawski, A., and Polacka, B.
Acta Physiol. Polon. 23: 811-7 (1963)
CA 61: 5105c

D 1434 The effects of a solvent mixture on the absorption and fluorescence spectra of dyes.
Kawski, A., Polacka, B., and Czyz, P.
Acta Physiol. Polon. 23: 705-14 (1963)
CA 61: 186d

D 1435 Triplet excitons and delayed fluorescence in anthracene crystals.
Kepler, R.G., Caris, J.C., Avakian, P., and Abramson, E.
Phys. Rev. Letters 10: 400-2 (1963)
CA 59: 4697a

D 1436 Fluorometric determination of calcium in blood serum.
Kepner, B.L., and Hercules, D.M.
Anal. Chem. 35: 1238 (1963)

D 1437 Rise of phosphorescence of organic luminophors.
Khalupovskii, M.D.
Izv. Akad. Nauk SSSR, Ser. Fiz. 27: 644-6 (1963)
CA 59: 8233e

D 1438 Temperature effect on the fluorescence spectra of n-paraffin solutions of some derivatives of pyrene.
Khesina, A.Ya.
Opt. i Spektroskopiya Akad. Nauk SSR, Otd. Fiz.-Mat. Nauk, Sb. Statei, 1: 43-51 (1963)
CA 59: 5955a

D 1439 Role of the triplet state of aromatic amines in the photodehydration of alcohols at 77°K.
Kholmogorov, V.E., Baranov, E.V., and Terenin, A.N.
Dokl. Akad. Nauk SSSR 152: 1399-402 (1963)
CA 60: 4988c

D 1440 Identification reaction for poly(vinyl alcohol).
Khosla, M.M.L., and Balakrishna, K.J.
"LABDEV" (Kanpur, India) No. 1, 49 (1963)
CA 60: 5641e

D 1441 Radioluminescence of organic substances. I. Luminescence yield of some substances in nonluminescent solvents.
Kilin, S.F., Kouyrzina, K.A., and Rozman, I.M.
Opt. i Spektroskopiya Akad. Nauk SSR, Otd. Fiz.-Mat. Nauk, Sb. Statei, 1: 147-51 (1963)
CA 59: 8277h

D 1442 Correction for anomalous fluorescence peaks caused by grating transmission characteristics.
King, R.M., and Hercules, D.M.
Anal. Chem. 35: 1099-1100 (1963)
CA 59: 5950a

D 1443 Effect of temperature on the phosphorescence of organic materials.
Kislyak, G.M., and Lisenko, G.M.
Izv. Akad. Nauk SSSR, Ser. Fiz. 27: 717-19 (1963)
CA 59: 14760a

D 1444 The law of damping of phosphorescence of organic substances.
Kislyak, G.M., and Lisenko, G.M.
Ukr. Fiz. Zh. 8: 772-8 (1963)
CA 59: 12319c

D 1445 Some luminescence properties of organic dyes.
Kislyak, G.M., and Lisenko, G.M.
Ukr. Fiz. Zh. 8: 900-6 (1963)
CA 60: 133c

D 1446 Crystal field splitting in $CaF_2:Nd^{+3}$.
Kiss, Z.I.
J. Chem. Phys. 38: 1476-80 (1963)

D 1447 Fluorometric microdetermination of thallium with Rhodamine B.
Kisser, W.
Arch. Toxikol. 20: 108-13 (1963)
CA 59: 8117c

D 1448 Dependence of the quenching of fluor-
escence by foreign substances on the
viscosity of solution. II.
Kitanskaya, L.A., and Seveshnikov, B.Ya.
Opt. i Spektroskopiya, Akad. Nauk SSSR,
Otd. Fiz.-Mat. Nauk, Sb. Statei 1: 60-5
(1963)
CA 59: 5900h

D 1449 Fluorescence determination of nio-
bium with lumogallion.
Klimov, V.V., and Didkovskaya, O.S.
Zavodsk. Lab. 29: 147-8 (1963)
CA 59: 19e

D 1450 Absorption and emission spectroscopy
of pyrene at 20° and 4°K.
Klimova, L.A.
Opt. i Spektroskopiya 15: 344-56 (1963)
CA 60: 1244d

D 1451 Reflectance and emission measurements on optically bleached fabrics and
their relation to the degree of soil removal.
Kling, A.
Fette, Seifen, Anstrichmittel 65: 285-8
(1963)
CA 59: 4089a

D 1452 Effect of scattered light on measure-
ment of luminescence spectra.
Klochkov, V.P.
Izv. Akad. Nauk SSSR, Ser. Fiz. 27: 17-18
(1963)
CA 59: 131f

D 1453 Temperature effect on the quantum
yield of the luminescence of vapors of
organic compounds.
Klochkov, V.P.
Izv. Akad. Nauk SSSR, Ser. Fiz. 27: 570-5
(1963)
CA 59: 8256c

D 1454 Temperature dependence of the
quantum yield of fluorescence in solu-
tions.
Klochkov, V.P., and Korotkov, S.M.
Opt. i Spektroskopiya Akad. Nauk SSSR, Otd.
Fiz.-Mat. Nauk, Sb. Statei, 1: 51-7
(1963)
CA 59: 5955b

D 1455 Temperature dependence of the intensity and shape of the absorption and
fluorescence spectra of anthracene
derivatives in the vapor state.
Klochkov, V.P., and Makushenko, A.M.
Opt. i Spektroskopiya 15: 52-60 (1963)
CA 59: 9474h

D 1456 Effect of foreign gases on the spectra
and fluorescence yield of vapors of
anthracene derivatives.
Klochkov, V.P., and Makushenko, A.M.
Opt. i Spektroskopiya 15: 237-44 (1963)
CA 59: 13498a

D 1457 Relation between the degree of polari-
zation of fluorescence and the wave-
length of the luminescence studied.
Klochkov, V.P., and Neporent, B.S.
Opt. i Spektroskopiya 14: 812-14 (1963)
CA 59: 10902e

D 1458 Fluorescence of solution scintillators
with naphthalene as a solvent.
Kobayashi, S., and Hayakawa, S.
Japan J. Appl. Phys. 2: 281-8 (1963)
CA 59: 8271h

D 1459 Hydrogen bonding between uranyl
nitrate herahydrate and Mg phthalocya-
mine as revealed by the luminescence
spectrum.
Kobyshev, G.I., Lyalin, G.N., and Terenin,
A.N.
Opt. i Spektroskopiya 15: 837-8 (1963)
CA 60: 7586a

D 1460 Flow chamber for the differential
microfluorimeter of Chance and
Legalluis. Preliminary work with
glass-grown ascites cells.
Kohen, E.
Biochim. Biophys. Acta 75: 139-42 (1963)
CA 59: 9175h

D 1461 Fluorescence studies.
Kok, B.
Natl. Acad. Sci. — Natl. Res. Council,
Misc. Publ. No. 1145, 45-55 (1963)
CA 60: 14745f

D 1462 Measurement of fluorescent lifetime.
Kokubun, H.
Bunko Kenkyu 11: 147-55 (1963)
CA 60: 12760d

D 1463 Rate constant for the thermal excitation of sodium in mixtures of sodium vapors with argon and nitrogen.
Kondrat'ev, V.N.
Dokl. Akad. Nauk SSSR 153: 1108-10 (1963)
CA 60: 10057a

D 1464 Nature of ultraviolet cell luminescence.
Konev, S.V., Lyskova, T.I., and Bobrovich, V.P.
Biofizika 8: 433-40 (1963)
CA 59: 9001d

D 1465 Extraction-fluorimetric determination of europium and terbium.
Kononenko, L.I., Lauer, R.S., and Poluektov, N.S.
Zh. Analit. Khim. 18: 1468-74 (1963)
CA 60: 11366g

D 1466 Fluorescence detection of bacterial enzymes which metabolize amino acids.
Kononenko, A.P.
Zh. Mikrobiol., Epidemiol. i Immunobiol. 40: 117-21 (1963)
CA 60: 16236h

D 1467 Fluorescent reactions with thiooxin.
Korenman, I.M., Sheyanova, F.R., and Starodulova, N.I.
Tr. po Khim. i Khim. Tekhnol. 1963: 243-5 (1963)
CA 61: 4959e

D 1468 Electron spectra of N-methylphthalimide vapors.
Korotkevich, V.T., Zelinskii, V.V., and Borisevich, N.A.
Izv. Akad. Nauk SSSR, Ser. Fiz. 27: 576-9 (1963)
CA 59: 9476f

D 1469 Triplet-level energy migration in benzophenone crystals.
Korsunskii, V.M., and Faidysh, A.N.
Dokl. Akad. Nauk SSSR 150: 771-4 (1963)
CA 59: 5905d

D 1470 Fluorescent light emitted by aqueous solutions (of sodium salicylate) in a γ-radiation field.
Kosek, S., and Zagorski, Z.P.
Inst. Nucl. Res. (Warsaw), Rept. No. 488, 5 pp. (1963)
CA 61: 11493c

D 1471 Optical investigations on CdS-doped AgBr crystals at low temperatures.
Koswig, H.D.
Czech. J. Phys. 13: 197-200 (1963)
CA 59: 14758d

D 1472 Vibrational analysis of the fluorescence spectra of a frozen naphthalene solution in pentane and of naphthalene vapors.
Kovner, M.A., and Krainov, E.P.
Opt. i Spektroskopiya 15: 562-5 (1963)
CA 60: 3630g

D 1473 Simple apparatus for estimation of concentration of fluorescent substances in eluates from chromatographic columns.
Koziol, J.
Prace Zakresu Towaroznawst Chem., Wyzsza Szkola Ekon. Poznan., Zeszyty Nauk Ser. I 10: 79-83 (1963)
CA 60: 4451f

D 1474 A substrate for the fluorometric determination of lipase activity.
Kramer, D.N., and Guilbault, G.G.
Anal. Chem. 35: 588-9 (1963)

D 1475 Analytical properties of fluorexon.
Kreingol'd, S.V., and Bozhevol'nov, E.A.
Tr. Vses. Nauchn.-Issled. Inst. Khim. Reaktivov No. 25, 358-73 (1963)
CA 61: 6362c

D 1476 Determination of the instability constant of the complex of 8-(p-toluene sulfonamido)quinoline with zinc.
Kreingol'd, S.V., Bozhevol'nov, E.A., and Serebryakova, G.V.
Tr. Vses. Nauchn.-Issled. Inst. Khim. Reaktivov No. 25, 422-6 (1963)
CA 61: 82a

D 1477 Causes of the difference in the half-widths of the absorption and luminescence bands of crystal phosphors.
Kristofel, N.N., Rebane, K.K., Sil'd, O.I., and Khizhnyakov, V.V.
Opt. i Spektroskopiya 15: 569-72 (1963)
CA 60: 3631e

D 1478 Enhancement of fluorescence yield of rare earth ions by heavy water.
Kropp, J.L., and Windsor, M.W.
J. Chem. Phys. 39: 2769-70 (1963)
CA 60: 1244c

D 1479 Electronic energy levels of E_r^{3+} (^{4}fu) in LaF$_3$.
Krupke, W.F.
U.S. Dept. Comm., Office Tech. Serv., AD 427,079, 33 pp. (1963)
CA 61: 1400a

D 1480 Absorption and fluorescence spectra of E_r^{3+} ($4f^{11}$) in LaF$_3$.
Krupke, W.F.
J. Chem. Phys. 39: 1024-30 (1963)
CA 59: 5951e

D 1481 Threshold for electron radiation damage in ZnSe.
Kulp, B.A., and Detweiler, R.M.
Phys. Rev. 129: 2422-4 (1963)
CA 58: 8471d

D 1482 The absorption and fluorescence spectra of trivalent europium in silicate glasses.
Kurkjian, C.R., Gallagher, P.K., Sinclair, W.R., and Sigety, E.A.
Phys. Chem. Glasses 4: 239-46 (1963)
CA 60: 7768c

D 1483 Effect of the meduin on the optical characteristics of some 5-membered heterocyclic rings.
Kutsyna, L.M., and Ogurtsoua, L.A.
Izv. Akad. Nauk SSSR, Ser. Fiz. 26: 739-44 (1963)
CA 59: 14758g

D 1484 Effect of impurities on the fluorescence of anthracene crystals containing tetracene.
Lacey, A.R., Lyons, L.E., and White, J.W.
J. Chem. Soc. 1963: 3670-5 (1963)
CA 59: 4698a

D 1485 Acridine: a low-temperature investigation of its enigmatic spectral characteristics.
Ladner, S., and Becker, R.S.
J. Phys. Chem. 67: 2481-6 (1963)
CA 59: 14765b

D 1486 Vibrational spectra and structure of the cycloheptatriene molecule.
la Hau, C., and de Ruyter, H.
Spectrochim. Acta 19: 1559-66 (1963)
CA 59: 9470e

D 1487 A mirrored test tube for fluorescence analysis.
Laikin, M.
Appl. Spectry. 17: 26-7 (1963)
CA 58: 11934a

D 1488 New double-beam fluorometer.
Laikin, M.
Rev. Sci. Instr. 34: 773-7 (1963)
CA 60: 1084h

D 1489 Singlet-triplet transitions in the tetraphenyls.
La Paglia, S.R.
Spectrochim. Acta 18: 1295-8 (1963)
CA 59: 132f

D 1490 Spectral properties of optical bleaching agents.
Lapshina, Z.K.
Bumazhn. Prom. 38: 10-11 (1963)
CA 59: 6609d

D 1491 Spectroscopic indications of the heterogeneity of chlorophyll in vivo.
Lavorel, J.
Colloq. Intern. Centre Natl. Rech. Sci. (Paris) 119: 161-74 (1963)
CA 61: 3344h

D 1492 Intramolecular electronic energy transfer in 4-(1-naphthylmethyl)benzophenone.
Leermakers, P.A., Byers, G.W., Lamola, A.A., and Hammond, G.S.
J. Am. Chem. Soc. 85: 2670-1 (1963)
CA 59: 13455d

D 1493 Crystalline luminescence of cell-nuclear chemicals under ionizing radiation.
Lehman, R.L.
U.S. At. Energy Comm. UCRL-10,724, 77 pp. (1963)
CA 59: 7123e

D 1494 Luminescence of crystalline cell-nuclear chemicals under ionizing radiation.
Lehman, R.L., and Wallace, R.W.
U.S. At. Energy Comm. UCRL-10961, 62 pp. (1963)
CA 61: 141b

D 1495 Optical maser action in europium
benzoylacetonate.
Lempicki, A., and Samuelson, H.
Phys. Letters 4: 133-5 (1963)
CA 58: 12063d

D 1496 Stimulated processes in organic
compounds.
Lempicki, A., and Samuelson, H.
Appl. Phys. Letters 2: 159-61 (1963)
CA 59: 1205a

D 1497 Electron transfer reactions of excited
perylene.
Leonhardt, H., and Weller, A.
Ber. Bunsenges. Physik. Chem. 67: 791-5
(1963)
CA 60: 4969f

D 1498 Fluorescence and phosphorescence
spectra of β-methyl-naphthalene in
normal and isoparaffin solutions at
77°K.
Levshin, V.L., and Mamedov, K.I.
Vestn. Mosk. Univ., Ser. III, Fiz., Astron.
18: 30-6 (1963)
CA 59: 5953h

D 1499 Fluorescence of silicon carbide.
Leushkina, G.V., Zvyagin, V.I., Lobanov,
E.M., and Dutov, A.G.
Izv. Akad. Nauk Uz. SSR, Ser. Fiz.-Mat.
Nauk 7: 98-9 (1963)
CA 60: 15314c

D 1500 Luminescence of aluminum electrodes
in the process of a.c. electrolytic oxi-
dation.
Leweski, T.
Acta Physiol. Polon. 23: 215-20 (1963)
CA 60: 11473h

D 1501 Nonradiative electronic energy transfer in organic liquids at high "donor"
concentrations.
Lipsky, S.
U.S. At. Energy Comm. AFCRL-63-610,
33 pp. (1963)
CA 61: 14134a

D 1502 Evidence for triplet-triplet transfer
from benzene to biacetyl in cyclohex-
ane solution.
Lipsky, S.
J. Chem. Phys. 38: 2786-7 (1963)
CA 59: 12335f

D 1503 Influence of an external electric field
on the fluorescence of molecules. I.
Theory.
Liptay, W.
Z. Naturforsch. 18a: 705-18 (1963)
CA 59: 7057f

D 1504 Thermodynamic data for quinone-
quinol equilibrium from electronic ab-
sorption and emission spectra.
Loeber, G.
Z. Chem. 3: 359 (1963)
CA 59: 14655f

D 1505 The low-temperature fluorescence of
some quinol derivatives.
Loeber, G.
Z. Chem. 3: 399-400 (1963)
CA 60: 3629h

D 1506 Effect of solvent on the electronic
spectra of α- and β-naphthylamines.
Loeber, G.
Z. Chem. 3: 437-8 (1963)
CA 60: 6360a

D 1507 Measurement of absorption and fluor-
escence spectra at lower temperatures.
Loeber, G.
Z. Physik. Chem. 223 (1/2): 90-8 (1963)
CA 59: 13470e

D 1508 2,3-Diaminonaphthalene as a reagent
for the determination of milligram to
submicrogram amounts of selenium.
Lott, P.F., Cukor, P., Moriber, G., and
Solga, J.
Anal. Chem. 35: 1159 (1963)

D 1509 Extinction of the fluorescence of
Rhodamine B solutions.
Lucatu, E.
J. Chim. Phys. 60: 1381-4 (1963)
CA 60: 11508g

D 1510 Interaction between electronic and
vibrational states of F- and F_A-centers
in KCl.
Luty, F., and Pick, H.
J. Phys. Soc. Japan 18: 240-2 (1963)
CA 59: 5899e

D 1511 Quenching of fluorescence of
carotenoid adsorbates.
Lyalin, G.N., Kobyshev, G.I., and Terenin,
A.N.
Dokl. Akad. Nauk SSSR 150: 407-10 (1963)
CA 59: 5409a

D 1512 Adsorption and fluorescence of fat-
soluble fluorescent dyes on Class I
and Class II Saccharomyces cerevisiae.
Lycette, R.M., and Hedrick, L.R.
J. Bacteriol. 85: 1-6 (1963)
CA 58: 4821h

D 1513 Do oxygen and water vapor fluoresce?
MacAdam, D.L.
J. Opt. Soc. Am. 53: 397-8 (1963)
CA 58: 12081f

D 1514 Fluorescent substance adhesion to
glass.
Machida, M., and Ishihara, M.
Japan 20,224 (1963)
CA 61: p4048f

D 1515 The theory of shape, width, and shift
of luminescence lines of impurity
centers in ionic crystals.
Malkin, B.Z.
Fiz. Tverd. Tela 5: 3088-94 (1963)
CA 60: 3631c

D 1516 Free energy of nonequilibrium polari-
zation systems. II. Homogeneous and
electrode systems.
Marcus, R.A.
J. Chem. Phys. 38: 1858-62 (1963)
CA 58: 10743d

D 1517 Fluorescence in mixed crystals of
Ce:PrCl$_3$.
Margolis, J.S., Stafsudd, O., and Young,
E.Y.
J. Chem. Phys. 38: 2045-6 (1963)
CA 59: 3412e

D 1518 Solvent effects on the fluorescence
spectrum of the ion-pair (naphthol-
triethylamine system).
Mataga, N., Kawasaki, Y., and Torihashi,
Y.
Bull. Chem. Soc. Japan 36: 358-9 (1963)
CA 58: 12081e

D 1519 Solvent effects on the electronic
spectra of naphthalene.
Mataga, N.
Bull. Chem. Soc. Japan 36: 1502-4 (1963)
CA 60: 4966h

D 1520 The electronic spectra and electronic
structures of amino-substituted ben-
zenes.
Mataga, N.
Bull. Chem. Soc. Japan 36: 1607-18 (1963)
CA 60: 8773d

D 1521 Inhibition properties of oximes. II.
Quenching of fluorescence of luminol
by means of oximes.
Matkovic, J., Weber, K., and Palla, L.
Arhiv Hig. Rada Toksikol. 14: 95-106
(1963)
CA 61: 1406d

D 1522 Fluorescence in ketone solutions of
europium and terbium salts.
Matovich, E., and Suzuki, C.K.
J. Chem. Phys. 39: 1442-4 (1963)
CA 59: 14762e

D 1523 Fluorescence in solid-state and electronic absorption in solutions of sali-
cylidene amines and benzylidene
amines.
Matsushita, H.
Nippon Kagaku Zasshi 84: 373-8 (1963)
CA 59: 8278e

D 1524 Fluorescence in the solid-state and
electronic absorption in solutions of
N-(2-hydroxy-α-naphthylmethylene)
amines.
Matsushita, H.
Nippon Kagaku Zasshi 84: 378-84 (1963)
CA 59: 8278g

D 1525 Remover for fluorescent and tempor-
ary camouflage paint systems.
Matuska, A., and Koury, J.
U.S. 3,115,471, Dec. 24, 1963
CA 60: p7030d

D 1526 Excitation in radio-frequency dis-
charges.
Mavrodineanu, R., and Hughes, R.C.
Spectrochim. Acta 19: 1309-17 (1963)
CA 59: 7091c

D 1527 Electron donor properties of zinc
phthalocyanine.
McCartin, P.J.
J. Am. Chem. Soc. 85: 2021 (1963)
CA 59: 5954a

D 1528 External heavy-atom spin-orbital coupling effect. IV. Intersystem crossing.
McGlynn, S.P., Daigre, J., and Smith, F.J.
J. Chem. Phys. 39: 675-9 (1963)
CA 59: 7088h

D 1529 Total luminescence of organic molecules of petrochemical interest. I. Naphthalene, phenanthrene, and 1,2,4,5-tetramethyl benzene.
McGlynn, S.P., Neely, B.T., and Neely, C.
Anal. Chim. Acta 28: 472-9 (1963)
CA 59: 4548f

D 1530 Molecular vibrations in the exciton theory for molecular aggregates. V. Electronic spectra of weakly-coupled systems.
McRae, E.G.
Australian J. Chem. 16: 295-314 (1963)
CA 59: 4631e

D 1531 Vibronic interactions in the exciton theory for molecular crystals.
McRae, E.G.
J. Chem. Phys. 39: 2974-82 (1963)
CA 60: 7565c

D 1531a Detection of cyanide.
Meditsch, J.
Eng. Quim. (Rio de Janeiro) 15: 15 (1963)
CA 60: 15140b

D 1532 Thermoluminescence in quartz.
Medlin, W.L.
J. Chem. Phys. 38: 1132-43 (1963)
CA 58: 7515d

D 1533 Effect of solvent viscosity on excitation-energy transfer between unlike molecules.
Melhuish, W.H.
J. Phys. Chem. 67: 1681-3 (1963)
CA 59: 7009h

D 1534 Spectrofluorometric identification of phenothiazine drugs.
Mellinger, T.J., and Keeler, L.E.
Anal. Chem. 35: 554-8 (1963)

D 1535 Fluorescence microscopy in dermatology.
Mescon, H., and Grots, I.A.
J. Invest. Dermatol. 41: 181-96 (1963)
CA 60: 12528a

D 1536 Fluorescence lifetime of the europium dibenzoylmethides.
Metlay, M.
J. Chem. Phys. 39: 491-2 (1963)
CA 59: 10866f

D 1537 True and parasitic luminescence of durene.
Meyer, Y., and Astier, R.
J. Phys. 24: 1089-94 (1963)
CA 60: 14023h

D 1538 Studies on the processes of histogenesis in dental tissue by the method of tetracycline marking in rodents and carnivores. II. Site and aspects of the fluorescence of enamel.
Miani, A., and Brusati, R.
Boll. Soc. Ital. Biol. Sper. 39: 1207-9 (1963)
CA 60: 4580h

D 1539 The concentration dependence of the luminescence of F and F' centers in KCl.
Miehlich, A.
Z. Physik 176: 168-90 (1963)
CA 60: 4913f

D 1540 Photopolarization in anthracene.
Milanez, C.S., and Murphy, V.
Anais Acad. Brasil. Cienc. 35: 153-7 (1963)
CA 60: 11480h

D 1541 Immunofluorescent methods in the diagnosis of infectious diseases.
Miller, J.N., Boak, R.A., Carpenter, C.M., and Fazzen, F.
Am. J. Med. Technol. 29: 25-32 (1963)
CA 59: 4365b

D 1542 Vibrational fluorescence of carbon monoxide.
Millikan, R.C.
J. Chem. Phys. 38: 2855-60 (1963)
CA 59: 1195d

D 1543 Polymers containing unsaturated aromatic chains and Group V elements.
Minnesota Mining and Manufg. Co.
Ger. 1,145,799, Mar. 21, 1963
CA 59: 1776h

D 1544 The quenching action of pyridine and quinoline on the fluorescence of naphthalene derivatives.
Miwa, T., and Koizumi, M.
Bull. Chem. Soc. Japan 36: 1619-29 (1963)
CA 60: 7586h

D 1545 Electroluminescent and fluorescent material.
Mujagawa, S., and Sakamoto, N.
Japan 4326 (1963)
CA 60: p11489f

D 1546 Electronic structure of excited N,N-dimethyl-α-naphthylamine.
Mataga, N.
Bull. Chem. Soc. Japan 36: 620-1 (1963)
CA 60: 2458f

D 1547 Optical studies. III. Fluorescent brightening agents on textiles: elementary optical theory and its practical implications.
Morton, T.H.
J. Soc. Dyers Colourists 79: 238-42 (1963)
CA 59: 6565a

D 1548 Probabilities of radiation and non-irradiation transitions in molecules of organic dyes.
Morozov, Yu.V.
Biofizika 8: 165-171 (1963)
CA 59: 3432h

D 1549 Peculiarities of the luminescence of Acridine Orange and of Rhodamine B dimers.
Morozov, Yu.V.
Biofizika 8: 331-4 (1963)
CA 59: 5955f

D 1550 Effect of temperature on phycocyanin and chlorophyll properties of a cyanophyceae in vivo.
Moyse, A., and Guyon, D.
Colloq. Intern. Centre Natl. Rech. Sci. (Paris) 119: 407-19 (1963)
CA 61: 2188b

D 1551 Luminescence spectra of diketonic vat dyes in Lavsan and in solutions.
Murmukhametov, R.N., and Bondareva, L.V.
Zh. Fiz. Khim. 37: 1143-8 (1963)
CA 59: 5953d

D 1552 Luminescence of naphthyl and anthryl derivatives of ethylene.
Nagornaga, L.L., Nurmukhametov, R.N., Malkes, L.Ya., and Shubina, L.V.
Izv. Akad. Nauk SSSR, Ser. Fiz. 27: 748-53 (1963)
CA 59: 14761b

D 1553 Fluorescence spectra of organic compounds in solution. I. The positions of the O,O-bands of the fluorescence spectra.
Nakamizo, M., and Kanda, Y.
Spectrochim. Acta 19: 1235-48 (1963)
CA 59: 4698e

D 1554 Polarization of green emission from a durene crystal.
Nakamura, T., and Hayakawa, S.
J. Phys. Soc. Japan 18: 793-6 (1963)
CA 59: 1193a

D 1555 The methodology of the measurement and representation of fluorescence spectra.
Nebbia, G.
Proc. Colloq. Spectros. Intern., 10th, Univ. Maryland 1962: 605-29 (1963)
CA 61: 9065b

D 1556 Spectrofluorimeter for the visible spectral region.
Nevinskii, A.A., and Belen'kii, B.G.
Zavodsk. Lab. 29: 1391-2 (1963)
CA 60: 3631b

D 1557 Excitation of the Nd^{+3} fluorescence in $CaWO_4$ by recombination radiation in GaAs.
Newman, R.
J. Appl. Phys. 34: 437 (1963)
CA 58: 12080g

D 1558 Franck-Condon factor surface for the $I_2 (B^3 \Pi_{O_u}{}^+ - X' \Sigma_g{}^+)$ band system.
Nicholls, R.W.
J. Chem. Phys. 38: 1029-30 (1963)
CA 59: 126g

D 1559 Direct determination of exciton interactions for triplet states of organic crystals.
Nieman, G.C., and Robinson, G.W.
J. Chem. Phys. 39: 1298-1307 (1963)
CA 59: 7073e

D 1560 Fluorescence of solutions of some
    substituted polyenes.
    Nikitina, A.N., Ter-Sarkisyan, G.S.,
       Mikhailov, B.M., and Minehenkova, L.E.
    Opt. i Spektroskopiya 14: 655-63 (1963);
       Opt. Spect. (USSR) (English Transl.)
       14: 347-50 (1963)
    CA 59: 4696c

D 1561 Stereospecific preparation of epoxy
    ketones by photochemical oxygenation.
    Nikon, A., and Mendelson, W.L.
    J. Am. Chem. Soc. 85: 1894-5 (1963)
    CA 59: 4003g

D 1562 Fluorescent substance.
    Nishigaki, S., and Kobayoshi, K.
    Japan 22,263, 1961 (1963)
    CA 60: 2426h

D 1563 Paper-chromatographic investigations
    of fluorescent compounds in lolium.
    Nitzcshe, W.
    Z. Pflanzenzuecht. 49: 101-6 (1963)
    CA 61: 9779d

D 1564 Fluorometric and spectrophotometric
    determination of aluminum in indus-
    trial water.
    Noll, C.A., and Stefaneli, L.J.
    Anal. Chem. 35: 1914-16 (1963)

D 1566 Luminescence spectra of solutions of
    indigo and some of its derivatives at
    77°K.
    Normukhametov, R.N., Shigorin, D.N., and
       Kozlov, Yu.I.
    Izv. Akad. Nauk SSSR, Ser. Fiz. 27: 686-9
       (1963)
    CA 59: 9476h

D 1567 Spectrofluorometric study of the com-
    plexation of indium ions with 2-methyl-
    8-quinolinol in absolute ethyl alcohol.
    Ohnesorge, W.E.
    Anal. Chem. 35: 1137-42 (1963)
    CA 59: 8342d

D 1568 Fluorescence of metal chelate com-
    pounds of 8-quinolinol. III. The mono-
    (8-quinolinol)-aluminum species in
    absolute ethanol.
    Ohnesorge, W.E., and Capotosto, A.
    J. Inorg. Nucl. Chem. 24: 829-38 (1963)
    CA 58: 8614g

D 1569 A histochemical study of folic acid.
    Okamoto, K.
    Bitamin 28: 49-58 (1963)
    CA 61: 16525e

D 1570 A toxic metabolite from ascochyta
    fabae having antibiotic activity.
    Oku, H., and Nakanishi, T.
    Phytopathology 53: 1321-5 (1963)
    CA 60: 9629a

D 1571 Phosphorescence lifetime studies in
    some organic crystals at low tempera-
    tures.
    Olness, D., and Sponer, H.
    J. Chem. Phys. 38: 1779-82 (1963)
    CA 58: 10870e

D 1572 Fluorescence spectra of alkaline
    earth derivatives of azo compounds.
    Olsen, R.L.
    Dissertation, Iowa State University (1963)
    CA 58: 10869f

D 1573 O,O'-Dihydroxyazo compounds as
    fluorometric reagents for magnesium.
    Olsen, R., and Diehl, H.
    Anal. Chem. 35: 1142 (1963)

D 1574 α-Centers in alkali halide crystals.
    Onaka, R., Fujita, I., and Fokoda, A.
    J. Phys. Soc. Japan 18: 263-8 (1963)
    CA 59: 5899d

D 1575 Pigment production by Pseudomonas
    aeruginosa on glutamic acid medium
    and gel filtration on the culture fluid
    filtrate.
    Osawa, S., Yabunchi, E., Narano, Y.,
       Kakata, M., Kosono, Y., Takashina, K.,
       and Tanahe, T.
    Japan. J. Microbiol. 7: 87-95 (1963)
    CA 60: 16234h

D 1576 Photochemical and spectral properties
    of chlorophyllin.
    Oster, G., Bellin, J.S., and Broyce, S.B.
    NASA, Doc. N63-23644, 50 pp. (1963)
    CA 60: 10974b

D 1577 Electron ejection and fluorescence in
    aqueous β-naphthol solutions.
    Ottolenghi, M.
    J. Am. Chem. Soc. 85: 3557-62 (1963)
    CA 60: 133e

D 1578 The use of β-thiopropionic acid for
stabilizing the fluorescence of adreno-
lutin and noradrenolutin.
Palmer, J.F.
J. Pharm. Pharmacol. 15: 777-8 (1963)
CA 60: 373f

D 1579 Delayed fluorescence from naphtha-
lene solutions.
Parker, C.A.
Spectrochim. Acta 19: 989-94 (1963)
CA 59: 3443d

D 1580 Delayed fluorescence of 3, 4-benzpy-
rene solutions.
Parker, C.A.
Nature 200: 331-2 (1963)
CA 60: 1245c

D 1581 Sensitized P-type delayed fluores-
cence.
Parker, C.A.
Proc. Roy. Soc. 276: 125-35 (1963)
CA 59: 13499c

D 1582 Delayed fluorescence of pyrene in
ethanol.
Parker, C.A., and Hatchard, C.G.
Trans. Faraday Soc. 59: 284-95 (1963)
CA 59: 1195f

D 1583 Tempering effect in the fluorescence
spectrum of thin anthracene films.
Perkampus, H.H., and Pohl, L.
Z. Physik. Chem. 39: 397-401 (1963)
CA 60: 14024e

D 1584 Linear emission and absorption spec-
tra of phthalocyanine in frozen crystal-
line solutions.
Personov, R.I.
Opt. i Spektroskopiya 15: 61-71 (1963)
CA 59: 9475b

D 1585 Determination of the types of symme-
try of some vibrations of naphthalene.
Pesteil, L., Vergnes, J., and Pesteil, P.
J. Chim. Phys. 60: 492-500 (1963)
CA 59: 2301g

D 1586 Absorption spectra at low tempera-
ture of some aromatic molecules in
solution.
Pesteil, L., Troisplis, R., and Pesteil, P.
J. Chim. Phys. 60: 1294-1300 (1963)
CA 60: 10066f

D 1587 Fluorescent lifetime of terbium in the
presence of other rare-earth ions.
Peterson, G.E., and Bridenbaugh, P.M.
J. Opt. Soc. Am. 53: 301-2 (1963)
CA 58: 10867c

D 1588 Effects of rare-earth-ion substitution
on the fluorescent lifetime of terbium
in the compound terbium hexa-
antipyrine tri-iodide.
Peterson, G.E., and Bridenbaugh, P.M.
J. Opt. Soc. Am. 53: 494-5 (1963)
CA 59: 9435g

D 1589 Relaxation processes in Tb^{+3} using
pulsed excitation.
Peterson, G.E., and Bridenbaugh, P.M.
J. Opt. Soc. Am. 53: 1129-38 (1963)
CA 60: 2461f

D 1590 Double photon excitation in organic
crystals.
Peticolas, W.L., Goldsborough, J.P., and
Rieckhoff, K.E.
Phys. Rev. Letters 10: 43-5 (1963)
CA 58: 10867a

D 1591 Double-photon excitation of organic
molecules in dilute solution.
Peticolas, W.L., and Rieckhoff, K.E.
J. Chem. Phys. 39: 1347-8 (1963)
CA 59: 14762b

D 1592 Synthesis of fluorescent copolymers
from styrene-maleic anhydride co-
polymers.
Petit, J., and Strzelecki, L.
Compt. Rend. 257: 2654-6 (1963)
CA 60: 5650e

D 1593 Phosphorus-containing polymers. VI.
Synthesis of polyphosphites and poly-
phosphonites based on glucose.
Petrov, K.A., Nifant'ev, E.E., Lysenko,
T.N., and Suzanskii, A.I.
Vysokomolekul. Soedin. 5: 712-18 (1963)
CA 60: 12120f

D 1594 Spectra and quantum yields of the
fluorescence of some coumarin derivatives.
Petrovich, P.I., and Borisevich, N.A.
Izv. Akad. Nauk SSSR, Ser. Fiz. 27: 703-7
(1963)
CA 59: 9477h

D 1595 Ultraviolet fluorescence of white-rat
    liver cells during embryonic growth.
    Pil'shchik, E.M., and Nikoaeva, M.V.
    Dokl. Akad. Nauk SSSR 148: 199-201 (1963)
    CA 58: 10568f

D 1596 Temperature dependence of the quan-
    tum yield of the phosphorescence of
    organophosphors.
    Pilipovich, V.A., and Tursunov, N.I.
    Izv. Akad. Nauk SSSR, Ser. Fiz. 27: 641-3
    (1963)
    CA 59: 7060d

D 1597 Investigation of photoexcited triplet
    states of polyatomic molecules by the
    electron paramagnetic resonance
    (e.p.r.) method and by means of phos-
    phorescence.
    Piskunov, A.K., Nurmukhametov, R.N.,
    Shigorin, D.N., Moromtsev, V.I., and
    Ozorova, G.A.
    Izv. Akad. Nauk SSSR, Ser. Fiz. 27: 634-7
    (1963)
    CA 59: 7085b

D 1598 Dependence of absorption and fluores-
    cence spectra of complex molecules in
    solution on temperature.
    Piterskaya, I.V., and Bakhshiev, N.G.
    Izv. Akad. Nauk SSSR, Ser. Fiz. 27: 623-7
    (1963)
    CA 59: 9476g

D 1599 Multistage radiative transitions in
    CaF_2:Er^{+++}.
    Pollack, S.A.
    J. Chem. Phys. 38: 2521-9 (1963)
    CA 58: 13311c

D 1600 Fluorescence and absorption spectra
    of biological dyes.
    Porro, T.J., Dadik, S.P., Green, M., and
    Morse, H.T.
    Stain Technol. 38: 37-48 (1963)
    CA 59: 5638g

D 1601 Decay of the triplet state.
    Porter, G., and Hilpern, J.W.
    NASA, Doc. N64-13759, 39 pp. (1963)
    CA 61: 12789h

D 1602 Emission spectra of the chromium
    (III) hexaurea complex.
    Porter, G.B., and Schlaefer, H.L.
    Z. Physik. Chem. 37: 109-14 (1963)
    CA 59: 12315g

D 1603 Vibrational structure in the $^2E_g \rightarrow {}^4A_{2g}$
    phosphorescence of tris-trimethylene-
    diaminechromium (III) ion.
    Porter, G.B., and Schlaefer, H.L.
    Z. Physik. Chem. 38: 227-33 (1963)
    CA 60: 2461h

D 1604 The fluorescence and the emission
    spectra of anisole.
    Prakash, S., and Singh, N.L.
    Indian J. Phys. 37: 59-66 (1963)
    CA 59: 9479d

D 1605 Mechanism of the conversion of CdF_2
    from an insulator to a semiconductor.
    Prener, J.S., and Kingsley, J.D.
    J. Chem. Phys. 38: 667-71 (1963)
    CA 58: 6300b

D 1606 The fluorescence spectrum of naphtha-
    lene crystals between 2° and 100°K.
    Proebstl, A., and Wolf, H.C.
    Z. Naturforsch. 18: 724-34 (1963)
    CA 59: 7087d

D 1607 Temperature dependence of the sensi-
    tized fluorescence in naphthalene
    crystals, between 2° and 100°K.
    Proebstl, A., and Wolf, H.C.
    Z. Naturforsch. 18: 822-8 (1963)
    CA 59: 5900g

D 1608 Carbazole and phenanthrene phosphor-
    escence at liquid-oxygen temperature.
    Pyatnitshii, B.A., and Ulasenko, A.I.
    Izv. Akad. Nauk SSSR, Ser. Fiz. 27: 647-50
    (1963)
    CA 59: 7061b

D 1609 Fluorescence of Sm^{+3} in CaF_2.
    Rabbiner, N.
    Phys. Rev. 130: 502-6 (1963)
    CA 58: 10870c

D 1610 Fluorescence of Dy^{3+} in CaF_2.
    Rabbiner, N.
    Phys. Rev. 132: 224-7 (1963)
    CA 59: 10901e

D 1611 Thorium phosphate phosphors.
    Ranby, P.W., and Hobbs, D.Y.
    J. Electrochem. Soc. 110: 280-4 (1963)
    CA 58: 10835e

D 1612 Fluorescent substances.
    Ranby, W., and Palowkar, R.
    Ger. 1,155,553, Oct. 10, 1963
    CA 60: 4942e

D 1613 Energy transfer in naphthalene-
　　tetracene solid solutions.
　　Reed, C.W., and Lipsett, F.R.
　　J. Mol. Spectry. 11: 139-61 (1963)
　　CA 59: 13451f

D 1614 Calculation of radiation transfer in
　　x-type fluorescence.
　　Renaud, M.
　　Compt. Rend. 256: 3837-40 (1963)
　　CA 59: 3457h

D 1615 Factors affecting photosynthetic lum-
　　inescence.
　　Reporter, M.C., and Strehler, B.L.
　　Studies Microalgae Photosyn. Bacteria,
　　Collection Papers 1963: 281-90 (1963)
　　CA 61: 7367b

D 1616 Fluorescence spectra of 3,4-benzopy-
　　rene in aqueous media.
　　Reske, G., and Stauff, J.
　　Z. Naturforsch. 18: 773-4 (1963)
　　CA 60: 131f

D 1617 Edge emission and Zeeman effects in
　　CdS.
　　Reynolds, D.C., and Litton, C.W.
　　Phys. Rev. 132: 1023-9 (1963)
　　CA 59: 13498g

D 1618 Fluorescence and absorption spectra
　　of azulene in frozen crystalline solu-
　　tions.
　　Rezevich, Z.S.
　　Opt. i Spektroskopiya 15: 357-61 (1963)
　　CA 60: 129f

D 1619 Fluorescence of chickens and eggs
　　following the feeding of benzopyrene
　　crystals.
　　Rigdon, R.H., and Neal, J.
　　Texas Rept. Biol. Med 21: 558-66 (1963)
　　CA 60: 16256a

D 1620 Measurement of the amplification co-
　　efficient of a fluorescent line of iodine
　　vapor excited by the 5461-A line of
　　mercury.
　　Rivoire, G., and Dupeyrat, R.
　　Compt. Rend. 256: 2575-7 (1963)
　　CA 58: 13305d

D 1621 The influence of urea on the fluores-
　　cence of aqueous dye solution.
　　Rohatagi, K.K., and Singhal, G.S.
　　J. Phys. Chem. 67: 2844-6 (1963)
　　CA 60: 7584e

D 1622 Fluorescence in two-pigment systems.
　　Rosenberg, J.L., and Bigat, T.
　　Natl. Acad. Sci. — Natl. Res. Council,
　　Misc. Publ. No. 1145, 122-3 (1963)
　　CA 60: 14823f

D 1623 Measuring fluorescent pigments.
　　Roux, C.
　　Farbe 10: 91-100 (1963)
　　CA 61: 5102h

D 1624 Identification of some corticosteroids
　　under Wood's light.
　　Russu, C., and Cruceanu, I.
　　Rev. Chim. (Bucharest) 14: 48-9 (1963)
　　CA 59: 10415d

D 1625 Identification of certain corticoster-
　　oids in Wood's light.
　　Russu, C. and Cruceanu, I.
　　Farmacia (Bucharest) 11: 251-5 (1963)
　　CA 59: 11824h

D 1626 Zeeman and uniaxial stress spectra of
　　$CaF_2(Eu^{++})$.
　　Runciman, W.A., and Stager, C.V.
　　J. Chem. Phys. 38: 279-80 (1963)
　　CA 58: 10863c

D 1627 Fluorescence and absorption spectra
　　of azulene in frozen crystalline solu-
　　tions.
　　Ruziewicz, Z.
　　Bull. Acad. Polon. Sci., Ser. Sci., Math.,
　　Astron. Phys. 11: 79-83 (1963)
　　CA 59: 2301g

D 1628 Escape probability of the trapped elec-
　　trons in phosphorescent zinc sulfide at
　　various temperatures.
　　Saddy, J.
　　J. Physiol. (Paris) 24: 199-202 (1963)
　　CA 59: 7087b

D 1629 Fluorescence and lifetimes of Eu
　　chelates.
　　Samelson, H., and Lempecki, A.
　　J. Chem. Phys. 39 (1): 110-12 (1963)
　　CA 59: 2290g

D 1630 Fluorescence spectral study of wave-
　　length shifters for scintillation plastics.
　　Sandler, S.R., and Tsou, K.C.
　　J. Chem. Phys. 39: 1062-7 (1963)
　　CA 59: 7089e

D 1631 Semiconductivity and color of pyran-
    threne.
Sano, M., and Akamatsu, H.
Bull. Chem. Soc. Japan 36: 1695-6 (1963)
CA 60: 6314e

D 1632 The relative yields of fluorescence
    and phosphorescence of biacetyl in
    fluid solutions.
Sandros, K., and Almgren, M.
Acta Chem. Scand. 17: 552-3 (1963)
CA 58: 13303h

D 1633 The electronic structure of complexes
    with a d^3 configuration in ligand field
    theory.
Sartori, G., Cervone, E., and
    Cancellieri, P.
Atti Accad. Nazl. Lincei, Rend., Classe
    Sci. Fis., Mat. Nat. 35: 226-32 (1963)
CA 61: 10184h

D 1634 Comparison of spectrophotometric
    and spectrophotofluorometric methods
    for the determination of malonaldehyde.
Sawicki, E., Stanley, T.W., and Johnson, H.
Anal. Chem. 35: 199-205 (1963)

D 1635 Fluorometric determination of
    malonaldehyde.
Sawicki, E., Stanley, T.W., and Johnson, H.
Chemist-Analyst 52: 4 (1963)

D 1636 Histones using fluorescent and iso-
    topic labeling.
Savage, R.E.
Dissertation, University of Wisconsin
    (1963)
CA 60: 13477g

D 1637 Spectral and photochemical properties
    of water-soluble chlorophyll analogs.
Savkina, I.G., and Eustigneev, V.B.
Biofizika 8: 335-43 (1963)
CA 59: 4896f

D 1638 Absorption and fluorescence spectra
    of water-soluble analogs.
Savkina, I.G., and Evstigneev, V.B.
Izv. Akad. Nauk SSSR, Ser. Fiz. 26: 782-6
    (1963)
CA 60: 131f

D 1639 Fluorescent components of ion ex-
    change resins.
Saxby, M.J.
Chem. Ind. (London) 1963: 1725 (1963)
CA 60: 1245d

D 1640 Fluorescence of binary systems of
    crystalline aromatic hydrocarbons.
Schmillen, A., and Ledger, R.
Z. Naturforsch. 18: 1-9 (1963)
CA 58: 12081d

D 1641 Investigations of the luminescence of
    doped single crystals of aromatic
    hydrocarbons.
Schmillen, A., and Kohlmannsperger, J.
Z. Naturforsch. 18: 627-32 (1963)
CA 59: 5904a

D 1642 Polarization of electron bonds of aro-
    matic compounds. VII. Indole, inda-
    zole, benzimidazole, bentriazole, and
    carbazole.
Schuett, H., and Zimmerman, H.
Ber. Bunsenges. Physik. Chem. 67: 54-62
    (1963)
CA 58: 13301a

D 1643 Direct determination of fluorescent
    substances in thin layer chromato-
    grams.
Seiler, N., Werner, G., and Wiechmann,
    M.
Naturwissenschaften 50: 643 (1963)
CA 60: 26a

D 1644 Fluorometric determination of sele-
    nium in plants and animals with
    3,3'-diaminobenzidine. Caution notice.
Serlin, I.
Anal. Chem. 35: 2221 (1963)

D 1645 Quasi-line electron vibration spectra
    of porphine and dihydroporphine.
Sevchenko, A.N., Solov'ev, K.N.,
    Shkivman, S.F., and Sarzhevskaya,
    M.V.
Dokl. Akad. Nauk SSSR 153: 1391-4 (1963)
CA 60: 12794a

D 1646 Spectrofluorometric determination of
    some rare earth elements.
Sevchenko, A.N., and Kuznetsova, V.V.
Redkozem. Elementy, Akad. Nauk SSSR,
    Inst. Geokhim. Analit. Khim. 1963:
    358-61 (1963)
CA 61: 2474d

D 1647 Fluorescent reaction of scandium
    with 8-hydroxyquinoline.
Shcherbov, D.P., and Lovchi, A.K.
Redkozem. Elementy, Akad. Nauk SSSR,
    Inst. Geokhim. Analit. Khim. 1963:
    362-6 (1963)
CA 61: 2475e

D 1648 A polish containing a fluorescent
brightener.
Shimizu, H., and Kimura, S.
Japan 7630, 1960 (1963)
CA 61: 10925b

D 1649 Ultraviolet fluorescence of muscles
and muscle proteins.
Shtrankfel'd, I.G.
Biofizika 8: 690-5 (1963)
CA 60: 5957f

D 1650 Radiation-free transitions and fluor-
escence capacity in molecules. Deri-
vatives of the diphenylmethyl cation
and other torsionable aromatic systems.
Simon, Z.
Rev. Chim., Acad. Rep. Populaire
Roumaine 8: 211-6 (1963)
CA 61: 5105d

D 1651 Double-photon excitation of fluores-
cence in anthracene single crystals.
Singh, S., and Stoicheff, B.P.
J. Chem. Phys. 38: 2032-3 (1963)
CA 59: 4694g

D 1652 Molecular excitation of water by
γ-irradiation.
Sitharamarao, D.N., and Duncan, J.F.
J. Phys. Chem. 67: 2126-32 (1963)
CA 59: 12320c

D 1653 Effect of hydrogen peroxide on fluor-
escence color of estrone in concentra-
ted sulfuric acid.
Smoezkiewiczowa, A., and Sioda, R.
J. Pharm. Pharmacol. 15: 486 (1963)
CA 59: 7800a

D 1654 Investigation of spectra of solid solu-
tions of anthracene in crystals of a
series of polyphenyls at 20.4°K.
Solov'ev, A.V.
Fiz. Probl. Spektroskopii, Akad. Nauk
SSSR, Materialy 13-go Soveshch.,
Leningrad, 1960, 2: 205-7 (1963)
CA 59: 12318h

D 1655 Spectral luminescent properties of
benzoporphyrins.
Solov'ev, K.N., Shkirman, S.F., and
Kachura, T.F.
Izv. Akad. Nauk SSSR, Ser. Fiz. 27: 767-71
(1963)
CA 60: 1245f

D 1656 The luminescence characteristics of
some group III-VI compounds.
Springford, M.
Proc. Phys. Soc. (London) 82: 1020-8
(1963)
CA 60: 1246f

D 1657 The sectoral distribution of lumines-
cent centers in synthetic quartz.
Starodubtsev, S.V., Vakhidov, S.A., and
Tsinober, L.I.
Kristallografiya 8: 770-3 (1963)
CA 60: 4913f

D 1658 Phosphorescence of deoxyribonucleic
acid (DNA).
Stauff, J., and Mennigmann, H.D.
Z. Naturforsch. 18b: 852 (1963)
CA 60: 7062d

D 1659 Phosphorescence in cis-stilbene.
Stegemeyer, H., and Perkampus, H.H.
Z. Physik. Chem. 39: 125-8 (1963)
CA 60: 6312c

D 1660 The ultraviolet fluorescence of pro-
teins. I. The influence of pH and
temperature.
Steiner, R.F., and Edelhoch, H.
Biochim. Biophys. Acta 66: 391-55 (1963)
CA 59: 9863h

D 1661 Fluorescence of some 17α-substituted
steroids in concentrated HCl.
Steinetz, B.G., Beach, V.L., Dubnick, B.,
Meli, A., and Fujimoto, G.
Steroids 1: 395-408 (1963)
CA 59: 6470b

D 1662 Versatile technique for measuring
fluorescence decay times in the nano-
second region.
Steingraber, O.J., and Berlman, I.B.
Rev. Sci. Instr. 34: 524-9 (1963)
CA 60: 11508d

D 1663 Triplet-triplet annihilation and delayed
fluorescence in molecular aggregates.
Sternlicht, H., Nieman, G.C., and Robin-
son, G.W.
J. Chem. Phys. 38: 1326-35 (1963)
CA 58: 8525b

D 1664 Errata: Triplet-triplet annihilation and delayed fluorescence in molecular aggregates; and comments concerning ruby-laser-induced fluorescence in anthracene crystals.
Sternlicht, H., Nieman, G.C., and Robinson, G.W.
J. Chem. Phys. 39: 1610-11 (1963)
CA 59: 13998b

D 1665 The kinetics of excimer formation in fluid media.
Stevens, B., Ban, M.I., and Walker, M.S.
U.S. Dept. Comm., Office Tech. Serv.
AD 430,510, 50 pp. (1963)
CA 61: 15548h

D 1666 Excimer fluorescence of some naphthalene derivatives in the molten state.
Stevens, B., and Dickinson, T.
J. Chem. Soc. 1963: 5492-6 (1963)
CA 60: 133g

D 1667 Simultaneous quenching of molecular fluorescence by oxygen and biacetyl in solution.
Stevens, B., and Dubois, J.T.
Trans. Faraday Soc. 59: 2813-19 (1963)
CA 60: 4924h

D 1668 Delayed fluorescence from microcrystalline aromatic hydrocarbons.
Stevens, B., and Hutton, E.
Proc. Phys. Soc. (London) 81: 893-7 (1963)
CA 58: 12080b

D 1669 Delayed fluorescence in aromatic hydrocarbon vapor.
Stevens, B., Walker, M.S., and Hutton, E.
Proc. Chem. Soc. 1963: 62 (1963)
CA 58: 12081c

D 1670 Phosphorescence and delayed fluorescence lifetimes of pyrene in liquid paraffin.
Stevens, B., and Walker, M.S.
Proc. Chem. Soc. 1963: 181 (1963)
CA 59: 9479g

D 1671 Spectroscopic studies with the ruby optical maser.
Stoicheff, B.P.
Proc. Colloq. Spectros. Intern., 10th, Univ. Maryland 1962: 399-415 (1963)
CA 61: 12825e

D 1672 Decay behavior of liquid organic scintillators.
Stolz, W.
Wiss. Z. Tech. Univ. Dresden, Beih. 12: 1167-71 (1963)
CA 61: 2678a

D 1673 Thin film analysis by X-ray fluorescence.
Stone, R.R., and Potts, K.T.
Norelco Reptr. 10: 94-7 (1963)
CA 59: 9631g

D 1674 Optical spectrum of bivalent vanadium in octahedral coordination.
Sturge, M.D.
Phys. Rev. 130: 639-46 (1963)
CA 58: 10870a

D 1675 Strain-induced splitting of the ^2E state of V^{2+} in MgO.
Sturge, M.D.
Phys. Rev. 131: 1456-8 (1963)
CA 59: 7090b

D 1676 Electronic spectra and hydrogen bonding. II. Anthrols.
Suzuki, S., and Baba, H.
J. Chem. Phys. 38: 349-53 (1963)
CA 58: 4059d

D 1677 Dependence of the quenching of fluorescence by foreign substances on the viscosity of the solution.
Sveshnikov, B.Y., Selivanenko, A.S., Shirokov, V.I., and Kiyanskaya, L.A.
Opt. i Spektroskopiya 14: 45-8 (1963)
CA 58: 12080f

D 1678 Electronic spectra of aromatic molecular crystals. II. Crystal structure and spectra of perylene.
Tanaka, J.
Bull. Chem. Soc. Japan 36: 1237-49 (1963)
CA 60: 123f

D 1679 Delayed fluorescence spectrum of pyrene solutions at low temperatures.
Tanaka, C., Tanaka, J., Hutton, E., and Stevens, B.
Nature 198: 1192 (1963)
CA 59: 13496f

D 1680 Quantitative immunofluorescein titration of human and bovine gamma globulins.
Tengerdy, T.R.
Anal. Chem. 35: 1084-6 (1963)

D 1681 Quasilinear phosphorescence spectra of phenanthrene solutions.
Teplyakov, P.A.
Opt. i Spektroskopiya 15: 645-50 (1963)
CA 60: 4972d

D 1682 Phosphorescence spectra of phenanthrene in heptane and in magnesium oxide.
Teplyakov, P.A., and Grosul, V.P.
Ukr. Fiz. Zh. 8: 864-9 (1963)
CA 59: 9478b

D 1683 Phosphorescence spectra of fluorene in heptane and hexane.
Teplyakov, P.A., and Trusov, V.V.
Ukr. Fiz. Zh. 8: 1280-2 (1963)
CA 60: 8796a

D 1684 Bound excitons in CaP.
Thomas, D.G., Gershenzon, M., and Hopfield, J.J.
Phys. Rev. 131: 2397-403 (1963)
CA 59: 9428h

D 1685 Energy levels of Tb^{+++} in $LaCl_3$ and other chlorides.
Thomas, K.S., Singh, S., and Dieke, G.H.
J. Chem. Phys. 38: 2180-90 (1963)
CA 58: 12075g

D 1686 Some benzofuran and naphthofuran derivatives used as scintillators in solution.
Tibu, M., Viscrian, I., Arventiev, B. Offenberg, H., and Nicolaescue, T.
Analele Stiint. Univ. "A.I. Cuza," Iasi, Sect. I, 9: 261-70 (1963)
CA 60: 2434b

D 1687 Dependence of the effective excitation energy of vapor molecules on the frequency of exciting radiation.
Tolkachev, V.A.
Izv. Akad. Nauk SSSR, Ser. Fiz. 27: 584-7 (1963)
CA 59: 10885f

D 1688 Average energy of the molecules of rarefied fluorescent vapor.
Tolkachev, V.A., and Borisevich, N.A.
Opt. i Spektroskopiya 15: 306-9 (1963)
CA 59: 14764d

D 1689 Temperature dependence of the fluorescence yield of complex molecules in the vapor phase and the activation energy of radiationless transitions.
Tolkachev, V.A., and Borisevich, N.A.
Opt. i Spektroskopiya, Akad. Nauk SSSR, Otd. Fiz.-Mat. Nauk, Sb. Statei 1: 16-21 (1963)
CA 59: 5954c

D 1690 Relation between the mean energy of excited vapor molecules and the frequency of the absorbed radiation.
Tolkachev, V.A., and Borisevich, N.A.
Opt. i Spektroskopiya, Akad. Nauk SSSR, Otd. Fiz.-Mat. Nauk, Sb. Statei 1: 22-8 (1963)
CA 61: 8249f

D 1691 Energy transfer by resonance in KCl/Ag, Pb single crystals.
Tomura, M., and Nishimura, H.
J. Phys. Soc. Japan 18: 277-81 (1963)
CA 59: 5952h

D 1692 Fluorometric microdetermination of carbohydrates.
Towne, J.C., and Spikner, J.E.
Anal. Chem. 35: 211-4 (1963)

D 1693 Separation of two light-induced electron-spin-resonance signals in several algal species.
Trehorne, R.W., Brown, T.E., and Vernon, L.P.
Biochim. Biophys. Acta. 75: 324-32 (1963)
CA 60: 3274d

D 1694 Acenaphthene and biphenyl phosphorescence spectra.
Trusov, V.V., and Teplyakov, P.O.
Ukr. Fiz. Zh. 8: 1353-7 (1963)
CA 60: 11509c

D 1695 Investigation by fluorescence on the mechanism of the proteolytic reactions in monaqueous media. I. The dissociation reaction of strong acids.
Urban, W., and Weller, A.
Ber. Bunsenges. Physik. Chem. 67: 787-91 (1963)
CA 60: 1151e

D 1696 Modification of the Zeiss spectrophotometer for high-sensitivity fluorimetric determinations. Medicalbiological value of the 5-hydroxytryptamine determinations.
Urinceanu, R., and Uluitu, M.
Acad. Rep. Populaire Romine, Studii Cercetari Fiziol. 8: 649-61 (1963)
CA 60: 13558c

D 1697 Method of determining fluorescent substances in tissues and biological fluids and its use for the study of the permeability of the tissue-blood barriers.
Utevskayer, L.B.
Akad. Nauk SSSR, Inst. Biol. Fiz. Sb. Rabot 1963: 209-15 (1963)
CA 60: 11035f

D 1698 Fluorescence properties of tetrahydrofolate and related compounds.
Uyeda, K., and Rabinowitz, J.C.
Anal. Biochem. 6: 100-9 (1963)
CA 59: 15536d

D 1699 Reflectance fluorescence spectra of aromatic compounds in potassium bromide pellets.
Van Duuren, B.L., and Bardi, C.E.
Anal. Chem. 35: 2198-202 (1963)
CA 60: 6361e

D 1700 Determination of the mean life of the first excited state of some complex aromatic compounds in aerated solution and the absolute quantum yield of fluorescence of biacetyl in solution.
Van Hermert, R.L.
U.S. Dept. Comm., Office Tech. Serv. AD 418,562, 71 pp. (1963)
CA 61: 11497c

D 1701 A comparison of the intensities of emission of Eu^{+++} and Tb^{+++} in tungstates and molybdates.
Van Uitert, L.G.
J. Electrochem. Soc. 110: 46-51 (1963)
CA 58: 5154a

D 1702 Unusual infrared fluorescence in $PrCl_3$ (Nd^{+3}) crystals.
Varsanyi, F.
Phys. Rev. Letters 11: 314-16 (1963)
CA 60: 131b

D 1703 Blue-violet fluorescence of fluorite.
Vasil'kova, N.N.
Mineral'n. Syr'e, Vses. Nauchn.-Issled. Inst. Mineral'n. Syr'ya 7: 55-61 (1963)
CA 59: 15033h

D 1704 Fluorescence of several complexes of manganese chloride and some alkaloids.
Velasevic, K.B., Slivic, S., and Nikolic, K.
Compt. Rend. 257: 3855-7 (1963)
CA 60: 8796h

D 1705 Fluorescent salts of popaverine with zinc and cadmium halides.
Velasevic, K.B., Slevic, S., and Buric, I.
Compt. Rend. 257: 4163-6 (1963)
CA 60: 10079a

D 1706 Fluorescence spectra of various natural antiamebics and therapeutic derivatives.
Viel, C.
Compt. Rend. 256: 4770-3 (1963)
CA 59: 4979a

D 1707 Fluorescent materials from Ephestia kuehniella Zeller. IV. Synthesis of erythropterin, ekapterin, and lepidopterin.
Viscontini, M., and Stierlin, H.
Helv. Chim. Acta 46: 51-6 (1963)
CA 58: 13954d

D 1708 Ultrahigh-frequency phase fluorometer.
Volkov, S.V., Limereva, C.C., and Shikorov, V.I.
Izv. Akad. Nauk SSSR, Ser. Fiz. 27: 558-61 (1963)
CA 59: 13496h

D 1709 Fluorescence of adenine and inosine nucleotides.
Walaas, E.
Acta Chem. Scand. 17: 461-3 (1963)
CA 59: 1873d

D 1710 Fluorescence of magnesium, calcium, and zinc 8-quinolinol complexes.
Watanabe, S., Frantz, W., and Trottier, D.
Anal. Biochem. 5: 345-59 (1963)
CA 58: 11948h

D 1711 Optical properties of chromophoremacromolecule complexes: absorption and fluorescence of acridine dyes bound to polyphosphates and DNA.
Weill, G., and Calvin, M.
Biopolymers 1: 401-17 (1963)
CA 60: 8263c

D 1712 Fluorescence of naphthacene vapor.
Williams, R., and Goldsmith, G.J.
J. Chem. Phys. 39: 2008-11 (1963)
CA 59: 10900e

D 1713 Phosphorimetry as a means of chemical analysis. The analysis of aspirin in blood serum and plasma.
Winefordner, J.D., and Latz, H.W.
Anal. Chem. 35: 1517-1522 (1963)

D 1714 Solvents for phosphorimetry.
Winefordner, J.D., and St. John, P.A.
Anal. Chem. 35: 2211 (1963)

D 1715 Energy-transfer studies by spectrophotofluorometric method.
Wilkinson, F., and Dubois, J.T.
J. Chem. Phys. 39: 377-83 (1963)
CA 59: 3442d

D 1716 XVIIth scientific meeting of the protein foundation (Cambridge, Mass., 1962). Application of studies of fluorescence in structural determinations in proteins.
Winkler, M.H.
Vox Sanguinis 8: 113 (1963)
CA 59: 6633h

D 1717 Fluorescence of europium thenoyltrifluoroacetonate. I. Evaluation of laser threshold parameters.
Winston, H., Marsh, O.J., Susuki, C.K., and Telk, C.L.
J. Chem. Phys. 39: 267-71 (1963)
CA 59: 3442f

D 1718 Trace determination by optical luminescence in alkaline earth oxides.
Witzmann, H., and Piesche, L.
Reinststoffe Wiss. Tech., Intern. Symp. 1, Dresden 1961: 439-48 (1963)
CA 60: 15116h

D 1719 Optical maser action in a Eu^{+++}-containing organic matrix.
Wolff, N.E., and Pressley, R.J.
Appl. Phys. Letters 2: 152-4 (1963)
CA 59: 1204g

D 1720 Absorption and fluorescence spectra of several praseodymium-doped crystals and the change of covalence in the chemical bonds of the praseodymium ion.
Wong, E.Y., Stafsudd, O.M., Johnston, D.R.
J. Chem. Phys. 39: 786-93 (1963)
CA 59: 4697g

D 1721 Energy levels of Yb^{3+} in garnets.
Wood, D.L.
J. Chem. Phys. 39: 1671-3 (1963)
CA 59: 9456b

D 1722 The fluorescence of some tetracyanoethylene complexes.
Wyant, R.E., Poziomek, E.J., and Poirier, R.H.
Anal. Chim. Acta 28: 496-8 (1963)
CA 59: 4696g

D 1723 A fluorescence test on carpeting.
Wylezich, A.
Spinner Weber Textilveredl. 81: 1334-5 (1963)
CA 60: 12162h

D 1724 Fluorescence of styrene homopolymers and copolymers.
Yanari, S.S., Bovey, F.A., and Lumry, R.
Nature 200: 242-4 (1963)
CA 60: 1889c

D 1725 The energy distribution in the continuous absorption spectrum of xenon.
Yankov, V.V.
Opt. i Spektroskopiya 14: 29-34 (1963)
CA 58: 7504g

D 1726 Vibronics in gadolinium compounds.
Yatsiv, S., Adato, I., and Goren, A.
Phys. Rev. Letters 11: 108-10 (1963)
CA 59: 10885a

D 1727 Structural transformations of bovine pancreatic ribonuclease in solution: a study of polarization of fluorescence.
Young, D.M., and Potts, J.T., Jr.
J. Biol. Chem. 238: 1995-2002 (1963)
CA 58: 14390g

D 1728 Delayed fluorescence of aromatic hydrocarbons.
Zander, M.
Naturwissenschaften 50: 327-8 (1963)
CA 59: 130h

D 1729 Experiments with the optical excitation of mercury vapor for the production of isolated fluorescence.
Zito, R., and Schraeder, A.E.
U.S. Dept. Comm., Office Tech. Serv. AD 412,951, 22 pp. (1963)
CA 60: 8797c

D 1730 An enrichment of zircon and monazite in the contract zone of the Rattlesnake granite.
Zimmerle, W.
Neues Jahrb. Mineral., Abhandl. 100: 164-84 (1963)
CA 60: 6650b

D 1731 Fluorescent agents for detergents.
Zussman, H.W.
J. Am. Oil Chemists' Soc. 40: 695-8 (1963)
CA 60: 5753h

D 1732 Action of nuclear radiation on the luminescence of silicon carbide.
Zvyagin, V.I., Lobanov, E.M., Dutov, A.G., Leushkina, G.V., and Maksumov, R.
Radiats. Effekty v Tverd. Telakh, Akad. Nauk Uz. SSR, Inst. Yadern. Fiz. 1963: 50-5 (1963)
CA 60: 132e

D 1733 The effect of photosynthesis inhibitors on oxygen evolution and fluorescence of illuminated chlorella.
Zweig, G., Tamas, I., and Greenberg, E.
Biochim. Biophys. Acta 66: 196-203 (1963)
CA 58: 11685h

D 1734 Luminescence properties of some purines and pyrimidines. A study by fluorescence spectrophotometry of the sites of protonation and of the types of lowest excited singlet states.
Acta Chem. Scand. 17: 921-9 (1963)
CA 61: 5955d

D 1735 Solvent effects on the absorption and fluorescence spectra of naphthylamines and isomeric aminobenzoic acids.
Bull. Chem. Soc. Japan 36: 654-62 (1963)
CA 59: 5955g

D 1736 Fluorescence of nitric oxide. I. Determination of the mean lifetime of the $A^2\Sigma^+$ state. II. Vibrational energy transfer between $NDA^2\Sigma^+$ (γ =3, 2, and 1) and $N_2X^1\Sigma^+_g$ (v =O).
Trans. Faraday Soc. 59: 1270-47 (1963)
CA 59: 9478c

D1737 Spectroscopic investigations concerning fluorimetric steroid analysis.
Abraham, R., and Staudinger, H.
Z. Klin. Chem. 2: 16-21 (1964)
CA 61: 14976f

D 1738 An automatic recording polarization spectrofluorimeter.
Ainsworth, S., and Winter, E.
Appl. Opt. 3: 371-83 (1964)
CA 61: 5103h

D 1739 Fluorescent white dyes: calculation of fluorescence from reflectivity values.
Allen, E.
J. Opt. Soc. Am. 54: 505-15 (1964)
CA 61: 1361g

D 1740 Absolute fluorescent quantum efficiency of sodium salicylate.
Allison, R., Burns, J., and Tuyzolino, A.J.
J. Opt. Soc. Am. 54: 747-51 (1964)
CA 61: 6543c

D 1741 Kininase inhibition by a fluorescent substance prepared from liver.
Amundsen, E., Waaler, B.A., Dedichen, J., Laland, P., Laland, S., and Thorsdalen, N.
Nature 203: 1245-8 (1964)
CA 61: 14965h

D 1742 Fluorometric determination of uranium with Rhodamine B.
Andersen, N.R., and Hercules, D.M.
Anal. Chem. 36: 2138-41 (1964)

D 1743 Anomolous storage of light in phosphors.
Antonov-Romanovshii, V.V., Vinokurov, L.A., and Fok, M.V.
Opt. i Spektroskopiya 16: 279-84 (1964)
CA 60: 15277b

D 1744 Spectral characteristics of fluorescent chelates.
Argauer, R.J.
Dissertation, University of Maryland (1964)
CA 61: 11494g

D 1745 Fluorescent compounds for calibration of excitation and emission units of a spectrofluorometer.
Argauer, R.J., and White, C.E.
Anal. Chem. 36: 368171 (1964)
CA 60: 12794b

D 1746 Effect of substituent groups on fluorescence of metal chelates.
Argauer, R.J., and White, C.E.
Anal. Chem. 36: 2141-4 (1964)

D 1747 Fluorescence spectrum of uranyl ion in neutron-irradiated cesium uranyl nitrate.
Asundi, R.K., and Dixit, R.M.
Current Sci. (India) 33: 38-9 (1964)
CA 60: 11509a

D 1748 Preparation of di-cesium uranyl-nitrate, $Cs_2UO_2(NO_3)_4$.
Asundi, R.K., and Dixit, R.M.
Current Sci.(India) 33: 332-3 (1964)
CA 61: 5174e

D 1749 Optical studies of biochemical events in the electric organ of electrophorus.
Aubert, X., Chance, B., and Keynes, R.D.
Proc. Roy. Soc. (London), B 160: 211-45 (1964)
CA 61: 2227d

D 1750 Fluorescence and energy transfer in $Y_2O_3:Eu^{+3}$.
Axe, J.D., and Weller, P.F.
J. Chem. Phys. 40: 3066-9 (1964)
CA 60: 15314h

D 1751 Fluorescence of GaAs under intense electron excitation.
Babcock, R.V.
J. Appl. Phys. 35: 3354-7 (1964)
CA 61: 15549f

D 1752 The prototropic equilibrium constants of 3,4-benzocinnoline in electronically excited states.
Bollard, R.E., and Edwards, J.W.
Spectrochim. Acta 20: 1275-81 (1964)
CA 61: 6530h

D 1753 The relationship between ultraviolet fluorescence and the protein content of milk. A preliminary investigation.
Bakalor, S.
Australian J. Dairy Technol. 19: 29 (1964)
CA 61: 13803f

D 1754 Universal intermolecular interactions and their effect on the position of the electronic spectra of molecules in two-component solutions. VII. Theory (general case of an isotropic solution).
Bakhshiev, N.G.
Opt. i Spektroskopiya 16: 821-32 (1964)
CA 61: 9043h

D 1755 Investigation of the fluorescence and photochemical primary processes in a vacuum-ultraviolet of NH_3, N_2H_4, PH_3, and reactions of the electron-excited radicals $NH^+(^1\pi)$, $NH^+(^3\pi)$, and $PH^+(^3\pi)$.
Becker, K.H., and Welge, K.H.
Z. Naturforsch. 19a: 1006-15 (1964)
CA 61: 11495b

D 1756 Fluorescent detection and spectro-fluorometric characterization and estimation of carbazoles and polynuclear carbazoles separated by thin layer chromatography.
Bender, D.F., Sawicki, E., and Wilson, R.M., Jr.
Anal. Chem. 36: 1011-17 (1964)
CA 61: 2487d

D 1757 Electric and magnetic properties of polymers with conjugated double bonds. II. Fluorescence of poly (phenylacetyl-enes).
Benderskii, V.A., and Stunzhas, P.A.
Vysokomolekul. Soedin. 6: 1104-10 (1964)
CA 61: 8430f

D 1758 Radiationless intermolecular energy transfer. I. Singlet → singlet transfer.
Bennett, R.G.
J. Chem. Phys. 41: 3037-40 (1964)
CA 61: 15528h

D 1759 Concentration dependence of the energy transfer in anthracene mixed crystals.
Benz, K.W., and Wolf, H.C.
Z. Naturforsch. 19a: 177-81 (1964)
CA 60: 15314a

D 1760 Improved fluorescence decay-time measuring apparatus.
Berlman, I.B., and Steingraber, O.J.
Trans. Nucl. Sci. 11: 27-8 (1964)
CA 61: 10280b

D 1761 Phosphorescence in nucleotides and nucleic acids.
Bersohn, R., and Isenberg, I.
J. Chem. Phys. 40: 3175-80 (1964)
CA 61: 878h

D 1762 Cellular localization of 5-hydroxy-
tryptamine in the rat pineal gland.
Bertler, A., Falck, B., and Owman, C.
Kgl. Fysiograf. Sallskap. Lund, Forh. 33:
13-16 (1963)
CA 60: 13645a

D 1763 Photodissociation of H_2, N_2, O_2, NO,
CO, H_2O, CO_2, and NH_3 in extreme
vacuum-ultraviolet.
Beyer, K.D., and Welge, K.H.
Z. Naturforsch. 19: 19-28 (1964)
CA 60: 10093g

D 1764 Quenching and temperature depend-
ence of fluorescence in rare-earth
chelates.
Bhaumik, M.L.
J. Chem. Phys. 40: 3711-15 (1964)
CA 61: 2615h

D 1765 Relaxation in europium chelates.
Bhaumik, M.L.
J. Chem. Phys. 41: 574-5 (1964)
CA 61: 10199h

D 1766 Effect of impurities upon the fluores-
cence spectra of rare-earth chelates.
Bhaumik, M.L., Yamakawa, K.A., and
Tannenbaum, I.R.
Proc. Conf. Rare Earth Res., 3rd, Clear-
water, Fla., 1963: 491-7 (1964)
CA 61: 2615h

D 1767 Geochemical petroleum prospecting.
Biederman, E.W., Jr., and Heinze, B.
U.S. 3,149,068, Sept. 15, 1964
CA 61: P15907b

D 1768 Delayed fluorescence of pyrene solu-
tion (Correction).
Birks, J.B.
J. Phys. Chem. 68: 439-40 (1964)
CA 60: 10078a

D 1769 Excimer fluorescence spectra of
aromatic liquids.
Birks, J.B., and Aladekomo, J.B.
Spectrochim. Acta 20: 15-21 (1964)
CA 60: 3634h

D 1770 A comparison of the scintillation and
photofluorescence spectra of organic
solutions.
Birks, J.B., Braga, C.L., and Lumb, M.D.
Brit. J. Appl. Phys. 15: 399-404 (1964)
CA 60: 12794c

D 1771 Excimer fluorescence. III. Lifetime
studies of 1:2-benzanthracene derivatives in solution.
Birks, J.B., Dyson, D.J., and King, T.A.
Proc. Roy. Soc. (London), Ser. A 277:
270-8 (1964)
CA 60: 3635d

D 1772 Excimer fluorescence. IV. Solution
spectra of polycyclic hydrocarbons.
Birks, J.B., and Christophorou, L.G.
Proc. Roy. Soc. (London), Ser. A 277:
571-82 (1964)
CA 60: 6362c

D 1773 Excimer flourescence. V. Influence
of solvent viscosity and temperature.
Birks, J.B., Lumb, M.D., and Munro, I.H.
Proc. Roy. Soc. (London), Ser. A 280:
289-97 (1964)
CA 61: 3831f

D 1774 Fluorescent properties of some
europium-activated phosphors.
Bril, A., and Wanmaker, W.L.
J. Electrochem. Soc. 111: 1363-8 (1964)
(Eng.)
CA 62: 141c

D 1775 Peculiarities of luminescence of
amylase in the crystalline state.
Bobrovich, V.P., and Konev, S.V.
Dokl. Akad. Nauk SSSR 155: 197-200 (1964)
CA 60: 14025c

D 1776 Evaluation of the bleaching effect of
fluorescent materials.
Bocharov, V.G.
Vestn. Mosk. Univ., Ser. III, Fiz., Astron.
19: 52-5 (1964)
CA 61: 14838a

D 1777 Extension of the limits of applicability
of the E.V. Shpol'skii effect.
Bogomolov, S.G., and Silant'ev, B.Y.
Spektroskopiya, Metody i Primenenie,
Akad. Nauk SSSR, Sibirsk. Otd. 1964:
200-2 (1964)
CA 62: 127h

D 1778 Anti-Stokes fluorescence of molecules.
Borisevich, N.A., Gruzinskii, V.V., and
Tolkachev, V.A.
Opt. i Spektroskopiya 16: 171-4 (1964)
CA 60: 14024f

D 1779 Luminescence of complexes of acrid-
ine orange with nucleic acids.
Borisova, O.F., and Tumerman, L.A.
Biofizika,9: 537-44 (1964)
CA 61: 14941a

D 1780 Fluorescence and phosphorescence
of hexafluoroacetone vapor.
Bowers, P.G., and Porter, G.B.
J. Phys. Chem. 68: 2982-5 (1964)
CA 61: 12810e

D 1781 Modifications of histidine residues
leading to the appearance of visible
fluorescence.
Brand, L., and Shaltiel, S.
Biochim. Biophys. Acta 88: 338-51 (1964)
CA 61: 16330b

D 1782 The dependence of electronic energy
transfer efficiency on exciting wavelength.
Braun, C.L.
Dissertation, University of Minnesota
(1964)
CA 62: 7240c

D 1783 Stimulated emission processes in
some rare earth chelates.
Brecher, C., Lempicki, A., and Samelson,
H.
Nature 202: 580-1 (1964)
CA 61: 3840h

D 1784 Spectrographic analysis and the fluor-
escence of zircons from Madagascar.
Briere, Y., and Kurylenko, C.
Cahiers Phys. No. 165: 215-21 (1964)
CA 62: 4596g

D 1785 A study of the calcium molybdate-
rare earth niobate systems.
Brixner, L.H.
J. Electrochem. Soc. 111: 690-7 (1964)
CA 61: 190a

D 1786 Calcium orthovanadate.
Brixner, L.H., Flournoy, P.A., and
Babcock, K.
J. Electrochem. Soc. 111: 873-4 (1964)
CA 61: 6469d

D 1787 γ-Excitation of the singlet and triplet
states of naphthalene in solution.
Brochlehurst, B., Porter, G., and Yates,
J.M.
J. Phys. Chem. 68: 203-5 (1964)
CA 60: 7584g

D 1788 An estimate of the effective sizes of
chlorophyll a aggregates in vivo as de-
termined from emission spectra.
Brody, S.S.
J. Theoret. Biol. 7: 352-9 (1964)
CA 62: 4366e

D 1789 Geometrical rearrangement of aggre-
gated chlorophyll in vivo during the
greening process.
Brody, S.S.
Nature 204: 470-1 (1964) (Eng.)
CA 62: 4366f

D 1790 Fluorescence lifetimes of ruby.
Brown, G.C., Jr.
J. Appl. Phys. 35: 3062-3 (1964)
CA 62: 155b

D 1791 Measurement of relaxation-times of
paramagnetic ions in crystals: ruby.
Brown, G.C., Jr.
US Dept. of Comm. Bull., AD 602186, 42 pp.
(1964)
CA 62: 2391e

D 1792 Limitation of the linear intensity-
concentration approximation in electron
probe microanalysis.
Brown, J.D.
Advan. X-Ray Anal. 7: 340-52 (1964)
CA 62: 1063g

D 1793 Infrared quantum counter action in
Er-doped fluoride lattices.
Brown, M.R., and Shand, W.A.
Phys. Rev. Letters 12: 367-9 (1964)
CA 61: 2615f

D 1794 Infrared quantum counter action in
HO-doped fluoride lattices.
Brown, M.R., and Shand, W.A.
Phys. Rev. Letters 11: 219-20 (1964)
CA 61: 10187d

D 1795 Energy transfer in the fluorescence of
iodine excited by the sodium D lines.
Brown, R.L., and Klemperer, W.
J. Chem. Phys. 41: 3072-89 (1964)
CA 61: 15549c

D 1796 The spectral and photochemical pro-
perties of chlorophyllin.
Broyde, S.B.
Dissertation, Polytechnic Institute of
Brooklyn, New York (1964)
CA 60: 14009g

D 1797 Effect of glycolytic poisons on the ultraviolet fluorescence of cells.
Brumberg, E.M., and Brumberg, I.E.
Biofizika 9: 748-50 (1964)
CA 62: 5773g

D 1798 Stainability with Rhodamine B and the inherent fluorescence of the cells of domestic and tropical orchids. I. Metachromatic effects in staining with Rhodamine B.
Burian, K.
Protoplasma 58: 551-60 (1964)
CA 62: 8111d

D 1798a Stainability with Rhodamine B and the inherent fluorescence of the cells of domestic and tropical orchids. II. Inherent fluorescence of normal and pathological origin.
Burian, K.
Protoplasma 58: 561-78 (1964)
CA 62: 8111d

D 1799 Nonadiabatic transitions associated with atomic-molecular collisions. Quenching of the resonance fluorescence of mercury.
Bykhovskii, V.K., and Nikitin, E.E.
Opt. i Spektroskopiya 16: 201-7 (1964)
CA 60: 12773h

D 1800 Radiative transfer of energy between rare earth ions.
Cabezas, A.Y., and Deshazer, L.G.
Appl. Phys. Letters 4: 37-9 (1964)
CA 60: 6310c

D 1801 Fluorescence of nitric oxide. III. Determination of the rate constants for the predissociation spontaneous radiation of NO $C^2\pi$ (v = O).
Callear, A.B., and Smith, I.S.M.
Discussions Faraday Soc. No. 37, 96-111 (1964)
CA 62: 3543d

D 1802 Interferometric phase-shift technique for measuring short fluorescent lifetimes.
Carbone, R.J., and Longaker, P.R.
Appl. Phys. Letters 4: 32-4 (1964)
CA 60: 11508c

D 1803 Spectrofluorometric study of (2-methyl-8-quinolinato) zinc (II) chelates in absolute ethyl alcohol.
Carter, D.A., and Ohnesorge, W.E.
Anal. Chem. 36: 327-30 (1964)
CA 60: 11508f

D 1804 Fluorescence spectra of high-molecular-weight aromatic components of Noriisk crude oil.
Chakrviani, M.K., and Usharauli, O.A.
Tr. Inst. Khim., Akad. Nauk Gruz. SSR 17: 85-102 (1964)
CA 62: 8903a

D 1805 Fluorescence emission of mitochondrial reduced diphosphopyridine nucleotide (DPNH) as a factor in the ultraviolet sensitivity of visual receptors.
Chance, B.
Proc. Natl. Acad. Sci. U.S. 51: 359-61 (1964)
CA 60: 12253b

D 1806 Novel chemiluminescent electron transfer reaction.
Chandross, E.A., and Sonntag, F.I.
J. Am. Chem. Soc. 86: 3179-80 (1964)
CA 61: 11495g

D 1807 Spectra and energy levels of Eu^{+3} in Y_2O_3.
Chang, N.C., and Gruber, J.B.
J. Chem. Phys. 41: 3227-34 (1964)
CA 61: 15550a

D 1808 Sensitized fluorescence in vapors of alkali metals. I. Energy transfer in potassium-potassium collisions.
Chapman, G.D., Krause, L., and Brockman, I.H.
Can. J. Phys. 42: 535-47 (1964)
CA 60: 10080a

D 1809 Stimulating effect of certain compounds on the quenching of chlorophyll fluorescence by nitro compounds.
Chernyuk, I.N., and Dilung, I.I.
Dokl. Akad. Nauk SSSR 156: 149-51 (1964)
CA 61: 5956h

D 1810 Use of tetracycline drugs to mark advanced fry and fingerling brook trout (Salvelinus fontinalis).
Choate, J.
Trans. Am. Fisheries Soc. 93: 309-11 (1964)
CA 62: 6849h

D 1811 Presence of two light-dependent enzymatic processes accompanying
fluorescence decay during photosynthesis in isolated wheat chloroplasts.
Chow, P.C., Shien, K.F., and Tang, P.S.
Sci. Sinica (Peking) 13: 1532-4 (1964)
CA 62: 5581f

D 1812 The presence of two light-dependent
enzymatic processes accompanying
fluorescence decay in wheat leaves
during photosynthesis.
Chow, P.C., Lin, S., Ch'u, C., Hsien, K.,
and Tang, P.S.
Chih Wu Hsueh Pao 12: 82-7 (1964)
CA 62: 5580e

D 1813 Polarization of charge-transfer bands.
Chowdhury, M., and Goodman, L.
J. Am. Chem. Soc. 86: 2777-81 (1964)
CA 61: 5095b

D 1814 Energy transfer in charge-transfer
complexes. III. Intersystem crossing.
Christodouleas, N.D., and McGlynn, S.P.
J. Chem. Phys. 40: 166-74 (1964)
CA 60: 4975d

D 1815 Phosphorescence properties of pyri-
doxal 5-phosphate.
Churchich, J.E.
Biochim. Biophys. Acta 79: 643-6 (1964)
CA 61: 5959d

D 1816 Luminescence properties of murami-
dase and reoxidized muramidase.
Churchich, J.E.
Biochim. Biophys. Acta 92: 194-7 (1964)
(Eng.)
CA 62: 2983d

D 1817 Microscopic study of vulcanizates
containing pale fillers.
Clamroth, R., and Palla, H.
Kautschuk, Gummi, Kunststoffe, Plasto-
mere, Elastomere, Duromere 17:
253-62 (1964)
CA 61: 7205g

D 1818 Fluorimetry
Conrad, A.L.
Optical methods. Treatise Anal. Chem.
I.M. Kolthoff and P.J. Elving, editors,
Interscience, 3057-78 (1964)
CA 60: 9875g

D 1819 Photoionization and absorption cross
sections of O_2 and N_2 in the 600 to
1000 A region.
Cook, G.R., and Metzer, P.H.
J. Chem. Phys. 41: 321-36 (1964)
CA 61: 5120e

D 1820 Fluorescence methods for histochem-
ical identification of monoamines. II.
Identification of the fluorescent pro-
ducts from dopamine and formalde-
hyde.
Corrodi, H., and Hillarp, N.A.
Helv. Chim. Acta 47: 911-18 (1964)
CA 61: 3002f

D 1821 Fluorescence and the structure of
proteins. III. Effects of denaturation
on fluorescence of insulin and ribonuclease.
Cowgill, R.W.
Arch. Biochem. Biophys. 104: 84-92 (1964)
CA 60: 8251b

D 1822 Triplet-triplet polarization measure-
ments in mixed crystals.
Craig, D.P., and Fischer, G.
Proc. Chem. Soc. 1964: 176-84 (1964)
CA 61: 7833e

D 1823 Absorption and fluorescence of crys-
talline naphthalene.
Craig, D.P., and Wolf, H.C.
J. Chem. Phys. 40: 2057-9 (1964)
CA 61: 176e

D 1824 Excretion of colored ultraviolet-
absorbing substances (flavonols or
tannins) by marine algae.
Craigie, J.S., and McLachlan, J.
Can. J. Botany 42: 23-33 (1964)
CA 60: 11052a

D 1825 Recombination scheme and intrinsic
gap variation in GaAs, Px semi-
conductors from electron beam and
p-n diode excitation.
Dusano, D.A., Fenner, G.E., and Carlson,
R.O.
Appl. Phys. Letters 5: 144-6 (1964)
CA 61: 15495e

D 1826 Anisotropy of fluorescence and phos-
phorescence. Naphthalene and some
halogen derivatives.
Czekalla, J., Liptay, W., and Doellefeld,
E.
Ber. Bunsenges. Physik. Chem. 68: 80-90
(1964)
CA 60: 14024a

D 1827 Fluorescence detection of the chemical relaxation of the reaction of lactate
dehydrogenase with reduced nicotinamide adenine denucleotide.
Czerlinski, G.H., and Schreck, G.
J. Biol. Chem. 239: 913-21 (1964)
CA 60: 8282g

D 1828 Application of fluorescence polariza-
tion to the antigen-antibody reaction.
Theory and experimental method.
Dandliker, W.B., Schapiro, H.C., Meduski,
J.W., Alonso, R., Feigen, G.A., and
Hamrick, J.R., Jr.
Immunochemistry 1: 165-91 (1964)
CA 62: 3247g

D 1829 X-ray fluorescence spectrometric
analysis of Fe(III), Co(II), Ni(II), and
Ca(II) chelates of 8-quinolinol.
Daugherty, K.E., Robinson, R.J., and
Mueller, J.I.
Anal. Chem. 36: 1869-70 (1964)
CA 61: 8888d

D 1830 Optical radiation from nitrogen and
air at high pressure excited by ener-
getic electrons.
Davidson, G., and O'Neil, R.
J. Chem. Phys. 41: 3946-55 (1964)
CA 62: 2371f

D 1831 A simple method of quenching the
fluorescence of fluorescent agents on
dyed textiles.
Dawson, P.R.
J. Soc. Dyers Colourists 80: 430-1 (1964)
CA 61: 9625a

D 1832 Luminescence excitation spectra and
recombination radiation of diamond in
the fundamental absorption region.
Dean, P.J., and Male, J.C.
Proc. Roy. Soc. (London), Ser. A 277:
330-47 (1964)
CA 60: 6363a

D 1833 Fluorescence of zinc and cadmium
sulfides, and manganese-amine complexes.
de la Garanderie, H.P.
Ann. Phys. 9: 649-78 (1964)
CA 62: 8537c

D 1834 The combination of chymotrypsin and
chymotrypsinogen with fluorescent
dyes.
Deranleau, D.A.
Dissertation, University of Washington,
Seattle (1964)
CA 61: 8565c

D 1835 Stimulated and fluorescent optical
emission in ruby from 4.2 to 300°K.
Zero-field splitting and mode structure.
D'Haenens, I.J., and Asawa, C.K.
Intern. Aerospace Abstr. 4: 1625 (1964)
CA 62: 8533c

D 1836 Fluorescence of transition metal ions
in crystals.
DiBartolo, B.
US Dept. of Comm. Bull., AD603459,
74 pp. (1964)
CA 62: 3543b

D 1837 Absorption fluorescence, and energy
levels of HO^{+3} in hexagonal $LaCl_3$.
Dieke, G.H., and Pandey, B.
J. Chem. Phys. 41: 1952-69 (1964)
CA 61: 12811b

D 1838 The fluorescence spectrum of dices-
ium uranyl nitrate at 77°K.
Dixit, R.M.
Proc. Indian. Acad. Sci., Sect. A 60: 90-8
(1964)
CA 62: 8536f

D 1839 See D 841a

D 1840 Correction of luminescence spectra
and calculation of quantum efficiencies
using computer techniques.
Drushel, H.V., and Companion, A.L.
J. Chem. Phys. 40: 1205-7 (1964)
CA 60: 7583f

D 1841 The theory of pseudo-Stark splitting of
R lines in ruby spectrum.
Druzhinin, V.V., and Cherepanov, V.I.
Fiz. Tverd. Tela 6: 2495-2501 (1964)
CA 61: 14053a

D 1842 Lifetimes of excited states in solution
by the quenching method.
Dubois, J.T., and Van Hemert, R.L.
J. Chem. Phys. 40: 923-5 (1964)
CA 60: 6362g

D 1843 Mechanism of retarded fluorescence
of some aromatic hydrocarbons in the
case of solidified solutions.
Dupuy, F., Lochet, R., and Rousset, A.
Compt. Rend. 258: 4223-7 (1964)
CA 61: 3830h

D 1844 The use of fluorescence in studies of
the structure and interactions of pro-
teins.
Edelhoch, H., and Steiner, R.F.
Electron. Aspects Biochem., Proc. Intern.
Symp. Ravello, Italy 1963: 7-22 (1964)
CA 61: 14931e

D 1845 Luminescence analysis.
Eisenbrand, J.
Z. Naturwiss.-Med. Grundlagenforsch. 2:
132-57 (1964)
CA 62: 15407e

D 1846 Stimulation of the fluorescence of
malvin on filter paper.
Eisenbrand, J., and Hett, O.
Z. Lebensm.-Untersuch.-Forsch. 125:
385-90 (1964)
CA 62: 3373b

D 1847 The measurement of ionization con-
stants of electronically excited species
from aminonaphthols.
Ellis, D.W., and Rogers, L.B.
Spectrochim. Acta 20: 1709-20 (1964)
CA 61: 15410g

D 1848 Vanishing first- and second-order
intramolecular heavy-atom effects on
the $(\pi * \rightarrow n)$ phosphorescence in carbo-
nyls.
El-Sayed, M.A.
J. Chem. Phys. 41: 2462-7 (1964)
CA 61: 12809f

D 1849 Effect of fluorescent substance on the
chemiluminescence of lucigenin.
Erdey, L., Takocs, J., and Buzas, I.
Acta Chim. Acad. Sci. Hung. 39: 295-300
(1964)
CA 60: 9878h

D 1850 Triplet-triplet energy transfer be-
tween identical molecules in solid solu-
tions at 90°K.
Ermolaev, V.L.
Opt. i Spektroskopiya 16: 548 (1964)
CA 61: 191e

D 1851 Use of triplet-singlet transfer for the
study of the internal degradation of
electronic energy in organic molecules.
Ermolaev, V.L., and Sveshnikova, E.B.
Opt. i Spektroskopiya 16: 587-93 (1964)
CA 61: 5104g

D 1852 The width of luminescence spectra and
Stokes losses for transitions from
fluorescent and phosphorescent levels
of aromatic compounds.
Ermolaev, V.L.
Opt. i Spektroskopiya 16: 704-5 (1964),
Opt. Specty. (USSR) (English Transl.)
16: 383 (1964)
CA 61: 2615d

D 1853 Fluorometer for the determination of
catechol amines in biological fluids.
Esikov, A.D.
Adrenalin i Noradrenalin, Akad. Nauk
SSSR, Lab. Neiro-Gumoral'n.
Regulyatsii 1964: 305-8 (1964)
CA 62: 10795d

D 1854 Conformational dependence of the
fluorescence of copolymers of tyrosine
and glutamic acid in aqueous solution.
Fastman, G.D., Norland, K., and Pesce, A.
Biopolymers, Symp. No. 1, 325-31 (1964)
CA 60: 12260e

D 1855 On the mechanism of fluorescence
quenching. Tyrosine and similar com-
pounds.
Feitelson, J.
J. Phys. Chem. 68: 391-7 (1964)
CA 60: 7587h

D 1856 Substituent effects on intramolecular
energy transfer. II. Fluorescence
spectra of europium and terbium
β-diketone chelates.
Filipescu, N., Sager, W.F., and Serabin,
F.A.
J. Phys. Chem. 68: 3324-46 (1964)
CA 62: 141e

D 1857 Identification of antioxidants and accelerators in rubber mixtures by mixtures by spectrophotometry and chromatography.
Fiorenza, A., Bonomic, G., and Piacentini, R.
Rev. Gen. Caoutchouc Plastiques 41: 995-9 (1964)
CA 62: 14905e

D 1858 Lifetime of the N_2 ($C^3\pi_u$), N_2^+ ($B^2\Sigma_u^+$), NH ($A^3\pi$), NH ($C^1\pi$), and PH ($^3\pi$) electron states.
Fink, E., and Welge, K.H.
Z. Naturforsch. 19a: 1193-1201 (1964)
CA 62: 120g

D 1859 Concentration change of the fluorescence of aromatic hydrocarbons in micellar colloidal solution.
Foerster, T., and Selinger, B.
Z. Naturforsch. 19: 38-41 (1964)
CA 60: 9950g

D 1860 Spectra of chlorophyll a in various media.
Frackowiak, D.
Bull. Acad. Polon. Sci., Ser. Sci., Math., Astron. Phys. 12: 119-24 (1964)
CA 61: 11494e

D 1861 Recording microfluorimeter of high sensitivity and low noise for measurements in living tissues.
Frank, M., Gurdarelli, G., and Crescenzi, G.S.
Ann. Chim. 54: 788-94 (1964)
CA 62: 5559g

D 1862 The structure of anionic polymers by measurement of fluorescent polarization.
Frey, M., Wahl, P., and Benoit, H.
J. Chim. Phys. 61: 1005-17 (1964)
CA 62: 2838e

D 1863 Luminescence of cadmium iodide activated lead iodide.
Frumar, M.
Collection Czech. Chem. Commun. 29: 672-8 (1964)
CA 60: 10032b

D 1864 Photoinduced electron transfer in dye-sulfhydryl protein complex.
Fujimuri, E.
Nature 201: 1183-5 (1964)
CA 60: 14638b

D 1865 Radioactive self-luminous compounds. V. Trapping center and cross section of excitation.
Fujimura, R., Ato, Y., and Oishi, M.
Nagoya Kogyo Gijutsu Shikensho Hokoku 13: 176-83 (1964)
CA 61: 9063a

D 1866 Vibronic coupling. II. Spectra of dimers.
Fulton, R.L., and Gouterman, M.
J. Chem. Phys. 41: 2280-6 (1964)
CA 61: 12805e

D 1867 Determination of the molecular weight of solid and liquid polymers by fluorescence.
Gachkovskii, V.F.
USSR 161,974, 1962 (1964)
CA 61: 8435h

D 1868 Fluorescent method for determining the molecular weight of solid and liquid polymers.
Gachkovskii, V.F.
USSR 163,010, 1962 (1964)
CA 61: 8435h

D 1869 Interaction between chlorophyll a and fatty alcohol molecules in mixed monomolecular films.
Gaines, G.L., Bellamy, W.D., and Tweet, A.G.
J. Chem. Phys. 41: 538-42 (1964)
CA 61: 6541h

D 1870 Monomolecular films of methyl chlorophyllide.
Gaines, G.L., Jr., Bellamy, W.D., and Tweet, A.G.
J. Chem. Phys. 41: 2572-3 (1964)
CA 62: 2373h

D 1871 Absorption and fluorescence of europium (III) in aqueous solutions.
Gallagher, P.K.
J. Chem. Phys. 41: 3061-9 (1964)
CA 61: 15549a

D 1872 Two-step energy transfer in solution.
Gallagher, P.K., Heller, A., and Wasserman, E.
J. Chem. Phys. 41: 3921-4 (1964) (Eng.)
CA 62: 2377a

D 1873 Effects of conformation and environ-
  ment on the fluorescence of proteins
  and polypeptides.
  Gally, J.A., and Edelman, G.M.
  Biopolymers, Symp. No. 1, 367-81 (1964)
  CA 60: 13485b

D 1874 Laser oscillations in Nd-doped yttrium
  aluminum, yttrium gallium, and gadoli-
  nium (gallium) garnets.
  Geusic, J.E., Marcos, H.M., and van
  Uitert, L.G.
  Appl. Phys. Letters 4: 182-4 (1964)
  CA 61: 3840a

D 1875 A proposed method for identifying
  states of N atoms produced by disso-
  ciative recombination of N_2^+ ions and
  electrons.
  Ghosh, S.N., and Sharma, A.
  Sci. Cult. (Calcutta) 30: 501-2 (1964)
  CA 62: 15445d

D 1876 Fluorescence in vivo of some marine
  algae.
  Girand, G.
  Proc. Intern. Seaweed Symp., 4th,
    Biarritz, France 1961: 326-30 (1964)
  CA 61: 6047c

D 1877 Electronic spectra of phthalimides in
  the crystal state.
  Gladchenko, L.F.
  Dokl. Akad. Nauk Belorussk. SSR 8: 29-32
    (1964)
  CA 60: 14018h

D 1878 Rhodamine 3B as a fluorescent re-
  agent for indium.
  Glovadskii, Y., Golovina, A.P., Levshin,
    L.V., and Mittsel, Y.A.
  Zh. Analit. Khim. 19: 693-6 (1964)
  CA 61: 8888a

D 1879 Electronic-vibrational spectra of some
  arylethylenes at 77°K.
  Gobov, G.V., Nurmukhametov, R.N., and
    Nagornaya, L.L.
  Zh. Fiz. Khim. 35: 1142-7 (1964)
  CA 61: 9064a

D 1880 Fluorescence bands and chlorophyll a
  forms.
  Goedheer, J.C.
  Biochim. Biophys. Acta 88: 304-17 (1964)
  CA 61: 14946h

D 1881 Fluorescence and photoconductivity
  of exciton in CuCl crystal.
  Goto, T., and Veta, M.
  J. Phys. Soc. Japan 19: 774-5 (1964)
  CA 61: 14007a

D 1882 Analysis of the emission spectra of
  benzyl radicals and some of their
  deuterated isotopes.
  Grajcar, L., and Leach, S.
  J. Chim. Phys. 61: 1523-30 (1964)
  CA 62: 12618h

D 1883 Avidin. V. Quenching of fluorescence
  by dinitrophenyl (DNP) groups.
  Green, N.M.
  Biochem. J. 90: 564-8 (1964)
  CA 60: 9504c

D 1884 Crystal-field splitting of trivalent
  thulium and erbium levels in yttrium
  oxide.
  Gruber, J.B., Krupke, W.F., and
    Poindexter, J.M.
  J. Chem. Phys. 41: 3363-77 (1964)
  CA 62: 1219e

D 1885 Effect of the intermolecular interac-
  tions in solutions on the mean duration
  of fluorescence of acriflavine.
  Grudzinski, H., and Heldt, J.
  Acta Phys. Polon. 25: 391-400 (1964)
  CA 62: 1216e

D 1886 On the shift (of the maxima) of the
  α-phosphorescence and fluorescence
  spectrum of fluorescein in boric acid.
  Grzywacz, J., and Pohoski, R.
  Z. Naturforsch. 19a: 440-4 (1964)
  CA 61: 3830f

D 1887 Dependence of the fluorescence-
  polarization degree of trypoflavine on
  the emission wavelength.
  Grzywacz, J., and Pohoski, R.
  Z. Naturforsch. 19a: 1621-2 (1964)
  CA 62: 11310d

D 1888 Effect of polymerization on the emis-
  sion spectrum of trypoflavine in methyl
  methacrylate.
  Grzywacz, J.
  Z. Naturforsch. 19a: 1622-3 (1964)
  CA 62: 11310f

D 1889 Polarization and relaxation time as a means for frequency selection from the emission of neodymium-containing calcium tungstate lasers.
Guers, K.
Z. Naturforsch. 19a: 515-6 (1964)
CA 61: 5115d

D 1890 Fluorometric determination of lipase, acylase, alpha- and gamma-chymotrypsin and inhibitors of these enzymes.
Guilbault, G.G., and Kramer, D.N.
Anal. Chem. 36: 409-12 (1964)

D 1890a New direct method for measuring dehydrogenase activity.
Guilbault, G.G., and Kramer, D.N.
Anal. Chem. 36: 2497-8 (1964)

D 1891 Fluorescent sample for focusing on electron-microscope beam.
Hall, L.C.
Anal. Chem. 36: 2515 (1964)
CA 62: 4794c

D 1892 Luminescence of europium hexafluoro-acetylacetonate.
Halverson, F., Brinen, J.S., and Leto, J.R.
J. Chem. Phys. 40: 2790-2 (1964)
CA 60: 15315a

D 1893 The effect of molecular shape on the fluorescent properties of 9,10-diphenyl-anthracene.
Hamilton, T.P.S.
Photochem. Photobiol. 3: 153-6 (1964)
CA 61: 6898f

D 1894 Polarized crystal spectra of $Ni(NH_3)_4(NCS)_2$ and $Ni(NH_3)_4(NO_2)_2$.
Hare, C.R., and Ballhausen, C.J.
J. Chem. Phys. 40: 792-5 (1964)
CA 60: 6351h

D 1895 Stimulated emission from Y_2O_3:Nd^{+3}
Hoskins, R.H., and Soffer, B.H.
Appl. Phys. Letters 4: 22-3 (1964)
CA 60: 7585e

D 1896 Association and self-quenching of proflavine in water.
Haugen, G.R., and Melhuish, W.H.
Trans. Faraday Soc. 60: 386-94 (1964)
CA 60: 14010e

D 1897 Fluorescent indicator for a confirmatory test for the aluminum ion.
Haworth, D.T., Starshak, R.J., and Surak, J.G.
J. Chem. Educ. 41: 436-7 (1964)
CA 61: 10019a

D 1898 Fluorescence of $Nd(4f^3)$ by irradiation into the chromium absorption bands of the system Al_2O_3:Cr, Nd.
Heitmann, W., Moeller, A., and Schultz, G.V.
Phys. Letters 10: 26-7 (1964)
CA 61: 5104h

D 1899 Organic liquid scintillators. VI. Substituted distyrylbenzenes: scintillation properties and spectra of absorption and fluorescence.
Heller, A.
J. Chem. Phys. 40: 2839-51 (1964)
CA 60: 15396f

D 1900 Color and fluorescence of cyclic Si compounds. IV. Determination of the orientation of substituting groups on siloxene by bridge formation.
Hengge, E., and Grupe, H.
Ber. 97: 1783-8 (1964)
CA 61: 9063d

D 1901 Spectra and structure of the free HSiCl and HSiBr radicals.
Herzberg, G., and Verna, R.D.
Can. J. Phys. 42: 395-432 (1964)
CA 60: 10063c

D 1902 Infrared transmission and fluorescence of doped gallium arsenide.
Hill, D.E.
Phys. Rev. 133: 866-72 (1964)
CA 60: 4960c

D 1903 Effect of Nd environment on its adsorption and emission characteristics in glass.
Hirayama, C., and Lewis, D.W.
Phys. Chem. Glasses 5: 44-51 (1964)
CA 60: 12989g

D 1904 Determination of the efficiency of triplet energy migration in benzophenone crystals.
Hockstrasser, R.L.
J. Chem. Phys. 40: 1038-40 (1964)
CA 60: 6362h

D 1905 Polarized emission and triplet-triplet absorption spectra of aromatic hydrocarbons in benzophenone crystals.
Hochstrasser, R.M., and Lower, S.K.
J. Chem. Phys. 40: 1041-6 (1964)
CA 60: 6362h

D 1906 Experimental evidence for localized excitons in the spectra of charge-transfer complex molecular crystals.
Hochstrasser, R.M., Lower, S.K., and Reid, C.
J. Chem. Phys. 41: 1073-8 (1964)
CA 61: 6503a

D 1907 Spectral effects of strong exciton coupling in the lowest electronic transition of perylene.
Hochstrasser, R.M.
J. Chem. Phys. 40: 2559-64 (1964)
CA 60: 14015e

D 1908 Fluorescence and energy transfer in pure solvents and liquid mixed systems of organic scintillators.
Hoefer, G.
Z. Physik 181: 44-57 (1964)
CA 62: 2471b

D 1909 Poly-p-xylylidenes.
Hoeg, D.F., Lusk, D.I., and Goldberg, E.P.
J. Polymer Sci. Pt. B 2: 697-701 (1964)
CA 61: 8418e

D 1910 Fluorescent yield W_{KL} of the L shell.
Hohmuth, K., and Winter, G.
Phys. Rev. Letters 10: 58-9 (1964)
CA 61: 5104b

D 1911 Effect of magnetic ordering on the fluorescence of MnF_2.
Holloway, W.W., Jr., and Kestigian, M.
Phys. Rev. Letters 13: 235-7 (1964)
CA 61: 11493a

D 1912 Double-photon excitation of fluorescence in anthracene.
Jannuzzi, M., and Polacco, E.
Phys. Rev. Letters 13: 371-2 (1964)
CA 61: 12767d

D 1913 Temperature dependence of the width and position of the $^2E \rightarrow {}^4A_2$ fluorescence lines of Cr^{3+} and V^{2+} in MgO.
Imbusch, G.F., Yen, W.M., Schawlow, A.L., Sturge, M.D., and McCumber, D.E.
Phys. Rev. 133: 1029-34 (1964)
CA 60: 6362b

D 1914 Rhodamine dyes and related compounds. X. Flourescence of solutions of alkyl- and arylalkylrhodamines.
Ioffe, I.S., and Gofman, I.A.
Zh. Obshch. Khim. 34: 2039-41 (1964)
CA 61: 10200a

D 1915 Delayed fluorescence in deoxyribonucleic acid (DNA) acridine dye complexes.
Isenberg, I., Leslie, R.B., Baird, S.L., Rosenbluth, R., and Bersohn, R.
Proc. Natl. Acad. Sci. U.S. 52: 379-87 (1964)
CA 61: 14943b

D 1916 Excited state pK's. I. Azobenzene and azoxybenzene.
Jaffe, H.H., Beveridge, D.L., and Jones, H.L.
J. Am. Chem. Soc. 86: 2932-4 (1964)
CA 61: 12794c

D 1917 Electric dipole nature of pyrazine 3700A. T $\rightarrow$ S emission.
Jamattona, B., and Goodman, L.
J. Chem. Phys. 40: 2042-3 (1964)
CA 61: 188c

D 1918 Lifetime measurements of some excited states of nitrogen, nitric oxide, and formaldehyde.
Jeunehomme, M., and Duncan, A.B.F.
J. Chem. Phys. 41: 1692-9 (1964)
CA 61: 10191g

D 1919 Selected absorption and fluorescence studies of certain transition metal chelates and complexes.
Jones, D.E.
Dissertation, Purdue University, Indiana (1964)
CA 61: 11479d

D 1920 An experimental study of the phosphorescence of polyatomic molecules in viscous media.
Jones, T.H.
Dissertation, University of Minnesota (1964)
CA 62: 4754h

D 1921 Spectrum of azulene. III. Flourescence intensities in azulene and azulene-d_8.
Johnson, G.D., Logan, L.M., and Ross, I.G.
J. Mol. Spectry. 14: 198-200 (1964)
CA 62: 1216b

D 1922 Fluorescence yields of the L_{II} and L_{III} shells in heavy elements.
Jopson, R.C., Khan, J.M., Mark, H., Swift, C.D., and Williamson, M.A.
Phys. Rev. 133: 381-4 (1964)
CA 60: 3634f

D 1923 Phosphorescence decay of X-ray irradiated KCl:Tl phosphors.
Joshi, R.V.
J. Phys. Chem. Solids 25: 135-9 (1964)
CA 60: 8749d

D 1924 Interpretation of the fluorescence spectra of vegetable oils.
Jung, L., and Morand, P.
Ann. Fals. Expert. Chim. 57: 17-25 (1964)
CA 61: 11493e

D 1925 Vibronic spectroscopy: detailed polarization of absorption, fluorescence, and phosphorescence in four benzene derivatives.
Kalantar, A.H.
Dissertation, Cornell University, New York (1964)
CA 61: 5102g

D 1926 Vibronic spectroscopy: detailed polarization of absorption, fluorescence, and phosphorescence in four benzene derivatives.
Kalantar, A.H., and Albrecht, A.C.
Ber. Bunsenges. Physik. Chem. 68: 361-76 (1964)
CA 61: 10179f

D 1927 The theory of vibronic interactions in four C_6H_6 derivatives.
Kalantar, A.H., and Albrecht, A.C.
Ber. Bunsenges. Physik. Chem. 68: 377-89 (1964)
CA 61: 10177c

D 1928 Effect of F-Cl ration on crystal growth and some fluorescent characteristics of calcium halophosphate phosphor.
Kamiya, S.
Denki Kagaku 32: 432-6 (1964)
CA 62: 7222a

D 1929 Dependence of emission properties of Sb-activated Ca halophosphate phosphors on activator concentration and temperature.
Kamiya, S., and Masuda, M.
Denki Kagaku 32: 679-84 (1964)
CA 62: 7221h

D 1930 Stimulated emission of neodymium ion in inorganic glasses.
Kan, F.H., Chiang, C.H., and Tsai, Y.S.
K'o Hsueh T'ung Pao 1964: 54-6 (1964)
CA 61: 10215d

D 1931 Optical properties and spectra of rare earth oxides in inorganic glass. III. Fluorescence spectrum of rare earth ions.
Kan, F.H., Chiang, C.H., Tsai, Y.S., and Hsiao, K.Y.
K'o Hsueh T'ung Pao 1964: 52-4 (1964)
CA 61: 11492h

D 1932 Singlet-triplet absorption spectra of several carbonyl compounds.
Kanda, Y., Kaseda, H., and Matsumura, T.
Spectrochim. Acta 20: 1387-96 (1964)
CA 61: 9059c

D 1933 Protein-sulfhydryl reagents. I. Synthesis of benzimidazole; derivatives of maleimide; fluorescent labeling of maleimide.
Kanaoka, Y., Sekine, T., Machida, M., Soma, Y., Tanizawa, K., and Ban, Y.
Chem. Pharm. Bull. (Tokyo) 12: 127-34 (1964)
CA 60: 15856g

D 1934 Determination of the absolute transition momentum direction.
Kawski, A.
Z. Naturforsch. 19: 159-60 (1964)
CA 60: 10057b

D 1935 Effect of polar molecules on electronic spectrum on 4-amino-phthalimide.
Kawski, A.
Acta Physiol. Polon. 25: 285-90 (1964)
CA 61: 12792e

D 1936 Intermolecular exchange of excitation energy in fluorescein-plexiglas solutions.
Kawski, A.
Bull. Acad. Polon. Sci., Ser. Sci., Math., Astron. Phys. 12: 173-8 (1964)
CA 61: 12106e

D 1937 Emission properties of tetracene-pyrene and pentacene-pyrene mixed crystals.
Kawaoka, K., and Kearns, D.R.
J. Phys. Chem. 41: 2095-7 (1964)
CA 61: 12809e

D 1938 The photoluminescence of aluminum oxide layer fluorescein luminophors.
Kawski, A., Korba, M., Czyz, P., Malinowski, S., and Szymkowiak, H.
Z. Naturforsch. 19a: 1328-9 (1964)
CA 62: 2376e

D 1939 β-Phosphorescence of Plexiglas luminophors at room temperature.
Kawski, A., and Pohoski, R.
Nature 201: 1116-7 (1964)
CA 60: 12760d

D 1940 The photoluminescence of some coumarin derivatives in poly(vinyl alcohol).
Kawski, A., Pohoski, R., and Sliwicki, E.
Z. Naturforsch. 19a: 1330-1 (1964)
CA 62: 2376e

D 1941 The polarization of the fluorescence of organic scintillation crystals on α-ray excitation.
Kayser, J.
Z. Physik 178: 445-56 (1964)
CA 61: 2615c

D 1942 Location of charge resonance states in aromatic molecular crystals with nondimeric structures.
Kearns, D.R.
J. Chem. Phys. 41: 581-2 (1964)
CA 61: 12721h

D 1943 Radiationless intermolecular energy transfer. III. Determination of phosphorescence efficiencies.
Kellogg, R.E., and Bennett, R.G.
J. Chem. Phys. 41: 3042-5 (1964)
CA 61: 15529b

D 1944 Temperature effect on triplet-state lifetimes in solid solutions.
Kellogg, R.E., and Schwenken, R.P.
J. Chem. Phys. 41: 2860-3 (1964)
CA 61: 14055g

D 1945 Chemiluminescent indicator titration of cadmium with potassium ferricyanide.
Kenny, F.
Anal. Chem. 34: 529-32 (1964)

D 1946 Complex investigation of the absorption centers and phosphorescence centers in phosphorescing alkali halide crystals.
Khalilov, A.K.
Izv. Akad. Nauk Azerb. SSR, Ser. Fiz.-Mat. i Tekhn. Nauk 1964: 95-104 (1964)
CA 61: 11444d

D 1947 Spectrum of Er^{3+} in single crystals of Y_2O_3.
Kisliuk, P., Krupke, W.F., and Gruber, J.B.
J. Chem. Phys. 40: 3606-10 (1964)
CA 61: 2515g

D 1948 Control of finish on fibrous materials with a fluorescing substance.
Klein, E.
U.S. 3,188,060, 1959 (1964)
CA 60: 14673g

D 1949 Enhancement of fluorescence yield of chelated lanthanide ions by Lewis bases.
Kleinerman, M., Hovey, R.J., and Hoffman, D.O.
J. Chem. Phys. 41: 4009-10 (1964)
CA 62: 8536h

D 1950 Pressed and extracted cocoa butter.
Kleinert, J.
Rev. Intern. Chocolat. 19: 142-53 (1964)
CA 61: 4605b

D 1951 Modified process for the identification of amino acids on a one-dimensional paper chromatogram.
Klembala, M., and Szekacs, J.
Orv. Hetilap 105: 1658-60 (1964)
CA 62: 30b

D 1952 Lead fluorescence of higher plants at room temperature.
Klochkova, M.P., and Moshkov, B.S.
Biofizika 9: 469-76 (1964)
CA 61: 9767f

D 1953 Nature of the self-activated blue luminescence center in cubic ZnS:Cl single crystals.
Koda, T., and Shionoya, S.
Phys. Rev. 136: 541-55 (1964)
CA 61: 12808h

D 1954 Correlation of π-electron absorption
bands with the aid of the method of
fluorescence polarization.
Koerberg, W., and Zanker, V.
Z. Angew. Phys. 17: 398-404 (1964)
CA 62: 8535f

D 1955 Pyridine nucleotide compartmentali-
zation in glass-grown ascites cells.
Kohen, E.
Exptl. Cell Res. 35: 303-16 (1964)
CA 62: 908e

D 1956 Enhancement of acridone fluorescence
by pyridine.
Kokubun, H., and Kobayashi, M.
Z. Physik. Chem. 41: 245-7 (1964)
CA 61: 9065c

D 1957 Duration of fluorescence of chloro-
phyll at various concentrations of it in
adsorbed state and in a green leaf.
Komissarov, G.G., Nekrasov, L.I., and
Kobozev, N.I.
Dokl. Akad. Nauk SSSR 154: 950-2 (1964)
CA 60: 13578b

D 1958 The influence of formaldehyde in
quantum fluorescence of tryptophan and
its derivatives.
Konev, S.V., and Chernitskii, E.A.
Biofizika 9: 520-2 (1964)
CA 61: 9700f

D 1959 Energy levels and crystal-field calcu-
lations of neodymium in yttrium alum-
inum garnet.
Konigstein, J.A., and Geusic, J.E.
Phys. Rev. 136: 711-6 (1964)
CA 61: 14038d

D 1960 Energy levels and crystal-field calcu-
lations of Er^{+3} in yttrium aluminum
garnet.
Konigstein, J.A., and Geusic, J.E.
Phys. Rev. 136: 726-8 (1964)
CA 61: 14038c

D 1961 Extraction-fluorometric determina-
tion of samarium and europium in
rare earth oxide mixtures.
Kononenko, L.I., Poluektov, N.S., and
Nikonova, M.P.
Zavodsk. Lab. 30: 779-83 (1964)
CA 61: 8891f

D 1962 Development of microorganisms and
fluorescence on poultry dipped in
water containing iron.
Kraft, A.A., and Ayres, J.C.
J. Food Sci. 29: 218-23 (1964)
CA 61: 6273h

D 1963 Vibrational spectra of aromatic compounds. XIX. Calculation and interpretation of vibrational spectra of
naphthalene and some deuteronaphtha-
lenes.
Krainov, E.P.
Opt. i Spektroskopiya 16: 763-7 (1964)
CA 61: 5090g

D 1964 Resorfufin acetate as substrate for
determination of hydrolytic enzymes at
low enzyme and substrate concentrations.
Kramer, D.N., and Guilbault, G.G.
Anal. Chem. 36: 1662-3 (1964)

D 1965 Ruthenium 2,2'-bipyridine complexes
as fluorescent oxidation-reduction indicators.
Kratochuil, B., and Zatke, D.A.
Anal. Chem. 36: 527-29 (1964)

D 1966 Determination of absolute values for
the effective cross sections of the
second kind collisions in the sensitized
fluorescence of sodium and mercury
vapors.
Kraulina, E.
Opt. i Spektroskopiya 17: 464-6 (1964)
CA 61: 15363e

D 1967 Absorption and fluorescence spectra
of some azomethine derivatives of ben-
zidine and of its 2,2'- and 3,3'-dichloro
derivatives.
Krasovitskii, B.M., Smelyakov, V.B., and
Nurmukhametov, R.N.
Opt. i Spektroskopiya 17: 558-64 (1964)
CA 62: 4791e

D 1968 Ruthenium complexes of ligands con-
taining the ferroin group as fluorescent
precipitation indicators for iodimetry.
Kratochvil, B., and White, M.C.
Anal. Chim. Acta 31: 528-33 (1964)
CA 62: 3398a

D 1969 The construction of a sensitive fluorimeter.
Krebs, V., and Smetana, R.
Chem. Listy 58: 230-1 (1964)
CA 60: 14785g

D 1970 Fluorescence changes in porphyridium exposed to green light of different intensity. A new emission band at 693 μ and its significance to photosynthesis.
Krey, A., and Govindjee
Proc. Natl. Acad. Sci. U.S. 52: 1568-72 (1964)
CA 62: 10829h

D 1971 Polarization of near-ultraviolet absorption and phosphorescence of aromatic ketones.
Krishna, V.G.
J. Mol. Spectry. 13: 296-304 (1964)
CA 61: 3826c

D 1972 Energy levels of Er^{3+} in LaF_3 and coherent emission at 1.61 μ.
Krupke, W.F., and Gruber, J.B.
J. Chem. Phys. 41: 1225-32 (1964)
CA 61: 9066d

D 1973 Relationship between luminescence and semiconducting properties of some synthetic polymers.
Kryszewski, M., Kurczewska, H., and Szymanski, A.
J. Polymer Sci., Pt. C 4: 1417-27 (1964)
CA 60: 5657c

D 1974 Phosphorescence in the ethyl-alcohol solutions of volatile oils.
Kucharski, J., and Moscicki, W.
Acta Physiol. Polon. 25: 299-300 (1964)
CA 61: 11579e

D 1975 Luminescence from zone-refined anthracene at 4° and the Davydov splitting.
Lacey, A.R., and Lyons, L.E.
J. Chem. Soc. 1964: 5393-400 (1964)
CA 62: 3545a

D 1976 Effects of acceptor concentration gradients in GaAs junctions on the energy of the fluorescent peak.
Lucovsky, G., and Varga, A.J.
J. Appl. Phys. 35: 3419 (1964)
CA 62: 3506c

D 1977 The effect of viscosity on the internal deactivation modes of the triplet state.
Ladner, J.S., and Becker, R.S.
J. Am. Chem. Soc. 86: 4205-6 (1964)
CA 62: 3525h

D 1978 The absorption and fluorescence spectra of 1,10-phenanthroline and related compounds.
Langmiur, M.E.L.
Dissertation, Purdue University, Indiana (1964)
CA 61: 11494h

D 1979 A new phosphoroscope.
Langouet, L.
Bull. Soc. Sci. Bretagne 37: 11-15 (1962) pub. 1964
CA 61: 2614b

D 1980 Anomalous anti-Stokes fluorescence of aniline.
LaPaglia, S.R.
Trans. Faraday Soc. 60: 1210-3 (1964)
CA 61: 5105a

D 1981 Attempts to detect fluorescence from rare earth-doped GaAs and InP.
Lasher, G.J., Stern, F., and Weiser, K.
Injection Laser Study, IBM Corp., Watson Res. Center 1963: 48 (1964)
Sci. Tech. Aerospace Rept. 2: 1295 (1964)
CA 62: 106h

D 1982 Heterogeneity of chlorophyll in vivo. II. Polarization and fluorescence action spectra.
Lavorel, J.
Biochim. Biophys. Acta 88: 20-36 (1964)
CA 61: 11063e

D 1983 The first triplet state of benzene.
Leach, S., and Lopez-Delgado, R.
J. Chim. Phys. 61: 1636-42 (1964)
CA 62: 12621b

D 1984 Infrared fluorescence in gaseous CO_2 and N_2O caused by active nitrogen.
Legay, F., and Barchewitz, P.
Mem. Soc. Roy. Sci. Liege, Collection in 8° 9:74-8 (1964)
CA 61: 5090a

D 1985 Luminescence and absorption studies on sapphire with flash light excitation.
Lehmann, H.W., and Gunthard, H.H.
J. Phys. Chem. Solids 25: 941-50 (1964)
CA 61: 10152g

D 1986 Fluorescence studies on poly-α-amino acids. II. Conformation-dependent excimer emission band in poly-L-tyrosine and poly-L-tryptophan.
Lehrer, S.S., and Fasman, G.D.
Biopolymers 2: 199-203 (1964)
CA 61: 3325c

D 1987 Comparison of thick and thin sodium salicylate layers with terphenyl as fluorescence detectors for far ultraviolet spectroscopy.
Lemonnier, J.C., Priol, M., Quemerais, A., and Robin, S.
J. Phys. (Paris), Suppl. 25: 79A-82A (1964)
CA 62: 1217a

D 1988 Radiationless transitions and deuterium effect on luminescence of some aromatics.
Lim, E.C., and Laposa, J.D.
J. Chem. Phys. 41: 3257-9 (1964)
CA 62: 5983a

D 1989 Phosphorescence of organic phosphors with two metastable levels.
Lisenko, G.M., and Kislyak, G.M.
Ukr. Fiz. Zh. 9: 160-5 (1964)
CA 60: 15277d

D 1990 Double carrier injection and negative resistance in CdS.
Litton, C.W., and Reynolds, D.C.
Phys. Rev. 133: 536-41 (1964)
CA 60: 7549c

D 1991 Infrared-excited luminescence of fluorescent CaF_2-Dy^{+2} crystal.
Liu, S.H., Wu, H.L., Lin, K.H., Chen, S.C., and Wang, F.K.
K'o Hsueh T'ung Pao 1964: 56-8 (1964)
CA 61: 5104c

D 1992 Fluorescence-spectrometric investigations of the equilibrium anthrone-anthranol.
Loeber, G.
Acta Chim. Acad. Sci. Hung. 40: 9-16 (1964)
CA 61: 7752d

D 1993 Rotational relaxation in rigid media by polarized photoselection.
Lombardi, J.R., Raymonda, J.W., and Albrecht, A.C.
J. Chem. Phys. 40: 1148-56 (1964)
CA 60: 6230c

D 1994 Preparation and study of the spectral properties of europium tris(5-nitro-1-naphthalensulfonate).
Loriers, J., and Heindl, R.
Compt. Rend. 259: 4571-4 (1964)
CA 62: 15597h

D 1995 Green fluorescent pigment accumulated by a mutant of cellvibrio gilvus.
Love, S.H., and Hulcher, F.H.
J. Bacteriol. 87: 39-45 (1964)
CA 60: 4489c

D 1996 R fluorescence spectrum ground state splitting of ruby.
Lu, T.Y., Yu, W.Y., Tang, K.S., and Chang, C.J.
K'o Hsueh T'ung Pao 1964: 59-60 (1964)
CA 61: 6545d

D 1997 Rare earth chelates and the molecular approach to lasers.
Lyons, H., and Bhaumick, M.L.
Intern. Aerospace Abstr. 4: 1623 (1964)
CA 62: 8550g

D 1998 Diffusion theory of concentration extinction of fluorescence.
Machwe, M.K., Kishore, J., Krishman, K.G., and Chandhuri, K.D.
Current Sci. 33: 301 (1964)
CA 61: 5102f

D 1999 The fluorescence of polymers obtained by anionic polymerization.
Macionis, Z., and Erofeev, B.V.
Lietuvous TSR Aukstuju Mokyklu Mokslo Darbai, Chem. ir Chem. Technol. 5: 83-8 (1964)
CA 61: 8411b

D 2000 Behavior of zinc white pigments when sintered.
Magdanz, H., and Hering, I.
Farbe Lack 70: 603-11 (1964)
CA 61: 13528f

D 2001 Transients in glycolytic metabolism following electrical activity in electrophorus.
Maitra, P.K., Ghosh, A., Schoener, B., and Chance, B.
Biochim. Biophys. Acta 88: 112-19 (1964)
CA 61: 11069f

D 2002 Spectral investigations of mixed
     dibenzyl-stilbene crystal at 20.4 and
     4.2°K.
Malikhina, N.M., and Shpak, M.T.
Ukr. Fiz. Zh. 3: 172-8 (1964)
CA 61: 1406h

D 2003 Temperature dependence of photoiso-
     merization. III. Direct and sensitized
     photoisomerization of stilbenes.
Malkin, S., and Fischer, E.
J. Phys. Chem. 68: 1153-63 (1964)
CA 60: 15332c

D 2004 New spectrofluorimetric determina-
     tion of traces of boron.
Marcantonatos, M., Monnier, D., and
     Marcantonatos, A.
Helv. Chim. Acta 47: 709-10 (1964)
CA 61: 2463c

D 2005 Dependence of the decay and relative
     yield of slow fluorescence of benzene
     on the concentration of the solvent.
Marszolek, T.
Bull. Acad. Polon. Sci., Ser. Sci., Math.,
     Astron. Phys. 12: 73-8 (1964)
CA 61: 7848a

D 2006 Optical absorption spectra of the Pr^{+3}
     ion in a single crystal of $AlLaO_3$;
     fluorescence of the transition
     $^3P_0 \rightarrow {}^3H_4$.
Martin-Brunetiere, F., and Jansen, R.
Compt. Rend. 259: 2629-32 (1964)
CA 62: 4794f

D 2007 Fluorescent brighteners for synthetic
     fibers. IV. Bis(benzoxazolyl)ethylene
     brighteners containing substituted
     aminomethyl groups.
Maruyama, T., Kuroki, N., and Konishi, K.
Kogyo Kagaku Zasshi 67: 155-9 (1964)
CA 61: 5820h

D 2008 Fluorescent brighteners for synthetic
     fibers. V. Bis(benzoxazolyl)ethylene
     brighteners containing N-substituted
     aminoethylsulfamoyl groups.
Maruyama, T., Araki, I., Kuroki, N., and
     Konishi, K.
Kogyo Kagaku Zasshi 67: 159-63 (1964)
CA 61: 5821c

D 2009 Charge transfer and proton transfer
     reaction in the excited hydrogen-
     bonded complex in nonpolar solvents.
Mataga, N., and Kaifu, Y.
Mol. Phys. 7: 137-47 (1963-64)
CA 60: 14025e

D 2010 Electronic structures of carbazole and
     indole and the solvent effects on the
     electronic spectra.
Mataga, N., Tarihashi, Y., and Ezumi, K.
Theoret. Chim. Acta 2: 158-67 (1964)
CA 61: 176c

D 2011 Absorption, fluorescence, and phos-
     phorescence spectra of symmetrical
     carbocyanines at 77°K.
Mazzucato, U., Favaro, G., and
     Mazzucato, M.P.
Ric. Sci., Rend. Sez. A. 4: 501-8 (1964)
CA 62: 11302e

D 2012 Spectra and quantum states of euro-
     pium β-diketone chelates in polymeth-
     ylmethacrylate.
McAvoy, N., Filipescu, N., Kagan, M.R.,
     and Serafin, F.A.
J. Phys. Chem. Solids 25: 461-8 (1964)
CA 60: 15312d

D 2013 Infrared fluorescence of CO.
McCao, D.J., and Williams, D.
J. Opt. Soc. Am. 54: 326-30 (1964)
CA 61: 1394d

D 2014 Fluorometric method for the deter-
     mination of urea in blood.
McCleskey, J.E.
Anal. Chem. 36: 1646-8 (1964)

D 2015 Protein and bound coenzyme fluores-
     cence of lactic dehydrogenases.
McKay, R.H., and Kaplan, N.O.
Biochim. Biophys. Acta 79: 273-83 (1964)
CA 60: 14779d

D 2016 The resonance fluorescence in gases.
Mead, C.A.
J. Chem. Phys. 40: 606 (1964)
CA 60: 8796d

D 2017 See D 1531a

D 2018 Synthesis and fluorescence of some
trivalent lanthanide complexes.
Melby, L.R., Rose, N.J., Abramson, E.,
and Caris, J.C.
J. Am. Chem. Soc. 86: 5117-25 (1964)
CA 62: 1306h

D 2019 Measurement of quantum efficiencies
of fluorescence and phosphorescence
and some suggested luminescence
standards.
Melhuish, W.H.
J. Opt. Soc. Am. 54: 183-6 (1964)
CA 60: 10032a

D 2020 Factors influencing spectrofluorome-
try of phenothiazine drugs.
Mellinger, T.J., and Keeler, C.E.
Anal. Chem. 36: 1840-7 (1964)

D 2021 Isolation and spectral properties of
chlorophyll.
Mel'nikov, S.S., and Evstigneev, V.B.
Biofizika 9: 414-22 (1964)
CA 61: 9713b

D 2022 Spectrofluorimetric estimation of
catechol amines by means of a photoelectric assembly for studies of spectra of combination dispersion.
Men'shikov, V.V., and Esikov, A.D.
Vopr. Med. Khim. 10: 77-80 (1964)
CA 60: 13568b

D 2023 The fluorescence of the europium and
terbium dibenzoyl-melhides.
Metlay, M.
J. Electrochem. Soc. 111: 1253-5 (1964)
CA 61: 14055e

D 2024 Structure of the $^5D_0 \rightarrow {}^7F_2$ transition
in europium benzoylacetonate solution.
Meyer, Y., Poncet, H., and Verron, M.
Compt. Rend. 259: 103-6 (1964)
CA 61: 10296c

D 2025 Shell fluorescence in australorbis
glabratus and other aquatic snails ex-
posed to tetracyclines.
Michelson, E.H.
J. Parasitol. 50: 743-7 (1964)
CA 62: 9509a

D 2026 Infrared spectra of NF, NCl, and NBr
(also phosphorescence).
Milligan, D.E., and Jacox, M.E.
J. Chem. Phys. 40: 2461-6 (1964)
CA 60: 14015g

D 2027 Vibrational relaxation of carbon mono-
xide by ortho- and para-hydrogen.
Millikan, R.C., and Osburg, L.A.
J. Chem. Phys. 41: 2196-7 (1964)
CA 62: 140e

D 2028 Fluorometric analysis of pyruvic
acid with 4'-hydrazino-2-stilbazole.
Mizutani, S., Nakajima, T., Matsumoto,
A., and Tamura, Z.
Chem. Pharm. Bull. 12: 850-3 (1964)
CA 61: 11340e

D 2029 Flourescent microscopy of blood and
bone marrow cells in leukemia.
Molotilova, G.P.
Probl. Gematol. i Pereliv. Krovi. 9: 12-16
(1964)
CA 62: 9581f

D 2030 Fluorimetric determination of sub-
microgram amounts of boron by
dihydroxy-2,4-benzophenone.
Monnier, D., Marcantonatos, A., and
Marcantonatos, M.
Helv. Chim. Acta 47: 1980-6 (1964)
CA 62: 5880a

D 2031 Kinetics of fluorescence of chloro-
phyll in vivo in the first instants after
the beginning of illumination.
Morin, P.
J. Chim. Phys. 61: 674-80 (1964)
CA 61: 6052e

D 2032 Estimation of the probability of
singlet-singlet and singlet-triplet
transitions in molecules having a conjugated π-electron system.
Morozov, Y.V., and Kuklin, A.I.
Biofizika 9: 299-305 (1964)
CA 61: 7847d

D 2033 Anisotropy of the electronic spectra of
a single crystal of 1,12-benzperylene
($C_{22}H_{12}$).
Mukherjee, B.C., and Ganguly, S.C.
Proc. Phys. Soc. (London), 83: 93-7 (1964)
CA 60: 3635c

D 2034 Energy transfer from 3d to 4f elec-
trons in $LaAlO_3:Cr_1Nd$.
Murphy, J., Ohlmann, R.C., and Mazelsky,
R.
Phys. Rev. Letters 13: 135-7 (1964)
CA 61: 10176a

D 2035 Theory of the electronic spectra of aromatic hydrocarbon dimers.
Murrell, J.N., and Tanaka, J.
Mol. Phys. 7: 363-80 (1963-4)
CA 60: 14014d

D 2036 Intermolecular quenching of higher excited states.
Murty, N.R., and Rabinowitch, E.
J. Chem. Phys. 41: 599-601 (1964)
CA 61: 9065d

D 2037 Effect of the physical environment on excited states of amino-acids and proteins.
Nag-Chaudhuri, J., and Augenstein; L.
Biopolymers, Symp. No. 1, 441-52 (1964)
CA 60: 12253e

D 2038 The mechanism of zymogen activation.
Neurath, H.
Federation Proc. 23: 1-7 (1964)
CA 60: 12309f

D 2039 Nonadiabatic transtions involved in atomic collisions. Quenching of the resonance fluorescence of sodium vapors by argon.
Nikitin, E.E., and Bykhovskii, V.K.
Opt. i Spektroskopiya 17: 815-20 (1964)
CA 62: 7127a

D 2040 The optical properties and photochemcial behavior of certain anthraquinone dyes. Pyranthrone, flavanthrone, and Indanthrene Blue RS.
Nitzl, K., and Doerr, F.
Melliand Tertilber. 45: 893-7 (1964)
CA 61: 14814g

D 2041 Delayed fluorescence in pulse radiolysis of anthracene solutions in benzene.
Nosworthy, J.M., and Keene, J.P.
Proc. Chem. Soc. 1964: 114 (1964)
CA 61: 190d

D 2042 Spectroscopic study of dianthrylethylenes.
Nurmukhametov, R.M., Timofeyuk, G.M., Chaplina, I.M., and Nagormaya, L.L.
Zh. Fiz. Khim. 38: 2465-9 (1964)
CA 62: 6010h

D 2043 Application of the luminescence method to determine the state of stabilizing additives in a polymer.
Nurmukhametov, R.N., Bondareva, L.V., Shigorin, D.N., Mikhailov, N.V., and Tokarev, L.G.
Vysokomolekul. Soedin. 6: 1411-14 (1964)
CA 61: 13467a

D 2044 The variation of the quantum efficiency of sodium salicylate with thickness of material.
Nygaard, K.J.
Brit. J. Appl. Phys. 15: 597-9 (1964)
CA 60: 15314h

D 2045 Fluorescent vinyl products.
O'Brien, W.J.
U.S. 3,125,536, 1959 (1964)
CA 60: p14690b

D 2046 Lattice energy and transfer stimulated emission from $CeF_3:Nd^{+3}$.
O'Connor, J.R., and Hargreaves, W.A.
Appl. Phys. Letters 4: 208-9 (1964)
CA 61: 6559d

D 2047 Pressure effect on energy transfer in anthracene-tetracene mixed crystals.
Ohigashi, H., Shirotani, I., Inokuchi, H., and Minomura, S.
J. Phys. Soc. Japan 19: 1996-7 (1964)
CA 62: 11311e

D 2048 Fluorescence properties of europium dibenzoylmethide and its complexes with Lewis bases.
Ohlmann, R.C., and Charles, R.G.
J. Chem. Phys. 40: 3131-3 (1964)
CA 61: 9064h

D 2049 Europium chelate solution as a potential liquid optical maser.
Ohlmann, R.C., Riedel, E.P., Charles, R.G., and Feldman, J.M.
Intern. Aerospace Abstr. 4: 1625 (1964)
CA 62: 8552h

D 2050 Effects of sample cell position and dimensions on the fluorescence intensity-concentration curve for "perpendicular-type" fluorimeters.
Ohnesorge, W.E.
Anal. Chim. Acta 31: 484-7 (1964)
CA 62: 1063g

D 2051 Polarized phosphorescence of hexa-
chlorbenzene and paradichlorobenzene
crystals at 77°K.
Olds, D.W.
Dissertation, Duke University, North
Carolina (1964)
CA 61: 14055h

D 2052 Molecular orientation. Spectral de-
pendence of bifluorescence of chloro-
plasts in vivo.
Olson, R.A., Jennings, W.H., and Butler,
W.L.
Biochim. Biophys. Acta 88: 331-7 (1964)
CA 61: 14920b

D 2053 Comparison of the lightfastness of
polymeric and monomeric fluorescent
dyes.
Osawa, E., Kwihara, O., and Nozawa, T.
J. Soc. Dyers Colourists 80: 205-6 (1964)
CA 62: 1765h

D 2054 Spectral properties of chlorophyllin a.
Oster, G., Broyde, S.B., and Bellin, J.S.
J. Am. Chem. Soc. 86: 1309-13 (1964)
CA 60: 11503b

D 2055 Fluorescence methods in polymer
science.
Oster, G., and Nishijima, Y.
Fortschr. Hochpolymer Forsch. 3: 313-31
(1964)
CA 61: 5761d

D 2056 Fluorescent compounds.
Panchenkov, G.M., Zhorov, Y.M., and
Venhatachalam, K.A.
U.S.S.R. 162,266, 1962 (1964)
CA 61: 14451d

D 2057 Hydrolytic uranyl species and their
emission spectra.
Pant, D.D., and Bist, H.D.
Indian J. Pure Appl. Phys. 2: 223-8 (1964)
CA 61: 12806h

D 2058 Phosphorescence and delayed fluores-
cence from solutions.
Parker, C.A.
Advan. Photochem. (A. Noyes, G.S.
Hammond, and J.N. Pitts, editors,
Interscience) 2: 305-83 (1964)
CA 61: 6877e

D 2059 Transient effects in triplet-triplet
annihilation.
Parker, C.A.
Trans. Faraday Soc. 60: 1998-2008 (1964)
CA 62: 3505b

D 2060 P-type delayed fluorescence from
ionic species and aromatic hydrocar-
bons.
Parker, C.A., Hatchard, C.G., and Joyce,
T.A.
J. Mol. Spectry. 14: 311-19 (1964)
CA 62: 1219a

D 2061 Fluorescence decay of Pr^{+3}, Nd^{+3}, and
Sm^{+2} in $LaCl_3$.
Partlow, W.D., and Carlson, E.H.
J. Chem. Phys. 41: 3645-6 (1964)
CA 62: 7263d

D 2062 Choice of the primary emission for
excitation of the X-ray fluorescent
spectrum.
Pavlinskii, G.V., and Losev, N.F.
Zavodsk. Lab 30: 165-8 (1964)
CA 60: 11496b

D 2063 Spectroscopic investigation of the
mechanism of the intramolecular heavy
atom effect on the phosphorescence
process. I.
Pavlopoulos, T., and El-Sayed, M.A.
J. Chem. Phys. 41: 1082-92 (1964)
CA 61: 6542g

D 2064 A two-quantasome theory of chlorophyll a fluorescence in green plant
photosynthesis.
Pearlstein, R.M.
Proc. Natl. Acad. Sci. U.S. 52: 824-30
(1964)
CA 62: 822f

D 2065 Fluorescence spectra of thin films of
aromatic hydrocarbons.
Perkampus, H.H., and Pohl, L.
Z. Physik. Chem. 40: 162-88 (1964)
CA 61: 191f

D 2066 Correlation between optical rotation,
fluorescence, and biological activity of
pepsinogen.
Perlmann, G.E.
Biopolymers, Symp. No. 1, 383-7 (1964)
CA 60: 13509a

D 2067 Fluorescence studies on poly-α-amino acids. Models of protein conformation. III. Copolymers of tyrosine with glutamic acid or lysine.
Pesce, A., Bodenheimer, E., Norland, K., and Fasman, G.D.
J. Am. Chem. Soc. 86: 5669-75 (1964)
CA 62: 4242a

D 2068 Electronic spectra of fluoranthrene and acenaphthylene in solutions at low temperature.
Pesteil, L., Pesteil, P., and Laurent, F.
Can. J. Chem. 42: 2601-6 (1964)
CA 62: 2376h

D 2069 Relations between absorption and fluorescence bands of brighteners and fluorescent dyes.
Pestemer, M., Berger, A., and Wagner, A.
SVF (Schweiz. Ver. Faerbereifachleuten) Fachorgan Textilveredlung 19: 420-5 (1964)
CA 61: 10201b

D 2070 Reexamination of the luminescent properties of the mixed phenyl- and p-biphenylsilanes.
Peterson, E.A.
US Dept. of Comm. Bull., AD 603615, 208 pp., (1964)
CA 62: 4794b

D 2071 Study of relaxation processes in Nd using pulsed excitation.
Peterson, G.E., and Bridenbaugh, P.M.
J. Opt. Soc. Am. 54: 644-50 (1964)
CA 61: 5103c

D 2072 Determination of constants of dipole moments of excited molecules by the temperature displacement of fluorescence and absorption spectra.
Pikulik, L.G., and Gladchenko, L.F.
Dokl. Akad. Nauk Belorussk. SSR 8: 641-4 (1964)
CA 62: 7130d

D 2073 Anomalous dependence of the emission anisotropy on the viscosity of the solution.
Polacka, B.
Bull. Acad. Polon. Sci., Ser. Sci., Math., Astron. Phys. 12: 131-6 (1964)
CA 61: 11494d

D 2074 Polarization spectra of POPOP and α-NPO in polymethylmethacrylate.
Polacka, B.
Bull. Acad. Polon. Sci., Ser. Sci., Math., Astron. Phys. 12: 491-5 (1964)
CA 62: 8518g

D 2075 Spectrum of trivalent erbium·ion in the matrix of calcium fluoride.
Pollack, S.A.
J. Chem. Phys. 40: 2751-67 (1964)
CA 60: 15311e

D 2076 Quenching of the europium luminescence in the crystals of intracomplex compounds in the presence of other rare earths.
Poluektov, N.S., Kononenko, L.I., Vitkun, R.A., and Nikonova, M.P.
Opt. i Spektroskopiya 17: 73-7 (1964)
CA 61: 14054d

D 2077 Fluorescent determination of microquantities of europium.
Poluektov, N.S., Vitkun, R.A., and Kononenko, L.I.
Ukr. Khim. Zh. 30: 629-35 (1964)
CA 61: 10029a

D 2078 Luminescence of chromium (III) complexes in rigid glass.
Porter, G.B., and Schlaefer, H.L.
Z. Physik. Chem. 40: 280-300 (1964)
CA 61: 9062h

D 2079 Low-lying energy levels and comparison of laser action of U^{+3} in CaF_2.
Porto, S.P.S., and Yariv, A.
Intern. Aerospace Abstr. 4: 1623-4 (1964)
CA 62: 9963d

D 2080 The Berta fluorite deposit (San Cugat del Valles, Barcelona).
Pous, J.M., and Altaba, M.F.
Bol. Real Soc. Espan. Hist. Nat., Secc. Geol. 62: 229-39 (1964)
CA 62: 11552c

D 2081 The fluorite layer of Mina Berta (San Cugat del Valles, Barcelona). II. Spectrographic study of the principal mineral species and the intruding rock.
Pous, J.M., and Altaba, M.F.
Notas Commun. Inst. Geol. Minero Espana 74: 61-8 (1964)
CA 61: 13058g

D 2082 Spectrofluorometric measurement of
   phenothiazines.
   Ragland, J.B., and Kinross-Wright, V.J.
   Anal. Chem. 36: 1356-9 (1964)
   CA 61: 4971b

D 2083 Electron spin resonance (E.S.R.) and
   luminescence studies of excited states
   of nucleic acids.
   Rahn, R.O., Longworth, J.W., Eisinger, J.,
   and Shulman, R.G.
   Proc. Natl. Acad. Sci. U.S. 51: 1299-1304
   (1964)
   CA 61: 12821d

D 2084 Use of X-ray fluorescence analysis
   for rapid mill control assaying.
   Rawling, B.S., and Greaves, M.C.
   Australasian Inst. Mining Met. Proc.
   No. 211, 135-55 (1964)
   CA 62: 13826b

D 2085 Effect of the time of firing and flux
   content on the intensity of phosphorescence of Ba sulfide-Cu (borax) phosphors.
   Razdan, K.N.
   Indian J. Pure Appl. Phys. 2: 271-2 (1964)
   CA 61: 14055a

D 2086 Absorption and fluorescence spectra
   of Eu^{+2} in alkali halide crystals.
   Reisfeld, R., and Glasner, A.
   J. Opt. Soc. Am. 54: 331-3 (1964)
   CA 60: 14017c

D 2087 Photoreaction of 3,4-benzopyrene
   with proteins.
   Reske, G., and Stauff, J.
   Ber. Bunsenges. Physik. Chem. 68: 783-4
   (1964)
   CA 62: 9377d

D 2088 Studies of the triplet-singlet decay
   lifetime for certain organic crystals at
   low temperatures.
   Richardson, B.M.
   Dissertation, Duke University, North
   Carolina (1964)
   CA 62: 8537e

D 2089 Vibronic spectra of $SrF_2:Sm^{+2}$ and
   $BaF_2:Sm^{+2}$.
   Richman, I.
   Phys. Rev. 133: 1364-6 (1964)
   CA 60: 10072b

D 2090 H bonding effects on triplet energy
   transfer in solution.
   Richtol, H.H., and Klappmeier, F.H.
   J. Am. Chem. Soc. 86: 1255-6 (1964)
   CA 60: 15156b

D 2091 Fluorescence spectrum of the benzyl
   radical trapped in a crystalline methyl-
   cyclohexane matrix.
   Ripoche, J.
   Compt. Rend. 259: 1071-3 (1964)
   CA 62: 1218h

D 2092 Effect of the pumping power on the
   amplification coefficient of a fluorescence line of iodine vapors excited by
   the 5461-Å mercury line.
   Rivoire, G., Hervet, H., and Dupeyrat, R.
   Compt. Rend. 259: 2404-7 (1964)
   CA 62: 4794h

D 2093 Low-temperature emission spectrum
   of O_2^- in alkali halides.
   Rolfe, J.
   J. Chem. Phys. 40: 1664-70 (1964)
   CA 60: 10077f

D 2094 Spectral properties of rare earth
   oxide phosphors.
   Ropp, R.C.
   J. Electrochem. Soc. 111: 311-17 (1964)
   CA 60: 10080d

D 2095 Fluorescence studies of the photosyn-
   thetic induction period.
   Rosenberg, J.L., Bigat, T., and Dejaegere,
   S.
   Biochim. Biophys. Acta 79: 9-19 (1964)
   CA 60: 8351a

D 2096 Modulation of resonance fluorescence
   with lamor frequency of the ground
   state.
   Rosinski, K.
   Bull. Acad. Polon. Sci., Ser. Sci., Math.,
   Astron. Phys. 12: 497-502 (1964)
   CA 61: 14062b

D 2097 See D 2174a

D 2098 Luminescence and the triplet state.
   Rousset, A.
   J. Chim. Phys. 61: 1621-30 (1964)
   CA 62: 14062d

D 2099 Mechanism of the retarded fluorescence of the impurities in the crystals of aromatic molecules close to room temperature.
Rousset, Y.
Compt. Rend. 258: 4687-9 (1964)
CA 61: 5103c

D 2100 Donor-acceptor relative orientation for maximum triplet-triplet energy transfer.
Roy, J.K., and El-Sayed, M.A.
J. Chem. Phys. 40: 3442-3 (1964)
CA 61: 10198f

D 2101 Electronic states of benzene by photoselection. The polarization of luminescence in benzene, perdeutero-benzene, and perchlorobenzene.
Russell, P.G., and Albrecht, A.C.
J. Chem. Phys. 41: 2536-50 (1964)
CA 61: 12810b

D 2102 Room temperature operation of Eu chelate liquid laser.
Samuelson, H., Lempicki, A., Brecher, C., and Brophy, V.
Appl. Phys. Letters 5: 173-4 (1964)
CA 62: 4806a

D 2103 Laser phenomena in europium chelates. I. Spectroscopic properties of europium benzoylacetonate.
Samuelson, H., Lempicki, A., Brophy, V.A., and Brecher, C.
J. Chem. Phys. 40: 2547-53 (1964)
CA 60: 14042g

D 2104 Molecular orientation in quantasomes. II. Absorption spectra, hill activity, and fluorescence yields.
Sauer, K., and Park, R.B.
Biochim. Biophys. Acta 79: 476-89 (1964)
CA 61: 3419f

D 2105 Ultraviolet spectrum of chrysene.
Sauvageau, P., and Sandorfy, C.
Can. J. Chem. 42: 197-200 (1964)
CA 60: 6348h

D 2106 Aluminum, gallium, and indium chelates of salicylidene-O-aminophenol.
Saylor, J.H., and Ledbetter, J.W.
Anal. Chim. Acta 30: 427-33 (1964)
CA 61: 1250f

D 2107 Color reaction of pentoses with anthrone. III. Fluorescence formed by the reaction of pentoses with anthrone.
Sawamura, R., and Koyama, T.
Chem. Pharm. Bull. (Tokyo) 12: 706-9 (1964)
CA 62: 1084f

D 2108 Characterization of aromatic compounds by low-temperature fluorescence and phosphorescence. Application to air pollution studies.
Sawicki, E., and Johnson, H.
Microchem. J. 8: 85-101 (1964)
CA 61: 4967b

D 2109 Direct spectrophotofluorometric analysis of aromatic compounds on thin-layer chromatograms.
Sawicki, E., Stanley, T.W., and Johnson, H.
Microchem. J. 8: 257-84 (1964)
CA 62: 1082c

D 2110 Thin-layer chromatographic separation and analysis of polynuclear aza hetrocyclic compounds.
Sawicki, E., Stanley, T.W., Pfaff, J.D., and Elbert, W.C.
Anal. Chim. Acta 31: 359-75 (1964)
CA 61: 15354h

D 2111 Quenchofluorometric analysis for fluoranthenic hydrocarbons in the presence of other types of aromatic hydrocarbons.
Sawicki, E., Stanley, T.W., and Elbert, W.C.
Talanta 11: 1431-9 (1964)
CA 61: 12639b

D 2112 A spectrophotofluorometric method for the analysis of aromatic and aza hetrocyclic hydrocarbons on thin layer chromatograms.
Sawicki, E., Stanley, T.W., and Elbert, W.C.
Occupational Health Rev. 16: 8 (1964)

D 2113 Automatic triparametric recording in fluorometry of polynuclear hydrocarbons.
Schachter, M.M., and Haenni, E.O.
Anal. Chem. 36: 2045-7 (1964)
CA 61: 12610c

D 2114 Apparatus for determining the fluorescence fading functions.
Schaefer, F.P., and Roellig, K.
Z. Physik. Chem. 49: 198-222 (1964)
CA 61: 191g

D 2115 Copper-activated thorium phosphate phosphors.
Schmid, W.F., and Mooney, R.W.
J. Electrochem. Soc. 111: 668-73 (1964)
CA 61: 193d

D 2116 Structural influence of alcoholic solutions on phosphorescence.
Schmillen, A., and Tschampo, A.
Z. Naturforsch. 19a: 190-4 (1964)
CA 60: 15277f

D 2117 Absorption and re-emission of oxide crystals containing dissimilar foreign ions.
Schneider, W.
US Dept. of Comm. Bull., AD 606837, 44 pp. (1964)
CA 62: 8520g

D 2118 Fluorescence emission spectra and structure of humic and fulvic acids.
Seal, B.K., Roy, K.B., and Mukherjee, S.K.
J. Indian Chem. Soc. 41: 212-14 (1964)
CA 61: 3829c

D 2119 Detection of amino acids within the 10^{-10} mole scale. Separation of 1-dimethylamino-5-naphthalene-sulfonyl amino acids by thin-layer chromatography.
Seiler, N., and Wiechmann, J.
Experientia 20: 559-60 (1964)
CA 62: 7f

D 2120 Fluorescence of organic mixed excimers.
Selinger, B.K.
Nature 203: 1062-3 (1964)
CA 62: 6028g

D 2121 The effect of iron concentration of flavine synthesis by Nocardia erythropolis.
Shaposhnikov, V.N., and Finogenova, T.V.
Dokl. Akad. Nauk SSSR 156: 692-4 (1964)
CA 61: 4750b

D 2122 Scattering of light, fluorescence, and transient phenomena.
Shorygin, P.P., and Krushinskii, L.L.
Dokl. Akad. Nauk SSSR 154: 571-4 (1964)
CA 60: 14023d

D 2123 Ultraviolet luminescence of muscle in various functional states.
Shtrankfel'd, I.G.
Dokl. Akad. Nauk SSSR 155: 461-4 (1964)
CA 60: 14889b

D 2124 Luminescence of crystalline anthracene.
Shpak, M.T., and Sheremet, N.I.
Opt. i Spektroskopiya 17: 694-704 (1964)
CA 62: 7264d

D 2125 Effect of some metabolic poisons of the respiratory chain on the ultraviolet fluorescence of cells.
Shudel, M.S., Chernogryadskaya, N.A., Brumberg, V.A., Rozanov, Y.M., and Brumberg, E.M.
Dokl. Akad. Nauk SSSR 157: 447-50 (1964)
CA 61: 12431b

D 2126 Intensity dependence of laser-induced delayed fluorescence.
Silver, M., and Zahlan, A.B.
J. Chem. Phys. 40: 1458 (1964)
CA 60: 11509b

D 2127 Precipitation of submicrogram quantities of thorium by barium sulfate and application to fluorometric determination of thorium in mineralogical and biological samples.
Sill, C.W., and Willis, C.P.
Anal. Chem. 34: 622-30 (1964)

D 2128 Fluorescence of uranium in fluoride-carbonate melts.
Singer, E., and Cifkova, D.
Z. Anal. Chem. 202: 401-7 (1964)
CA 61: 8892g

D 2129 The construction of an objective fluorimeter for measuring fluorescence of melts and an apparatus for the determination of traces of uranium.
Singer, E., Ruzicka, B., and Turtenwald, J.
Chem. Listy 58: 224-30 (1964)
CA 60: 11352a

D 2130 Three-photon absorption in naphthalene
crystals by laser excitation.
Singh, S., and Bradley, L.T.
Phys. Rev. Letters 12: 612-4 (1964)
CA 61: 5105g

D 2131 Effect of purity and temperature on the
fluorescence of anthracene excited by
red light.
Singh, S., and Lipsett, F.R.
J. Chem. Phys. 41: 1163-4 (1964)
CA 61: 11493f

D 2132 2,2'-Dipyridyl complexes of rare
earths. I. Preparation and infrared
and some other spectroscopic data.
Sinha, S.P.
Spectrochim. Acta 20: 879-86 (1964)
CA 61: 5093a

D 2133 Fluorescence spectra of Eu(III)
phthalate and naphthalate and of Sa(III),
Eu(III), Tb(III), and Dy(III) dipyridyl
complexes.
Sinha, S.P., Klixbuell, C., Jorgensen, C.K.
and Pappalardo, R.
Z. Naturforsch. 19a: 434-9 (1964)
CA 61: 3831c

D 2134 The use of the background as a stand-
ard in X-ray fluorescence analysis.
Smagunova, A.N., Belova, R.A., Afonin,
V.P., and Losev, N.F.
Zavodsk. Lab. 30: 426-31 (1964)
CA 61: 3668a

D 2135 Correlation of the optical and electron
paramagnetic resonance spectra of
type I natural diamonds.
Sobolev, E.V., Bokii, G.B., Dvoryankin,
V.F., and Samsonenko, N.D.
Zh. Strukt. Khim. 5: 557-61 (1964)
CA 62: 151g

D 2136 Fluorescence and stimulated emission
from $Gd_2O_3:Nd^{+3}$ at room temperature
and 77°K.
Soffer, B.H., and Hoskins, R.H.
Appl. Phys. Letters 4: 113-4 (1964)
CA 60: 14023e

D 2137 Triplet-singlet luminescence from
methylated benzenes in the crystalline
state and in rigid glass solutions.
Sponer, H., and Kanda, Y.
J. Chem. Phys. 40: 778-87 (1964)
CA 60: 6362f

D 2138 Structure-dependent phosphorescence
and chemiluminescence of a protein.
Stauff, J., and Wolf, H.
Z. Naturforsch. 19b: 87-97 (1964)
CA 60: 15314d

D 2139 Fluorometric determination of corticosteroids in human blood; comparison of results with the Silber-Porter
method.
Steenburg, R.W., and Thomasson, B.H.
J. Clin. Endocrinol. Metab. 24: 875-83
(1964)
CA 61: 14974d

D 2140 Ultraviolet fluorescence of proteins.
Influence of conformation and environ-
ment.
Steiner, R.F., Lippoldt, R.E., Edelhoch,
H., and Frattali, V.
Biopolymers, Symp. No. 1, 355-66 (1964)
CA 60: 13485d

D 2141 Starch-primuline fluorescence.
Sterling, C.
Protoplasma 59: 180-92 (1964)
CA 62: 9447d

D 2142 Spectrophotometric determination of
enthalpies and entropies of photo-
association for dissolved aromatic
hydrocarbons.
Stevens, B., and Ban, M.I.
Trans. Faraday Soc. 60: 1515-23 (1964)
CA 61: 12709d

D 2143 Photodimerization in crystalline
9-cyanoanthracene.
Stevens, B., Dickinson, T., and Sharpe,
R.R.
Nature 204: 876-7 (1964)
CA 62: 4813c

D 2144 Fluorescence self-quenching in aro-
matic vapors: the effect of fluorescence
reabsorption.
Stevens, B., and McCartin, P.J.
Mol. Phys. 8: 597-606 (1964)
CA 62: 14060g

D 2145 A mechanism for triplet-state relaxa-
tion of aromatic molecules in a fluid
environment.
Stevens, B., and Walker, M.S.
Proc. Chem. Soc. 1964: 26-7 (1964)
CA 60: 10078h

D 2146 Kinetics of triplet-state relaxation. Correction.
Stevens, B., and Walker, M.S.
Proc. Chem. Soc. 1964: 109 (1964)
CA 61: 3803d

D 2147 The kinetics of phosphorescence and delayed fluorescence decay for aromatic hydrocarbons in liquid paraffin.
Stevens, B., and Walker, M.S.
Proc. Roy. Soc. (London), Ser. A 281: 420-36 (1964)
CA 61: 9065f

D 2148 The prosthetic group of succinic dehydrogenase. I. Fluorescence and enzymic hydrolysis of a flavine peptide.
Strom, R., and Cerletti, P.
Ital. J. Biochem. 11: 208-20 (1962)
CA 62: 6734g

D 2149 Fluorescent rare earths in semiconducting thiospinels.
Suchow, L., and Stemple, N.R.
J. Electrochem. Soc. 111: 191-5 (1964)
CA 60: 6315a

D 2150 Electronic spectra of substituted aromatic hydrocarbons. IV. Anthramines.
Suzuki, S., and Baba, H.
Bull. Chem. Soc. Japan 37: 519-24 (1964)
CA 61: 11479e

D 2151 Effect of temperature on the shape of lines in the quasilinear fluorescence spectrum of coronene.
Svishchev, G.M.
Opt. i Spektroskopiya 16: 364-5 (1964)
CA 60: 14023f

D 2152 Color centers in x-irradiated halophosphate crystals.
Swank, R.K.
Phys. Rev. 135: 266-75 (1964)
CA 61: 2577e

D 2153 Intermolecular energy transfer and concentration depolarization during fluorescence.
Szalay, L.
Ann. Physik 14: 221-8 (1964)
CA 62: 7220d

D 2154 Effect of the solvent on the fluorescence spectrum of trypaflavine and fluorescein.
Szalay, L., and Tombacz, E.
Acta Phys. Acad. Sci. Hung. 16: 365-9 (1964)
CA 61: 14054h

D 2155 Fluorescence of sanitary textile products.
Szekely, A.
Magy. Textiltech. 16: 490-6 (1964)
CA 62: 14873h

D 2156 Fluorescent photometer using a low-pressure mercury arc.
Tabata, T., and Shibazaki, T.
Bunseki Kagaku 13: 56-8 (1964)
CA 60: 9878c

D 2157 Synthesis of new monomers and polymers. IV. Synthesis and properties of poly(diphenyldiacetylenes).
Teyssie, P., and Korn-Girard, A.C.
J. Polymer Sci., Pt. A 2: 2849-58 (1964)
CA 61: 5777g

D 2158 Methodics of the precipitation processes. XIX. Fluorescence of sodium fluresceinate, Rhadamine B, and their mixtures, as a function of hydrogen ion concentration in water solutions.
Tezak, D., and Tezak, B.
Croat. Chem. Acta 36: 59-66 (1964)
CA 61: 15327b

D 2159 Pair spectra and edge emission in gallium phosphide.
Thomas, D.G., Gershenzon, M., and Trumbore, F.A.
Phys. Rev. 133: 269-79 (1964)
CA 60: 3634d

D 2160 Fluorescence responses of chlorophyll in vivo to treatment with acetone.
Thomas, J.B., and Flight, W.F.G.
Biochim. Biophys. Acta 79: 500-10 (1964)
CA 61: 3344e

D 2161 Methandrostenolone. Mechanism of hydrochloric acid-induced fluorescence.
Tishler, F., and Brody, S.M.
J. Pharm. Sci. 53: 161-4 (1964)
CA 60: 14562d

D 2162 An adapter for the SF-4 spectrophoto-
meter for sensitive fluorescence
measurements.
Tomana, M., and Stranskey, Z.
Chem. Listy 58: 978-98 (1964)
CA 61: 12601c

D 2163 Study of excited-state reactions of
β-naphthol by means of fluorescence
determination.
Trieff, N.M.
Dissertation, New York University, New
York (1964)
CA 60: 15191f

D 2164 An absolute spectrofluorometer.
Turner, G.K.
Science 146: 183-9 (1964)

D 2165 Phosphorescence spectra of biphenyl,
fluorene, acenaphthalene, and carbazole.
Trusov, V.V., and Teplyakov, P.O.
Opt. i Spektroskopiya 16: 52-7 (1964)
CA 60: 11508e

D 2166 Energy transfer in monomolecular
films containing chlorophyll a.
Tweet, A.G., Gaines, G.L., and Bellamy,
W.D.
Nature 202: 696-7 (1964)
CA 61: 5105e

D 2167 Fluorescence of chlorophyll a in
monolayers.
Tweet, A.G., Gaines, G.L., and Bellamy,
W.D.
J. Chem. Phys. 40: 2596-2600 (1964)
CA 60: 14025c

D 2168 Angular dependence of fluorescence
from chlorophyll a in monolayers.
Tweet, A.G., Gaines, G.L., and Bellamy,
W.D.
J. Chem. Phys. 41: 1008-10 (1964)
CA 61: 6507g

D 2169 Fluorescence quenching and energy
transfer in monomolecular films con-
taining chlorophyll.
Tweet, A.G., Bellamy, W.D., and Gaines,
G.L.
J. Chem. Phys. 41: 2068-77 (1964)
CA 61: 12810g

D 2170 Fluorescence of pure zinc sulfide in
correlation with deviation from stoich-
iometry.
Uchida, I.
J. Phys. Soc. Japan 19: 670-4 (1964)
CA 61: 1361f

D 2171 The action of heparin on the total
catechol amine content of whole blood
and its significance.
Unghvary, L., Farkas, K., Hovanyi, M.,
and Farkas, F.
Z. Rheumaforsch. 23: 204-6 (1964)
CA 61: 13792f

D 2172 Comparative study of the fluorescence
and emission spectra of p-dichloroben-
zene.
Upadhya, K.N., and Rai, J.N.
Indian J. Pure Appl. Phys. 2: 284-6 (1964)
CA 62: 140b

D 2173 Excited singlet and triplet states of
some aromatic hydrocarbons.
Uzhinov, B.M., Kuz'min, M.G., Morozov,
Y.V., and Berezin, I.V.
Vestn. Mosk. Univ., Ser. II, Khim. 19:
62-4 (1964)
CA 62: 6008e

D 2174 Fluorescence of 1,3,7,9-tetramethyl-
uric acid complexes of aromatic hydro-
carbons.
Van Duuren, B.L.
J. Phys. Chem. 68: 2544-53 (1964)
CA 61: 10201d

D 2174a The fluorescence emitted by slices
of rat liver and avian salt gland.
Van Rossum, G.D.V.
Biochim. Biophys. Acta 88: 507-16 (1964)
CA 62: 2925a

D 2175 Excitation of fluorescence with mono-
chromatic light in rare earth crystals.
Varsanyi, F.
Intern. Aerospace Abstr. 4: 1625 (1964)
CA 62: 8534g

D 2176 Neodymium fluorescence in the 5- to
6-μ region.
Varsanyi, F.
Phys. Rev. Letters 11: 193-4 (1964)
CA 61: 11494g

D 2177 The dependence of the duration of
scintillations from CsI-Tl and KI-Tl
phosphors on the activator concentration.
Vasil'eva, N.N.
Opt. i Spektroskopiya 16: 851-3 (1964)
CA 61: 6544c

D 2178 The fluorescence of fossils. III. Comparative chemical and fluorescence observations on the constitution of teeth
of carcharodon megalodon in natural
and experimental conditions.
Vialli, G.S.
Atti Ist. Geol. Univ. Pavia No. 15, 89-145
(1964)
CA 62: 14342d

D 2179 Visible fluorescence spectra of vari-
ous natural amoebicides and possible
synthetic therapeutic substitutes.
Viel, C.
J. Chim. Phys. 61: 1234-9 (1964)
CA 62: 3886e

D 2180 See D 2208a

D 2181 See D 2208b

D 2182 Excited electronic states of isotactic
polystyrene and poly(vinylnaphthalene).
Vola, M.T., Silbey, R., Rice, S.A., and
Jortner, J.
J. Chem. Phys. 41: 2846-53 (1964)
CA 61: 15530a

D 2183 Fluorescence of tyrosine and trypto-
phan residues incorporated into poly-
peptide chains.
Wada, A., and Veno, Y.
Biopolymers, Symp. No. 1, 343-53 (1964)
CA 60: 13476e

D 2184 A very luminous grating spectrograph
for the study of the Raman effect.
Wallart, F.
Compt. Rend. 258: 5390-3 (1964)
CA 62: 143h

D 2185 2,2'-p-Phenylenebis(4-methyl-5-
phenyloxazole) and other solutes for
scintillation counting. I.
Walker, D., and Waugh, T.D.
J. Heterocyclic Chem. 1: 72-3 (1964)
CA 60: 15850h

D 2186 Fluorescent properties of terbium-
activated alkaline earth borates.
Wanmaker, W.L., and Bril, A.
Philips Res. Rept. 19: 479-97 (1964)
CA 62: 14062e

D 2187 Absorption intensity and fluorescence
lifetimes of molecules.
Ware, W.R., and Baldwin, B.A.
J. Chem. Phys. 40: 1703-5 (1964)
CA 60: 10074d

D 2188 Zeeman effect of the $^{(2)}F_{(5/2)}$,
$E_{(5/2)} \rightarrow {}^{(2)}F_{(7/2)}$, $E_{(5/2)}$ transition in
CaF_2:Tm^{2+}.
Weakleim, H.A., and Kiss, Z.J.
J. Chem. Phys. 4(5): 1507-8 (1964)
CA 61: 9066b

D 2189 Proton-tranfer effects in the quench-
ing of fluorescence of tyrosine copoly-
mers.
Weber, G., and Rosenheck, K.
Biopolymers, Symp. No. 1, 333-41 (1964)
CA 60: 13475c

D 2190 Fragmentation of lovine serum albu-
min by pepsin. I. Origin of the acid
expansion of the albumin molecule.
Weber, G., and Young, L.B.
J. Biol. Chem. 239: 1415-23 (1964)
CA 60: 13481e

D 2191 Photoionization and photoionization-
induced ion-molecule reactions.
Weissler, G.L., et al.
NASA Accession No. N64-19633 Rept. No.
AD 435601, 7 pp. (1964)
CA 62: 1242d

D 2192 Two-photon absorption in crystalline
anthracene and naphthalene excited
with a xenon flash.
Weisz, S.Z., Zahlan, A.B., Gilreath, J.,
and Jarnagin, R.C.
J. Chem. Phys. 41: 3491-5 (1964)
CA 62: 1219b

D 2193 Spectroscopy of carbon vapor con-
densed in rare-gas matrixes at 4 and
20°K.
Weltner, I.W., Walsh, P.N., and Angell,
C.L.
J. Chem. Phys. 40: 1299-1305 (1964)
CA 60: 7590b

D 2194 Fluorometric analysis.
White, C.E., and Weissler, A.
Anal. Chem. 36: 140R-141R (1964)
CA 60: 15107d

D 2195 Fluorescence of solutions: a review.
Williams, R.T., and Bridges, J.W.
J. Clin. Pathol. 17: 371-94 (1964)
CA 62: 9369d

D 2196 Atomic fluorescence spectrometry as
a means of chemical analysis.
Winefordner, J.D., and Vickers, J.J.
Anal. Chem. 36: 161-5 (1964)
CA 60: 6199h

D 2197 Phosphorescence of oxygen-containing
luminophors. II.
Witzmann, H., Anderson, H., and Grieser,
G.
Z. Physik. Chem. 226: 333-42 (1964)
CA 62: 2376g

D 2198 Luminescence of the system
CaO.B_2O_3:Pb.
Witzmann, H., Herzog, G.H., and
Schlemmer, L.
Naturwissenschaften 51: 101-2 (1964)
CA 60: 15314d

D 2199 Emission properties of lead-
manganese-activated calcium borate
phosphors (CaO·xB_2O_3; Pb, Mn).
Witzmann, H., Buhrow, J., and Mueller, K.
Naturwissenschaften 51: 103 (1964)
CA 61: 1406h

D 2200 Tetracycline fluorescence in mouse
gastric carcinoma.
Yabe, Y., Warren, I.A., Berk, J.E., and
Kobernich, S.
J. Histochem. Cytochem. 12: 842-6 (1964)
CA 62: 9576f

D 2201 Effect of an electric field on the sharp
green fluorescence transitions of Er in
single-crystal fluorite.
Yaney, P.P.
Sci. Tech. Aerospace Rept. 2: 1934 (1964)
CA 62: 8535a

D 2202 Effect of a uniform electric field on
sharp fluorescence transitions in single
crystalline CaF_2 (Er^{+3}).
Yaney, P.P.
Dissertation, University of Cincinnati,
Cincinnati, Ohio (1964)
CA 61: 7819a

D 2203 Phenon-induced relaxation in excited
optical states of trivalent praseodym-
ium in LaF_3.
Yen, W.M., Scott, W.C., and Schawlow,
A.L.
Phys. Rev. 136: 271-83 (1964)
CA 61: 11474b

D 2204 Biochemical changes in tobacco in-
fected with colletotrichum destructivam.
I. Fluorescent compounds, phenols,
and associated enzymes.
Yu, L.M., and Hampton, R.E.
Phytochemistry 3: 269-72 (1964)
CA 60: 14846a

D 2205 Damping by oxygen of the fluorescence
of acriflavine on silica.
Zakharov, I.A., and Aleskovskii, V.B.
Izv. Vysshikh Uchebn. Zavedenii, Khim. i
Khim. Tekhnol. 7: 517-19 (1964)
CA 62: 140a

D 2206 Delayed fluorescence of aromatic
hydrocarbons.
Zander, M.
Ber. Bunsenges. Physik. Chem. 68: 301-7
(1964)
CA 61: 1406b

D 2207 Phosphorescence spectra of benzocar-
bazoles.
Zander, M.
Ber. 97: 2695-9 (1964)
CA 61: 14053d

D 2208 Calculation of intensity distribution in
the vibrational structure of electronic
transitions: the $B^3\Pi_{o+u}$ - $X^1\Sigma_{o+g}$
resonance series of molecular iodine.
Zare, R.N.
J. Chem. Phys. 40: 1934-44 (1964)
CA 60: 10079h

D 2208a Rate analysis of multiple-step excit-
ation in mercury vapor.
Zito, R.
Intern. Aerospace Abstr. 4: 1622 (1964)
CA 62: 8535d

D 2208b Effect of solvent and temperature on
the position of the electronic spectra of
the 3,4-benzopyrene molecule.
Zurauskiene, E.
Lietuvos Fiz. Rinkinys, Lietuvos TSR
Mokslu Akad., Lietuvos TSR Aukstosios
Mokyklos 4: 565-71 (1964)
CA 62: 14041h

SUPPLEMENT

E 1 Automatic measurement of light absorption and fluorescence on paper chromatograms.
Marsh, M., and Brown, J.
Anal. Chem. 25: 1865-9 (1953)

E 2 Fluorescence characteristics of 5-hydroxytryptamine (serotonin).
Udenfriend, S., Bodanski, D. F., and Weissbach, H.
Science 122: 972-3 (1955)

E 3 Estimation of 5-hydroxytryptamine (serotonin) in biological tissue.
Udenfriend, S., Weissbach, H., and Clark, C.T.
J. Biol. Chem. 215: 337-44 (1955)

E 4 Phosphorimetry, a new list of analyses.
Kiers, R.J., Britt, R.D., and Wentworth, W.E.
Anal. Chem., Feb. (1957)
CA 51: 56211

E 5 A spectrophotofluorometric study of organic compounds of pharmacological interest.
Udenfriend, S., Duggan, D.E., Vasta, B.M., and Brodie, B.B.
J. Pharmacol. Exptl. Therap. 120 (1): 26-37 (1957)
CA 51: 18473b

E 6 Fluorometric method for estimation of tyrosine in plasma and tissues.
Waalkes, T. P., and Undenfriend, S.
J. Lab. Clin. Med. 50 (5): 733-6 (1957)

E 7 Spectrophotofluorometry for pesticide determinations.
Hornstein, I.
J. Agr. Food Chem., 6: 32-4 (1958)
CA 52: 7606i

E 8 Fluorometric determination of kynurenic acid and xanthurenic acid in human urine.
Satoh, K., and Price, J.M.
J. Biol. Chem. 230(2): 781-9 (1958)
CA 52: 12053d

E 9 A new instrument for fluorescent analysis.
Howerton, H.K.
ISA J., Vol. 6, No. 10 (1959)

E 10 A method for the fluorometric assay of histamine in tissues.
Shore, P.A., Burkhalter, A., and Cohn, V.H.
J. Pharmacol. Exptl. Therap. 127: 182-6 (1959)

E 11 The Chesapeake Bay Institute study of the Baltimore Harbor.
Carpenter, J.H.
Proc. Ann. Conf., Maryland-Delaware Water Sewage Assoc. 33: 62-78 (1960)

E 12 Tracer for circulation and mixing in natural waters.
Carpenter, J.H.
Public Works, Vol. 91, No. 6 (1960)

E 13 New instrument for phosphorescent analysis.
Howerton, H.K.
J. Opt. Soc. Am. 50: 505 (1960)

E 14 Influence of scatter on fluorescence spectra of dilute solutions. I. Observations with two f/4 Czerny-Turner monochromators.
Price, J.M., Kaihara, M., and Howerton, H.K.
Pitts. Conf. Anal. Chem. Appl. Spect. (1960)

E 15 Measurements of turbulent diffusion in
Estuarine and inshore waters.
Pritchard, D.W., and Carpenter, J.H.
Bull. Intern. Assoc. Scientific Hydrology
20: 37-50 (1960)

E 16 Quinidine.
Gelfman, N., and Seligson, D.
Am. J. Clin. Pathol. 36: 390-2 (1961)

E 16a Photometric determination of boron in
magnesium alloys.
Gordievskii, A.V., and Ustyugov, G. P.,
Izvest. Vysshikh. Ucheb Zavedenii, Khim.
i Khim. Tekhnol. 4: 366-9 (1961)
CA 55: 25558li

E 17 Studies of the physical, immunological,
and biological properties of insulin
conjugated with fluorescein isothiocya-
nates.
Halikis, D.N., and Arguilla, E.R.
Diabetes 10: 142-7 (1961)

E 18 Gluocester-forced circulation of Babson
Reservoir.
Nickerson, H.D.
Sanitalk 9: 1-10 (1961)

E 19 A fluorometric method to determine
levels of histamine in human plasma.
Noah, J.W., and Brand, A.
J. Allergy 32: 236-40 (1961)

E 20 Standard methods of clinical chemistry.
Seligson, D.
Academic Press (1961)

E 21 A study of the factors affecting the aluminum oxide trihydroxyindole procedures
for analysis of catecholamine.
Anton, A.H., and Sayre, A.F.
J. Pharmacol. Exptl. Therap. 138, Dec.
(1962)

E 22 Formation of phenolic acids in brain
after administration of
3,4-dihydroxyphenylalanine.
Carlsson, A., and Hillays, N.A.
Acta Phys. Scand. 55: 95-100 (1962)

E 23 The selective sensitization of biacetyl
triplet state in the vapor phase.
Dubois, J.T.
J. Am. Chem. Soc. 84: 2902-4 (1962)

E 24 Fluorescence of tetracyclines in bone.
Absorption maximum, hydration shell,
and polarization effects.
Hattner, R., and Frost, H.M.
J. Surg. Res. 2: 262-7 (1962)
CA 62: 1896e

E 25 New fluorometric micromethod for the
determination of reserpine.
Jakovljevic, I.M., Fose, J.M., and Kuzel,
N.R.
Anal. Chem. 34: 410-3 (1962)

E 26 Isolation of dehydroepiandrosterone and
17 α-hydroxy-pregnenolone from the
polycystic ovaries of the Stein-
Leventhal syndrome.
Maheah, V.B., Greenblatt, D.P., and
Greenblatt, R.B.
J. Clin. Endocrinol. Metab. 22: 441-8
(1962)

E 27 Excitation of optical fluorescence spec-
tra of transition elements by means of
X rays.
Makovsky, J., Low, W., and Yatsiv, S.
Phys. Rev. Letters 2: 186-7 (1962)
CA 62: 9932e

E 28 Method of determining the composition
of piridine nucleotides in biological
materials.
Martelli, P., and Ricci, C.
Accad. Fis. Atti 11: 3-16 (1962)

E 29 Calibration of spectrofluorometers for
measuring corrected emission spectra.
Melhuish, W.H.
J. Opt. Soc. Am. 52: 1256-8 (1952)

E 30 α-Haloacrylic polymers. X. Quenching
effect of poly(α-chloroacrylate) and the
dehalogenated polymer on the fluorescence of a liquid scintillator.
Okamura, S., and Yamada, Y.
Doitai To Hoshasen 4: 331-42 (1962)
CA 62: 12620e

E 31 α-Haloacrylic polymers. XI. Quenching
effects of copolymer and mixture of
poly(ethyl α-chloroacrylate).
Okamura, S., and Yamada, Y.
Doitai To Hoshasen 4: 343-52 (1962)
CA 62: 12620g

E 32 α-Haloacrylic polymers. XII. Quenching the fluorescence of a liquid scintillator in a solvent by organic halides.
Okamura, S., and Yamada, Y.
Doitai To Hoshasen 4: 353-62 (1962)
CA 62: 12620h

E 33 A comparative study of various methods for the detection of formaldehyde.
Sawicki, E., Stanley, T.W., and Pfaff, J.
Chem. Anal. 51, Mar. (1962)

E 34 Determination of the secretion rate of aldosterone in normal man by use of 7-H^3-d-aldosterone and acid hydrolysis of urine.
Siegenthaler, W.E., Dowdy, A., and Leutscher, J.A.
J. Clin. Endocrinol. Metab. 22: 172-7 (1962)

E 35 Fluorescence of fresh and glycerinated muscle stained by Acridine Orange.
Takahashi, M.
Sapporo Igaku Zasshi 21: 85-9 (1962)
CA 62: 10812f

E 36 Fluorescence assay in biology and medicine.
Udenfriend, S.
Academic Press (1962)

E 37 Formation of highly fluorescent zinc phosphate in the presence of uranium and its application to the direct fluorimetry of uranium in aqueous media.
Alberti, G., and Saini, A.
Anal. Chim. Acta 28: 536-42 (1963)

E 38 Basal content and the induced biosynthesis of pyridine nucleotides in the rat liver under the influences of insulin.
Alertsen, A.R., Haugen, H.N., and Walass, E.
Acta Physiol. Scand. 57: 317-27 (1963)

E 39 Fluorescence methods for the determination of correction factors for fluorescence excitation and emission spectra.
Argauer, R., and White, C.E.
Eleventh Detroit Anachem Conf., Wayne State Univ. Paper No. 59 (1963)

E 40 One-hour subcutaneous ACTH test with determination of plasma corticosteroids.
Arner, B., Hedner, P., Karlefors, T., and Rerup, C.
Acta Med. Scand. 173: 91-7 (1963)

E 41 A sensitive method of high specificity for determination of urinary estrogens.
Barlow, J.J.
Anal. Biochem. 6: 435-50 (1963)

E 42 Fluorescent pesticide techniques.
Beckman, H.F., Bruce, R.B., and MacDougall, D.
Analytical Methods for Pesticides, Plant Growth Regulators, and Food Additives, Zweig, G., editor, Academic Press, N.Y., Vol. I, Ch. 8 (1963)

E 43 Study of thalidomide in organisms.
Beckmann, R.
Arzneimittel-Forsch. 13: 185-91 (1963)

E 44 Fluorimetric determination of formaldehyde.
Belman, S.
Anal. Chim. Acta 29: 120-6 (1963)

E 45 Photodissociation and fluorescence in the Schumann ultraviolet.
Beyer, K., and Welge, K.H.
US Dept. of Comm. Bull., AD 438456,15 pp. (1963)
CA 62: 2373a

E 46 On the luminescence properties of some purines and pyrimidines. A study by fluorescence spectrophotometry of the sites of protonation and of the types of lowest excited singlet states.
Borresen, H.C.
Acta Chem. Scand. 17: 921-9 (1963)

E 47 Study of the Momose fluorometric determination of blood glucose.
Bourne, B.B.
Clin. Chem. 9 (120): 460 (1963)

E 48 Introduction of fluorescence into proteins by treatment with N-bromosuccinimide.
Brand, L., and Shaltiel, S.
Israel J. Chem. 1: 51-2 (1963)

E 49 Lower limits of organic reactions.
Brandt, R., and Cheronis, N.D.
Mikrochim. Acta 1963: 467-73 (1963)

E 50 Simplified method for the determination
of reduced triphosphopyridine nucleotide by means of enzymic cycling.
Brown, E., and Clarke, D.L.
J. Lab. Clin. Med. 61: 889-92 (1963)

E 51 Differential fluorometry in catecholamine determination. A simplified
method of calculation.
Brunjes, S., and Wybenga, D.
Clin. Chem. 9: 626-30 (1963)

E 52 Chemiluminescence of phenazine methosulfate in the presence of hydrogen
peroxide induced by reductants including reduced nicotinanide adenine dinucleotide (NADH) and ascorbic acid.
Chayet, C., Steele, R.H., and
Breckinridge, B.S.
Biochem. Biophys. Res. Commun. 10:
390-395 (1963)

E 53 Fluorometric assay of yohimbine.
Chiang, H.C., and Chen, W.F.
J. Pharm. Sci. 52: 808-9 (1963)

E 54 Light — an essential factor in the trihydroxyindole spectrofluorometric
assay of norepinephrine.
Chin, L., Picchioni, A.L., and Childs, R.F.
J. Pharm. Sci. 52: 907-9 (1963)

E 55 Direct proportionality of urinary excretion rate and serum level of tetracycline in human subjects.
Chulski, T., Johnson, R.H., Schlagel, C.A.,
and Wagner, J.G.
Nature 198: 450-3 (1963)

E 56 Fluorescent analysis of α-tubocurarine
hydrochloride.
Cohen, E.N.
J. Lab. Clin. Med. 61: 338-45 (1963)

E 57 Quantitative determination of α-tubocur-
arine in body tissues and fluids.
Cohen, E.N.
J. Lab. Clin. Med. 62: 979-84 (1963)

E 58 Fluorometric assay of glutathione.
Cohn, V.H., and Lyle, T.
Federation Proc. 22(2), Paper 1564 (1963)

E 59 Joint occurrence of a lichen depsidone
and its probable depside precursor.
Culberson, C.F.
Science 143: 255-6 (1963)

E 60 Correction of luminescence spectra and
calculation of quantum efficiencies
using computer techniques.
Drushel, H.V., Sommers, A.L., and Cox,
R.C,.
Anal. Chem. 35: 2166 (1963)

E 61 Singlet-singlet energy transfer in fluid
solutions.
Dubois, J.T., and Cox, M.
J. Chem. Phys., 38(10): 2536-41 (1963)

E 62 An improved bioassay for blood ACTH.
Ducommun, P., Ducommun, S., Kraiser,
J., Jobin, M., and Fortier, C.
Proc. Can. Fed. Biol. Sci. 6: 20 (1963)

E 63 Paper chromatographic differentiation of
some important phenothiazine encount-
ered in toxicology.
Eagleson, D.A.
Am. J. Clin. Pathol. 39: 648-51 (1963)

E 64 Stationary phase as color reagent in
glass paper chromatography of estrogens.
Epstein, E., and Zak, B.
Clin. Chem. 9: 70-8 (1963)

E 65 Testosterone analysis by glass paper
chromatography.
Epstein, E., and Zak, B.
Eleventh Detroit Anachem Conf., Wayne
State Univ., Paper No. 46 (1963)

E 66 Fluorescent tracers for dispersion
measurements.
Feuerstein, D.L., and Selleck, R.E.
J. Sanit. Eng. Div., Am. Soc. Civil Engrs.
89: 1-21 (1963)

E 67 Tooth fluorometer.
Forziati, A.F., Kumpula, J.W., and
Barone, J.J.
J. Am. Dental Assoc. 67: 663-9 (1963)

E 68 Fluorometry as a method of determining protein content of milk.
Fox, K.K., Holsinger, V.H., and Pallansch, M.J.
J. Dairy Sci. 26: 302-9 (1963)

E 69 Colorimetric and fluorometric reactions of digitalis.
Frerejacque, M., and DeGrave, P.
Ann. Pharm. Franc. 21: 509-28 (1963)

E 70 Fluorometric determination of biliverdin in serum bile and urine.
Garay, E.A.R., and Argevich, T.C.
J. Lab. Clin. Med. 62: 141-7 (1963)

E 71 Fluorometric determination of total phospholipids in rat tissues.
Harris, R.A., and Gambal, D.
Anal. Biochem. 5: 479-88 (1963)

E 72 Fluorometric analysis of 4-chloro-2-oxybenzoic-n-butylamide.
Haussler, A., and Hajdu, P.
Arzneimittel-Forsch. 13: 16-7 (1963)

E 73 Separation of pyridoxine, pyridoxal, and pyridoxamine by a sulfonic acid ion exchange resin.
Hedin, P.A.
Agr. Food Chem. 11: 343-5 (1963)

E 74 Measurement of protein in millimicrogram amounts by quenching of dye fluorescence.
Hiraoka, T., and Glick, D.
Anal. Biochem. 5: 497-504 (1963)

E 75 Utility of 2,4-bis-N,N[1]-di(carboxymethyl)aminomethyl fluorescein in the fluorometric estimation of aluminum, alkaline earths, cobalt copper, nickel, and zinc in micromolar concentrations.
Hoelzl-Wallach, D.F., and Steck, T.L.
Anal. Chem. 35: 1035-44 (1963)

E 76 Fluorometric determination of oxytetracycline in biological material.
Ibsen, K.H., Saunders, R.L., and Urist, M.R.
Anal. Biochem. 5: 505-14 (1963)

E 77 The fluorimetric determination of rhenium in ores with Rhodamine 6J.
Ivankova, A.I., and Shcherbov, D.P.
Ind. Lab. (USSR) (English Trans.) 29: 843-5 (1963)

E 78 Fractionation and fluorometric quantitation of digoxin and its metabolites from human urine.
Jellinek, M., and Willman, V.L.
Federation Proc. 22(2), Paper 152 (1963)

E 79 Detection of quanidine compounds on paper chromatograms.
Jones, A.S., and Thompson, T.W.
J. Chromatog. 10: 248-9 (1963)

E 80 Fluorometric determination of albumin in cerebrospinal fluid and serum.
Kaplan, A., and Johnstone, M.A.
Clin. Chem. 9(126): 461 (1963)

E 81 The structure of the phenylalanine hydroxylation cofactor.
Kaufman, S.
Proc. Natl. Acad. U.S. 50: 1085-93 (1963)

E 82 The identification of 3-methoxy-anthranilic acid, additional tryptophan metabolite, in human urine.
Kido, R., Tsuji, T., and Matsumura, Y.
Biochem. Biophys. 13: 428-30 (1963)

E 83 Paper chromatography of flavin analogs.
Kimmich, G.A., and McCormick, D.B.
J. Chromatog. 12: 394-400 (1963)

E 84 The use of lumogallion IREA for the fluorescent determination of niobium.
Klimov, V.V., and Didkovskaya, O.S.
Ind. Lab. (USSR) (English Trans.) 29: 128-9 (1963)

E 85 Assay of disulfide and sulfhydryl content of proteins and peptides by fluorescence quenching.
Klinman, N., and Karush, F.
Federation Proc. 22 (2), 1123 (1963)

E 86 Determination of chlordiazepoxide and of a metabolite of lactum character in plasma of humans, dogs, and rats by a specific spectrofluorometric micro method.
Koechlin, B.A., and D'Arconte, L.
Anal. Biochem. 5: 195-207 (1963)

E 87 Testosterone production rates in normal adults.
Korenman, S.G., Wilson, H., and Lipsett, M.R.
J. Clin. Invest. 42: 1753-60 (1963)

E 88 Change of the character of bond
strengths in the Nb-N system.
Korsunskii, M.I., and Genkin, Y.E.
Izv. Akad. Nauk Kaz. SSR, Ser. Fiz.-Mat.
Nauk 1963: 70-5 (1963)
CA 62: 1216g

E 89 Spectrophotofluorometric determination
of o, o-diethyl-2-pyrazinyl phosphoro-
thioate and its oxygen analog in soil
and plant tissues.
Kugenagi, U., and Terriere, L.C.
Agr. Food Chem. 11: 293-7 (1963)

E 90 A sensitive fluorometric assay for
morphine.
Kupferberg, H.J., Burkhatter, A., and
Way, E.L.
Federation Proc. 22(2), Paper 533 (1963)

E 91 Effect of impurities on the effectiveness
of liquid scintillators.
Kutsyna, L.M., Verkhovtseva, E.T., and
Poduzharlo, V.F.
Stsintillyatory i Stsintillyats. Maerialy
1963: 32-5 (1963)
CA 62: 11310a

E 92 Isolation of estrone, estradiol-17β, and
estriol from female human urine.
Lodany, S., and Finkelstein, M.
Steroids 2: 297-318 (1963)

E 93 Accumulation of oestrone in the allantoic
fluid.
Lunaas, T.
J. Endocrinol. 26: 401-6 (1963)

E 94 An outline of magnesium metabolism in
health and disease — a review.
MacIntyre, I.
J. Chronic Diseases 16: 201-15 (1963)

E 95 A method for the enzymic determination
of corticosteroids in extracts from
whole blood, plasma, and urine.
Margraf, H.W., Margraf, C.O., and
Werchselbaum, T.E.
Steroids 2: 143-54 (1963)

E 96 Use of sephadex G-25 for the separation
of catecholamines from plasma.
Marshall, C.S.
Biochim. Biophys. Acta 74: 158-9 (1963)

E 97 Chromatographic separation of estriol,
16-ketostradiol-17β, 16-epiestriol,
estradiol-17α, estradiol-17β, estrone,
and 2-methoxyestrone on the column of
partially esterified carboxylic acid
type ion exchange resin.
Matsumoto, K., and Seki, T.
Endocrinol. Japon. 10: 136-41 (1963)

E 98 Nd^{3+} fluorescence and stimulated
emission in oxide glasses.
Maurer, R.D.
Polytech. Inst. Brooklyn, Microwave Res.
Inst. Symp. Ser. 13: 435-449 (1963)
CA 62: 3778a

E 99 Urinary excretion of magnesium in man
following the ingestion of ethanol.
McCollister, R.J., Flink, E.B., and Lewis,
M.D.
Am. J. Clin. Nutr. 12: 415-20 (1963)

E 100 Purification and analysis of fluorescein
labeled antisera by column chromato-
graphy.
McDevitt, H.O., Peter, J.H., Polland,
L.W., Harter, J.G., and Coons, A.H.
J. Immunol. 90: 634-642 (1963)

E 101 Some catechol compounds other than
noradrenaline and adrenaline in
brains.
Montagu, K.
Biochem. J. 86: 9-11 (1963)

E 102 Effect of serotonin loading on histamine
release and blood flow of isolated per-
fused liver and lung.
Moore, T.C., Normell, L., and Eiseman, B.
Arch. Surgery 87: 42-3 (1963)

E 103 Browning of fish meat. XII. Fluores-
cent substance in flounder extracts.
Nagayama, F., and Ono, T.
J. Tokyo Univ. Fisheries 50: 31-6 (1963)
CA 62: 11070d

E 104 Quantitative fluorometric determination of anthranilic acid, 3-hydroxy- and
5-hydroxy-anthranilic acid in the urine.
Nakken, K.F.
Scand. J. Clin. Lab. Invest., 15: 78 (1963)

E 105 Determination of millimicrogram
amounts of protein.
Newmark, M.Z., and Wenger, B.S.
Federation Proc. 22(2), Paper 1899 (1963)

E 106 Luminescence in animals.
Nicol, J.A.C.
Endeavor 22: 37-41 (1963)

E 107 Simplified micromethod for measuring
histamine in human plasma.
Noah, J.W., and Brand, A.
J. Lab. Clin. Med. 62: 506-10 (1963)

E 108 Hydraulic model tests of estuarial
waste dispersion.
O'Connell, R.L., and Walter, C.M.
J. Sanit. Eng. Div., Am. Soc. Civil Engrs.
89: 3394 (1963)

E 109 The use of β-thiopropionic acid for the
analysis of mixtures of adrenaline and
noradrenaline in plasma by the fluorometric trihydroxyindole method.
Palmer, J.F.
West Indian Med. J. 13: 38-53 (1963)

E 110 Fluorescence determination of gibber-
ellic acid in cherries.
Parker, K.G., St. John, L.E., and Lisk,
D.J.
J. Assoc. Offic. Agr. Chemists 46: 986-8
(1963)

E 111 Fluorometric determination of
11-hydroxycorticosteroids in human
plasma.
Popens, Y., Silinsh, E., and Vitols, I.
Vopr. Med. Khim. 8: 628-34 (1962);
Federation Proc. Trans. Suppl. 22,
T957-60 (1963)

E 112 An initial inquiry into a photoelectric
device to detect menhaden marked with
fluorescent pigments.
Reintjes, J.W.
North Atlantic Fish Marking Symp., Spec.
Publ. No. 4 (1963)

E 113 Separation and determination of thiamine and pyrothiamine in biological
materials by chromatography on polyethylene powder.
Rindi, G., and Perri, V.
Anal. Biochem. 5: 179-86 (1963)

E 114 Stability assays of pharmaceutical preparations by quantitative paper chromatography.
Roberts, H.R., and Siino, M.R.
J. Pharm. Sci. 52: 370-5 (1963)

E 115 Application of fluorescence assay to the
identification of acidic components in
psoriatic lesions.
Roe, D.A.
J. Invest. Dermatol. 41: 319-24 (1963)

E 116 Fluorometric determination of trypsin.
Roth, M.
Clin. Chim. Acta 8: 574-8 (1963)

E 117 Fluorogenic substrates for B-D-galac-
tosidases and phosphatases derived
from fluorescein (3,6-dihydroxy-
fluoran) and its monomethyl ether.
Rotman, B., Zderic, J.A., and Edelstein,
M.
Proc. Natl. Acad. Sci. U.S. 50: 1-6 (1963)

E 118 A new fluorimetric method of plasma
cortisol assay with a study of pituitary-
adrenal function using metyrapone.
Rudd, B.T., Sampson, P., and Brooke, B.N.
J. Endocrinol. 27: 317-25 (1963)

E 119 Detection of alloxan by paper chroma-
tography.
Said, A., and Fleita, D.H.
Chem. Anal. 52: 109-10 (1963)

E 120 Thin-layer chromatographic separation
and fluorescence analysis of polynuclear azo hydrocarbons.
Sawicki, E., Stanley, T.W., Elbert, W.C.,
and Pfaff, J.D.
Eleventh Detroit Anachem Conf., Wayne
State Univ., Paper No. 48 (1963)

E 121 Comparison of spectrophotometric and
spectrophotofluorometric methods for
the determination of malonaldehyde.
Sawicki, E., Stanley, T.W., and Johnson, H.
Anal. Chem. 35, Feb. (1963)

E 122 Comparative study of some new meth-
ods for the detection of malonaldehyde.
Sawicki, E., Stanley, T.W., and Johnson, H.
Chem. Anal. 52, Jan. (1963)

E 123 Spectrophotofluorimetric determination
of formaldehyde and acrolein with
J-acid, comparison with other methods.
Sawicki, E., Stanley, T.W., and Pfaff, J.
Anal. Chim. Acta 28: 156-63 (1963)

E 124 Fluorescence-structure relationship for polynuclear hydrocarbons by automatic triparametric recording.
Schachter, M.M., and Haenni, E.O.
Eleventh Detroit Anachem Conf., Wayne State Univ., Paper No. 57 (1963)

E 125 Interaction of some fluorescent quaternary ammonium salts with muscle protein.
Schell, H.D.
Acad. Rep. Populare Romine, Studii Cercetari Biochim. 6: 545 (1963)

E 126 A new fluorometric technique for measuring serum high and low-density lipoproteins.
Searcy, R.L., Korotzer, J.C., and Bergquist, L.M.
Clin. Chim. Acta 8: 148-51 (1963)

E 127 Fluorometric method for the estimation of 4-hydroxy-3-methoxyphenylacetic acid (homovanillic acid) and its identification in brain tissue.
Sharman, D.F.
Brit. J. Pharmacol. 20: 204-13 (1963)

E 128 Microdetermination of calcium by aequorin luminescence.
Shimomura, O., Johnson, F.H., and Saiga, Y.
Science 140: 1339-40 (1963)

E 129 Blue fluorescence in crystals excited by the ruby optical maser.
Singh, S., and Stoicheff, B.P.
Polytech. Inst. Brooklyn, Microwave Res. Inst. Symp. Ser. 13: 385-403 (1963)
CA 62: 6028a

E 130 Fluorimetric determination of total catecholamines in urine.
Small, N.A.
Clin. Chim. Acta 8: 803-6 (1963)

E 131 Distribution of forms of lactic dehydrogenase within the developing rat kidney.
Smith, C.H., and Kissane, J.M.
Exptl. Bio. 8: 151-64 (1963)

E 132 The presence and distribution of tyramine in mammalian tissues.
Spector, S., Melmon, K., Lovenberg, W., and Sjoerdsena, A.
J. Pharmacol. Exptl. Therap. 140: 229-35 (1963)

E 133 Fluorometric and spectrophotometric determination of magnesium with O,O'-dihydroxyazobenzene.
Spielholtz, G.I., and Jensen, R.
Anal. Chem. 35: 1144 (1963)

E 134 Excretion of a tryptophan metabolite in rheumatoid arthritis.
Spiera, H.
Arthritis Rheumat. 6: 364-71 (1963)

E 135 Fluorometric determination of chlorotetracycline in premixes.
Spick, J., and Katz, S.E.
J. Assoc. Offic. Agr. Chemists 46: 434-7 (1963)

E 136 A simple fluorometric method for the routine determination of corticosteroids in small quantities of plasma.
Stahl, F., Hertline, I., and Knappe, G.
Acta Biol. Med. Ger. 10: 480-7 (1963)

E 137 A photo-induced chemiluminescence of riboflavin in water containing hydrogen peroxide. I. The primary photochemical phase.
Steele, R.H.
Biochemistry 2: 529-36 (1963)

E 138 A partial kinetic analysis of the chemiluminescence of phenazine methosulfate.
Steele, R.H., and Breckenridge, B.S.
Biochem. Biophys. Res. Comm. 10: 396-400 (1963)

E 139 Oestriol and pregnandiol estimations in urine as an aid in the examination of placental function.
Strand, A.
Acta Obstet. Gynecol. Scand. 42: 96-104 (1963)

E 140 Enzymatic formation of α-isopropyl malic acid, an intermediate in leucine biosynthesis.
Straussman, M., and Ceci, L.N.
J. Biol. Chem. 238: 2445-2452 (1963)

E 141 A quantitative fluorometric method for the determination of serpasil (reserpine) in feeds at the micro level.
Tishler, F., Sheth, P.B., and Giaimo, M.B.
J. Assoc. Offic. Agr. Chemists 46: 448-51 (1963)

E 142 Salicylism.
 Tschetter, P.N.
 Diseases Children 106: 134-46 (1963)

E 143 Fluorescence of purines and pyrimidines.
 Udenfriend, S.
 Eleventh Detroit Anachem Conf., Wayne State Univ., Paper No. 47 (1963)

E 144 A new method for the determination of dopamine (3-hydroxytyramine).
 Uuspaa, V.J.
 Ann. Med. Exper. Biol. Fennial 41: 194-201 (1963)

E 145 A ninhydrin reaction giving a sensitive quantitative fluorescence assay for 5-hydroxytryptamine.
 Vanable, J.W.
 Anal. Biochem. 6: 393-403 (1963)

E 146 Clinical significance of trace elements.
 Vollee, B.L.
 Mod. Medicine 8: 111-27 (1963)

E 147 Quantitative determination of 3,4-dihydroxymandelic acid (DOMA) in human urine.
 Wada, Y.
 Tohoku J. Exptl. Med. 79: 389-400 (1963)

E 148 Magnesium binding constants of a denosine triphosphate and some other compounds estimated by the use of fluorescence of magnesium-8-hydroxyquinoline.
 Watanabe, S., Trosper, T., Lynn, M., and Evenson, L.
 J. Biochem. (Tokyo) 54: 17-24 (1963)
 CA 60: 12267h

E 149 Fluorescent biological stains as markers for Drosophila.
 Wave, H.E., Henneberry, T.J., and Mason, H.C.
 J. Econ. Entomol. 56: 890-1 (1963)

E 150 Estimation of serotonin in biological material.
 Weissbach, H.
 Std. Methods Clin. Chem., Vol. 4, p. 197, Academic Press (1963)

E 151 Rapid appearance of injected fat in the gut of the rat.
 Wilkins, D.J.
 Proc. Soc. Exptl. Biol. Med. 112: 953 (1963)

E 152 A routine procedure for screening of phenylketonuria in the newborn.
 Wong, P., Inouye, T., and Hsia, D.Y.Y.
 Clin. Chem. 9(71): 444 (1963)

E 153 Ethyl esters of coumarin-4-acetic acids.
 Woods, L.L., and Sapp, J.
 J. Chem. Eng. Data 8: 235-6 (1963)

E 154 Plasma N-acetyl-β-glucosaminidase and β-glucuronidase in health and disease.
 Woolen, J.W., and Turner, P.
 Clin. Chem. 9(77): 446 (1963)

E 155 Inherited metabolic disorders. Errors of phenylalanine and tyrosine metabolism.
 Woolf, L.I.
 Adv. Clin. Chem., Vol. 6, pp. 97-230, H. Sobotka and C.P. Stewart, editors, Academic Press (1963)
 CA 60: 8462c

E 156 Spectrophotofluorometric determination of the dissociation constants of amides from the enzyme-reduced coenzyme complex of liver alcohol dehydrogenase.
 Woronick, C.L.
 Acta Chem. Scand. 17: 1789-91 (1963)

E 157 Aspects of the clinical chemistry of demethylimipramine in man.
 Yates, C.M., Todrick, A., and Tait, A.C.
 J. Pharm. Pharmacol. 15: 432-9 (1963)
 CA 59: 6881e

E 158 Method for the determination of phytoplankton chlorophyll and phaeophytin by fluorescence.
 Yentsch, C.S., and Menzel, D.W.
 Deep-Sea Res. 10: 221-31 (1963)

E 159 Histamine in human blood.
 Zachariae, H.
 Scand. J. Clin. Lab. Invest. 15: 173-8 (1963)

E 160 Use of dimedon for the detection of
       keto sugars by paper chromatography.
       Adachi, S.
       Anal. Biochem. 9: 224-7 (1964)

E 161 A method for the determination of
       estrogens in urine.
       Aizawa, Y., and Pincus, G.
       Steroids 4: 249-54 (1964)

E 162 Polarization ratios in tetracene-
       anthracene mixed crystals.
       Akon, C.D., and Craig, D.P.
       J. Chem. Phys. 41: 4000-1 (1964)
       CA 62: 8518h

E 163 Determination of submicrogram
       amounts of selenium in biological mat-
       erials.
       Allaway, W.H., and Cary, E.E.
       Anal. Chem. 36: 1359-62 (1964)

E 164 Determination of phenoxyacetic acids
       with J and phenyl J acids.
       Aly, O.M., and Faust, S.D.
       Anal. Chem. 36: 2200-2201 (1964)

E 165 Determination of antibody-hapten
       building constant by fluorescence en-
       hancement.
       Amkrant, A.A.
       Immunochemistry 1: 231-5 (1964)

E 166 Fluorometric determination of uranium
       with Rhodamine B.
       Anderson, N.R., and Hercules, D.M.
       Anal. Chem. 36: 2138-41 (1964)
       CA 62: 26d

E 167 Lead poisoning test studied.
       Anon.
       Lab. World 15: 802 (1964)

E 168 Dopamine in CNS and a method for its
       determination.
       Anton, A.H., and Sayre, D.F.
       Federation Proc. Abst. 23, Abst. 2339
       (1964)

E 169 Effect of the concentrational transfor-
       mation of the fluorescence spectrum of
       pyrene solutions at low temperatures.
       Arabidze, A.A.
       Opt. i Spektroskopiya 17: 633-5 (1964)

E 170 Effect of substituent groups on fluores-
       cence of metal chelates.
       Argauer, R.J., and White, C.E.
       Anal. Chem. 36: 2141-4 (1964)
       CA 62: 1218d

E 171 Use of the excitation spectrum to deter-
       mine the degree of dissociation of
       fluorescent metal chelates.
       Argauer, R.J., and White, C.E.
       Spectrochim. Acta 20: 1323-6 (1964)

E 172 5-Hydroxytryptophan decarboxylase
       activity in nerve endings of the rat
       brain.
       Arnaiz, G.R.D.L., and de Robertis, E.
       J. Neurochem. 11: 213 (1964)

E 173 Estimation of 5-hydroxytryptamine in
       human blood.
       Ashcroft, G.W., and Crawford, T.B.B.
       Clin. Chim. Acta 9: 364-9 (1964)

E 174 Measurement of the temperature of the
       ionosphere from the incidence of the
       twilight glow of aluminum monoxide.
       Authier, B., Blamont, J.E., and
       Carpenter, G.
       Ann. Geophys. 20: 342-5 (1964)
       CA 62: 14037d

E 175 Studies of coporporphyrin. VII. Adapt-
       ation of the Eriksen paper chromato-
       graphic method to the quantitative ana-
       lysis of the isomers in normal human
       urine.
       Aziz, M.A., Schwartz, S., and Watson, C.J.
       J. Lab. Clin. Med. 63: 585-9 (1964)

E 176 Prototropic equilibrium and fluores-
       cence of some 8-hydroxyquinoline
       derivatives.
       Ballard, R.E., and Edwards, J.W.
       J. Chem. Soc. 1964: 4868-74 (1964)
       CA 62: 5174b

E 177 Formation of 5-hydroxytryptophol from
       endogenous 5-hydroxytryptamine by
       isolated blood platelets.
       Bartholini, G., Pletscher, A., and
       Bruderer, H.
       Nature 203: 1281-3 (1964)

E 178 Fluorometric estimation of magnesium
      in serum and urine.
      Batsakis, J.G., Madera-Orsini, F., Stiles,
      D., and Briere, R.O.
      Am. J. Clin. Pathol. 42: 541-6 (1964)

E 179 Octacoordinate chelate of lanthanides.
      Two series of compounds.
      Bauer, H., Blanc, J., and Ross, D.L.
      J. Am. Chem. Soc. 86: 5125-31 (1964)
      CA 62: 3636a

E 180 Ultraviolet-induced radiophotolumines-
      cence in silver-activated metaphosphate
      glasses.
      Becker, K.
      Z. Naturforsch. 19a: 1233-4 (1964)
      CA 62: 141c

E 181 Catalytic synthesis of a new luminophor
      in the presence of the natural alumino-
      silicate gumbrin.
      Bekauri, N.G., Shuikin, N.I., Shakarashvili,
      T.S., Topuridze, L.F., and Goderzish-
      vili, K.G.
      Tr. Inst. Khim., Akad. Nauk Gruz. SSR 17:
      145-58 (1964)
      CA 62: 7220h

E 182 Fluorometer for chemical dosimetry.
      Berlinguette, G.E., and Tate, P.A.
      Rev. Sci. Instr. 35: 1725-6 (1964)
      CA 62: 4874g

E 183 A micro method for blood serotonin.
      Berman, J.L., and Hsia, D.Y.
      Clin. Chem. 10: 641 (1964)

E 184 A fluorescence study of specific and
      nonspecific dye-protein interactions.
      Berns, D.S., and Singer, S.J.
      Immunochemistry 1: 209-17 (1964)
      CA 62: 3242e

E 185 Studies on the inhibition of dihydrofo-
      late reductase by the folate antagonists.
      Bertino, J.R., Booth, B.A., Bieber, A.L.,
      Cashmore, A., and Sartorelli, A.C.
      J. Biol. Chem. 239: 479-485 (1964)

E 186 Simplified determination of blood
      adenosine triphosphate using the fire-
      fly system.
      Beutler, E., and Baluda, M.C.
      Blood 23: 688-98 (1964)

E 187 Fluorometric assay of α-chymotrypsin.
      Bielski, B.H.J., and Freed, S.
      Anal. Biochem. 7: 192-8 (1964)

E 188 Qualitative and quantitative analysis of
      amino acids.
      Boulton, A.A., and Bush, I.E.
      Biochem. J. 92: 11 (1964)

E 189 Application and modification of the
      Momose-Ohkura fluorometric deter-
      mination of blood glucose.
      Bourne, B.B.
      Clin. Chem. 10: 1121-30 (1964)

E 190 Reproducibility of R_f and correlation of
      chromatographic patterns on paper and
      thin layer plates.
      Brodasky, T.F.
      Anal. Chem. 36: 996-9 (1964)

E 191 The significance of lactate dehydro-
      genase isozymes in abnormal skeletal
      muscle.
      Brody, I.A.
      Neurology 14: 1091-9 (1964)

E 192 Fluorometric determination of urinary
      metanephrine and normetanephrine.
      Brunjes, S., and Wybenga, D.
      Clin. Chem. 10: 1-12 (1964)

E 193 Time of travel of soluble contaminants
      in streams.
      Buchanan, T.J.
      J. Sanit. Eng. Div., Proc.Am. Soc. Civil
      Engrs. 90: 1-12 (1964)

E 194 Plasma cortisol and corticosterone
      response to infused corticotrophin in
      normal subjects.
      Cameron, E.A., and Kilborn, J.R.
      Clin. Chim. Acta 10: 308-13 (1964)

E 195 Method for the fluorometric determina-
      tion of 3-methoxytyramine in tissues
      and the occurrence of this amine in
      brain.
      Carlsson, A., and Waldeck, B.
      Scand. J. Clin. Lab. Invest. 16: 133-8
      (1964)

E 196 Fluorescent method for determining
      fibrinolytic activity.
      Caviezel, V.O.
      Schweiz. Med. Wochschr. 94: 555-6 (1964)

E 197 Bioluminescence-production of light by organisms.
Chase, A.M.
Photophysiology, Vol. II, pp. 389-421, A.C. Giese, editor, Academic Press N.Y. (1964)

E 198 Sensitive fluorescence reaction for vitamins D and dihydrotachysterol.
Chen, P.S., Terepka, A.R., and Lane, K.
Anal. Biochem. 8: 34-42 (1964)

E 199 Photoinactivation of L-glutamate dehydrogenase in a spectrophotofluorometer.
Chen, R.F.
Biochem. Biophys. 17: 141-5 (1964)

E 200 Experiments on determination of melphalan by fluorescence. Interaction with protein and various solutions.
Chirigos, M.A. and Mead, J.A.R.
Anal. Biochem. 7: 259-268 (1964)

E 201 A procedure for the direct reading of fluorescent spots on thin-layer chromatography plates using the Turner fluorometer.
Connors, W.M., and Boak, W.K.
J. Chromatog. 16: 243-5 (1964)

E 202 The estimation of 5-hydroxytryptamine in human blood.
Contractor, S.F.
Biochem. Pharmacol. 13: 1351-7 (1964)

E 203 The fluorometric determination of acetylcholine.
Cooper, J.R.
Biochem. Pharmacol. 13: 795-7 (1964)

E 204 Direct extraction of corticosterone from rat adrenal gland under an applied electrical field.
Cortes, J.M., and Peron, F.G.
Federation Proc. Abst. 23 (3), Abst. 1055 (1964)

E 205 Application of isotopic dilution analysis to the fluorometric determination of selenium in plant materials.
Cukor, P., Walzcyk, J., and Lott, P.F.
Anal. Chim. Acta 30: 473 (1964)

E 206 Fluorimetric determination of adrenal corticosteroids: observations on interfering fluorogens in human plasma.
Daly, J.R., and Spencer-Peet, J.
J. Endocrinol. 30: 255-63 (1964)

E 207 Determination of serotonin in blood using an ion-exchange resin.
Davis, V.E., Huff, J.A., Brown, H., and Alfrey, C.P.
Clin. Chim. Acta 9: 419-26 (1964)

E 208 Determination of serotonin in tissues using an ion-exchange resin.
Davis, V.E., Huff, J.A., and Brown, H.
Clin. Chim. Acta 9: 427-33 (1964)

E 209 Calcein, colmagite, and O,O'-dihydroxyazobenzene titrimetric, colorimetric, and fluorometric reagents for calcium and magnesium.
Diehl, H.
G. Frederick Smith Co., Ohio (1964)

E 210 Identification and quantitative determination scofolin and scopoletin in tobacco plants treated with 2,4-dichlorophenoxyacetic acid.
Dieterman, L.J., Lin, C.Y., Rohrbaugh, L., Thiesfeld, V., and Wender, S.H.
Anal. Biochem. 9: 139 (1964)

E 211 Carbonyl addition to nicotinamide adenine dinucleotide in frozen solution.
Dolin, M.I., and Jacobson, K.B.
J. Biol. Chem 239: 3007-16 (1964)

E 212 Simplified method for the fluorometric measurement of free cortisol and corticosterone in urine.
Dorner, G., and Stahl, F.
Acta Biol. Med. Ger. 12: 606-11 (1964)

E 213 Micro-detection of β-phenethylbiquanide.
Durfee, D.A., Bailey, R.E., and Beck, J.H.
Federation Proc. Abst. 23, Abst. 2344 (1964)

E 214 Chromatographic assay of estrogen in pregnancy urine.
Epstein, E., and Zak, B.
Clin. Chem. 10: 637 (1964)

E 215 Thiamine retention as influenced by processing method, storage time, and temperature, and type of container.
Everson, G.J., Chang, J., Leonard, S., Luh, B.S., and Simone, M.
Food Technol. 18: 84-6 (1964)

E 216 Single extraction method for the simultaneous determination of serotonin, dopamine, and norepinephrine in brain.
Fleming, R.M., Clark, W.G., and Clark, P.T.
Federation Proc. Abst. 23, Abst. 2340 (1964)

E 217 Experience with a simple procedure for the determination of plasma and urine free 11-hydroxycorticosteroids.
Gantt, C.L., Maynard, D.E., and Hamwi, G.G.
Metab., Clin. Eptl. 13: 1327-32 (1964)

E 218 Enzymic determination of free myoinositol in human cerebrospinal fluid and plasma.
Garcia-Bunuel, L., and Garcia-Bunuel, V.M.
J. Lab. Clin. Med. 64: 461-8 (1964)

E 219 Fluorescein labeled clot assay of plasmathrombolytic activity.
Genton, E., and Fletcher, A.P.
Federation Proc. Abst. 23, Abst. 6395 (1964)

E 220 Assay of plasma thrombolytic activity with fluorescein-labeled clots.
Genton, E., Fletcher, A.P., Alkjaersig, N., and Sherry, S.
J. Lab. Clin. Med. 64: 313 (1964)

E 221 Modified trihydroxyindole procedure for plasma catecholamines using stannous chloride and temperature control.
Gerst, E.C., and Steinsland, O.S.
Federation Proc. Abst. 23, Abst. 2337 (1964)

E 222 Quantitative microdetermination of enzymes in the sweat gland. II. Dehydrogenases in patients with cystic fibrosis and in control subjects.
Gibbs, G.E., and Reimer, K.
J. Pediat. 65: 540-1 (1964)

E 223 Fluorometric determination of corticostersone and cortisol in 0.02-0.05 milliliters of plasma on submilligram samples of adrenal tissue.
Glick, D., Von Redlick, D., and Levine, S.
Endocrinology 74: 653-5 (1964)

E 224 Fluorometric analysis of amidase and alkaline phosphatase in neonatal rat thymocytes.
Greenberg, L.J., and Cole, L.J.
Nature 201: 1001 (1964)

E 225 Separation and identification of pyridoxal and pyridoxal-5-phosphate by paper chromatography.
Hakanson, R.
J. Chromatog. 13: 263-5 (1964)

E 226 Photoluminescence of lanthanide complexes. II. Enhancement by an insulating sheath.
Halverson, F., Brinen, J.S., and Leto, J.R.
J. Chem. Phys. 41: 157-63 (1964)

E 227 On the molecular mechanism of bioluminescence. I. The role of long chain aldehyde.
Hastings, J.W., Gibson, Q.H., and Greenwood, C.
Proc. Natl. Acad. Sci. U.S. 52: 1529-35 (1964)

E 228 A simple method for quantitative estimation of tetracycline antibiotics.
Hayes, J.E., and Dubuy, H.G.
Anal. Biochem. 7: 322-7 (1964)

E 229 Progesterone levels in intact and ovariectomized pregnant guinea pigs.
Heap, R.B., and Deamesly, R.
J. Endocrinol. 30: ii-iii (1964)

E 230 A fluorescence assay of progesterone.
Heap, R.B.
J. Endocrinol. 30: 293-305 (1964)

E 231 Spectrofluorometric method for measuring 6-amino-nicotinamide in pyridine nucleotides of rat kidney.
Herken, H., and Neuhoff, V.
Naunyn-Schmiedebergs Arch. Exptl. Pathol. Pharmakol. 247: 187-201 (1964)

E 232 Fluorometric assay of sialic acid in brain gangliosides.
Hess, H.H., and Rolde, E.
J. Biol. Chem. 239: 3215-20 (1964)

E 233 Urinary estrogens in non-pregnant human subjects measured by modification of Bauld's method.
Hobkirk, R., and Nilsen, M.
Steroids 3: 453-470 (1964)

E 234 Fluorometric demonstration of tryptophan in dentin and bone protein.
Hoerman, K.C., and Mancewicz, S.A.
J. Dental Res. 43: 276-80 (1964)

E 235 Screening newborn infants for phenylketonuria.
Hsia, D.Y., Berman, J.L., and Slatis, H.M.
J. Am. Med. Assoc. 188: 203 (1964)

E 236 Blood phenylalanine levels of newborn infants (phenylketonuria).
Irwin, H.R., Notrica, S., and Fleming, W.
Calif. Med. 101: 331-3 (1964)

E 237 Quantitation of estrone, estradiol, and estriol on thin layer chromatograms by a photogrammetric procedure.
Jacobsohn, G.M.
Anal. Chem. 36: 275-9 (1964)

E 238 Fluorometric study of antihistamines.
Jensen, R.E., and Pflaum, R.T.
J. Pharm. Sci. 53: 835-7 (1964)

E 239 Fluorometric method for the determination of amprolum in feeds.
Kanora, J., and Szalkowski, C.P.
J. Assoc. Offic. Agr. Chemists 47: 209-13 (1964)

E 240 Assay method for disulfide groups by fluorescence quenching.
Karush, F., Klinman, N.R., and Marks, R.
Anal. Biochem. 9: 100-14 (1964)

E 241 Determination of penicillins and chlorotetracycline in premixed and mixed feeds.
Katz, S.E., and Helrich, K.
Pest. Rev. 7: 74-95 (1964)

E 242 Drugs. in feeds. Fluorometric determination of chlortetracycline in mixed feeds.
Katz, S.E., and Spock, J.
J. Assoc. Offic. Agr. Chemists 47: 203-8 (1964)

E 243 Fluorometric determination of chlortetracycline in low level mixed feeds.
Katz, S.E., and Spock, J.
J. Assoc. Offic. Agr. Chemists 47: 1157-61 (1964)

E 244 Xanthine dehydrogenase: Differences in activity among Drosophila strains.
Keller, E.C., and Glassman, E.
Science 143: 40-1 (1964)

E 245 Tracer studies with Rhodamine B in a 3.2 kilometer reach of the upper Ohio River.
Kisiel, C.C., Shapiro, M.A., Fiche, J.F., Morgan, P.V., and Spear, R.D.
Verh. Int. Ver. Limnol. XV: 265-75 (1964)

E 246 Phenylketonuria — a review of some deficits in our information.
Kleinman, D.S.
Pediatrics 33: 123-4 (1964)

E 247 Interferences by formaldehyde forming drugs in the determination of urinary catecholamines.
Klotz, M.O., Richter, H., and Meuffels, M.
Clin. Chem. 10: 372-3 (1964)

E 248 Simplified procedure for the estimation of testosterone production rates.
Korenman, S.G., Davis, T.E., Wilson, H., and Lipsett, M.D.
Steroids 3: 203-7 (1964)

E 249 Separation of uranium, thorium, and the rare earth elements by anion exchange.
Korkisch, J., and Arrhenius, G.
Anal. Chem. 36: 850-4 (1964)

E 250 Ion exchange determination of uranium in ferrous alloys.
Korkisch, J., and Hazan, I.
Anal. Chem. 36: 2464-6 (1964)

E 251 Determination of small amounts of aldosterone and corticosterone in the incubation medium of the adrenal gland of rats.
Kraus, M., and Popp, M.
Physiol. Bohemoslov. 13: 457-61 (1964)

E 252 Nicotinamide coenzyme concentrations in mammary biopsy samples from ketotic cows.
Kronfeld, D.S., and Raggi, F.
Biochem. J. 90: 219-24 (1964)

E 253 A sensitive fluorometric assay for morphine in plasma and brains.
Kupferberg, H., Burkhalter, A., and Way, E.L.
J. Pharmacol. Exptl. Therap. 145: 247-51 (1964)

E 254 Fluorometric identification of submicrogram amounts of morphine and related compounds on thin-layer chromatograms.
Kupferberg, H., Burkhalter, A., and Way, E.L.
J. Chromatog. 16: 558-9 (1964)

E 255 Determination of pyrrolidine and piperidine.
Langemann, H.M., and Honegger, C.G.
Anal. Biochem. 8: 529-31 (1964)

E 256 Simplified method for the determination of adrenaline and noradrenaline in urine.
Lauber, K.
Z. Klin. Chem., 2(3): 76-9 (1964)
CA 63: 5972b

E 257 Histamine release from human leukocytes by ragweed pollen antigen.
Lichtenstein, L.M., and Osler, A.G.
J. Exptl. Med. 120: 507-30 (1964)

E 258 Benzopyrene and other polynuclear hydrocarbons in charcoal-broiled meat.
Lijinsky, W., and Shubik, P.
Science 145: 53-5 (1964)

E 259 A critical study of pyridine nucleotide concentrations in normal fed, normal fasted, and diabetic rat liver.
Lindall, A.W., and Luzarow, A.
Metab., Clin. Exptl. 13: 259-71 (1964)
CA 61: 6175e

E 260 Effect of ischemia on known substrates and cofactors of the glycolytic pathway in brain.
Lowry, O.H., and Passonneau, J.V.
J. Biol. Chem. 239: 18-30 (1964)

E 261 The relationships between substrates and enzymes of glycolysis in brain.
Lowry, O.H., and Passonneau, J.V.
J. Biol. Chem. 239: 31-42 (1964)

E 262 Electrophoretic separation (and fluorescence measurement) of urobiline.
Lozzio, B.B., Gorodisch, S., and Royer, M.
Clin. Chim. Acta 9: 78-81 (1964)

E 263 The potential of fluorescence for pesticide residue analysis.
MacDougall, D.
Pest. Rev. 5: 119-29 (1964)

E 264 Fluorometric method for guthion.
MacDougall, D.
Analytical Methods for Pesticides, Plant Growth Regulators, and Food Additives, G. Zweig, editor, Vol. II, Ch. 20, Academic Press N.Y. (1964)

E 265 A paper chromatographic method for estrogen determination.
Mahesh, V.B.
Steroids 3: 647-61 (1964)

E 266 Fluorometric method for the enzymic determination of glycolytic intermediates.
Maitra, P.K., and Estabrook, R.W.
Anal. Biochem. 7: 472-84 (1964)

E 267 Simplified estimation of leucine aminopeptidase (LAP) activity.
Martinek, R.G., Berger, L., and Broida, D.
Clin. Chem. 10: 1087-97 (1964)

E 268 Rapid screening test for adrenal cortical function.
Mattingly, D., Dennis, P.M., Pearson, J., and Cope, C.L.
Lancet 2: 1046-9 (1964)

E 269 The mechanism of chemiluminescence: A new chemiluminescent reaction.
McCapro, F., and Richardson, D.G.
Tetrahedron Letters 43: 3167-72 (1964)

E 270 The hydrolysis of amino acyl-β-naphthylamide by plasma aminopeptidases.
McDonald, J.K., Reilly, T.J., and Ellis, S.
Biochem. Biophys. Res. Commun. 16(2): 135-40 (1964)
CA 61: 4780h

E 271 Determination of purity of fluorescein isothiocynates.
McKinney, R.M., Spillane, J.T., and Pearce, G.W.
Anal. Biochem. 7: 74-86 (1964)

E 272 Further studies of plasma tyrosine in patients with altered thyroid function.
Melmon, K.L., Rivlin, R., Oates, J.A., and Sjoerdsma, A.
J. Clin. Endocrinol. Metab. 24: 691 (1964)

E 273 A new in vitro test for the detection of antibody in sera of patients allergic to lolium multiflorum.
Millman, M., Wolter, G.H., Millman, S., and Rosen, R.
Ann. Allergy 22: 136-45 (1964)

E 274 Multidimensional chromatography of Arenes produced during combustion.
Mukai, M., Tebbens, B.D., and Thomas, J.F.
Anal. Chem. 36: 1126-30 (1964)

E 275 New fluorometric method for estimation of citrovorum factor.
Netrawaki, M.S., Radhakrishmamurty, R., and Sreenivasan, G.
Anal. Biochem. 8: 143 (1964)

E 276 Fluorometric determination of pyridine nucleotides.
Pande, S.V., Bhan, A.K., and Venkitasubramanian, T.A.
Anal. Biochem. 8: 446-62 (1964)

E 277 Quantitative fluorometric determination of panthenol in multivitamin preparations.
Panier, R.G., and Close, J.A.
J. Pharm. Sci. 53: 108-10 (1964)

E 278 Fluorimetry and spectrofluorimetry.
Parker, C.A., and Rees, W.T.
Trace Analysis of Semiconductor Materials, pp. 228-246, J.P. Cali, editor, Pergamon Press (1964)

E 279 Transglucurondiase activity of the liver: a dosage method suitable for fragments obtained with needle biopsy.
Perona, G., Frezza, M., Dalla Rosa, C., and DeSandre, G.
Clin. Chim. Acta 10: 513-20 (1964)

E 280 The enzymatic measurement of γ-amino butyric-α-keto-glutaric transaminase.
Pitts, F.N.
Federation Proc. Abst. 23, Abst. 1695 (1964)

E 281 Vitamins and other nutrients — pyridoxine determined fluorometrically as pyridoxal cyanide compound.
Polansky, M.M., Camarra, R.T., and Toepfer, E.W.
J. Assoc. Offic. Agr. Chemists 47: 827-8 (1964)

E 282 Fluorometric determination of glycosidose in the locust and other insects.
Robinson, D.
Comp. Biochem. Physiol. 12: 95 (1964)

E 283 Coproporphyrin analyses on random urine samples.
Rogers, J.
Clin. Chem. 10: 678 (1964)

E 284 A fluorometric ultra-micro method for the determination of leucine aminopeptidase in biological fluids.
Roth, M.
Clin. Chim. Acta 9: 448-53 (1964)

E 285 Determination of alloxan via paper chromatography in solutions, blood, and plasma.
Said, A., and El Naggar, G.M.
Chem. Anal. 53: 69-71 (1964)

E 286 The determination of methanol in biological fluids.
Sardesai, V.M., and Provido, H.S.
J. Lab. Clin. Med. 64: 977-82 (1964)

E 287 Separation and analysis of polynuclear aromatic hydrocarbons present in the human environment. III.
Sawicki, E.
Chem. Anal. 53: 88-91 (1964)

E 288 Application of thin-layer chromatography to the analysis of atmospheric pollutants and determination of benzopyrene.
Sawicki, E., Stanley, T.W., Elbert, W.C., and Pfaff, J.D.
Anal. Chem. 36: 497-502 (1964)

E 289 Catecholamines in the urine of patients
       receiving methyldopa (presinol).
       Schlossmann, K., Bock, K.D., and
           Kroneberg, G.
       Klin. Wochschr. 42: 440-3 (1964)

E 290 Urobilin in urine.
       Schmidt, N.A., and Scholtis, R.J.H.
       Clin. Chim. Acta 10: 574-6 (1964)

E 291 The fluorometric determination of
       mescaline and some β-phenylethyl-
       amines.
       Seiler, N., and Wiechmann, M.
       Z. Physiol. Chem. 337: 229-40 (1964)

E 292 Experimental human magnesium deple-
       tion. I. Clinical observations and
       blood chemistry alterations.
       Shils, M.E.
       Am. J. Clin. Nutr. 15: 133-143 (1964)

E 293 Binding and release of metaraminol
       (aramine).
       Shore, P.A., and Alpers, H.S.
       Federation Proc. Abst. 23, Abst. 1498
       (1964)

E 294 Fluorometric estimation of metaramin-
       ol and related compounds.
       Shore, P.A., and Alpers, H.S.
       Life Sci. 3: 551 (1964)

E 295 Precipitation of submicrogram quanti-
       ties of thorium by barium sulfate and
       application to fluorometric determina-
       tion of thorium in mineralogical and
       biological samples.
       Sill, C.W., and Willis, C.P.
       Anal. Chem. 36: 622-30 (1964)

E 296 Some studies on the effect of angioten-
       sin II on adrenocortical hormone secre-
       tion in hypophysectomized rats with
       renal pedicle ligation.
       Singer, B., Losito, C., and Salmon, S.
       Endocrinology 74: 325-32 (1964)

E 297 Fluorescent marking and migration of
       grasshoppers from sprayed plots.
       Smith, D.S., Holmes, N.D., Swailes, G.E.,
           and McDonald, S.
       J. Econ. Entomol. 57: 990-2 (1964)

E 298 Excretion of injected catecholamines
       by white rats at 23°C. and 2°C.
       Smith, L.C., and Dugal, L.P.
       Can. J. Physiol. Pharmacol. 42: 579-84
       (1964)

E 299 Acetyl CoA condensation with α-keto
       acids.
       Strassman, M., Ceci, L.N., and Silverman,
           B.E.
       Federation Proc. Abst. 23, Abst. 1270
       (1964)

E 300 Spectrophotofluorometric determination
       of low concentrations of amobarbital in
       plasma.
       Swagdis, J.E., and Flanagan, T.L.
       Anal. Biochem. 7: 147-51 (1964)

E 301 Spectrofluorometric determination of
       low concentrations of amobarbital in
       plasma.
       Swagdis, J.E., and Flanagan, T.L.
       Anal. Biochem. 7: 147-51 (1964)

E 302 The determination of coproporphyrin
       isomers.
       Sweeney, G.O., and Eales, L.
       Scand. J. Clin. Lab. Invest. 16: 250-1
       (1964)

E 303 Quantitative determination of meta-
       nephrine and normetanephrine in urine.
       Taniguchi, K., Kahimoto, Y., and
           Armstrong, M.D.
       J. Lab. Clin. Med. 64: 469-84 (1964)

E 304 Fluorometric measurement of alkaline
       phosphatase activity in single cells of
       human fibroblast cultures.
       Tierney, J.H.
       Federation Proc. Abst. 23, Abst. 2381
       (1964)

E 305 Automated fluorescence-monitored
       system for the separation of serum
       proteins on sephadex.
       Toporek, M., and Phillipp, L.J.
       Federation Proc. Abst. 23, Abst. 2396
       (1964)

E 306 Spectrofluorometric determination of
       β-glucuronidase activity.
       Verity, M.A., Caper, R., and Brown, W.J.
       Arch. Biochem. Biophys. 106: 386-93
       (1964)

E 307 Determination of plasma cortisol by a fluorometric method.
Vermeulen, A., and Van der Straeten, M.
J. Clin. Endocrinol. Metab. 24: 1188-94 (1964)

E 308 Determination of selenium in semiconductor materials by the fluorescence method.
Vladimirova, V.M., and Kuchmistaya, G.I.
Zavodsk. Lab. 30: 528-9 (1964)

E 309 Quantitation of stress by catecholamine analysis.
Von Euler, U.S.
Clin. Pharmacol. Therap. 5: 398-404 (1964)

E 310 The estimation of uranin (fluorescein sodium) in blood.
Wagatsuma, T., and Wright, H.P.
J. Clin. Pathol. 17: 271-2 (1964)

E 311 The simultaneous estimation of catecholamines and their metabolites.
Weil-Malherbe, H.
Z. Klin. Chem. 2: 161-7 (1964)

E 312 Catecholamine excretion in smokers and non-smokers.
Westfall, T.C., and Watts, D.T.
J. Appl. Physiol. 19: 40 (1964)

E 313 Fluorescent labeling of polystyrene latex for tracing in biological systems.
Wilkins, D.J.
Nature 202: 798 (1964)

E 314 Fluorescence of solutions: A review.
Williams, R.T., and Bridges, J.W.
J. Clin. Pathol. 17: 371 (1964)

E 315 The determination of the acid-nonextractable flavin in mitochondrial preparations from heart muscle.
Wilson, D.F., and King, T.E.
J. Biol. Chem. 239: 2683-2690 (1964)

E 316 Phosphorimetric determination of procaine, phenobarbital, cocaine, and chlorpromazine in blood serum, and cocaine and atropine in urine.
Winefordner, J.D., and Tin, M.
Anal. Chim. Acta 31: 341-7 (1964)

E 317 The use of rigid ethanolic solutions for the phosphorimetric investigation of organic compounds of pharmacological interest.
Winefordner, J.D., and Tin, M.
Anal. Chim. Acta 31: 239-45 (1964)

E 318 Micromethods for measuring phenylalanine and tyrosine in serum.
Wong, P.W.K., O'Flynn, M.E., and Inouye, T.
Clin. Chem. 10: 1098-1104 (1964)

E 319 See E 318

E 320 A microprocedure for the determination of 4-pyridoxic acid in urine.
Woodring, M.J., Fisher, D.H., and Storvick, C.A.
Clin. Chem. 10: 479-89 (1964)

E 321 Skin histamine and delayed skin reactions.
Zachariae, H.
Acta Allergol. 19: 336-50 (1964)

E 322 Histamine in delayed skin reactions; fluorometric determinations on patch tests.
Zachariae, H.
J. Invest. Dermatol. 42: 431-4 (1964)

E 323 Analytical methods for pesticides, plant growth regulators, and food additives.
Zweig, G.
Academic Press, N.Y. (1963-4)

E 324 Measurement of anti-tuberculosis drug isoniazid.
Peters, J.H.
Am. Rev. Respirat. Diseases 81: 485-97 (1960)

E 325 Studies on the metabolism of isoniazid.
Peters, J.H.
Am. Rev. Respirat. Diseases 81(4): 485-97 (1960)

E 326 Fluorimetric method for the determination of major spironolactone (aldactone) metabolite in human plasma.
Gochman, N., and Gantt, C.L.
J. Pharmacol. Exptl. Therap. 135(3): 114-7 (1962)

E 327 Delayed fluorescence of solid solutions of polyacenes. II. Kinetic considerations.
Azumi, T., and McGlynn, S.P.
J. Chem. Phys. 39: 1186 (1963)

E 328 Testosterone production rates in normal adults
Korenman, S.G., Wilson, H., and Lipsett, M.B.
J. Clin. Invest. 42(11): 649-52 (1963)

E 329 Immunologic tests for tuberculosis.
Parlett, R.C.
Off. J. Am. Med. Tech., 13: 288-92 (1963)

E 330 Fluorometric and spectrophotometric determination of magnesium with o, o'-dehydroxyazobenzene.
Spielholtz, G.I., and Jensen, R.
Anal. Chem. 35: 1144 (1963)

E 331 Energy transfer studies by spectrophotofluorometric methods.
Wilkenson, F., and Dubois, J.T.
J. Chem. Phys., 39: 337-9 (1963)

E 332 The effect of photosynthesis inhibitors in oxygen evolution and fluorescence of illuminated chlorella.
Zweig, G., Tamas, I., and Greenberg, E.
Biochim. Biophys. Acta 4: 196-205 (1963)

E 333 The distribution of dopamine and dopa in various animals and a method for their determination in diverse biological material.
Anton, A.H., and Sayre, D.F.
J. Pharmacol. Exptl. Therap. 145: 326-36 (1964)

E 334 Fluorescent compounds for calibration of excitation and emission units of spectrofluorometer.
Argauer, R.J., and White, C.E.
Anal. Chem. 36: 368-71 (1964)

E 335 Characterization of carbazole and polynuclear carbazoles in urban air and in air polluted by coal tar pitch fumes by thin-layer chromatography and spectrophotofluorometry.
Bender, D.F., Sawicki, E., and Wilson, R.M.
Air and Water Poll. 8: 633-43 (1964)

E 336 Fluorescent detection and spectrophotofluorometric characterization and estimation of carbazoles and polynuclear carbazoles separated by thin-layer chromatography.
Bender, D.F., Sawicki, E., and Wilson, R.M.
Air and Water Poll. 8: 625-33 (1964)

E 337 Application and modification of the Momose-Ohkura fluorometric determination of blood glucose.
Bourne, B.B.
Clin. Chem. 10(12): 515-8 (1964)

E 338 Photoinactivation of L-glutamate dehydrogenase in a spectrophotofluorometer.
Chen, R.F.
Biochem. Biophys. Res. Commun. 17: 141-5 (1964)

E 339 Studies of the prolonged biochemical effects of 3-methyl-cholanthrene and of its physiological disposition in the rat.
Dayton, P.G., Vrindten, P., and Perel, J.M.
Biochem. Pharmacol. 13: 143-52 (1964)

E 340 Direct determination of inhibitors in polymers by luminescence techniques.
Drushel, H.V., and Sommers, A.L.
Anal. Chem. 36: 836-40 (1964)

E 341 Assay of plasma thrombolytic activity with fluorescein-labeled clots.
Genton, E., Fletcher, A.P., Alkjaersig, N., and Sherry, S.
J. Lab. Clin. Med. 64: 313-20 (1964)

E 342 Fluorometric determination of lipase acylase, alpha- and gamma-chymotrypsin and inhibitors of these enzymes.
Guilbault, G.G., and Kramer, D.N.
Anal. Chem. 36: 409-12 (1964)

E 343 Phosphorescence of calcified tissues.
Hoerman, K.C., and Mancewicz, S.A.
Arch. Oral Biol. 9: 517-34 (1964)

E 344 Characteristics of insoluble protein of tooth and bone.
Hoerman, K.C., and Mancewicz, S.A.
Arch. Oral Biol., 9: 835-42 (1964)
CA 61: 2253h

E 344a Fluorometric demonstration of trypto-
phan in dentin and bone protein.
Hoerman, K.C., and Mancewicz, S.A.
J. Dental Res. 43: 276-80 (1964)
CA 61: 2253h

E 345 Fluorescence and phosphorescence.
Howerton, H.K.
American Inst. Co., Silver Spring, Md.,
Reprint 222 (1964)

E 346 Factors influencing spectrofluorometry
of phenothiazine drugs.
Mellinger, T.J., and Keeler, C.E.
Anal. Chem. 36(9): 1822-4 (1964)

E 347 Spectrofluorometric measurement of
phenothiazines.
Ragland, J.B., and Kenross-Wright, V.J.
Anal. Chem. 36: 1356 (1964)

E 348 Direct spectrophotofluorometric analysis of aromatic compounds on thin-
layer chromatograms.
Sawicki, E., Stanley, T.W., and Johnson, H.
Microchem. J. 8: 257-84 (1964)

E 349 Thin-layer chromatographic separation
and analysis of polynuclear aza heterocyclic compounds.
Sawicki, E., Stanley, T.W., Pfaff, J.D.,
and Elbert, W.C.
Anal. Chim. Acta 31: 359-75 (1964)

E 350 Quenchofluorometric analysis for
fluoranthenic hydrocarbons in the presence of other types of aromatic hydrocarbons.
Sawicki, E., Stanley, T.W., and Elbert,
W.C.
Talanta 11: 1433-41 (1964)

E 351 The application of thin-layer chromatographic and spectral procedures to the
analysis of aza heterocyclic hydrocarbons in complex mixtures.
Sawicki, E., Stanley, T.W., and Elbert, W.C.
Occupational Health Rev. 16(3): 8-16 (1964)

AUTHOR INDEX

D60, D332, D409, D410, D411, D412, D691, D747, D809, D810, D811, D1282, D1283, D1284
Chernikov, A. A. - C34
Chernitskii, E. A. - D1285, D1286, D1958
Chernogryadskaya, N. A. - D812, D1270, D2125
Chernyuk, I. N. - D1304, D1809
Cheroff, G. - C112
Cheronis, N. D. - D413, D1258, E49
Chestnut, A. - D187
Chevalier, N. - D813
Chiang, C. H. - D1930, D1931, E53
Chibisov, A. K. - D932, D933
Chieco-Bianchi, L. - D863
Chieriei, L. - C550
Child, K. J. - C561
Childs, R. F. - E54
Chin, L. - E54
Chiozzotto, M. - A1
Chirigas, M. A. - D880, E200
Chmutina, L. A. - D1428
Choate, J. - D1810
Choi, S. I. - D1419
Chomse, H. - A21, A22, B459, C302, D61, D380, D1089
Chopin, J. - C607
Chou, C. N. - A348
Choudhury, P. K. - B460
Chow, P. C. - D1811, D1812
Chowdhury, M. - D814, D1287, D1813
Christodouleas, N. D. - D1001, D1002, D1814
Christophorou, L. G. - D773, D774, D1239, D1240, D1772
Ch'u, C. - D1812
Chulski, T. - E55
Chung, K. I. - C685
Chung, N. T. - D213
Churckich, J. E. - D815, D1288, D1815, D1816
Ciais, A. - C797, D816
Ciferri, A. - C303
Cifkova, D. - D2128
Cingolani, E. - C608
Ciotti, M. M. - C834
Ciusa, W. - D62
Clais, A. - D256
Clamroth, R. - D1817
Clar, E. - B461
Claringbull, G. F. - A182
Clark, C. D. - D818
Clark, C. T. - E3

Clark, L. E. - A23
Clark, P. T. - E216
Clark, W. G. - E216
Clarke, D. L. - E50
Claus, C. J. - B633
Cleiren, A. P. D. M. - C823, D325, D326
Clement, J. - A639
Clemmons, J. J. - B295
Clerc, P. - A236
Clogston, A. M. - C823
Close, J. A. - E277
Clotten, R. - C566
Coche, A. - C304, C609, D1395
Cochran, A. J. - B358
Codeau, C. - D463
Coffer, H. E. - A525
Coffman, R. - C305
Cohen, E. N. - E56, E57
Cohen, G. - C35
Cohen, H. - A9
Cohen, P. - D807
Cohen, S. G. - B31, B462, B603
Cohen, V. H. - E10, E58
Cohen-Tannoudji, C. - D819
Cole, E. W. - A526
Cole, L. J. - E224
Collier, D. M. - B89
Collins, F. C. - A58
Collins, R. J. - C652, D63, D64
Collinson, A. J. L. - D1289
Colombo, C. - B239
Colonge, J. - C610
Companion, A. L. - D1840
Compton, D. M. J. - B463, C135, C306
Comte, M. - B10, B220
Conn, R. B. - C611
Connors, W. M. - E201
Conrad, A. L. - D1818
Conrads, H. - C559
Constantinescu, D. G. - D36
Constantzas, N. - C307
Contaxis, C. - D577
Contractor, S. F. - E202
Conway, J. G. - B466, B500, C37, D65
Conway, W. - B633
Cook, G. R. - D1819
Cook, J. R. - B33
Cook, R. J. - A508
Coombs, M. M. - C308
Coons, A. H. - D66, E100
Coop, W. H. - D820
Cooper, B. S. - D854
Cooper, C. D. - A349

Gibson, G. E. - A385
Gibson, O. - A554
Gibson, Q. H. - E227
Gickihorn, J. - A225
Gier, J. D. - B62
Gillchriest, W. C. - D116
Gilman, H. - C342
Gilmore, E. H. - A55, A384, A385, C655
Gilreath, J. - D2192
Gindina, R. I. - D461
Ginther, R. J. - A131
Giral, J. - A56
Girand, G. - D117, D1876
Girdzhiyauskaite, E. A. - C210, C482
Giurgea, M. - C656
Givner, M. L. - D23, D120
Gladchenko, L. F. - D1359, D1877, D2072
Glaess, H. E. - B586
Glaid, A. J. - C780
Glarum, S. N. - B63
Glaser, F. - B493
Glasner, A. - D2086
Glassman, E. - E244
Glazunov, P. Ya. - D1150
Glick, D. - D1360, E74, E223
Glier, R. - C39
Glocker, R. - B26, B232, C71, C343
Glovadskii, Y. - D1878
Glowacki, J. - D1361, D1362, D1363
Gobin, F. - D118
Gobov, G. V. - D876, D1030, D1364, D1879
Gobrecht, H. - A386, A555, A877
Gochkovskii, V. F. - A556
Gochman, N. - E326
Goderzishvili, K. G. - E181
Goedheer, J. C. - B272, D1880
Goehring, M. - A367
Gofman, I. A. - D1914
Goldberg, P. - D119
Goldenberg, M. - C35
Goldenson, J. - C68
Goldfien, A. - C344
Gol'dina, T. A. - A2
Gol'dman, A. G. - D1367
Goldsborough, J. P. - D1590
Goldsmith, G J. - D1712
Goldstein, J. - A557, C72
Goldzieher, J. W. - A387, B64, D23, D120
Golikova, L. E. - C345
Golob, S. I. - C74
Golovina, A. P. - D3, D727, D878, D1878
Gol'tsev, V. D. - C73
Gombay, L. - A558

Gomez, A. M. - B273
Gomon, G. O. - D121
Gompper, R. - B494, C346, C657
Gondo, Y. - D466, D510, D930
Goodban, A. E. - C863
Goodgame, D. M. L. - D464
Goodman, L. - D465, D532, D814, D962,
    D1287, D1813, D1917
Goodwin, R. H. - A57, A388, B65
Goon, E. - A559
Gopala, K. - D945
Goplen, B. P. - B495
Gorbacheva, N. A. - A409
Gorbenko-Germanov, D. S. - D879
Gordeivskii, A. V. - E16a
Gordon, H. T. - C658
Goren, A. - D1726
Gornall, A. G. - B274
Gorodisch, S. - E262
Gorskii, G. U. - D1017
Gotlib, Y. Y. - D1368
Goto, K. - A394, B276, C659, D1881
Gouaze, A. - D1370
Gough, K. - D854
Gouterman, M. - D881, D882, D1866
Govindachari, T. R. - D467
Govindjee, R. - D1371, D1970
Grabar, P. - C647
Grabenstetter, R. J. - A72
Grabowski, Z. R. - D468
Gracian, J. - D883
Gragneva, G. I. - D410
Graham-Bryce, I. J. - D122, D414
Grajcar, L. - C738, D469, D1882
Grand, S. - A58
Grandberg, I. I. - D1372
Grandi, F. - D1299
Grandolini, G. - C643
Grant, D. W. - C268, D1211
Grauer, R. C. - B624
Graulier, M. - C660
Gray, P. H. H. - C75
Gray, W. R. - D1373
Greaves, M. C. - D2084
Greech, H. F. - C795
Green, M. - D1600
Green, N. M. - D1883
Greenberg, E. - D1733, E332
Greenberg, L. J. - E224
Greenblatt, D. P. - E26
Greenblatt, R. B. - E26
Greene, J. W. - C513, C866
Greenough, K. F. - D470

Hoffman, W. - B459
Hofstadter, R. - B393
Hoggan, M. D. - D1107
Hohensee, F. - C366
Hohmuth, K. - D1402, D1910
Hojman, J. - A166
Holiday, E. R. - A73
Hollas, J. M. - D902
Holleman, H. C. A. - A74
Holloway, W. W. - D1403, D1404, D1911
Holmes, N. D. - E297
Holsinger, V. H. - E68
Holstein, T. - A102
Holzapfel, L. - D906
Holzbecher, Z. - A568, B78, B283, B284,
    B285, B286, C680, C681, D142, D903,
    D1405
Holzl, J. - C682
Honegger, C. G. - E255
Honig, J. M. - C93
Hood, G. M. - B484
Hoogenstraten, W. - A93, B287, C313
Hopfield, J. J. - C856, D63, D680, D904,
    D1127, D1128, D1406, D1684
Horak, J. - D166
Horhammer, L. - A397, B288
Hori, J. - D905
Horne, J. E. T. - A75
Hornstein, I. - E7
Horvai, R. - B473, C27, C621, D159, D514,
    D941
Horwitz, L. - B262, B289, C337
Hoshino, M. - A76
Hoskins, R. H. - D558, D1895, D2136
Hossfeld, R. L. - B566
Hotta, K. - C683, D309
Hovanyi, M. - D2171
Hovey, R. J. - D1949
Howard, A. J. - D907
Howerton, H. K. - D1060, E9, E13, E14,
    E345
Hoyaux, M. - D143
Hoyer, E. - C657
Hrabal, L. - C684
Hsia, D. Y. Y. - E152, E183, E235
Hsias, K. Y. - D1931
Hsien, K. - D1812
Hsu, J. C. - C685
Hsu, R. - D1172
Hubert-Habart, M. - C434, D1019
Huefner, S. - D398
Hueniger, M. - D1407
Huennekens, F. M. - A208, C527

Huff, J. A. - E207, E208
Huffman, R. E. - D1408
Hughes, J. - B358
Hughes, R. C. - D1526
Huke, F. B. - A569
Hulcher, F. H. - D1995
Human, J. P. E. - A77
Hummel, F. A. - C358, C821, D492, D634
Humphreys, W. G. - D1211
Humuller, F. L. - D197
Hunt, B. E. - C686
Hunt, G. R. - B508
Huraux, M. J. - C658
Hussain, F. - D1409
Hutchinson, C. A. - C368, D144, D493,
    D1410, D1411
Huttenrauch, R. - C366
Hutton, E. - D30, D299, D670, D908, D1668,
    D1669, D1679
Hynie, I. - D331

Ibsen, K. H. - E76
Ichimura, K. - D578, D633
Ichimura, Y. - C687, D145, D146, D147
Ide, H. - A78, A505, C688, C689, D909
Idelhoch, H. - D1660
Igarashi, Y. - C691, D578
Ignjatovic, N. - A570
Iguchi, K. - C3, C94, C369, C690, D910,
    D1412
Iida, S. - D1143
Iimori, H. - A476
Iimori, M. - A571
Iimori, S. - A79, A398, A571, A581
Iizuka, H. - A399
Ikeda, M. - D578
Ikekawa, N. - C370
Il'ina, A. A. - A135, A234
Imada, Y. - A627
Imai, T. - C691
Imbush, G. F. - D1913
Imoto, E. - C159
Imoto, M. - D1413
Inamura, Y. - B529
Indemans, A. W. M. - B79
Ingram, D. J. E. - D599
Inokuchi, H. - D2047
Inoue, G. - C371
Inouye, T. - E152, E318, E319

Jones, S. - A81
Jones, T. H. - D1920
Joop, N. - D364, D714, D715
Joos, G. - B45
Jope, E. M. - A73
Jopson, R. C. - D1418, D1922
Jorgensen, C. K. - C101, D2133
Jorder, H. - C693
Jortner, J. - D502, D1419, D2182
Joshi, B. D. - B423
Joshi, R. V. - D97, D1923
Joslet, C. - B518
Joussot-Dubien, J. - C787
Joyce, T. A. - D2060
Juboer, J. F. W. - C857
Judd, B. R. - C694, C695
Judge, D. L. - D644, D1420
Julliot, C. - D1271
Jung, L. - D1421, D1924
Jung, W. - D1264
Junga, I. G. - D272
Jupe, N. F. L. - D184

Kachura, T. F. - D1655
Kacprzak, F. - D512
Kael, K. - D331
Kagan, M. R. - D862, D2012
Kahimoto, Y. - E303
Kahn, J. M. - D1922
Kaibe, Y. - B329
Kaifu, Y. - B328, B546, B547, B548, D995,
    D997, D998, D2009
Kaihara, M. - D1060, E14
Kaiser, G. - C102
Kaiser, H. - B519
Kaiser, W. - D64, D459, D503, D504, D505,
    D1185, D1422
Kajigaeshi, S. - D511
Kakata, M. - D1575
Kakisawa, H. - D590
Kalant, H. - B274
Kalantar, A. H. - D926, D1925, D1926,
    D1927
Kalckar, H. M. - C422
Kalenichenko, Y. I. - A403
Kalimbet, A. Z. - D876
Kalle, K. - A240
Kallistratos, G. - D152, D927
Kallman, H. - A19, A82, A83, A219, A241,
    A242, A335, A379, A401, A402, A549,

A550, B58, B267, B268, B269, B292,
    B520, C63, C103
Kallmann, H. P. - C293, C339, C375, D155,
    D506, D654, D870, D928
Kalojanoff, A. - C497
Kamath, N. R. - B521
Kambe, R. - D560
Kamimova, H. - D1339
Kaminskii, M. G. - C696
Kaminsko, V. - D1361
Kamiya, S. - C154, D1928, D1929
Kamiyama, M. - D1423
Kan, F. H. - D1930, D1931
Kanaoka, V. - D1932, D1933
Kanda, E. - A475, B191
Kanda, K. - D1424
Kanda, S. - D153
Kanda, Y. - C376, C490, C697, D48, D507,
    D508, D509, D510, D511, D929, D930,
    D931, D1553, D1932, D2137
Kaneko, Y. - C104, C698
Kanora, J. - E239
Kantardzhyan, L. T. - C105, D722
Kantor, J. - D1425
Kantor, S. M. - D766
Kapilza, W. - D865
Kaplan, A. - E80
Kaplan, N. O. - C834, D2015
Kaptil'nyi, M. A. - D1097
Kara, R. - B352
Karagoonis, G. - D934
Karapetyan, G. O. - C699, C875
Karapetyan, N. V. - D1426, D1427
Karazin, V. A. - C578
Karczewska, H. - D1973
Kardos, F. - C372
Kargakin, A. V. - D154
Karishin, A. P. - C700
Karlefors, T. - E40
Karler, R. - C344
Karpukhin, O. N. - D1148
Karras, H. - C845, D1365, D1366
Karreman, G. - C106
Karrer, P. - A84, A130, B396, B397
Karush, F. - E85, E240
Karyakin, A. V. - A243, A303, A305, A403,
    B86, B187, C107, C377, C701, C702,
    D531, D932, D933, D1428
Kaseda, H. - D931, D1932
Kasha, M. - A404, B216, B647, C426
Kashiwagi, H. - B195
Kaskan, W. E. - A85, A86
Kasper, K. - B259, C108
Katagiri, H. - C378

Kisin, V. I. - D375
Kisliuk, P. - D874, D1947
Kislyak, G. M. - C389, C390, C717, C718,
    C719, D165, D521, D923, D946, D947,
    D1443, D1444, D1445, D1989
Kislyakova, T. E. - D948
Kiss, Z. J. - D949, D1446, D2188
Kissane, J. M. - A131
Kisser, J. - A247
Kisser, W. - D1447
Kitanskaya, L. A. - D1448
Kivalo, P. - A416
Kiyanskaya, L. A. - D522, D523, D1677
Klappmeier, F. H. - D2090
Klaseno, H. A. - C115, C720
Klasens, H. A. - A583, A685, B301
Kleber, W. - C721, C722
Klein, E. - C15, D1948
Klein, M. P. - D874
Kleinberg, A. V. - B302, B527
Kleinerman, M. - D950, D951, D1949
Kleinert, J. - B303, D1950
Kleinman, D. S. - E246
Klembala, M. - D1951
Klemperu, W. - D1795
Klick, C. C. - A131, A132
Klikorka, J. - D166
Klimov, A. I. - D291
Klimov, V. V. - D1449, D1450, E84
Klimova, L. A. - A454, B167, B612, B613,
    C210, C838, D1099, D1100
Kling, A. - D167, D524, D924, D952, D1451
Klingenberg, H. - A336
Klinman, N. R. - E85, E240
Klixbuell, C. - D2133
Klochkov, V. P. - A616, B304, B337, B528,
    B559, C116, D747, D953, D1452, D1453,
    D1454, D1455, D1456, D1457
Klochkova, M. P. - D1952
Kloss, H. G. - D525, D954
Klotz, M. D. - E247
Kluver, H. - B305
Klyuev, Yu. A. - C742, D454
Knappe, G. - E136
Knau, H. - C117
Knopfe, H. - C118
Knowles, F. - B242
Knox, W. J. - C807
Kobayashi, M. - D1956
Kobayashi, S. - D1458
Kobayoshi, K. - D1562
Kobernich, S. D. - D2200
Kobozev, N. I. - D1957
Kobuke, Y. - C868

Kobyshev, G. I. - D168, D955, D1459, D1511
Koch, L. - C391
Koch, R. - D1181
Kocheminovskii, A. S. - D169, D336, D711
Kochetkov, N. K. - C405
Kocik, J. - B92
Kockmirovskii, A. S. - D363
Kocourek, J. - C307
Koda, T. - D1953
Kodera, Y. - A584, B306
Kodryashov, P. I. - C500, C501
Koe, B. K. - A89
Koechlin, B. A. - E86
Koechlin, Y. - D1354
Koehler, M. - D1324
Koelmans, H. - C662, C723, D170
Koerber, W. - D1192, D1954
Koetitz, G. - D1365, D1366
Kofuya, Y. - A153, B193, B194, D319
Kogan, I. B. - C392
Kogarlitskaya, N. V. - D650
Kohen, E. - D1460, D1955
Kohlmannsperger, J. - D1641
Koizumi, M. - A408, A426, A585, A586,
    A606, B87, B93, B112, B293, B327,
    B328, B329, B546, B547, B548, C109,
    D156, D320, D868, D1544
Kok, B. - D1461
Kokoski, C. J. - C393
Kokoski, R. J. - C393
Kokubun, H. - B529, C119, C120, C326,
    C394, C724, D1462, D1956
Kolbel, H. - B94
Kolesnikov, G. S. - C395
Kolevatykh, G. V. - C396
Kollman, H. P. - D956
Kolobkov, V. P. - B414, B662, B663, B664,
    B665, B666, C261, C345, C537, C538,
    C539, C540, C909, C910, C912, D32,
    D37, D38, D335, D363
Komar, O. F. - C862
Komissarov, G. G. - D1957
Kondo, T. - B530, C121
Kondaraki, N. I. - C261.
Kondrat'ev, V. N. - D1463
Kondrat'eva, T. M. - D571
Konev, S. V. - C725, C726, D1285, D1286,
    D1464, D1775, D1958
Konig, E. - C58
Konig, K. - C727, C728, C729
Koniger, M. - C497
Konigstein, J. A. - D1959, D1960
Konishi, K. - D2007, D2008
Kononenko, A. P. - D1466

Mastroeni, P. - D1279
Masuda, H. - A323
Masuda, I. - C868, D319, D563
Masuda, M. - D1929
Masuda, T. - C144, C406
Masuko, K. - D994
Mataga, N. - A408, A426, A606, B93, B112,
    B327, B328, B329, B545, B546, B547,
    B548, C145, C146, C147, C418, C419,
    C420, D995, D996, D997, D998, D1518,
    D1519, D1520, D1546, D2009, D2010
Matejec, R. - C498
Mathe, G. - C148
Matheson, M. S. - B67
Mathur, R. M. - A257
Matkovic, J. - D1521
Matovich, E. - D1522
Matsumoto, A. - D2028
Matsumoto, K. - E97
Matsumura, T. - D931, D1932
Matsumura, Y. - E82
Matsunaga, K. - C689
Matsushita, A. - D1523, D1524
Matthews, I. G. - B549, C323, C421
Matthies, H. G. - C769
Mattingly, D. - E268
Mattisson, A. G. M. - D564
Mattoo, B. N. - B84, B514, B515, B516,
    B550
Matuska, A. - D1525
Matyas, Z. - C149
Maurer, R. C. - E98
Mavrodineanu, R. - B330, D1526
Maxwiell, E. S. - C422
May, I. - A389
Maycraft, G. W. - D818
Maynard, D. E. - E217
Mazelsky, R. - D2034
Mazurenko, Y. T. - D999
Mazzucato, M. P. - D2011
Mazzucato, U. - D2011
McAlister, A. - C216, C217
McAnally, J. S. - B113
McAvoy, N. - D862, D2012
McCao, D. J. - D2013
McCapro, F. - E269
McCartin, P. J. - D301, D1527, D2144
McCarty, M. - C757
McCleskey, J. E. - D2014
McClintock, R. M. - D1386
McClure, D. S. - A55, A101, A283, A385,
    A427, A450, A607, A654, B114, B370,
    B551, C204, C284, C305, D825, D843

McClure, W. E. - D1000
McCollister, R. J. - E99
McConnell, H. - A607, D306
McConnell, W. V. - D198
McCormick, D. B. - E83
McCoubrey, A. D. - A102, B115
McCully, C. R. - B586
McCumber, D. E. - D1913
McDevitt, H. O. - E100
McDonald, J. K. - E270
McDonald, S. - E297
McDougall, D. J. - A428
McElroy, W. D. - C829
McFarland, R. H. - A666, B382, D7
McGauock, E. H. - D209
McGlynn, S. P. - C423, D199, D565, D740,
    D741, D950, D951, D1001, D1002,
    D1217, D1218, D1219, D1528, D1529,
    D1814, E327
McGregor, L. L. - D101
McKay, R. H. - D2015
McKeag, A. H. - C686, D1005
McKernan, W. - D1106
McKinney, R. M. - E271
McLachlan, A. D. - D200
McLachlan, J. - D1824
McLachlan, L. A. - D1003
McLaughlin, J. A. - B116
McLaughlin, R. D. - B466, B500
McLeary, J. F. - D1004
McLeod, G. C. - C424
McMillan, W. G. - C86
McMullen, W. H. - A559
McNabb, W. M. - C72
McNutt, W. S. - C425
McOmie, J. F. W. - A6
McPherson, J. F. - B357
McPherson, S. - D1082
McRae, E. G. - C426, D201, D566, D567,
    D1530, D1531
McWeeny, R. - D568
Mead, C. A. - D2016
Mead, J. A. R. - E200
Meckelburg, E. - C758
Meditsch, J. - D2017
Medlin, W. L. - D569, D570, D1532
Meduski, J. W. - D1828
Medvedev, A. N. - C734
Medvedev, V. S. - A198
Medzvetskii, D. S. - D1377
Meek, E. S. - B421
Megill, L. R. - B51
Mehler, A. H. - C150

Meijer, G. - C427, D12, D202
Meinnel, J. - D562
Meisel, M. N. - D2, D404, D571
Meissner, D. - C343
Meixner, H. - A608
Melamed, T. - D1388
Melby, L. R. - D2018
Melhuish, H. W. - B117, B331, C428, D203,
    D572, D573, D1006, D1281, D1533,
    D1896, D2019, E29
Meli, A. - D1661
Melikadze, L. D. - D1007
Mellichamp, J. W. - D720, D721
Mellinger, T. J. - D1534, D2020, E346
Mellors, R. C. - A429
Melmon, K. - A132, E272
Mel'nikov, S. S. - D2021
Memming, R. - D574
Mende, L. - B336
Mendelson, W. L. - D1561
Menningmann, H. D. - D1658
Men'shikov, V. V. - D448, D2022
Menzel, D. W. - E158
Menzies, A. C. - D1008
Mercier, J. - B320
Merklin, J. F. - D982
Merner, R. R. - A609
Merola, G. V. - B325
Mescon, H. - D1535
Mesnard, P. - A610, B118, B326
Messner, D. - C71, C151
Metcalf, W. S. - A194, B117, B119, C428,
    D204, D205
Methke, E. - C430
Metlay, M. - D1536, D2023
Metze, R. - C429
Metzer, P. H. - D1819
Metzger, F. R. - B120, B121, B332, B333,
    B552
Metzger, J. F. - D1107
Meuffels, M. - E247
Meurs-Hoekstra, W. - D798
Meyer, H. - D90
Meyer, J. - A325
Meyer, K. O. - D420, D1295
Meyer, W. C. - D1009
Meyer, Y. - D67, D1537, D2024
Meyer-Berkhout, U. - B310
Meyers, F. H. - D701
Mezincesu, M. D. - B334
Miani, A. - D1538
Michelson, E. H. - D2025
Michon, F. - B12

Michul, C. - C152, C153
Middleton, S. - C759
Miehlich, A. - D1539
Migiridicyan, E. - C738
Mihul, C. - B553, C760, D206, D575
Mika, N. - D840, D1310
Mikhailov, B. M. - D1560
Mikhailov, N. V. - D2043
Mikhant'ev, B. I. - B554
Milanez, C. S. - D1540
Milch, R. A. - C431
Milikhina, N. M. - D2002
Miller, A. - C638
Miller, D. G. - C199, D1010
Miller, D. H. - D421
Miller, E. J. - B335
Miller, H. K. - B73
Miller, I. M. - B357
Miller, J. N. - D1541
Miller, J. R. - A208
Miller, O. B. - A105
Miller, P. H. - A256
Millich, F. - C761
Milligan, D. E. - D576, D2026
Millikan, R. C. - D1011, D1542, D2027
Millman, M. - E273
Millman, S. - E273
Millson, H. E. - A104
Milner, M. - A526, B122, B660
Miluyanchuk, V. S. - B300
Minchenkova, L. E. - D1075, D1560
Minczewski, J. - D1012
Mink, Pho Duc - D587
Minomura, S. - D2047
Minzghor, K. - D1010
Miras, C. - D577
Mironenko, L. A. - B522
Mirumyants, S. O. - D207, D208
Miskin, S. F. A. - C583
Mitchell, B. W. - D215
Mitchell, E. W. J. - D818
Mitchell, R. S. - C349, D209
Mitsumoto, T. - C762
Mittsel, Y. A. - D1878
Miura, K. - D578
Miwa, T. - D1544
Mix, M. - C763
Miyazaki, H. - C404, D535
Mizuno, H. - C154
Mizunoya, T. - B641
Mizushima, Y. - C691
Mizutani, S. - D2028
Moeller, A. - D1898

Pedrotti, L. S. - D247, D248, D617, D618
Peibst, H. - C454
Pekkarinen, L. - C796, D188
Pektor, V. - A118
Pellam, J. R. - B28
Pellegrini, R. - B470
Pende, G. - C177
Penner, S. E. - B63
Penova, F. D. - D32
Peregud, E. A. - D250
Perel, J. M. - E339
Perkampus, H. H. - A301, A466, B180, D1583, D1659, D2065
Perkins, R. W. - D556
Perlin, Y. E. - C178, D252, D253
Perlitz, H. - A60, A391
Perlman, G. E. - D2066
Peron, F. G. - E204
Perona, G. - E279
Perri, V. - E113
Perronnet, J. - B349
Perry, C. T. - D249
Pershina, E. V. - C179
Personov, R. I. - D254, D1099, D1584
Pesce, A. - D1854, D2067
Pesez, M. - A119, B137
Pesteil, L. - B138, B139, B352, B3668, C797, D255, D601, D1050, D1585, D1586, D2068
Pesteil, P. - A271, A272, A434, A435, A518, A635, A636, A639, B7, B138, B139, B140, B141, B142, B350, B351, B352, B418, B498, B574, B668, D256, D816, D1585, D1586, D2068
Pestemer, M. - D2069
Peter, H. - D257
Peter, J. H. - E100, E324, E325
Peterman, L. A. - B575
Peters, T. - C180
Peterson, E. A. - D2070
Peterson, G. E. - D1051, D1587, D1588, D1589, D2071
Peticolas, W. L. - D1590, D1591
Petit, J. - D1592
Petit, L. - A120
Petracek, F. J. - B576
Petrov, A. A. - A140, A207
Petrov, K. A. - D1593
Petrovich, P. I. - D1594
Petrovskaya, O. G. - D992
Pettit, G. D. - C384, C708, C710, C711, D158, D1108
Petzold, O. - A121

Peyton, W. T. - A431
Pfaffi, J. D. - D2110, E33, E120, E123, E288, E349
Pfahnl, A. - C181
Pfau, A. - D152, D927
Pfeiffer, L. - C182
Pflaum, R. T. - E238
Pfleidever, W. - C455
Phadke, R. - B375
Philips, N. V. - C183
Phillipp, L. J. - E305
Phillips, H. B. - A637
Phillips, J. P. - A638
Piacentini, R. - D1857
Picchioni, A. L. - E54
Pichot, L. - A639
Pick, H. - D1510
Piesche, L. - D1718
Pietsch, H. - C559, C672
Piette, L. H. - D1052
Pih, B. M. - C279
Piksis, A. - D639
Pikulik, L. G. - B662, B664, B666, C798, D258, D259, D260, D602, D1053, D1054, D1359, D2072
Piliporvich, V. A. - C456, C457, C458, D261, D603, D1055, D1596
Pilon, A. M. - C799, D611
Pil'shchik, E. M. - D1595
Pinard, P. - D425
Pincus, G. - E161
Piraux, M. - D105
Pirkl, J. - C46
Piskonov, A. K. - D1597
Piskurov, A. K. - D1102
Piterskaya, I. V. - D1598
Pitts, F. N. - E280
Pitts, J. N. - D1056
Plaza, G. R. - A122
Pletscher, A. - E177
Plitt, K. F. - D604, D1136
Plotner, G. - C314, D435
Plsko, E. - D553
Podlubnaya, E. T. - D605
Podual'naya, R. L. - D329
Poduzharlo, U. F. - E91
Pohaski, R. - D1886, D1887, D1939, D1940
Pohl, L. - D1583, D2065
Pohoski, R. - D1383
Poindepter, J. M. - D1884
Poirier, R. H. - D1722
Polacco, E. - D1912

Richter, G. - D346
Richter, H. - E247
Richtol, H. H. - D2090
Rickard, E. F. - D1347
Riddle, F. H. - A375
Ridgeway, G. J. - D1163
Riecker, R. E. - D1065, D1066
Rieckhoft, K. E. - D1590, D1591
Ried, W. - C196, C810, D275
Riedel, E. P. - D2049
Rieke, F. F. - D1067
Riess, W. - C824
Rigdon, R. H. - D1619
Riggs, J. L. - D1068
Rikhireva, G. T. - D191, D1134
Riley, H. P. - D620
Rinbach, H. W. - D276
Rindi, G. - E113
Rio, G. - B585, D1394
Ripoche, J. - D2091
Ritschl, R. - A440
Rivlin, R. - E272
Rivoire, G. - D1620, D2092
Robert, J. G. - C98
Roberts, H. R. - E114
Robertson, J. C. - C811
Robin, S. - D1987
Robinson, D. - E282
Robinson, F. M. - B357
Robinson, G. W. - C757, D446, D621, D851,
    D1025, D1559, D1663, D1664
Robinson, R. J. - A226, D1829
Rodaelli, P. - D113
Roder, I. - C468
Rodomakina, G. M. - D294
Rodriquez, C. F. - A441
Roe, D. A. - E115
Roellig, K. - D2114
Rogers, D. A. - C197
Rogers, J. - E283
Rogers, L. B. - B573, C363, C784, C800,
    D135, D263, D1057, D1847
Rogers, T. H. - A442
Rogland, J. B. - D614
Rohatgi, K. K. - D623, D1069, D1621
Rohde, F. - A644, A653
Rohrbaugh, L. - E210
Rojo, E. A. - D1151
Rolde, E. - E232
Rolfe, J. - D624, D2093
Rollefson, G. K. - A15, A351, A352, A517,
    B72
Romain, P. - B118, B326

Romano, P. M. - C177
Romanowski, R. D. - D1070
Romantsova, G. I. - C812
Ron, A. - D1071
Roncero, A. V. - C469
Roodenbeck, A. - D213
Ropp, R. C. - D2094
Roquitte, B. C. - D965
Rorem, E. S. - C813
Rosahl, D. - A567, A619, A620, A645
Rose, N. J. - D2018
Rosebrook, D. D. - D1265
Rosen, A. - D625
Rosen, P. - D844
Rosen, R. - E273
Rosenbaum, E. J. - A251
Rosenberg, B. - C814, D626
Rosenberg, J. L. - D277, D627, D669,
    D1622, D2095
Rosenberger, D. - D1072
Rosenblatt, G. M. - D796, D1262
Rosenbluth, R. - D1915
Rosenheck, K. - D628, D2189
Rosenkranz, G. - C114
Rosenthal, H. L. - B150
Rosenthal, J. E. - A339
Rosinski, J. - B586
Rosinski, K. - A646, D2096
Ross, A. M. - B452
Ross, D. L. - E179
Ross, I. G. - B508
Ross, J. G. - D1921
Ross, M. A. S. - B358
Ross, R. E. - C23
Rossi, G. - C593
Rossi, S. - B270, C340
Rossman, I. M. - D629
Roston, S. - A615
Roth, L. M. - B653
Roth, M. - E116, E284
Rothschild, S. - B151
Rothwell, P. - A167
Rotman, B. - E117
Rottgardt, K. H. J. - B359
Rousset, A. - A647, B361, B539, C50, C815,
    C816, D67, D1319, D1843, D2098
Rousset, Y. - D2099
Roux, C. - D1623
Roux, J. - A648, B152
Roux, M. - D1073
Rowbottom, J. - B38
Rowell, J. C. - A274
Rowland, R. E. - A649

Style, D. W. G. - A362, A465, A665
Subbarao, E. C. - D1388
Suchow, L. - D2149
Suciu, M. - D575
Sucov, E. - B292
Sudarsanam, V. - D467
Suga, K. - C782
Sugano, S. - D308, D503, D639
Sugiura, M. - D309
Suhrmann, R. - A301, A466, B180, C498
Suk, V. - A469
Sukar, S. C. - B625
Sukornick, B. - C775
Surak, J. G. - D1897
Surorov, V. S. - C411
Suryanarayana, V. - B626, C219
Susuki, C. K. - D1717
Suzanskii, A. I. - D1593
Suzuc, M. - D936
Suzuki, C. K. - D1522
Suzuki, S. - D383, D742, D1676, D2150
Suzushino, G. - A230
Svendsen, A. B. - A139
Sverdlov, Z. M. - C220
Sveshnikov, B. Y. - A140, A141, A142, A207,
    A467, B381, B533, B534, B627, B628,
    B629, C124, C407, C456, C458, C499,
    C500, C501, C508, D178, D179, D180,
    D210, D307, D494, D495, D522, D523,
    D579, D732, D733, D914, D1207, D1208,
    D1209, D1210, D1677
Sveshnikova, E. B. - D853, D1851
Sveshnikova, E. V. - D1335
Svirbely, W. J. - B181
Svishchev, G. M. - D2151
Svitashev, K. K. - D95
Swagdis, J. E. - E300, E301
Swailes, G. E. - E297
Swank, R. K. - A515, A637, D1093, D2152
Swann, C. P. - C221
Swann, R. V. - B182
Swanson, R. E. - A666, B382
Sweat, M. L. - B183
Sweeney, G. O. - E302
Sweep, G. - C51
Sweers, H. F. - D1321
Swenson, G. W. - D975
Swift, C. D. - D1418, D1922
Swindells, F. E. - B184, B631
Swings, P. - C87, C316
Switzer, J. L. - A143, A144, A145, A667,
    C502, D1119
Switzer, R. C. - A144, A145, A667, C502

Switzes, K. - D672
Syenes, I. - B383
Sykes, W. O. - D310
Symons, M. C. R. - D1120
Syuiyum, S. - B632
Szalay, L. - A146, C341, D1121, D1122,
    D1273, D2153, D2154, D2155
Szalay, S. - C148
Szalkowski, C. P. - E239
Szasz, K. - B383
Szczurek, T. - D760
Szegho, C. S. - B185
Szekacs, J. - D1951
Szekely, A. - D673
Szekely, R. - D673
Szendrei, M. E. - A522
Szent-Gyorgi, A. - B116, B384, C215, C223,
    C224
Szollosy, L. - B450
Szor, P. - A147
Szymanski, A. - D1973
Szymkowiak, H. - D1938

Tabak, S. V. - D1372
Tabata, T. - B204, B413, B657, D2156
Tagantsev, K. V. - C225
Tait, A. C. - E157
Takagi, K. - C161
Takahashi, M. - E35
Takahashi, Y. - D185
Takashina, K. - D1575
Takatori, K. - D1414
Takeda, J. - C378
Takei, K. - D1123
Takenoshita, Y. - D930
Taketomo, Y. - D674
Takeyama, H. - B294, C512
Takeyama, N. - B641, C864
Talman, E. L. - B130
Tamari, M. - D186
Tamas, I. - D1733, E332
Tamura, Z. - D2028
Tanabe, H. - B294
Tanahe, T. - D1575
Tanaka, C. - D1678, D1679
Tanaka, E. - D309
Tanaka, J. - D1679
Tanaka, Y. - D1408
Tang, K. S. - D1996

Tomicek, O. - A469
Tomick, E. G. - C561
Tomimatsu, Y. - C863
Tomiser, J. - A470
Tomita, G. - B641, C512, C864, D1135
Tomlinson, T. B. - C232
Tomotsu, T. - A600, B109, B110, C417, D560, D991
Tomura, M. - D1691
Toner, S. D. - D604, D1136
Toporek, M. - E305
Topuridze, L. F. - E181
Torgov, V. G. - D3
Torihashi, Y. - D995, D996, D1518
Torii, K. - C233
Torio, K. - B386
Torok, C. - A38
Toropova, G. P. - C865
Torp, A. - D857
Touchstone, J. C. - C513, C866
Tousey, R. - A163
Tovey, G. P. - D198
Towne, J. C. - D1692
Traill, R. J. - A473
Traisme, M. - D385
Trapeznikova, Z. A. - A152
Trautner, E. M. - A306
Travnicek, M. - A474
Tredale, T. - A540
Trehorne, R. W. - D1693
Treibs, W. - C867, D682
Trevalion, P. A. - D1120
Trevoux, P. - D1291, D1346
Trieff, N. M. - D2163
Trofimov, A. K. - B387
Troisplis, R. - D1586
Tronche, P. - B46
Troncoso, V. - C114
Tropenko, M. A. - D1138
Tropper, H. - D424
Troshchenko, A. T. - C514
Trosken, O. - A669
Trosper, T. - E148
Trottier, D. - D1710
Troxler, F. - C515
Trumbore, F. A. - D2159
Trusk, B. A. - D1254
Trusov, V. V. - D683, D1683, D1694, D2165
Tschampo, A. - D2116
Tschesche, R. - C516
Tschetter, P. N. - E142
Tsinober, L. I. - D1657
Tsirlin, Ya. A. - D297

Tsou, K. C. - D1630
Tsubomura, I. - C233
Tsuchida, S. - C258, C259
Tsuji, F. I. - B190
Tsuji, T. - E82
Tsujikawa, I. - A475, B191
Tsumo, S. - C146, C147
Tuboi, S. - C683
Tumerman, L. A. - C234, C579, D1075, D1252, D1779
Turkevich, H. - A660
Turkevich, J. - D317
Turnbull, J. H. - C336
Turner, G. K. - D2164
Turner, P. - E154
Tursunov, N. I. - D1596
Turtenwald, J. - D2129
Tuyzolino, A. J. - D1740
Tweet, A. G. - D1869, D1870, D2166, D2167, D2168, D2169
Tweit, R. C. - D27
Tyler, J. E. - A307

Uchida, K. - D320
Udenfriend, S. - B230, B388, C49, E2, E3, E5, E6, E36, E143
Uebersfeld, J. - B192
Uehara, Y. - A153, B193, B194, C868, D318, D319
Uehida, I. - D2170
Ueno, H. - A76
Ueta, M. - A154
Uhl, O. - A308
Uhlenbroek, J. H. - D321
Ui, H. - B126
Uibo, L. Y. - C869
Ujhelyi, S. - D530
Ujihara, T. - D584
Ulasenko, A. I. - D1608
Ulrich, H. - B339, B340, C774
Ulrich, W. F. - C870
Uluitu, M. - D1696
Umegaki, Y. - B195
Umreiko, D. S. - D337, D338, D339
Unger, H. - D865
Unghuary, L. - D2171
Unohara, N. - A476
Upadhya, K. N. - D2172
Uphaus, R. A. - C871

Ward, H. F. - B3, B210
Ward, H. T. - B660
Ward, J. C. - A465, A665
Ward, R. A. - A143, A667
Ward, W. H. - C863
Ware, W. R. - D695, D1160, D1161, D2187
Warhurst, E. - C622, C623
Warminsky, R. - A342
Warncke, J. - D340
Warner, E. R. - B55
Warren, I. A. - D2200
Wasselburg, K. - D275
Wasserman, E. - D1872
Wassink, E. C. - A314, B176, B197
Wasson, M. N. - D1162
Watanabe, K. - A163, A315, D211
Watanabe, S. - D1710, E148
Watanabe, T. - D156
Watson, C. J. - B603, E175
Watson, W. F. - A164, A478
Watts, D. T. - E312
Waugh, T. D. - B61, D2185
Wauk, M. T. - D851
Wave, H. E. - E149
Way, E. L. - E90, E253, E254
Wayo, S. J. - B465
Weakleim, H. A. - D2188
Weber, B. C. - C239
Weber, D. D. - D1163
Weber, G. - A165, A479, A480, A674, A675,
    B201, B649, C226, C240, C523, C853,
    C893, D341, D628, D2189, D2190
Weber, K. - A166, C241, D1269, D1521,
    D1839
Webster, R. - D696
Wedin, B. - C894
Weigl, J. W. - C241a
Weiglin, W. - D906
Weil, S. A. - A481
Weill, G. - C303, D1711
Weil-Malherbe, H. - D342, E311
Weimer, E. Q. - D348, D349
Weinberger, L. A. - A292
Weinreb, A. - B31, B462, C242, C274, D21,
    D344, D697, D698, D768, D1164
Weipert, E. A. - C342
Weir, D. S. - D699, D1165, D1166
Weis, C. G. - D700
Weiser, K. - D1981
Weiss, J. B. - D1167
Weissbach, H. - B388, E2, E3, E150
Weissler, A. - D1175, D2194
Weissler, G. L. - A593, D644, D1168, D1420,
    D2191

Weissman, S. I. - B198, C265, D1169
Weisz, S. Z. - D2192
Weitkamp, H. - C401
Welge, K. H. - D1228, D1755, D1763, D1858,
    E45
Weller, A. - A482, B199, B200, B400, B650,
    C243, D550, D524, D525, D969, D1170,
    D1345, D1497, D1695
Weller, P. F. - D1750
Wells, D. - D701
Wells, R. S. - C8
Weltner, W. - D1171, D2193
Wen, W. Y. - D1172
Wendel, G. - D346, D525, D1173
Wender, S. H. - E210
Wenger, B. S. - E105
Weniger, S. - C90, C92
Wentworth, W. E. - C244, E4
Werchselbaum, T. E. - E95
Werner, G. - D1643
Werssermel, K. - C245
Werth, G. - B401
Wesselborg, K. - C810
West, D. - A167
West, E. J. - C127
West, K. - B227
West, W. - A316, D1174
Westfall, T. C. - E312
Westphal, O. - D1183
Weyl, W. A. - A676
Whan, R. E. - D69, D347
Whellock, E. - C895
White, A. - C896
White, C. E. - A97, A168, A317, A318,
    A483, A484, B486, B651, C246, C247,
    C526, D348, D349, D350, D1175,
    D1745, D1746, D2194, E39, E170, E171,
    E334
White, E. A. D. - C232
White, M. C. - D1968
White, N. E. - A612
White, W. J. - B495, D1484
Whitehead, W. L. - A169
Whiteley, H. R. - C527
Wiberly, S. E. - A559
Wiechmann, J. - D2119
Wiechmann, M. - D1643, E291
Wieder, H. - D1323
Wieder, I. - C897, D703
Wightman, T. - D127
Wiig, E. O. - A72
Wildermann, L. - B519
Wiley, R. H. - B402, C248
Wilke, K. T. - C249, C250

SUBJECT INDEX

Flash (also see Pulsed Excitation) - D969

Flavanone - C607

Flavanthrone - D2040

Flavine - C533, C714, C903, D35, D102, D108, D564, D707, D1896, D2148, E83, E315

Flavine Adenine Dinucleotide - C470

Flavine Mononucleotide - C470

Flavone - B288, C36, D277

Flavonoid - B116

Flavonol - B515, C674, C707, D218, D1254, D1824

Flour - A88, A120

Fluoranthene - D2068, E350

Fluorene - B7, B138, C30, C438, D681, D746, D1028, D1030, D1309, D2165

Fluorescein - A216, A384, B57, B162, B253, B363, B369, B419, B421, B541, C202, C399, C536, D46, D359, D622, D632, D786, D866, D936, D973, D1069, D1192, D1361, D1400, D1886, D1936, D1938, D2158, E75, E100, E117, E219, E220, E271, E310, E341

Fluorescence, Anti-Stokes (see Anti-Stokes Shift, Stokes Shift)

Fluorescence, atomic - D2196

Fluorescence, delayed - A239, C280, C284, C490, C816, D300, D1044, D1109, D1217, D1218, D1219, D1237, D1581, D1582, D1668, D1669, D1670, D1679, D1728, D1768, D1843, D2005, D2041, D2058, D2060, D2099, D2126, D2147, D2206

Fluorescence, immuno (see Immunofluorescence)

Fluorescence, line - D874, D1303

Fluorescence, micro (see Microfluorescence)

Fluorescence, nuclear resonance - B552, C221

Fluorescence, recombination - D1095

Fluorescence, reflection - D1699

Fluorescence, resonance - A575, B120, B121, B172, B209, B256, B333, B437, C263, C412, C632, C876, D140, D1011, D1323, D1799, D2016, D2039, D2096

Fluorescence, secondary - A134, A151, B449, B473, B27, C29, C99, C621, D647

Fluorescence, sensitized - A164, A222, A303, A666, B187, B202, B264, B382, B468, B564, C626, D284, D307, D695, D908, D1045, D1321, D1795, D1808, D1966

Fluorescence Yield (see Quantum Yield)

Fluorescent Photography (see also Photography) - B179, B254

Fluorescent Substances (also see Phosphor) - A78, A96, A111, A112, A113, A114, A115, A153, A157, A185, A307, A459, A502, A503, A505, A513, A565, A573, A584, A588, A598, A622, A623, A625, A626, A627, A641, A651, A656, A667, A677, B3, B4, B9, B32, B62, B132, B151, B157, B193, B194, B205, B210, B241, B266, B306, B313, B323, B336, B386, B481, C3, C8, C54, C154, C183, C203, C259, C688, D1350, D1545, D1562, D1612, D2056, E334

Fluorexon - D961, D1475

Fluorine - B330, D276, D764, D848

Fluorite - B70, D1703, D2080, D2081

Fluorochrome (see Dye, Reagent) - A233, C882, C898

Fluorogen - E206

Fluoromeite - D813

Fluorometer (see Instrument)

Fluorotoluene - B626, C219

Fluorspar (see also Calcium, minerals) - A202, A640, C454, C845, D2075, D2079

Folic Acid - B621, D1569

Food Products - D62

Formaldehyde - A16, A196, A361, C485, D672, D857, D1082, D1413, D1820, D1918, D1958, E33, E44, E123, E247

Formate - C527

Formic Acid - A362, A465, C390, C717, C718, D102

Franck-Condon Principle - A3, D977, D981, D1558

Fulvalene - D600

Fulvic Acid - D2118

Fungus (also see specific fungus) - C698, D126

Furan - D578

Gadolinium - A492, C43, D1726

Galactopyranoside - C409

Galactosidase - C409, E117

Galactoside - D1113

Gallium - C36, C95, C444, C445, C446, C478, C662, C692, C833, D3, D125, D171, D329, D545, D650, D782, D1041, D1377, D1406, D1751, D1902, D1976, D2106

Hydrogen – A658, B172, B225, B278, C92,
  C873, D77, D1306, D1763, D2027
Hydrogen Bonding – B329, B546, C146, C147,
  D780, D995, D997, D998, D1459, D1676,
  D2009, D2090
Hydrogen Peroxide – B76, D1653
Hydrogenase (also see Dehydrogenase) –
  C834
Hydroxy (see also Hydroxyl) – D589
Hydroxy Naphthoic Acid – C301, D412, D897
Hydroxyanthraquinone – A403, C400
Hydroxyazo – C680
Hydroxyazobenzene – D1302
Hydroxycorticosteroid – E111, E217
Hydroxycorticosterone – B18
Hydroxyfluoran – C409
Hydroxyindole – C214
Hydroxyl (see also pH, Acid, Alcohol) –
  B233, C591, C592, D502, D804
Hydroxyphenazine – B180
Hydroxyquinoline – B46, B584, D1244,
  D1392, D1647, E148, E176
Hydroquinone – C672
Hydroxyquinoxaline – D537, D538
Hydroxytyramine (see Dopamine)
Hydroxytryptamine (see Seratonin)
Hydroxytryptophan – E172
Hydroxytryptophol – E177
Hypericin – B487

Ice (see also Water) – B67, C673
Identification (also see Technique) – B159
Imidazole – A323, A324, A325, A488, A489,
  A490, B270, C267
Imidazolone – A264
Imine – C757
Immunoelectrophoresis – C647
Immunofluorescence – D66, D1068, D1107,
  D1202, D1541
Indanone – D682
Indanthrene Blue RS – D2040
Indazole – D1642
Indicators – A108, A469, D88, D114, D225,
  D413, D1023, D1897, D1945, D1965,
  D1968, E149
Indigo – D229, D283, D1103, D1566
Indium – B186, B567, C168, C169, D3, D170,
  D545, D1567, D1878, D2106
Indole – C214, C335, C387, C505, C530,
  D686, D714, D724, D1286, D1292,
  D1642, D2010, E109
Indoleacetic Acid – A211
Indoxyl – C68, D175
Infrared – A236, A331, A504, B296, C17,

C112, C228, C384, C427, D1, D71, D98,
  D223, D388, D450, D536, D576, D798,
  D877, D1011, D1019, D1117, D1157,
  D1275, D1337, D1357, D1702, D1793,
  D1794, D1902, D1984, D1991, D2013,
  D2026, D2132, D2176
Inosine – D1709
Insecticide (also see Pesticide) – D1138
Insects – A358, B247, B653, D112
Inspection – D118
Instrument – A120, A180, A514, A542, A543,
  A569, A586, A644, A652, A671, B47,
  B55, B59, B89, B91, B113, B124, B182,
  B204, B274, B311, B334, B373, B391,
  B430, B438, B439, B440, B491, B528,
  B531, B539, B559, B598, B657, C12,
  C23, C31, C72, C148, C175, C176,
  C251, C311, C318, C374, C409, C453,
  C457, C560, C578, C579, C593, C595,
  C882, C847, C897, D15, D26, D100,
  D133, D184, D197, D216, D224, D231,
  D236, D290, D339, D373, D405, D448,
  D478, D525, D549, D552, D616, D652,
  D655, D659, D674, D735, D796, D844,
  D894, D919, D1046, D1097, D1124,
  D1197, D1341, D1342, D1473, D1460,
  D1488, D1556, D1696, D1708, D1738,
  D1760, D1839, D1853, D1861, D1969,
  D1979, D2050, D2113, D2114, D2129,
  D2156, D2162, D2164, D2184, E9, E13,
  E14, E67, E112, E182
Insulin – C216, D1821, E17, E38
Iodide – D523, D587, D660, D691
Iodimetry – D1968
Iodine – B111, C388, C851, D1262, D1620,
  D1795, D2092
Ion-Exchange Resins (also see Resin) –
  D453, D1343, D1639
Iridaceae – D620
Iron – B344, C714, D646, D727, D811,
  D1962, D1968, D2121, E250
Isoalloxazine – B480
Isomerization – D978
Isonicatinyl Hydrazide – B341
Isoquinoline – C653
Isothiocyanate – E17

Joshi Effect – B158

Kerosine – C129, C141, C152, C568, D291
Ketazine – D1183
Keto Compounds (see Ketone) – D331
Ketoestradiol – B19, E97

Ketone – A547, C308, C332, C371, C550,
  C654, D80, D115, D699, D1056, D1397,
  D1522, D1561, D1971
Kidney – D808
Kininase – D1741
Kojic Acid – D584
Krypton – A167, D282
Kymurenic Acid – E8

Lamp – A46, A74, A114, A115, A238, A579,
  C372, C471, C684, C858, C897, D82,
  D143
Lanthanum – B184, B466, B631, D706,
  D1184, D1291, D1301, D1346, D1949,
  D2018, E179, E226
Laser – D1422, D1423, D1664, D1717,
  D1889, D1997, D2079, D2102, D2103,
  D2126
Lead – A555, D35, D1852, D1863, E267
Lecithin – B365
Lepidopterin – D1152
Leucine Aminopeptidase (LAP) – E267, E270,
  E284
Leucovorin – B621
Leukemia – D1317, D2029
Leukocyte – D1279, D1317
Lifetime (also see Decay) – A52, A53, A54,
  A85, A207, A356, A517, A637, A644,
  B82, B174, B233, B292, B438, B439,
  B441, B456, B457, C24, C25, C42, C45,
  C690, C847, C882, D22, D26, D335,
  D394, D398, D421, D434, D447, D495,
  D550, D569, D592, D597, D622, D670,
  D708, D765, D873, D974, D1067, D1093,
  D1114, D1192, D1268, D1314, D1462,
  D1571, D1771, D1790, D1802, D1842,
  D1885, D1918, D2187
Light Sources (also see Lamp, Energy
  Source) – D140
Lignin – A187, A246, A247
Lipase – D1474, D1890, E342
Lipid – D1087, E71, E126
Liqueur – D605
Lithium – A318, D916, D1155, D1200, D1365
Liver – D2097
Lolium – D1563
Luciferesceine – A299
Luciferin – B190, C829
Lucigenin – C702, D793, D1849
Lucite – D1006
Lumazine – D190
Luminol – D1521
Luminophor (see also Phosphor) – B381
Lumocupferron – D783

Lumogallion – D782, D1449, E84
Lymphate – B355
Lysergic Acid – A200, C515
Lysine – D2067

Magnesium – A12, B50, B154, B155, B489,
  C691, C821, D276, D284, D285, D308,
  D319, D362, D439, D569, D580, D639,
  D1302, D1573, D1710, E94, E99, E133,
  E148, E178, E209, E292, E330
Malachite Green – B101
Maleimide – D1933
Malic Acid – D1592, E140
Malonaldehyde – D1634, D1635, E121, E122
Malvin – D1846
Manganese – A249, B344, D151, D227, D379,
  D464, D492, D563, D1005, D1404,
  D1407, D1704, D1833, D1911
Maser – D918, D1108, D1127, D1178, D1188,
  D1416, D1417, D1495, D1671, D1719,
  D2049
Meat – A91, E258
Medium – D1483
Melanoidin – C48
Melatonin – D185
Melphalan – E200
Mematocides – D321
Mepacrine – A105

Naphthacene – B349, B362, B615, D1712
Naphthalene – A18, A45, A137, A197, A265,
  A283, A435, A450, A607, A624, A654,
  B54, B106, B114, B399, B403, B404,
  B405, B406, B407, B443, B496,
  B551, B574, B616, C38, C207, C266,
  C267, C362, C368, C490, C549, C576,
  C634, C690, C700, C767, C771, C899,
  C913, C914, D6, D57, D77, D95, D116,
  D135, D144, D255, D256, D283, D292,
  D306, D331, D357, D364, D365, D414,
  D415, D418, D428, D429, D493, D562,
  D568, D705, D746, D749, D768, D902,
  D908, D963, D1001, D1035, D1062,
  D1086, D1310, D1333, D1385, D1391,
  D1458, D1472, D1498, D1519, D1529,
  D1544, D1552, D1579, D1585, D1606,
  D1607, D1613, D1666, D1787, D1823,
  D1826, D1963, D1994, D2130, D2133,
  D2165, D2192
Naphthalenesulfonic – D1035

Panthenol - E277

Paper - C679, D54, D127, D137, D353, D436

Paper Chromatography (see Chromatography, paper)

Paracyclophane - D1071

Paraffin - B226, C624, D845, D846, D1438, D1498, D1670, D2147

Particle Size - B586

Penicillin - B271, E241

Pentacene - D1937

Pentane - D1472

Pentose - D2107

Pepsin - D2190

Pepsinogen - D2066

Peptide - A170, D791, D1293, D1373, D1873, D2148, D2183, E85

Perimidine - A488, A489, A490

Permanganate - D1120

Peroxidases - B164

Perylene - D254, D876, D898, D1497, D1678, D1907

Pesticide (see individual compound, Insectiside) - C687, E7, E42, E263, E264, E323

Petroleum (see Oil) - A169, A245, A285, A354, A378, B371, B562, C447, C754, C760, D294, D1326, D1529, D1767

pH - A45, A57, A388, B203, B648, C326, C788, C896, D205, D441, D497, D502, D589, D628, D632, D667, D804, D1154, D1660, D2158

Phaeophytin - E158

Pharmaceutical (see Drug)

Phenanthrene - A31, A475, C211, C697, C899, D29, D300, D677, D681, D714, D741, D1044, D1050, D1232, D1319, D1376, D1529, D1608, D1681, D1682

Phenanthroline - D1265, D1978

Phenanthrylboric Acid - C506

Phenobarbital - E316

Phenol -A548, C915, D136, D625, D1292, D1364, D2204, E22

Phenothiazines - D614, D1534, D2020, D2082, E63, E346, E347

Phenoxazine - C768, D917

Phenoxyacetic Acid - E164

Phenyl - D29, D546

Phenylacetylene - D1757

Phenylalanine - C563, E22, E81, E155, E236, E318, E319

Phenylene - C612

Phenylenediamine - C142, C631

Phenylethylamine - E291

Phenylfluorone - D289

Phenylhydrazine - A478, C550

Phenylketonuria - E152, E235, E236, E246

Phenylnaphthacene - B244

Pheophytin - A366, B261, B487, C402, C650, D176

Pheylsilane - D2070

Phosphatase - E304, E224

Phosphate (also see Phosphorus, Phosphor) - D479

Phosphine - D464, D665

Phosphor - A19, A21, A22, A42, A59, A60, A92, A126, A131, A132, A140, A141, A142, A152, A154, A176, A186, A249, A270, A279, A280, A289, A290, A294, A311, A313, A315, A339, A341, A342, A359, A390, A391, A401, A405, A406, A407, A409, A411, A412, A449, A453, A458, A462, A474, A521, A527, A539, A550, A560, A571, A579, A582, A586, A594, A660, A681, A685, B2, B14, B153, B184, B287, B290, B293, B297, B324, B379, B420, B428, B429, B459, B471, B475, B497, B511, B512, B513, B520, B522, B578, B602, B609, B631, B632, B655, C65, C80, C105, C110, C111, C112, C113, C118, C125, C151, C157, C158, C161, C173, C174, C193, C195, C271, C290, C302, C321, C334, C358, C383, C384, C389, C456, C545, C646, C684, C685, C686, C703, C704, C708, C711, C716, C734, C753, C765, C778, C819, C858, C868, C869, C887, C892, C902, D119, D131, D132, D182, D183, D193, D202, D210, D240, D247, D248, D261, D271, D276, D279, D324, D325, D326, D346, D376, D377, D391, D396, D406, D425, D437, D438, D450, D461, D477, D479, D492, D515, D516, D527, D561, D563, D690, D694, D702, D708, D916, D942, D944, D1015, D1034, D1048, D1049, D1055, D1063, D1076, D1092, D1095, D1096, D1126, D1127, D1128, D1142, D1209, D1289, D1388, D1398, D1437, D1471, D1477, D1510, D1515, D1517, D1539, D1557, D1596, D1599, D1609, D1610, D1611, D1617, D1626, D1628, D1685, D1743, D1837, D1928, D1929, D1989, D2006, D2034, D2046, D2061, D2071, D2085, D2089, D2094, D2115, D2136, D2159, D2170, D2177, D2186, D2197, D2198, D2199, D2201, D2202

Phosphorescence, sensitized - A363, A472, B252, B477, C478, B637

Phosphorescent Screen - D1084

Phosphorus - C410, D132, D276, D315, D762

Photochemical - D468

Photoconduction - D645

Photodecomposition - D649, D691

Photodissociation - D757, D1228

Photography- A23, A171, A210, A316, A333, A340, A593, B262, C337, C566, D436

Photoinactivation - E338

Photolysis - C676, C677, D520, D788, D789

Photooxidation - D926, D1009, D1111

Photosynthesis (also see Chlorophyll) - A309, A314, B197, C254, D500, D1371, D1427, D1733, D1811, D1812, D1970, D2064, E332

Photosynthetic - D402, D1353, D2095

Phthaladehyde - C888

Phthalate - D2133

Phthalazinedione - A156

Phthalic - C322, D923

Phthalic Acid - C315, C625, D947

Phthalimide - A616, B20, B128, B304, B337, B414, B665, C288, C537, D208, D258, D409, D602, D635, D1081, D1468, D1877, D1935

Phthalocyanine - A34, A35, A48, A49, A50, A51, A380, A381, A556, B489, C64, C228, D387, D473, D967, D984, D1459, D1527, D1584

Phycocyanin - A376, B6, D1233, D1550

Phycoerythrin - A376, D1285

Piazrelenol - D1043

Pigment - A96, A143, A144, A145, A172, A227, A364, A441, A566, B40, B41, B136, B203, B313, B485, B488, B536, B537, C24, C45, C66, C340, C433, C762, C870, C894, D381, D501, D512, D585, D598, D672, D717, D1073, D1119, D1125, D1135, D1167, D1277, D1296, D1353, D1356, D1370, D1575, D1622, D1623, D1995, D2000, E112

Pituitary - E118

Plant - A195, A300, A526, A657, B122 (soybean), B248 B322, B325, B377, B509, B525, B536, B537, B645, B660, C56, C213, C393, C439, C468, C469, C682, C747, C863, C873, D130, D157, D174, D189, D228, D232, D948, D1000, D1010, D1070, D1134

Plasma- A615, D1713, E6, E19, E86, E95, E96, D107, E109, E111, E118, E136, E154, E206, E217, E218, E219, E220, E221, E223, E253, E272, E285, E300, E301, E307, E326, E341

Plastic - A365, C67, D1038, D1167, D1327, D1330, D1347

Platinum - D580, D1013

Plexiglas - D938, D1936, D1939

Plutonium - B466, D544

Polarization - A125, A133, A141, A174, A197, A231, A292, A395, A415, A479, A480, A495, A551, A552, A590, A674, A675, B26, B52, B66, B142, B175, B177, B201, B256, B315, B339, B363, B415, B416, B483, B544, B617, B649, B667, C28, C202, C252, C262, C271, C299, C330, C341, C351, C381, C385, C386, C414, C501, C520, C522, C571, C574, C632, C667, C740, C745, C785, C828, C830, C893, C899, C906, C911, D30, D34, D56, D63, D72, D87, D103, D123, D149, D178, D179, D180, D183, D201, D286, D341, D361, D364, D365, D366, D387, D395, D397, D422, D474, D486, D490, D496, D532, D540, D541, D542, D560, D574, D591, D593, D603, D635, D714, D715, D740, D741, D760, D761, D815, D818, D840, D876, D881, D885, D888, D889, D896, D915, D917, D919, D937, D953, D962, D1027, D1053, D1081, D1083, D1121, D1122, D1226, D1285, D1286, D1309, D1310, D1330, D1332, D1333, D1363, D1368, D1383, D1384, D1430, D1457, D1516, D1554, D1642, D1727, D1738, D1813, D1822, D1828, D1862, D1887, D1889, D1894, D1895, D1905, D1925, D1926, D1941, D1954, D1971, D1982, D1993, D2051, D2074, D2101, D2153, E24, E162

Pollution, air (see Air Pollution)

Pollution, water (see Water Pollution)

Polyacene - C473, D1217, D1218, D1219, E327

Polyacetylene - B427

Polyacrylonitrile (see Acrylonitrile Polymer)

Polyadenylic Acid - C491

Polyatomic Molecules - C441

Polycyclic Compounds (also see Aromatic Compounds) - B453, C846, D511

Polyene - B667, D87, D1364, D1560

Polyethylene - C522, D1026, D1033

Polymer - A408, A516, B74, B77, B208, B426, B427, C143, C224, C512, C651, C742, D10, D35, D234, D401, D411, D427, D457, D534, D567, D604, D628,

B329, B335, B541, B548, B570, B613,
B634, B635, B636, C21, C116, C199,
C326, C407, C419, C540, C624, C744,
C745, C784, C798, C854, C905, C912,
D165, D169, D206, D258, D263, D363,
D410, D424, D453, D498, D535, D572,
D677, D683, D686, D712, D810, D958,
D1017, D1236, D1433, D1434, D1441,
D1518, D1519, D1533, D1714, D1735,
D1773, D1908, D2005, D2009, D2010,
D2154, D2181

Spectra Compensation (see Corrected Spectra)

Spectrophotofluorometer (see Instrument)

Spermatid - D812

Spinal - A445, E80, E218

Sputum - D330

Standard - A76, C746, C776, C912, D134,
D203, D441, D731, D2019, E334

Standardization (also see Corrected Spectra) - D1008, D1043, D1281

Staphylococci - D458

Starch - D2141

Stark Effect - C90

Statistical Method - D993

Steridine - C516

Stern-Volmer Law - A15

Steroid (also see Hormones, Aromatic Compounds, Drug, etc.) - A229, A387, A420,
A421, A498, A621, B17, B18, B19, B64,
B76, B126, B137, B424, B624, C26, C134,
C160, C365, C366, C484, C513, D27,
D483, D484, D701, D1133, D1661, D1737,
E25

Sterols - D13

Stilbamidine - A418

Stilbazole - D84, D85, D86, D2028

Stilbene - A32, A271, A272, A327, A418,
A609, C13, C69, C222, C255, C314,
C532, C565, C755, C803, D83, D435,
D640, D709, D746, D843, D900, D971,
D1659, D2002, D2003

Stilbestrol - D386

Stokes Shift (also see Anti-Stokes Shift) -
B557, D1035, D1045, D1852

Streptomycin - D451, D1338

Stress - D1076

Strontium - C708, C709, C892, D830, D1179,
D1216

Structure - D964

Styrene (also see Polystyrene) - A613, A614,
C610, D411, D754, D1187, D1282, D1283,
D1592, D1724

Substitution Effect (also see Theory) - D625,
D1310

Sulfate - A129, A386, C438, C862

Sulfobenzic Acid - D678

Sulfonamide - D672

Sulfonic Acid - A253, A584, C26, C46, C475,
D709, D1244

Sulfur - A29, A367, B577, D53, D470, D957,
E240

Sulfur Dioxide (also see Sulfur) - C894,
D470, D895

Sulfuric Acid - A205, A420, A421, A659,
B183, C866, D23, D25, D577, D800,
D1653

Surface Activity - D986

Tannin - D1824

Tanshinone - D590

Technique - A17, A28, A146, A317, B74,
B84, B159, B230, B240, B258, B275,
B356, B373, B493, B586, C32, C258,
C436, C457, C730, C746, C756, C822,
C826, C859, C879, D198, D279, D288,
D296, D345, D369, D380, D400, D441,
D449, D453, D472, D555, D627, D644,
D655, D711, D729, D731, D829, D835,
D867, D936, D970, D985, D992, D1057,
D1091, D1131, D1227, D1268, D1269,
D1313, D1330, D1507, D1514, D1535,
D1555, D1890a, E112, E350

Teeth (also see Enamel, Dentin) - B278,
D595, D1538, D2178, E67, E344

Telurium - C477, D7, D11, D151, D297,
D366, D476, D543, D679, D832

Temperature - A29, A32, A42, A52, A86,
A232, A238, A298, A302, A419, A471,
A508, A631, A657, B13, B180, B304,
B374, B382, B580, B582, B665, C537,
C576, C581, C582, C584, C654, C673,
C720, C744, C791, C817, C859, C910,
D19, D32, D40, D73, D98, D161, D193,
D258, D259, D260, D273, D278, D297,
D335, D360, D361, D365, D382, D402,
D406, D475, D499, D523, D592, D596,
D602, D611, D677, D698, D713, D732,
D733, D760, D872, D900, D921, D923,
D957, D967, D970, D1054, D1055,
D1077, D1131, D1178, D1208, D1210,
D1249, D1253, D1438, D1443, D1453,
D1454, D1455, D1550, D1598, D1607,